# ACCRETION

## A Collection of Influential Papers

**ADVANCED SERIES IN ASTROPHYSICS AND COSMOLOGY**

*Series Editors:* Fang Li Zhi and Remo Ruffini

---

Volume 1 Cosmology of the Early Universe
*eds Fang Li Zhi and Remo Ruffini*

Volume 2 Galaxies, Quasars and Cosmology
*eds Fang Li Zhi and Remo Ruffini*

Volume 3 Quantum Cosmology
*eds Fang Li Zhi and Remo Ruffini*

Volume 4 Gerard and Antoinette de Vaucouleurs: A Life for Astronomy
*eds M Capaccioli and H. G. Corwin, Jr.*

Advanced Series in Astrophysics and Cosmology – Volume 5

# ACCRETION
## A Collection of Influential Papers

Edited by:
**A. Treves & L. Maraschi**
*University of Milan*
*Italy*
**M. Abramowicz**
*International School for Advanced Studies*
*Trieste, Italy*

World Scientific
*Singapore • New Jersey • London • Hong Kong*

*Published by*

World Scientific Publishing Co. Pte. Ltd.,
P O Box 128, Farrer Road, Singapore 9128
*USA office:* 687 Hartwell Street, Teaneck, NJ 07666
*UK office:* 73 Lynton Mead, Totteridge, London N20 8DH

The editors and publisher are grateful to the authors and publishers who have granted permission to reproduce the reprinted papers in this volume.

While every effort has been made to contact the publishers of reprinted papers prior to publication, we have not always been successful. Where we could not contact the publishers, we have acknowledged the source of the material. Proper credit will be given to these publishers in future editions of this work after permission is granted.

**ACCRETION**
**A Collection of Influential Papers**



ISBN 981-02-0077-3

Printed in Singapore by JBW Printers and Binders Pte. Ltd.

*To the memory of V.F. Shvartsman*

# Preface

The theory of accretion has its foundations in the late thirties with the publication of the paper by Hoyle and Lyttleton (2), the 50th anniversary of which falls this year. At first the relations with actual astronomical situations were rather indirect, and the main emphasis was on mathematical aspects (see e.g. Bondi 3).

The situation changed drastically in the sixties with the discovery of Quasars and X-ray sources. It was immediately suspected, and later clarified, that in both cases the energy source is accretion, thus establishing that accretion itself was a major mechanism of energy production in the Universe. At present the theory of accretion enters in the study of various classes of objects, like cataclysmic variables (close binary systems containing a white dwarf), X-ray binaries (close binaries containing a neutron star or a black hole), and possibly the still mysterious gamma ray bursters. In the extragalactic domain accretion is recognized as the basic mechanism to explain the luminosity of active galactic nuclei (Seyfert nuclei, Quasars, BL Lac Objects, etc).

It is natural that the theory of accretion has gained a prominent role in the standard astrophysical curriculum. Basic elements can be found in various textbooks of Astronomy (see e.g. "Relativistic Astrophysics" by Zel'dovich and Novikov 1971, and "Black Holes, White Dwarfs and Neutron Stars" by Shapiro and Teukolsky, 1983), and at least one book fully devoted to the field has recently appeared ("Accretion Power in Astrophysics", by Frank, King and Raine, 1985). Still direct reference to the original papers is not only advisable from a didactical point of view, but is largely useful for research.

The main motivation of this book is to collect together a number of papers, which, in our opinion, have been most influential in the development of accretion theory. We start by reproducing the two seminal papers by Hoyle and Lyttleton (2), and by Bondi (3). They calculate the accretion rate of a gravitating body moving in a cold fluid (2), and spherical accretion of a hot fluid (3).

The following six papers were fundamentally important in establishing accretion as a basic mechanism for energy production in Astrophysics. The two short papers by Salpeter (4) and Zel'dovich (5), written soon after the discovery of quasars, proposed accretion onto a massive black hole ($10^8$ $M_\odot$) as the energy source of these objects.

In the sixties Kraft (and also Krzeminski, Robinson, Smak and others) discovered that cataclysmic variables are binary stellar systems which consist of a white dwarf surrounded by an accretion disk and an orbiting normal companion. We have chosen Kraft's paper (6) as representative of this discovery. Later Osaki (11) suggested that light curves of a special class of cataclysmic variables, called dwarf novae, which exhibit short (a few days) outbursts separated by long quiescent periods (a few weeks or months) would be explained by intrinsic disk instabilities. Observations of accretion disks in cataclysmic variables still provide today the major tests for the theory of disk accretion.

The paper of Shklovsky (7) proposed a model of Sco X–1 (the first X–ray source to be discovered), in terms of a close binary system containing an accreting neutron star, which is now the canonical picture of most galactic X–ray sources. A first attempt to account for the spectrum of Sco X–1 in the X-ray band was given in the paper by Zel'dovich and Shakura (9) in terms of spherical accretion onto a neutron star. Prendergast and Burbidge (8) recognized that the basic picture of cataclysmic variables could be adapted to binary X–ray sources, indicating the fundamental importance of accretion disks for the modelling of these systems. Lynden-Bell (10) suggested that accretion through a disk should be present not only in close binary systems where orbital motion is the natural source of angular momentum, but also in active galactic nuclei (AGN), where the dynamics of mass transfer is dominated by the central massive black hole. The observation, though still indirect, of this kind of disk, is one of the basic discoveries of the last decade in the field of AGN.

In the third part of the book we report on a number of papers which present the theory of spherical and disk accretion in a form which appears now to be standard, and some improvement to this standard theory is contained in the last part.

The limitation in size has forced us to neglect a number of important contributions. Moreover in various sectors we had to make a difficult choice among various, almost simultaneous papers with similar results. Our choice has not considered priority as the main criterion, but mostly that the paper was influential in the later development of accretion theory. This of course is a rather subjective issue.

Our recent review on accretion theory (1), which serves as an introductory chapter to the present collection, is meant to provide a simplified and self–consistent account of the subject, and may it be of use to locate in perspective the chosen papers.

The book ends with an extended reference list. This was built up from all the references of the papers reproduced here. It should contain most of the relevant papers on accretion, and various ones on affine fields. We hope that the list will be useful.

Finally we would like to express our gratitude to the International Center for Theoretical Physics for help in the reproduction of the papers, to Prof. G. Sedmak for allowing the use of the library of the Osservatorio Astronomico di Trieste, and to Prof. D. Sciama for useful discussions on this collection of papers.

# CONTENTS

Preface . . . . . . . . . . . . . . . . . . . . . . . . . . . . . . . . . . . . . . . . . . . . . . . . . . vii

**Chapter 1 INTRODUCTORY PAPER**

[1] A. Treves, L. Maraschi and M. Abramowicz, "Basic Elements of the Theory of Accretion," *Publ. Astron. Soc. Pac.* **100** (1988) 427 . . . . . . . . . . .3

**Chapter 2 THE SEMINAL PAPERS**

[2] F. Hoyle and R. A. Lyttleton, "The Effect of Interstellar Matter on Climatic Variation," *Proc. Camb. Phil. Soc.* **35** (1939) 405 . . . . . . . . . . . . . .31

[3] H. Bondi, "On Spherically Symmetrical Accretion," *Mon. Not. R. astr. Soc.* **112** (1952) 195 . . . . . . . . . . . . . . . . . . . . . . . . . . . . . . . . . .42

[4] E. Salpeter, "Accretion of Interstellar Matter by Massive Objects," *Astrophys. J.* **140** (1964) 796 . . . . . . . . . . . . . . . . . . . . . . . . . . . . . . . . . .52

[5] Ya. B. Zel'dovich, "The Fate of a Star and the Evolution of Gravitational Energy Upon Accretion," *Sov. Phys. —Dokl.* **9** (1964) 195 . . . . .57

[6] R. P. Kraft, "Binary Stars Among Cataclysmic Variables — I. U Geminorum Stars (Dwarf Novae)," *Astrophys. J.* **135** (1961) 408 . . . . . . . . . .60

[7] I. S. Shklovsky, "On the Nature of the Source of X-Ray Emission of Sco XR-1," *Astrphys. J.* **148** (1967) L1 . . . . . . . . . . . . . . . . . . . . . . . . .79

[8] K. H. Prendergast and G. R. Burbidge, "On the Nature of Some Galactic X-Ray Sources," *Astrophys. J.* **151** (1968) L83 . . . . . . . . . . . . . . . . .83

[9] Ya. B. Zel'dovich and N. I. Shakura, "X-Ray Emission Accompanying the Accretion of Gas by a Neutron Star," *Sov. Astron. —AJ.* **13** (1969) 175 . . . . . . . . . . . . . . . . . . . . . . . . . . . . . . . . . . . . . . . . . . . . .89

[10] D. Lynden-Bell, "Galactic Nuclei as Collapsed Old Quasars," *Nature* **223** (1969) 690 . . . . . . . . . . . . . . . . . . . . . . . . . . . . . . . . . . . . . .98

**Chapter 3 THE STANDARD ACCRETION THEORY**

# ACCRETION

## A Collection of Influential Papers

# Chapter 1 INTRODUCTORY PAPER

PUBLICATIONS OF THE ASTRONOMICAL SOCIETY OF THE PACIFIC
100: 427–451, April 1988

# BASIC ELEMENTS OF THE THEORY OF ACCRETION*

A. TREVES AND L. MARASCHI
Dipartimento di Fisica dell'Università di Milano, Italy

AND

M. ABRAMOWICZ
Scuola Internazionale Superiore di Studi Avanzati, Trieste, Italy

*Received 1988 January 8*

## ABSTRACT

The paper reviews what we consider to be the essential aspects of the theory of accretion. A general overview is given in Section I. The treatments of accretion of cold gas onto a moving star by Hoyle and Lyttleton and of spherical adiabatic accretion by Bondi are summarized in Section II. The formation of spectra in accretion flows onto neutron stars and white dwarfs is described in Section III, with brief mention of the effects of a magnetic field anchored to the star. Spherical accretion onto black holes is treated in Section IV; in particular we underline the differences between classes of models leading to different efficiencies and spectra. We stress the role of a magnetic field entangled in the infalling plasma and the importance of the Comptonization process. The effects of radiation pressure in spherical accretion are then discussed.

Thin accretion disks in the standard $\alpha$ approximation are introduced in Section V. The stability conditions for thermal and viscous processes are examined and the bivalued region of the $\Sigma, \dot{\mathfrak{M}}$ plot is discussed in view of a possible limit cycle behavior. Thick disks and their stability are briefly treated in Section VI.

Section VII offers threads between theory and observations mentioning the astrophysical systems where accretion plays a major role (cataclysmic variables, X-ray binaries, Active Galactic Nuclei) pointing out successes and weaknesses of the present models.

*Key words:* mass accretion–mass exchange–circumstellar disks

### I. Introduction

Accretion is the process by which a gravitating body aggregates matter from its surroundings. Its importance was first recognized in connection with the physics of the solar system (formation of planets). A few years later Hoyle and Lyttleton (1939) examined the possible change in luminosity of a Sun-like star due to its passage through an interstellar gas cloud, with a view to explaining the Earth's climatic variations. This fundamental paper contains the first derivation of the accretion rate for a star moving through cold gas.

The next important step in the development of accretion theory was the paper by Bondi (1952), who considered the infall of gas described by a polytropic equation of state onto a static gravitating body. The accretion rate was calculated and a full analytic solution for the fluid flow was given. The equations formulated by Bondi are also relevant for the case of outflow, i.e., for the theory of stellar (or solar) winds which, from the late fifties, were the subject of an enormous amount of research (e.g., Parker 1963; Holzer and Axford 1970).

Renewed interest for accretion was related in the sixties to two fundamental astrophysical discoveries: quasi-stellar radio sources (quasars) with extremely large luminosities ($L \sim 10^{12}\ L_{\odot}$) produced in comparatively small regions ($r \sim 1$ pc) and X-ray sources, with large powers emitted predominantly at very high temperatures ($\sim 10^7$ K). Salpeter (1964) and Zel'dovich (1964) were the first to examine the possibility of producing the luminosity of quasars by accretion of matter onto a collapsed object of large mass, which must necessarily be a black hole (though the word black hole was not yet in use when these papers were written). Shklovsky (1967) was the first to propose that a bright galactic X-ray source (Scorpius X-1) could be a binary system containing a neutron star accreting mass from its companion.

It was immediately apparent that, in more quantitative studies, the boundary conditions both at large distances from the collapsed object and at its surface would play an essential role. If the accreting matter possesses angular momentum, this has to be removed before the matter can

*One in a series of invited review articles currently appearing in these *Publications*.

fall. The simplest situation in this respect is that of a disk in which matter slowly spirals inward, as viscosity between differentially rotating rings transports angular momentum outward. The role of an accretion disk in binary X-ray sources and the similarity with the cataclysmic variables were pointed out by Prendergast and Burbidge (1968). The importance of disk accretion around a massive black hole was recognized by Lynden-Bell (1969), with special reference to the center of our Galaxy. Shakura (1972), Pringle and Rees (1972), and Shakura and Sunyaev (1973) gave a detailed discussion of the accretion disk with a computation of the emission spectrum. Disk accretion has the important characteristic that the properties of the disk itself are largely independent of the nature of the central object, except at the inner boundary. In order to reach this inner boundary a fixed fraction (50%) of the potential energy gained by the gas in approaching the gravitating body must be radiated away, thus fixing the efficiency of the process.

The effect of the inner boundary conditions is the following: if the central object is a white dwarf or a neutron star, the accreting gas must be slowed down and brought to rest at the surface of the star. Again the efficiency of the process is fixed and the mechanisms by which the slowing down occurs determine the emitted spectrum.

The presence of a strong surface magnetic field on a neutron star ($10^9$–$10^{12}$ G) or a white dwarf ($10^7$ G) introduces a further feature into the models. It defines a region (magnetosphere) where the accretion flow is dominated by magnetic pressure. Thus, even if an accretion disk may be present outside this region, within it the flow will be quasi-radial along the field lines. Obviously the radiation processes are strongly affected.

The case of accretion onto a black hole at first sight seems simpler. Two alternative approximations can be adopted: either a disk, which extends to the innermost stable orbit, or a spherically symmetric flow. In the former case the efficiency is fixed, as mentioned above, while in the second case it is the main unknown of the problem.

For spherical accretion onto black holes, the efficiency of conversion of gravitational potential energy into radiation is extremely small, if the accretion rate is low (Shapiro 1973*a*), but it may be substantially enhanced if the accreting plasma is even weakly magnetized, leading to internal heating by dissipative processes as proposed by Shvartsman (1971) and Meszaros (1975). Here and in the following the efficiency $\eta$ is defined by $\eta \equiv L/\dot{\mathfrak{M}}c^2$.

Profound changes in the accretion flows summarized above are introduced when the accretion luminosity becomes close to the Eddington luminosity for which radiation pressure balances gravity (see eq (4.10))

$$L_E = \frac{4\pi G \mathfrak{M} m_p c}{\sigma_T} \quad . \tag{1.1}$$

This is also called the critical luminosity and, associated with it, there is a critical accretion rate

$$\dot{\mathfrak{M}}_{crit} = L_E/c^2 = \frac{4\pi G \mathfrak{M} m_p}{c\sigma_T} \quad . \tag{1.2}$$

In the case of white dwarfs and neutron stars no stationary solution exists if the accretion rate is larger than $\dot{\mathfrak{M}}_{crit}/\eta$, while in the case of a black hole the radiation efficiency $\eta$ can be reduced so that the hole is able to swallow mass at an arbitrarily high rate with a limited energy output. The evolution of accretion theory is sketched in Figure 1.

The aim of this review is to present the basic elements of the theory of accretion in a simple and unified way. We will concentrate on results of general interest without going into the fine details of specific astrophysical models, but merely indicating which systems may be described by a certain class of accretion models. A special effort is made to explain in elementary terms the spectra deriving from different choices in the description of the accretion flow.

The plan of the paper is the following: in Section II, approximate formulae for the accretion rate are obtained through elementary considerations; next the theory of spherical accretion of a polytropic gas is introduced. The effects of the termination of the flow at the surface of a neutron star or white dwarf are discussed in Section III, where the interaction with a magnetic field anchored to the star is also examined. In Section IV, spherical accretion onto black holes is described considering, in particular, the role of the magnetic field in the accreting plasma and the issue of radiation pressure. Disk accretion is discussed in Section V, and the basic ideas of thick accretion disks are presented in Section VI. Section VII introduces the relations between the theoretical models and the observational world. Some concluding remarks are given in Section VIII. A summary of most-frequently-used symbols is reported in Table I.

Some of the subjects treated in this paper can be found

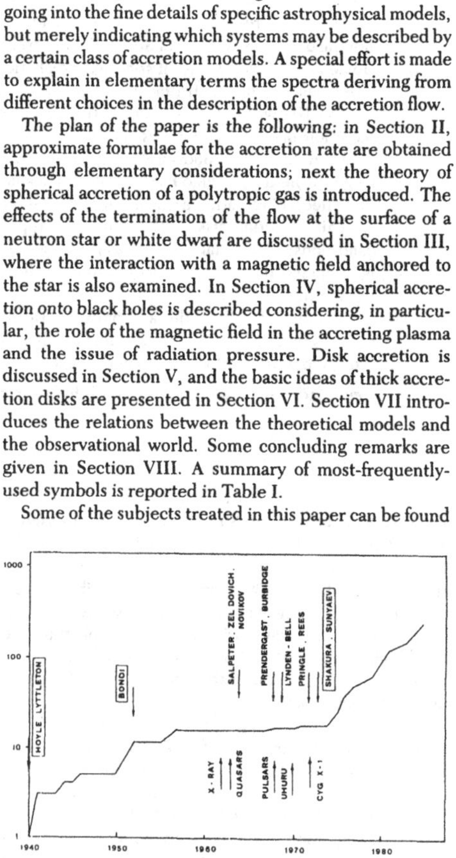


FIG. 1–Number of authors who published, prior to a given date, articles in the *Monthly Notices of the Royal Astronomical Society*. The basic papers and observational discoveries are marked with arrows (from Abramowicz and Marsi 1987).

TABLE I

LIST OF THE PRINCIPAL SYMBOLS USED IN THE PAPER

| | |
|---|---|
| $B$ = magnetic field | $T_{eff}$ = effective temperature |
| $D$ = height of shock | $T_s$ = shock temperature |
| $F$ = radiation flux from surface | $V_s$ = sound velocity |
| $f$ = radial function defined in 5.12 | $V_k$ = Keplerian velocity |
| $H$ = half thickness of the disk | $V_R$ = radial velocity in the disk |
| $L_E$ = Eddington luminosity | $V_\varphi$ = azimuthal velocity |
| $l$ = specific angular momentum | $W_B$ = magnetic energy density |
| $\mathfrak{M}$ = mass of the accreting body | $W_G$ = gravitational energy density |
| $\dot{\mathfrak{M}}$ = accretion rate | $\alpha$ = viscosity parameter |
| $\dot{\mathfrak{M}}_c$ = critical accretion rate | $\beta$ = ratio of gas pressure to total pressure |
| $P$ = pressure | $\gamma$ = polytropic index |
| $Q^+$ = viscous heat production rate | $\Gamma$ = specific heating rate |
| $Q^-$ = viscous heat loss | $\kappa$ = opacity coefficient |
| $q$ = horizontal heat flux | $\eta$ = radiation efficiency |
| $R_{in}$ = inner edge of the disk | $\Lambda$ = specific cooling rate |
| $r_A$ = Alfven radius | $\mu$ = mean molecular weight |
| $r_c$ = accretion radius | $\nu$ = kinematic viscosity |
| $r_s$ = sonic radius | $\rho$ = density |
| $r_{st}$ = star radius | $\Sigma$ = surface density of the disk |
| $r_G$ = gravitational radius | $\tau_T$ = Thomson optical depth |
| $r_{tr}$ = trapping radius | |

in textbooks such as *Relativistic Astrophysics* by Zel'dovich and Novikov (1971) and *Black Holes, White Dwarfs, and Neutron Stars* by Shapiro and Teukolsky (1983). A more extended and detailed account of the fundamentals of accretion theory in astronomy can be found in the recently published volume *Accretion Power in Astrophysics* by Frank, King, and Raine (1985).

## II. Basics of Accretion

### A. *Accretion of a Cold Gas*

Consider a star of mass $\mathfrak{M}$ moving with velocity $v_\infty$ in a cold gas of density $\rho_\infty$ (Hoyle and Lyttleton 1939). In the rest frame of the star at large distance the gas streams with a uniform velocity $-v_\infty$. The cold particles constituting the gas will follow Keplerian orbits in the gravitational field. They are deflected on the rear side of the object and particles with a given impact parameter intersect the star trajectory at a certain distance $r$ from the star. Suppose that at $r_c$ they collide with the particles having the symmetric trajectory and that through the collisions the transverse momentum can be dissipated. If the remaining parallel velocity component is lower than the escape velocity from the star at $r$ the particles will be gravitationally captured. Using Newtonian mechanics one can calculate that, with the above hypothesis, capture occurs for impact parameters smaller than

$$r_c = 2G\mathfrak{M}/v_\infty^2 \ . \tag{2.1a}$$

Correspondingly, the accretion rate is given by

$$\dot{\mathfrak{M}} = \rho_\infty v_\infty \pi r_c^2 = 4\pi\rho_\infty G^2\mathfrak{M}^2/v_\infty^3 \ . \tag{2.1b}$$

The two formulae (2.1a and 2.1b) play a fundamental role in the theory of accretion.

Consider now the other limit of a star at rest in a fluid of temperature $T$ (Bondi 1952). The influence of the gravitational field will be strong in the region where the sound speed $v_{s\infty} = (\partial p/\partial\rho)_\infty^{1/2}$, is smaller than the escape velocity, i.e., for

$$r < r_c = 2G\mathfrak{M}/v_{s\infty}^2 \ . \tag{2.2}$$

Therefore, the critical radius for accretion is formally similar to that of the previous problem, provided the sound velocity is used instead of the bulk velocity of the cold fluid. The accretion rate will be roughly given by

$$\dot{\mathfrak{M}} = 4\pi r_c^2 \rho v_{s\infty} = 4\pi\rho_\infty G^2\mathfrak{M}^2/v_{s\infty}^3 \ . \tag{2.3}$$

In the general case of a star moving with velocity $v$ in a fluid where the sound speed is $v_s$, it can be argued that the accretion rate should be given by

$$\dot{\mathfrak{M}} \simeq \frac{4\pi(G\mathfrak{M})^2}{(v_\infty^2 + v_{s\infty}^2)^{3/2}} , \qquad (2.4)$$

which reduces to equations (2.1) and (2.3) for $v_s = 0$ and $v_\infty = 0$, respectively.

B. *Spherical Accretion: Bondi Theory*

A rigorous treatment of the flow dynamics around a Newtonian point mass, taking into account pressure forces, was given by Bondi in 1952. The simplest formulation of the stationary spherical problem is specified by the equations

$$\dot{\mathfrak{M}} = 4\pi r^2 v \qquad \text{mass conservation} , \qquad (2.5)$$

$$v\frac{dv}{dr} = -\frac{1}{\rho}\frac{dp}{dr} - \frac{G\mathfrak{M}}{r^2} \qquad \text{Euler equation} , \qquad (2.6)$$

and

$$P = P(\rho) \qquad \text{equation of state} . \qquad (2.7)$$

Bondi considered a polytropic equation of state

$$P \propto \rho^\gamma \qquad (2.8)$$

with

$$1 \leq \gamma \leq 5/3 .$$

If we introduce the local sound velocity, the equations reduce to the form

$$\frac{v^2}{2} + \frac{1}{\gamma - 1} v_s^2 - \frac{G\mathfrak{M}}{v} = B \qquad (2.9)$$

$$v = \frac{\dot{\mathfrak{M}}}{4\pi\rho_\infty r^2}\left(\frac{v_{s\infty}}{v_s}\right)^{\frac{2}{\gamma-1}} , \qquad (2.10)$$

where $B$ is an integration constant and $\infty$ labels quantities at large distance with respect to $r_c$ (eq. (2.2)).

The first equation represents an ellipse and the second a hyperbola in the $v, v_s$ plane. The two points of intersection correspond to the existence of two solutions, one subsonic and one supersonic (see Zel'dovich and Novikov 1971). They have been studied in detail by Bondi (see also Frank *et al.* 1985). Here we consider only those where the velocity at infinity tends to zero, excluding the cases of outflows which are relevant for stellar winds. The condition implies that the constant $B$ in equation (2.9) equals $(1/\gamma - 1)v_{s\infty}^2$. Moreover, we suppose that the velocity increases monotonically toward the central star, which corresponds to the case of maximal accretion rate. In this case a sonic point is present at

$$r_s = \frac{1}{4}(5 - 3\gamma)\frac{G\mathfrak{M}}{v_{s\infty}^2} , \qquad (2.11)$$

and the accretion rate is uniquely determined by the equation

$$\dot{\mathfrak{M}} = \pi G^2 \mathfrak{M}^2 \frac{\rho_\infty}{v_{s\infty}^3}\left[\frac{2}{5 - 3\gamma}\right]^{\frac{5-3\gamma}{2\gamma-1}} . \qquad (2.12)$$

This expression for $\dot{\mathfrak{M}}$ differs from that obtained from elementary considerations (eq. (2.3)) only by a small numerical factor. Within the sonic radius the fluid falls essentially freely.

Particularly interesting is the explicit form of the solution in the case $\gamma = 5/3$ (nonrelativistic monoatomic gas). From equation (2.12) the accretion rate is given by

$$\dot{\mathfrak{M}} = \pi \frac{(G\mathfrak{M})^2}{v_{s\infty}^3} \rho_\infty . \qquad (2.13)$$

The sonic point is at $r = 0$, therefore the solution is everywhere subsonic. For $r \ll r_c = G\mathfrak{M}/v_{s\infty}^2$, the velocity is

$$v = \frac{1}{\sqrt{2}}\left(\frac{G\mathfrak{M}}{r}\right)^{1/2} , \qquad (2.14)$$

i.e., one-half of the free-fall velocity, and the density and pressure are

$$\rho = \rho_\infty\left(\frac{2r}{r_c}\right)^{-3/2} , \qquad (2.15)$$

$$P = P_\infty\left(\frac{2r}{r_c}\right)^{-5/2} . \qquad (2.16)$$

The temperature $T$ can be expressed as $p/nk$, where $n$ is the particle density, i.e., $T = p\mu m_p/k\rho$, where $\mu$ is the mean molecular weight. For ionized hydrogen $\mu = 1/2$, and

$$T = \frac{3}{20}\frac{m_p}{k}\frac{G\mathfrak{M}}{r} . \qquad (2.17)$$

Note that the temperature in this limiting case does not depend on the boundary conditions.

## III. Spherical Accretion onto White Dwarfs and Neutron Stars

In the spherical adiabatic approximation the accretion flow onto a compact star (neutron star or white dwarf) can be described by the solutions discussed in the previous section up to some distance from the star. The infall of the plasma is halted at the surface of the star and the solutions must be matched with appropriate boundary conditions. Since practically all the energy release occurs at this boundary, the microscopic processes by which the particles are decelerated determine the radiation spectrum.

Two extreme approximations are possible. In the first the energy loss of protons occurs through Coulomb collisions in the atmosphere of the star which requires, for a proton incident on a neutron star, a depth of 2–30 g cm$^{-2}$ (the kinetic energy of the proton is $\sim$ 100 MeV). The energy is uniformly dissipated and thermalized in this thick layer, which radiates roughly as a blackbody. The associated temperature is

$$T_{eff} = (L/\sigma 4\pi r_{st}^2)^{1/4} \ , \qquad (3.1)$$

which is the minimum temperature at which the radiation can emerge.

At the other extreme, if the proton momenta are randomized in a time much shorter than the time scale for energy loss, the transition from the supersonic flow to the settling atmosphere occurs through a strong adiabatic shock discontinuity. The temperature below the shock is then

$$T_s = \frac{3}{8}\frac{G\mathfrak{M}m_p}{kr_{st}} \ , \qquad (3.2)$$

which is the maximum temperature at which radiation can be emitted. For a typical neutron star $T_{eff} = 1 \times 10^7 (L_{37})^{1/4}$ K and $T_s = 10^{12}$ K, while for a typical white dwarf $T_{eff} = 6 \times 10^5 (L_{37})^{1/4}$ K and $T_s = 10^9$ K, where $L_{37}$ is the luminosity in units of $10^{37}$ erg $s^{-1}$.

More realistic calculations along the first line of approach were first carried out in the case of a neutron star by Zel'dovich and Shakura (1969) and further expanded by Alme and Wilson (1973). It is found that a large fraction of the power is emitted as blackbody radiation from the inner layers of the atmosphere; however, the upper layers are heated to higher temperatures and cool off mainly through inverse Compton scattering with the outgoing blackbody photons. This produces a high-energy tail on the blackbody distribution, which increases in importance with decreasing accretion rate (see Fig. 2).

The second type of approximation has been treated in detail in the case of a neutron star by Shapiro and Salpeter (1975). The adiabatic shock hypothesis requires very efficient plasma instabilities, so that protons are randomized in a time comparable to the inverse of the plasma frequency rather than to the inverse of the Coulomb collision frequency. This is a rather extreme assumption since, in the case of a neutron star, the ratio between the two frequencies is of the order of $10^{12}$. The extent of the hot emission zone below the shock is determined by the time required for the ions to transfer all their energy to the electrons, which cool rapidly via Compton scattering of the photons. The computed spectra show a blackbody component analogous to that computed by Zel'dovich and Shakura (1969); however, a large fraction of the total power is emitted in a high-energy component, which extends to the MeV region, due to Comptonization by relativistic electrons (see Section IV) immediately behind the shock (see Fig. 3).

On a white dwarf the incident protons are much less energetic (100 keV). Therefore, the depth in which their energy is deposited is much smaller and the shock approximation is usually adopted (Hoshi 1973; Aizu 1973). A

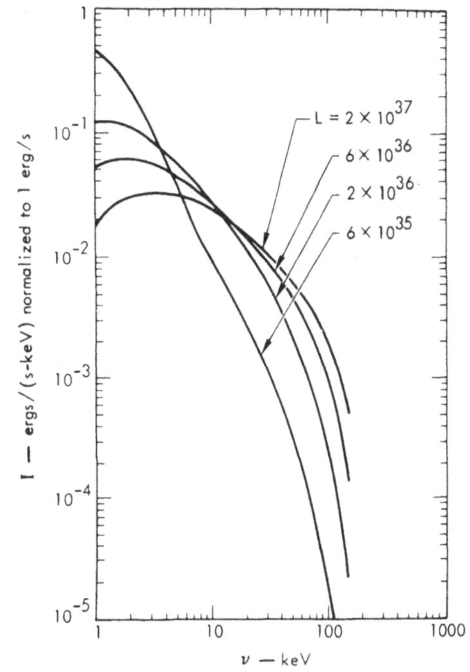


FIG. 2—X-ray spectra from accretion of a nonmagnetized neutron star of $0.5\,\mathfrak{M}_\odot$, in the hypothesis of no shock formation (from Alme and Wilson 1973).

particularly simple description is given by Fabian, Pringle, and Rees (1976). The layer below the shock (which, for reasonable values of the parameters, turns out to be smaller than the radius of the star) is assumed to be isothermal and homogeneous. The temperature and density are assumed to have the postshock values. The height of the shock $D$ is determined by the condition that the emissivity of the shocked gas is sufficient to radiate the deposited power. If the emission mechanism is bremsstrahlung, the height $D$ is given by the simple expression

$$D = 3 \times 10^7\, \dot{\mathfrak{M}}_{20} R_9^{1/2} \quad \text{cm} \ , \qquad (3.3)$$

where $\dot{\mathfrak{M}}_{20}$ is the accretion rate in units of $10^{20}$ g $s^{-1}$ and $R_9$ is the white-dwarf radius in units of $10^9$ cm. Note that the shock height increases with decreasing accretion rate. Inserting the same scaling, the temperature, as from equation (3.1), is given approximately by

$$T_s = 3.7 \times 10^8 (\mathfrak{M}/\mathfrak{M}_\odot) R_9^{-1} K \ . \qquad (3.4)$$

*Accretion on Stars Endowed with a Magnetic Field*

Neutron stars and white dwarfs may have large magnetic fields. On neutron stars typical values of the surface

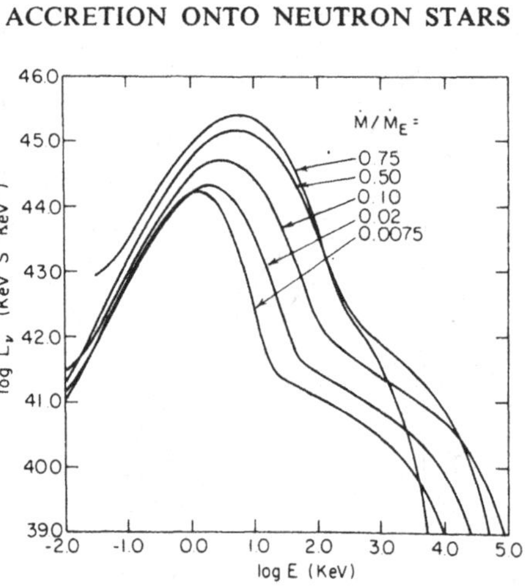


FIG. 3–Spectra from a 1.1 $\mathfrak{M}_\odot$ unmagnetized neutron star with shock formation (from Shapiro and Salpeter 1975).

field, $B_0 = 10^{12}$ G, are indicated by various estimates; on white dwarfs the surface field can reach $10^7$–$10^8$ G. Fields of this size affect the dynamics of accretion. The radius at which the influence of the field becomes important can be evaluated by equating the ram pressure of the infalling plasma $W_C$ to the magnetic pressure $W_B$. For spherical accretion

$$W_C = \frac{\dot{\mathfrak{M}}(G\mathfrak{M})^{1/2}}{4\pi\sqrt{2}} r^{-5/2} , \qquad (3.5)$$

and for a dipolar field

$$W_B = \frac{B_0^2}{8\pi}\left(\frac{r}{r_{st}}\right)^{-6} , \qquad (3.6)$$

where $r_{st}$ is the star radius.

The radius at which $W_C = W_B$, usually called the Alfven radius, is given by Davidson and Ostriker (1973), Lamb, Pethick, and Pines (1973), and Baan and Treves (1973) as

$$r_A = \left(\frac{B_0^2 r_{st}^6}{\sqrt{2G\mathfrak{M}}\dot{\mathfrak{M}}}\right)^{2/7} . \qquad (3.7)$$

For a neutron star with an accretion luminosity $L \sim 10^{37}$ erg s$^{-1}$, a typical value for a binary X-ray source, $r_A = 3 \times 10^8$ cm. Therefore, in a large volume around the neutron star accretion is dominated by the magnetic field. In the case of white dwarfs the Alfven radius turns out to be comparable to the white-dwarf radius.

Within the Alfven radius the accreting material is supposed to follow the magnetic-field lines. The magnetic surface, identified by the condition that its equatorial radius equals the Alfven radius, defines the path of the accretion flow. For a dipole field its shape resembles two funnels centered on the magnetic poles. Hence, the accreting material is channeled onto small areas around the polar caps. Elementary considerations yield for the radius of the polar cap the approximate formula

$$r_{pc} = r_{st}\left(\frac{r_{st}}{r_A}\right)^{1/2} . \qquad (3.8)$$

Considering again a neutron star with a luminosity of $10^{37}$ erg s$^{-1}$, one has that a fraction $f = (r_{st}/r_{pc})^2 \simeq 1\%$ of the star surface is hit by the accreting matter. The blackbody temperature evaluated by equation (3.1) is therefore raised by a factor 3, if energy transfer by conduction is neglected.

For the description of the accretion columns, the spherical approximation is still relevant, though limited to a fraction $f$ of the solid angle. For a neutron star with a high magnetic field the radiative processes are strongly dominated by cyclotron radiation in the semirelativistic regime $kT$ = 10–100 keV. The radiative transfer of the cyclotron radiation in the magnetic funnel is enormously complicated due to the anisotropy of emission and absorption processes and is much beyond the scope of this review. Suffice it to say that, while the general properties of accretion columns on white dwarfs seem recently understood (e.g., Lamb 1985), those of neutron-star accretion columns still present unsolved problems, a basic one being whether the outgoing radiation is beamed along the column or in a wide-angle hollow cone (see, however, Meszaros and Nagel 1985*a*,*b*).

## IV. Spherical Accretion onto Black Holes

It seems well established that the collapse of stars larger than a few solar masses would proceed through the Schwarzschild event horizon $r_G = 2G\mathfrak{M}/c^2$, giving rise to a black hole. The properties of the Schwarzschild metric are such that no radiation emitted below the horizon can reach the observer at infinity. Hence, if matter falling into a black hole releases its energy entirely within the horizon, the efficiency $\eta = L/\dot{\mathfrak{M}}c^2$ of the process for the external observer may be zero. However, in any realistic description of the accreting plasma, some fraction of the kinetic (gravitational) energy is radiated away before reaching the gravitational radius $r_G$. The computation of this fraction is one of the main issues in the problem of spherical black-hole accretion. It is clear that the most important region in this respect is just outside $r_G$, where $v \simeq c$; thus special and general relativistic corrections should be taken into account.

In the simplest, adiabatic, Newtonian approximation the expected temperature at $r_G$ is $T \simeq 10^{12}$ K, as given by equation (2.17). This is independent of the mass of the black hole.

Shapiro (1973*a*) solved under specific hypotheses a full set of relativistic equations, coupling the dynamical and thermodynamical behavior of the accreting plasma. This consists of adding to the dynamic equations (2.5, 2.6), rewritten in the fully relativistic form, the energy balance equation (e.g., Shvartsman 1971)

$$v\left(\frac{d\epsilon}{dr}-\frac{\epsilon+P}{n}\frac{dn}{dr}\right)=\Lambda-\Gamma\ , \qquad (4.1)$$

where $v$ is the velocity, $n$ is the scalar number density, $\epsilon$ is the proper internal-energy density, $P$ is the pressure, and $\Lambda$ and $\Gamma$ are the specific cooling and heating rates in erg $\mathrm{cm}^{-3}\,\mathrm{s}^{-1}$.

Shapiro (1973*a*) considered the case of $\Gamma = 0$ and bremsstrahlung as a cooling mechanism; he found that a solar-mass black hole accreting the interstellar medium radiates with an efficiency $\eta \simeq 10^{-10}$–$10^{-9}$. The efficiency increases for large accretion rates, since it is roughly proportional to the square of the density of the flow. However, it remains rather low $\eta \leq 10^{-4}$, with the equal sign corresponding to the critical accretion rate $\dot{\mathfrak{M}} \sim \dot{\mathfrak{M}}_c$. The radial dependence of the velocity closely approaches the free-fall case.

Subsequently, on the basis of this result, several authors, interested mainly in the computation of the emitted spectrum, have simply assumed velocity and density profiles similar to the free-fall case and have devoted attention to the energy-balance equation (4.1) only.

In a Newtonian scheme, if one supposes that the infall velocity scales as the free-fall velocity $v \propto r^{-1/2}$, and for the case of a completely ionized nonrelativistic hydrogen plasma, equation (4.1) reduces to the form (Meszaros 1975)

$$\frac{dT}{dr}=-\frac{T}{r}+\frac{4\pi m_p r^2(\Lambda-\Gamma)}{3k\dot{\mathfrak{M}}}\ . \qquad (4.2)$$

This is a single equation for $T$ and can be solved analytically or numerically if the heating and cooling rates are specified. The outgoing spectrum and luminosity can be computed, in the case of a transparent atmosphere, by integrating the spectral emissivities over the accreting volume.

A. *Role of the Magnetic Field in the Accreting Plasma*

Shvartsman (1971) pointed out that, if the infalling plasma is even weakly magnetized, the accretion process will amplify the field $B$. In fact, from the conservation of magnetic flux which is valid for highly conducting flows, it follows that the radial component of $B$ should increase as $r^{-2}$ and the magnetic energy density $W_B$ as $r^{-4}$. The buildup of the magnetic energy density occurs through the gravitational field, which is responsible for the compression of the plasma. Since the gravitational energy density $W_G$ scales as $r^{-5/2}$ (eq. (3.5)) at sufficiently small radii the magnetic energy density approaches the gravitational one. On energetic and stability grounds it is assumed that at small enough radii $W_G \simeq W_B$, giving

$$B = \dot{\mathfrak{M}}^{1/2}\, G\mathfrak{M}^{1/4} r^{-5/4}\ , \qquad (4.3)$$

independent of the initial value of $B$. For critical accretion rates equation (4.3) gives, at $r = r_G$, $B = 5 \times 10^7$ G for a 10 $\mathfrak{M}_\odot$ black hole and $B = 10^4$ G for $10^8\ \mathfrak{M}_\odot$. Alternative configurations are possible if one considers an initially ordered rather than chaotic magnetic field (see, e.g., Bisnovatyi-Kogan 1979).

Shapiro (1973*b*) showed that the dynamical effect of the $B$ field is small, as it does not impede the radial flow. However, the field has two main consequences. The most obvious is the activation of cyclotron (synchrotron) radiation from the thermal electrons as an emission process (cooling mechanism). The inclusion of this effect leads to radiative efficiencies far exceeding those obtained in the case when only bremsstrahlung is considered (Shapiro 1973*a*). Moreover, it shifts the frequency at which maximum emission occurs from values of the order of $kT$ = 1–10 MeV, appropriate for bremsstrahlung, to the thermal synchrotron frequency

$$\nu_s=\frac{3}{4}\frac{eB}{\pi mc}\left(\frac{kT}{mc^2}\right)^2\ , \qquad (4.4)$$

which falls in the far-infrared region of the spectrum.

A more subtle effect, which was clearly recognized by Meszaros (1975), is that the equipartition assumption implies a complicated physical situation, with instabilities and field reconnection preventing excessive growth of $W_B$ and leading to turbulent motion. A certain amount of dissipation must be associated with these processes, which is equivalent to saying that some fraction of the gravitational energy is transformed into heat, giving rise to a heating term $\Gamma$ in equations (4.1, 4.2). With this assumption $\Gamma$ can be written as a fraction of the gravitational energy density divided by the infall time, i.e.,

$$\Gamma \simeq \frac{1}{8\pi}\frac{G\mathfrak{M}\dot{\mathfrak{M}}}{r^4}\ . \qquad (4.5)$$

Equation (4.1) with a magnetic field given by equation (4.3) and a heating term given by equation (4.5) was studied by Meszaros (1975) and further considered by Ipser and Price (1977, 1982, 1983), Maraschi *et al.* (1979), and Maraschi, Roasio, and Treves (1982).

For low accretion rate, $\dot{\mathfrak{M}} \leq 10^{-7}\ \dot{\mathfrak{M}}_c$, even in the presence of turbulent heating, the radiative processes are inefficient and the results are qualitatively similar to those of Shapiro (1973*a*): the gas heats up quasi-adiabatically and the efficiency is low. At intermediate accretion rates the efficiency increases and opacity effects come into play. The optical depth for synchrotron self-absorption in the field given by equation (4.3) is for $\nu \lesssim \nu_s$ (Ipser and Price 1982)

$$\tau_{syn} \sim 2 \times 10^5 \left(\frac{\mathfrak{M}}{\mathfrak{M}_\odot}\right)^{1/2} \left(\frac{\dot{\mathfrak{M}}}{\dot{\mathfrak{M}}_c}\right)^{1/2} T_{10}^{-5} \left(\frac{r}{r_G}\right)^{3/4} \left(\frac{\nu}{\nu_s}\right)^{-5/3}, \quad (4.6)$$

where $T_{10}$ is the temperature in units of $10^{10}$ K.

For $\dot{\mathfrak{M}} > 10^{-4}\dot{\mathfrak{M}}_c$, $\tau_{syn} > 1$ and the effects of self-absorption are essential. The efficiency reaches $\sim 10^{-2}$ and the emission peak occurs at the transparency frequency, which can be $10^2 - 10^3\,\nu_s$.

B. *Comptonization*

Another essential effect at large accretion rates is related to the Thomson optical depth, which for free-fall is

$$\tau_T = \int_{r_G}^{\infty} \sigma_T n_e dr = \frac{\dot{\mathfrak{M}}}{\dot{\mathfrak{M}}_c}\left(\frac{r}{r_G}\right)^{-1/2}. \quad (4.7)$$

The outgoing shychrotron photons which initially have $h\nu \ll m_e c^2$ gain energy from the hot electrons through repeated Compton scatterings if $\tau_T$ approaches 1. This process is usually called Comptonization (Rybicki and Lightman 1979; Pozdnyakov, Sobol', and Sunyaev 1983). The average energy gain per scattering is

$$\Delta h\nu/h\nu = 4kT/mc^2 \quad \text{(a)}$$
$$\Delta h\nu/h\nu = 16\,(kT/mc^2)^2 \quad \text{(b)}, \quad (4.8)$$

in the nonrelativistic (a) and relativistic (b) limits, respectively. It has been shown both analytically and numerically (Shapiro, Lightman, and Eardley 1976; Pozdnyakov *et al.* 1983) that although the probability of repeated scattering decreases exponentially, it gives rise to an exponentially increasing energy gain. The combination of the two effects yields a power-law energy distribution $F_\nu \propto \nu^{-\alpha}$ up to $h\nu \simeq 3kT$ for initially monochromatic photons which propagate through even modest optical depths of hot gas. Here we report two expressions for the index of the power law derived (a) analytically in the nonrelativistic limit (Sunyaev and Titarchuk 1980), and (b) numerically with relativistic corrections (Pozdnyakov *et al.* 1983; Zdziarski 1985).

$$\alpha = \left(\frac{9}{4} + \frac{\pi^2}{3(\tau_T + \frac{2}{3})^2 \frac{kT}{m_e c^2}}\right)^{1/2} - \frac{3}{2} \quad (4.9a)$$

$$\alpha = -\ln P/\ln\left[1 + 4\frac{kT}{mc^2} + 16\left(\frac{kT}{mc^2}\right)^2\right] \quad (4.9b)$$

$$P = \frac{3}{4}\tau_T \quad (\tau_T < 1)$$

$$P = 1 - \frac{3}{4\tau_T} \quad (\tau_T > 1)$$

Comptonization is crucial in all problems in which the accretion rate is nearly critical. Specific model computations including the effects mentioned above were carried out by several authors (e.g., Takahara *et al.* 1981; Maraschi, Roasio, and Treves 1982; Ipser and Price 1982). Maximum temperatures of $10^9$–$10^{10}$ K are found and the results agree in finding high efficiencies ($\eta \sim 10^{-1}$) and power-law spectra extending from the synchrotron self-absorption frequency to $3\,kT$. The derived spectral slope depends on the optical depth, but not too strongly. For a range of two decades $\dot{\mathfrak{M}} = (10^{-2} - 1)\dot{\mathfrak{M}}_c$ in accretion rate, $\alpha$ varies between 1 and 0.5, due to the decrease of the temperature at high optical depths. Examples of temperature profiles and spectra are given in Figures 4 and 5.

C. *Spherical Accretion with Radiation Pressure*

The radiation force on a free electron is $f_{rad} = (\sigma_T/c)\,F$ where $F$ is the radiation flux. Since the electric field strongly couples protons and electrons in the accreting plasma, the resultant force on a volume element of completely ionized hydrogen is

$$\left(\frac{G\mathfrak{M}m_p}{r^2} - \frac{\sigma_T}{c}\frac{L(r)}{4\pi r^2}\right)\frac{\rho}{m_p}. \quad (4.10)$$

At large distance from the star the outgoing luminosity is constant and the vanishing of the resultant force, which is independent of $r$, defines the Eddington luminosity

$$L_E = \frac{4\pi G\mathfrak{M}_p c}{\sigma_T}. \quad (4.11)$$

Luminosities of this order are actually observed in accreting sources and it is interesting to discuss how the dynamics are modified. In the case of neutron stars or white dwarfs the accretion rate fixes the outgoing luminosity. In the simplest model (Thomson opacity only) it is found that

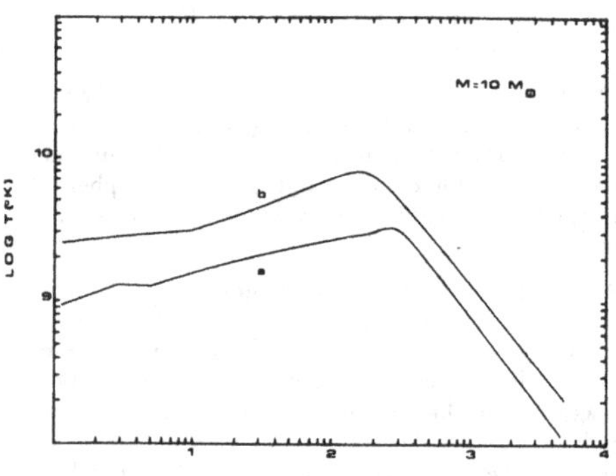


FIG. 4–Temperature profiles around a black hole of $10\,\mathfrak{M}_\odot$ accreting at a rate of $3 \times 10^{18}$ g s$^{-1}$ (a), and $\dot{\mathfrak{M}} = 3 \times 10^{17}$ g s$^{-1}$ (b) $\langle r \rangle = 3r_G$ (from Maraschi *et al.* 1982).

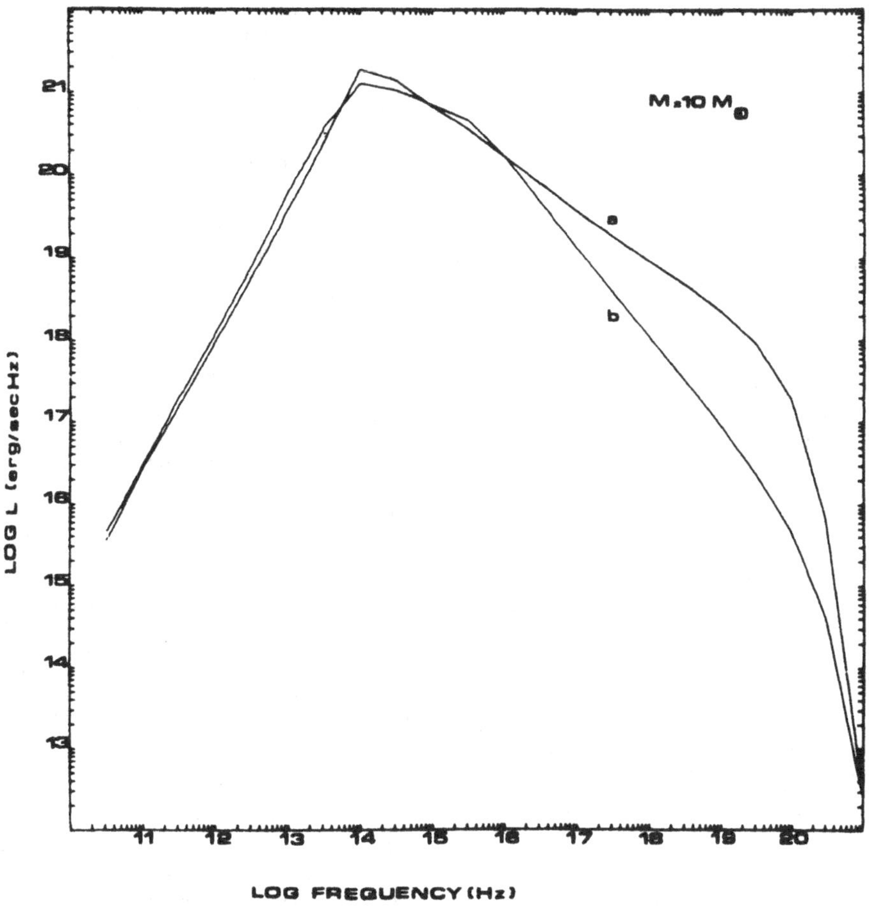


FIG. 5–Spectra from an accreting black hole of 10 $\mathfrak{M}_{\odot}$, for an accretion rate $\dot{\mathfrak{M}} = 3 \times 10^{18}$ g s$^{-1}$ (a) and $\dot{\mathfrak{M}} = 3 \times 10^{17}$ g s$^{-1}$ (b) $\langle r \rangle = 3r_G$ (from Maraschi *et al.* 1982).

the simplest model (Thomson opacity only) it is found that for $L = \lambda L_E$ with $\lambda \leq 1$ the flow occurs at velocities everywhere lower than free-fall. The infall velocity has a maximum at $r > r_{st}$, and decreases significantly toward the surface of the star (see Fig. 6).

For $\lambda > 1$ no stationary solution exists. Time-dependent solutions have been explored by Cowie, Ostriker, and Stark (1978) and Klein, Stockman, and Chevalier (1980). In the latter investigation, which includes supercritical accretion, an extended envelope is shown to form around the neutron star in 10 to 100 ms.

The case of supercritical accretion onto a black hole is different because of the different inner boundary conditions. The inflow velocity at the horizon as measured by static observers must equal the speed of light. The full hydrodynamic treatment is rather complicated (Thorne, Flammang, and Zytkow (1981); Flammang (1982, 1984)), but confirms the validity of a simple physical concept, the trapping radius (eq. (4.13)), introduced by Rees (1978) and studied in a simple hydrodynamic model by Begelman (1978, 1979). At high accretion rates the radiation transfer, in a first approximation, can be treated diffusively. The photon diffusion velocity at $r$ is $\sim c/\tau(r)$. At the same time photons are convected inward with veloc-

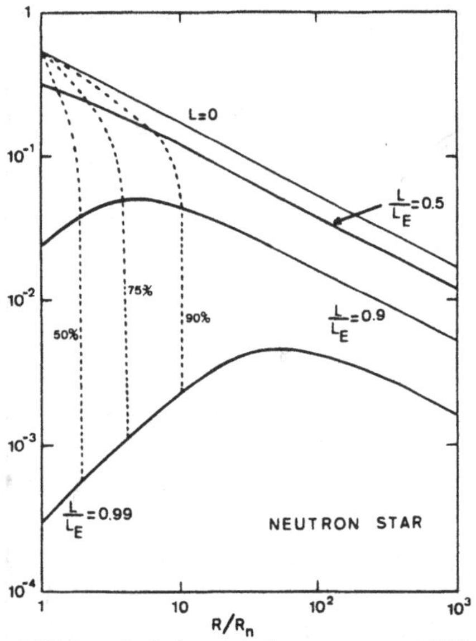


FIG. 6–Velocity profiles for the accretion flow on a neutron star of 1 $\mathfrak{M}_\odot$ and radius $r_n = 10^6$ cm, for different outgoing luminosities (from Maraschi *et al.* 1978). The abscissae are measured in star radii.

ity $v(r)$. For

$$v(r) > c/\tau(r) \ , \tag{4.12}$$

convection overwhelms diffusion, so that the radiation emitted by the gas is trapped within it and falls toward the black hole. Assuming that $v(r)$ is given by free-fall, i.e., $v(r) = c(r/r_G)^{-1/2}$, and taking expression (4.7) for the optical depth, one finds that equation (4.12) is valid when $r > r_{tr}$, where

$$r_{tr} = \frac{\dot{\mathfrak{M}}}{\dot{\mathfrak{M}}_c} r_G \ . \tag{4.13}$$

It can be shown that the luminosity reaching the observer at infinity, $L_\infty$, apart from minor corrections, is simply the luminosity produced at $r \geq r_{tr}$. If this is some fraction $\lambda$ of the gravitational power at $r_{tr}$,

$$L_\infty = L(> r_{tr}) = \lambda \frac{G\mathfrak{M}\dot{\mathfrak{M}}}{r_{tr}} \ , \tag{4.14}$$

and using the definition of equation (4.13) one finds

$$L_\infty = \lambda \frac{4\pi G \mathfrak{M} m_p c}{\sigma_T} = \lambda L_E \ . \tag{4.15}$$

Hence, whatever the accretion rate, the emerging luminosity is limited to a fraction of the Eddington luminosity. This is shown with a rigorous semianalytic treatment of Comptonization in a spherical flow by Colpi (1988). The effect is increased if $e^+ - e^-$ pair production contributes to the opacity (Lightman, Zdziarski, and Rees 1987).

The determination of the accretion rate, given the external boundary conditions is a different problem and requires a discussion of the critical points in the flow (see, e.g., Turolla, Nobili, and Calvani 1986).

D. *Two-Temperature Accretion*

In the previous section it was assumed that the accreting plasma was in thermal equilibrium, i.e., that the distribution functions of electrons and protons are described by Maxwellians with the same temperature. The relaxation times through Coulomb collisions and their scaling with distance in the spherical case are discussed, e.g., by Colpi, Maraschi, and Treves (1984). Under reasonable astrophysical conditions, the time required to achieve thermal equilibrium between the ions and electrons can be longer than the dynamical time scale (i.e., the free-fall time scale). Therefore, electrons and ions can be assumed to behave as independent fluids. This approach has been used both in spherical and disk models (e.g., Colpi *et al.* 1984; Colpi, Maraschi, and Treves 1986; Shapiro, Lightman, and Eardley 1976; Rees *et al.* 1982). Because protons are inefficient radiators they reach temperatures of the order of the virial temperature (eq. (2.17)) and dominate the pressure of the fluid, while the electrons, due to strong radiative losses, attain much lower temperatures. Therefore, the dynamical properties of the flow are determined by the protons and the radiative properties by the electrons. This situation may occur at accretion rates below the critical one and gives rise to interesting models, where the high ion temperature can be used to confine jets or to produce high-energy radiation (gamma rays).

It should be noted, however, that the coupling between electrons and ions could be much more efficient than the Coulomb one, as one can argue on the basis of collective plasma effects, which should be present in turbulent magnetic fields (Begelman and Chiueh 1987). In this case the temperatures of the two populations would be close, and the single temperature description would be realistic.

## V. Accretion Disks

In many astrophysical situations the accreted matter has substantial angular momentum. This completely changes the physical picture of accretion discussed in the previous sections. The most important process governing the accretion of rotating matter is the action of viscous stresses. Gas orbiting a compact central object gradually loses its angular momentum through viscous stresses (friction) to gas further out. With friction, energy is also dissipated and the inner gas spirals further inward. Viscous stresses not only drive accretion by transporting mass inward and angular momentum outward but also convert the gravitational energy of matter into heat. The heat diffuses toward the top and bottom surfaces of the flow where it is radiated away. Radiation from the hot innermost parts can be absorbed by much cooler outer regions and heat them up, forming a tenuous corona and blowing off a wind. If the central accreting object has a rigid surface a shock is formed close to it (see Section III for the spherical case). Black-hole accretion, on the other hand, is highly supersonic in the innermost part. Presence of a strong magnetic field is another complication.

There is no complete, satisfactory theoretical model of this rather-complex hydrodynamical situation. Advanced numerical codes are still not powerful enough to cope with it, and our present understanding of accretion of rotating matter is based on a very approximate, semianalytic approach. In particular, we refer to the papers of Prendergast and Burbidge (1968), Shakura (1972), Pringle and Rees (1972), Shakura and Sunyaev (1973), and Pringle (1981). We shall now present the basic assumptions using cylindrical coordinates: $R$ (horizontal), $Z$ (vertical), and $\varphi$ (azimuthal). The geometrical half-thickness of the flow $H = H(R)$, the angular velocity $\Omega$, and the surface density of matter

$$\Sigma(R) = 2\int_0^H \rho(R,Z)dZ \ , \tag{5.1}$$

appear in some of these assumptions.

GEOMETRY. The flow is axially symmetric and plane symmetric: its properties do not depend on $\varphi$ and depend on $Z$ only through $|Z|$. The vertical thickness is very small, $H \ll R$.

KINEMATICS. Azimuthal rotation agrees with the circular Keplerian motion

$$V_\varphi = \Omega R = V_k = (G\mathfrak{M}/R)^{1/2} \ . \tag{5.2}$$

Only this motion contributes to shear $\sigma = 1/2\, R(d\Omega/dR)$. This expression for the shear is valid when the angular velocity is independent of $Z$. (For a general definition see, e.g., Landau and Lifshitz 1959.) The horizontal component of velocity is very subsonic

$$V_R < V_s = \left(\frac{\partial p}{\partial \rho}\right)^{1/2} .$$

The vertical component is negligible

$$V_Z \ll V_R \ll V_\varphi \ .$$

DYNAMICS. The horizontal pressure gradient is negligible. At the inner edge of the accretion flow $R = R_{in}$ the viscous torque vanishes and the accretion rate is much smaller than the critical one $\dot{\mathfrak{M}} \ll \dot{\mathfrak{M}}_c$.

DISSIPATION. The horizontal heat flux is negligible. The $Z$-averaged kinematic viscosity,

$$\langle\nu\rangle = \frac{1}{\Sigma(R)}\int_0^H \rho(R,Z)\nu(R,Z) \ , \tag{5.3}$$

is given by the formula

$$\langle\nu\rangle = \alpha V_s H \ , \tag{5.4}$$

where $\alpha$ = const is a phenomenological viscosity coefficient. The opacity is dominated by electron scattering:

$$\kappa = \kappa_{es} = 0.4\ \mathrm{g\ cm^{-2}} \ . \tag{5.5}$$

These assumptions define the standard $\alpha$ model of thin accretion disks: "$\alpha$" because the assumption of equation (5.4) is crucial (and restrictive!) and "thin disk" because with $H/R \ll 1$ the flow shape resembles a thin disk. The action of viscous stresses is so important for the theory of accretion disks that our discussion must start from this point.

Consider a particular cylindrical surface $R = R_0$, crossing the accretion flow. The rates of mass and angular-momentum flow across this surface are $\dot{\mathfrak{M}}(R_0)$ and $\dot{J}(R_0)$:

$$\dot{J}(R_0) = \dot{\mathfrak{M}}(R_0)l(R_0) - g(R_0) \ , \tag{5.6}$$

with $l$ being the specific angular momentum and $g(R_0)$ the viscous torque acting through $R = R_0$:

$$g(R_0) = 2\pi\int R_0^3\left(\frac{d\Omega}{dR}\right)_{R_0}\rho(R_0,Z)\nu(R_0,Z)dZ$$

$$= 2\pi R_0^3\left(\frac{d\Omega}{dR}\right)_{R_0}\Sigma(R_0)\langle\nu\rangle(R_0) \ . \tag{5.7}$$

When $(d\Omega/dR) > 0$ the viscous torque transports the angular momentum outward. At the inner edge $R = R_{in}$ the torque vanishes, $g(R_{in}) = 0$, and therefore

$$\dot{J}(R_{in}) = \dot{\mathfrak{M}}(R_{in})l(R_{in}) \ . \tag{5.8}$$

In the stationary accretion disk the rate of mass and angular-momentum flow across any $R = R_0$ cylinder must be the same, they do not depend on $R$.

For a stationary disk,

$$\dot{\mathfrak{M}}(R) = \dot{\mathfrak{M}}_0 = \text{const}, \quad \dot{J}(R) = \dot{J}_0 = \text{const} \ . \tag{5.9}$$

From equations (5.6), (5.8), and (5.9) it follows that in the stationary accretion disk

$$\dot{\mathfrak{M}}_0[l(R) - l(R_{in})] = 2\pi R^3\left(\frac{d\Omega}{dR}\right)\Sigma\langle\nu\rangle \ . \tag{5.10}$$

When rotation is given by the Keplerian law, (eq. (5.2)), then the last formula reads

$$3\pi\langle\nu\rangle\Sigma = \dot{\mathfrak{M}}f \tag{5.11}$$

$$f = \frac{l(R) - l(R_{in})}{l(R)} = 1 - \left(\frac{R_{in}}{R}\right)^{1/2} . \tag{5.12}$$

The viscous stresses acting inside a mass $M$ dissipate the kinetic energy and transform it to heat at the rate (Landau and Lifshitz 1959)

$$\dot{E} = 4\int_0^M \sigma^2 \nu dM = \int\int 2\pi R^3 \left(\frac{d\Omega}{dR}\right)\nu\rho dR dZ \ . \quad (5.13)$$

For a mass inside an infinitesimal shell between $R = R_0$ and $R = R_0 + dR$ the last formula gives

$$Q^+(R) = \frac{1}{2\pi R}\frac{d\dot{E}}{dR} = \left(R\frac{d\Omega}{dR}\right)^2 \int \nu\rho dZ = \left(R\frac{d\Omega}{dR}\right)^2 \Sigma\nu$$

$$Q^+(R) = \frac{1}{2\pi R}\left[l(R) - l(R_{in})\right]\frac{d\Omega}{dR} \ . \quad (5.14)$$

The quantity $Q^+$ is the surface heat-generation rate due to viscous processes. When rotation is Keplerian it is equal to

$$Q^+(R) = \frac{3G\mathfrak{M}\dot{\mathfrak{M}}}{4\pi R^3} f \ . \quad (5.15)$$

All the equations basic to the accretion disk theory are collected in Table II and will be referred to with brackets. They are written in a simplified form apt to further generalization. Most of them are self-explanatory. For example, equation [2] approximates the hydrostatic equilibrium condition

$$\frac{1}{\rho}\frac{dP}{dZ} = \frac{d}{dz}\left[-\frac{G\mathfrak{M}}{(R^2+Z^2)^{1/2}}\right] \ . \quad (5.16)$$

In a similar way equation [7] approximates, for the optically thick case,

$$Q^- = -\frac{c}{3}\frac{d(aT^4)}{d\tau} \ , \quad \tau = \int K\rho dZ \ . \quad (5.17)$$

Table I gives the list of all the relevant accretion-disk structure quantities. Tables I and II contain all the information relevant to the standard α models of thin accretion disks.

There are 20 equations for 24 physical quantities. Therefore, four physical quantities are independent. It is convenient to use

$$\mathfrak{M}, \dot{\mathfrak{M}}, R, \alpha \quad (5.18)$$

as the four independent quantities.

All the equations for the standard disk model in Table II, with the exception of equation [11], are linear in their logarithmic version. Even equation [11] is linear in the two asymptotic cases in which either the gas pressure

$$P_{gas} = \frac{k}{m_p}\frac{T}{\langle\mu\rangle} \ ,$$

or the radiation pressure $P_{rad} = 1/3\, aT^4$ dominates. In the first case $\beta \simeq 1$; in the second case $\beta \simeq 0$ with β defined by

$$\beta = \frac{P_{gas}}{P} = \frac{P_{gas}}{P_{gas} + P_{rad}} \ . \quad (5.19)$$

Therefore, in the two asymptotic cases it is possible to write down an analytic solution for the standard accretion-disk structure. This was done first by Shakura and Sunyaev (1973). We present the Shakura-Sunyaev solution in a slightly modified version, better suited to astrophysical applications.

For the case of $P_{rad} \ll P_{gas}$:

$$\Sigma = 7.08 \times 10^4 \alpha^{-4/5} f^{3/5} (R/R_C)^{-3/5} \times (\mathfrak{M}/\mathfrak{M}_\odot)^{1/5} (\dot{\mathfrak{M}}/\dot{\mathfrak{M}}_c)^{3/5} \ [\mathrm{g/cm^2}] \quad (5.20)$$

$$H/R = 1.15 \times 10^{-2} \alpha^{-1/10} f^{1/5} (\mathfrak{M}/\mathfrak{M}_\odot)^{1/10} \times (\dot{\mathfrak{M}}/\dot{\mathfrak{M}}_c)^{1/5} \quad (5.21)$$

$$\rho = 1.03 \times 10\alpha^{-7/10} f^{2/5} (R/R_C)^{-33/20} \times (\mathfrak{M}/\mathfrak{M}_\odot)^{-7/10} (\dot{\mathfrak{M}}/\dot{\mathfrak{M}}_c)^{2/5} \ [\mathrm{g/cm^3}] \quad (5.22)$$

$$P = 3.01 \times 10^{17} \alpha^{-9/10} f^{4/5} (R/R_C)^{-51/20} \times (\mathfrak{M}/\mathfrak{M}_\odot)^{-9/10} (\dot{\mathfrak{M}}/\dot{\mathfrak{M}}_c)^{4/5} \ [\mathrm{dyn/cm^2}] \quad (5.23)$$

$$T = 3.53 \times 10^8 \alpha^{-1/5} f^{2/5} (R/R_C)^{-9/10} \times (\mathfrak{M}/\mathfrak{M}_\odot)^{-1/5} (\dot{\mathfrak{M}}/\dot{\mathfrak{M}}_c)^{2/5} \ [\mathrm{K}] \quad (5.24)$$

$$\frac{V}{c} = 3.5 \times 10^{-5} \alpha^{4/5} f^{-3/5} (R/R_C)^{-2/5} \times (\mathfrak{M}/\mathfrak{M}_\odot)^{-1/5} (\dot{\mathfrak{M}}/\dot{\mathfrak{M}}_c)^{2/5} \quad (5.25)$$

$$F = 8.3 \times 10^{25} f (R/R_C)^{-3} (\mathfrak{M}/\mathfrak{M}_\odot)^{-1} (\dot{\mathfrak{M}}/\dot{\mathfrak{M}}_c) \ [\mathrm{erg/cm^2 s}] \ . \quad (5.26)$$

For the case $P_{gas} \ll P_{rad}$:

$$\Sigma = 4.24\,\alpha^{-1} f^{-1} (R/R_C)^{3/2} (\dot{\mathfrak{M}}/\dot{\mathfrak{M}}_c)^{-1} \ [\mathrm{g/cm^2}] \quad (5.27)$$

$$\frac{H}{R} = 0.74 f (R/R_C)^{-1} (\dot{\mathfrak{M}}/\dot{\mathfrak{M}}_c) \quad (5.28)$$

$$\rho = 9.51 \times 10^{-6} \alpha^{-1} f^{-2} (R/R_C)^{3/2} \times (\mathfrak{M}/\mathfrak{M}_\odot)^{-1} (\dot{\mathfrak{M}}/\dot{\mathfrak{M}}_c)^{-2} \ [\mathrm{g/cm^3}] \quad (5.29)$$

$$P = 2.33 \times 10^{15} \alpha^{-1} (R/R_C)^{-3/2} \times (\mathfrak{M}/\mathfrak{M}_\odot)^{-1} \ [\mathrm{dyn/cm^2}] \quad (5.30)$$

$$T = 3.10 \times 10^7 \alpha^{-1/4} (R/R_C)^{-3/8} \times (\mathfrak{M}/\mathfrak{M}_\odot)^{-1/4} \ [\mathrm{K}] \quad (5.31)$$

$$\frac{V}{c} = 0.53\,\alpha f (R/R_C)^{-5/2} (\dot{\mathfrak{M}}/\dot{\mathfrak{M}}_c) \quad (5.32)$$

$$F = 8.3 \times 10^{25} f (R/R_C)^{-3} (\mathfrak{M}/\mathfrak{M}_\odot)^{-1} \times (\dot{\mathfrak{M}}/\dot{\mathfrak{M}}_c) \ [\mathrm{erg/cm^2 s}] \ . \quad (5.33)$$

The boundary between the two regions is within the radii derived in the two regimes

$$\frac{R}{R_C} = 6.4 \times 10^3 \alpha^{2/11} f^{-16/11} (\mathfrak{M}/\mathfrak{M}_\odot)^{2/11} \times (\dot{\mathfrak{M}}/\dot{\mathfrak{M}}_c)^{16/11} \quad (5.34)$$

$$\frac{R}{R_C} = 78.6\,\alpha^{2/21} f^{-16/21} (\mathfrak{M}/\mathfrak{M}_\odot)^{2/21} (\dot{\mathfrak{M}}/\dot{\mathfrak{M}}_c)^{16/21} \ . \quad (5.35)$$

The inner region is that dominated by radiation pressure and its extension increases with the accretion rate. To complete the solution one should compute the case $P_{gas} \sim P_{rad}$ numerically.

**TABLE II**

List of the Accretion Disk Structure Equations

| | | |
|---|---|---|
| [1] | $\Sigma = 2H\rho$ | definition of surface density |
| [2] | $\frac{H}{R} = \frac{V_s}{V_k}$ | vertical hydrostatic equilibrium |
| [3] | $V_s^2 = \frac{P}{\rho}$ | sound velocity |
| [4] | $\dot{M} = 2\pi R\Sigma V$ | mass conservation |
| [5] | $Q^+ = \frac{1}{2\pi R} l \frac{d\Omega}{dR} f\dot{M} \overset{*}{=} \frac{3GM\dot{M}f}{4\pi R^3}$ | viscous heat production rate |
| [6] | $\pi\nu\Sigma = \frac{1}{2}\dot{M}fl\left(\frac{d\Omega}{dR}\right)^{-1} \overset{*}{=} \frac{1}{3}\dot{M}f$ | angular momentum balance |
| [7] | $Q^- = \frac{4acT^4}{3\rho H k}$ | heat loss (optically thick case) |
| [8] | $F = \frac{1}{2}Q^-$ | radiation flux |
| [9] | $Q^+ = Q^- + q$ | heat balance |
| [10] | $\nu = \alpha V_s H$ | viscosity law |
| [11] | $P = \frac{\rho}{\bar{\mu} m_p}\rho T + \frac{1}{3}aT^4$ | equation of state |
| [12] | $f = \frac{\epsilon - l(R_{in})}{l} \overset{*}{=} 1 - \left(\frac{R_{in}}{R}\right)^{1/2}$ | radial function |
| [13] | $\Omega = \frac{l}{R^2}$ | angular velocity |
| [14] | $V_\varphi = \frac{l}{R}$ | azimuthal velocity |
| [15] | $R_{in} = R_{in}(\ldots) \overset{*}{=} \frac{6GM}{c^2}$ | inner edge |
| [16] | $V_\varphi = V_\varphi(\ldots) \overset{*}{=} V_k$ | azimuthal velocity |
| [17] | $q = q(\ldots) \overset{*}{=} O$ | horizontal heat flux |
| [18] | $\alpha = \alpha(\ldots) \overset{*}{=} \text{const}$ | viscosity law |
| [19] | $\kappa = \kappa(\ldots) \overset{*}{=} \kappa_{es}$ | opacity |
| [20] | $V_k = \left(\frac{GM}{R}\right)^{1/2}$ | Keplerian velocity |

The symbol $\overset{*}{=}$ indicates equality valid only in the case of the standard disk model. Other equalities are more general. The location of the inner edge in the table refers to the black hole accretion (onto a nonrotating hole).

A. *Single Disk with Fixed Accretion Rate*

We shall discuss, as an example, two disk models with $\mathfrak{M} = 10\,\mathfrak{M}_\odot$ and $\mathfrak{M} = 10^8\,\mathfrak{M}_\odot$ central black holes, both having $\alpha = 0.1$ and $\dot{\mathfrak{M}}/\dot{\mathfrak{M}}_c = 0.1$. The functions $T(R)$ and

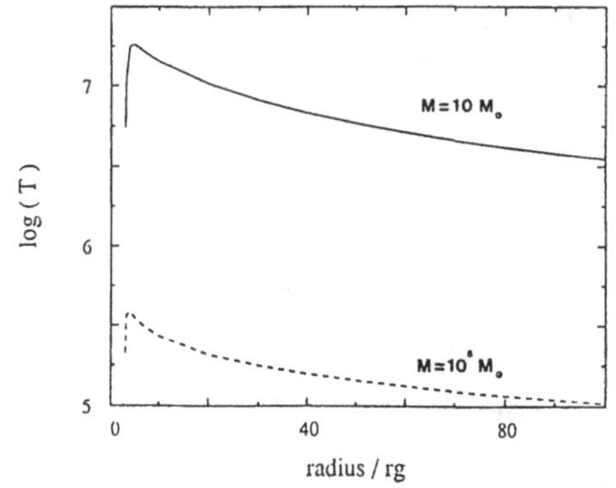


FIG. 7(a)–Equatorial temperature (K) profile for $\dot{\mathfrak{M}} = 0.1\,\dot{\mathfrak{M}}_c$, $\alpha = 0.1$, $\mathfrak{M} = 10\,\mathfrak{M}_\odot$ (upper curve) $\dot{\mathfrak{M}} = 0.1\,\dot{\mathfrak{M}}_c$, $\alpha = 0.1$, $\mathfrak{M} = 10^8\,\mathfrak{M}_\odot$ (lower curve).

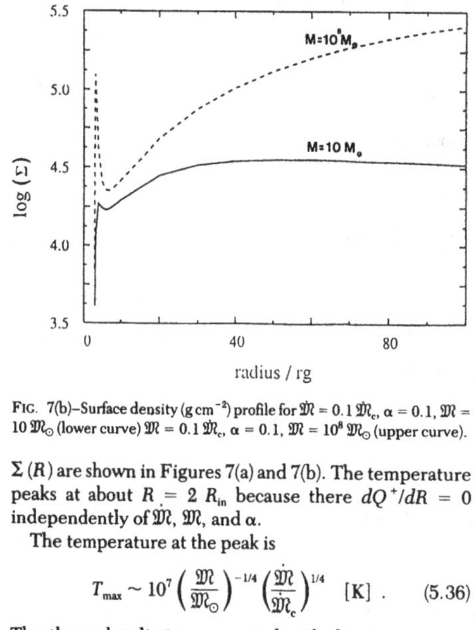


FIG. 7(b)–Surface density (g cm$^{-2}$) profile for $\dot{\mathfrak{M}} = 0.1\,\dot{\mathfrak{M}}_c$, $\alpha = 0.1$, $\mathfrak{M} = 10\,\mathfrak{M}_\odot$ (lower curve) $\dot{\mathfrak{M}} = 0.1\,\dot{\mathfrak{M}}_c$, $\alpha = 0.1$, $\mathfrak{M} = 10^8\,\mathfrak{M}_\odot$ (upper curve).

$\Sigma\,(R)$ are shown in Figures 7(a) and 7(b). The temperature peaks at about $R = 2\,R_{in}$ because there $dQ^+/dR = 0$ independently of $\mathfrak{M}$, $\dot{\mathfrak{M}}$, and $\alpha$.

The temperature at the peak is

$$T_{max} \sim 10^7 \left(\frac{\mathfrak{M}}{\mathfrak{M}_\odot}\right)^{-1/4} \left(\frac{\dot{\mathfrak{M}}}{\dot{\mathfrak{M}}_c}\right)^{1/4} \; [\mathrm{K}] \; . \qquad (5.36)$$

The thermal radiation connected with this temperature has maximal frequency $\nu_{max} = kT_{max}/h$ which equals

$$h\nu_{max} \simeq \left(\frac{\mathfrak{M}}{\mathfrak{M}_\odot}\right)^{1/4} \left(\frac{\dot{\mathfrak{M}}}{\dot{\mathfrak{M}}_c}\right)^{1/4} \; [\mathrm{keV}] \; . \qquad (5.37)$$

Therefore, optical and X-ray thermal photons can be produced by accretion disks around not-very-massive black holes ($\mathfrak{M} \simeq 10\,\mathfrak{M}_\odot$). Accretion onto supermassive black holes ($\mathfrak{M} \sim 10^8\,\mathfrak{M}_\odot$) can produce optical and ultraviolet photons. These and the following estimates refer to the optically thick case.

The temperature at the outer parts of the disk ($R/R_{in} \to \infty$) is given by

$$T_{out}(R) \simeq 5 \times 10^7 \left(\frac{\mathfrak{M}}{\mathfrak{M}_\odot}\right)^{-1/4} \left(\frac{\dot{\mathfrak{M}}}{\dot{\mathfrak{M}}_c}\right) \left(\frac{R}{R_{in}}\right)^{3/4} .. \qquad (5.38)$$

At a given radius one can assume the Planck formula (blackbody radiation) for the thermal spectrum, $S_\nu(\nu, T)$, with the temperature corresponding to this radius: for small frequencies (Rayleigh Jeans) $S_\nu \sim \nu^2$, and for the high ones $S_\nu \sim \nu^3 e^{-h\nu/kT}$. Integrating this over the whole disk one gets the approximate shape of the standard spectrum. The contributions from different disk regions form two parts, one called blackbody disk spectrum, which behaves like $\nu^{1/3}$, and the other one called modified blackbody with the behavior like $\nu^0$ (e.g., Novikov and Thorne 1973). In real situations the spectrum can be strongly influenced by the detailed radiative processes, which occur at the optically thin part of the disk (e.g., corona).

### B. *Sequences of the Disk Models, with Varying Accretion Rate*

Figure 8(a) shows the relation between $\dot{\mathfrak{M}}$ and $\Sigma$ for a sequence of models with $R = 5\,R_G$, $\alpha = 1$, and $\mathfrak{M} = 10^5$, $10^6$, $10^7$, $10^8\,\mathfrak{M}_\odot$. For the high accretion rates all the curves converge to a unique line, which asymptotically agrees with the approximate solution (eq. (4.27)). This is because there is no mass dependence there. The turning point of the $\dot{\mathfrak{M}}(\Sigma)$ curves always corresponds to $\beta = 2/5$, because

$$\frac{d\ln\Sigma}{d\ln\dot{\mathfrak{M}}} = 5\,\frac{\beta - 2/5}{2 + 3\beta} \; . \qquad (5.39)$$

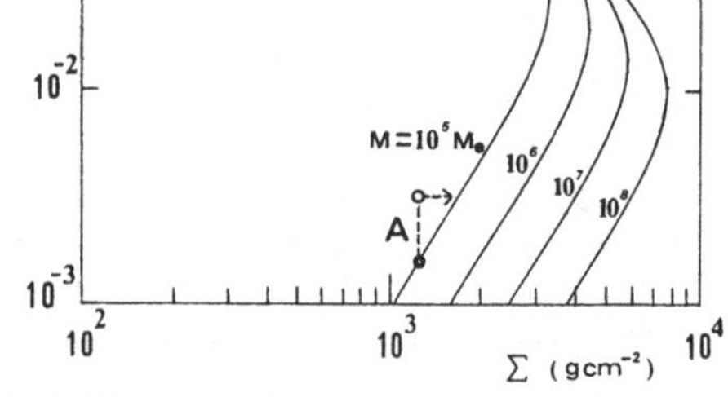

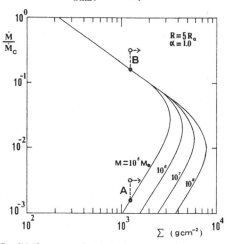

FIG. 8(a)–The sequence of standard disk accretion models for different masses of the black hole. $R = 5\,R_G$, $\alpha = 1.0$.

Its location in terms of $\dot{\mathfrak{M}}/\dot{\mathfrak{M}}_c$ depends very weakly on $\mathfrak{M}$ and $\alpha$: the turning point

$$\left(\frac{\dot{\mathfrak{M}}}{\dot{\mathfrak{M}}_c}\right) = 3.4\times10^{-3} f^{-1}\left(\frac{R}{R_G}\right)^{21/16}\alpha^{-1/8}\left(\frac{\mathfrak{M}}{\mathfrak{M}_\odot}\right)^{-1/8} . \quad (5.40)$$

C. *Stability*

We shall now show how the stability of standard models relates to the slope of the $\dot{\mathfrak{M}}(\Sigma)$ curve. Let us start with an argument which, although not quite formal, gives a quick insight into the nature of the problem (compare Bath and Pringle 1981). In Figure 8(a) the points A and B (open dots) lie above the equilibrium sequence. They can represent a perturbation of equilibrium models indicated by black dots. Thus, in both cases, the perturbed models have too high accretion rates, $\dot{\mathfrak{M}}$ (perturbed) > $\dot{\mathfrak{M}}$ (equilibrium). They are oversupplied with matter, and therefore their surface densities must increase, which shifts them to the right as indicated by arrows. This brings the model A, connected with a positive slope of the $\dot{\mathfrak{M}}(\Sigma)$ curve, back to equilibrium, indicating that the model A is stable. Model B goes further off the equilibrium curve and it is unstable. From the results of Piran (1978) one can obtain that

$$\frac{d\ln\dot{\mathfrak{M}}}{d\ln\Sigma} > 0 \quad (5.41)$$

is both sufficient and necessary for viscous stability. Piran introduced the phenomenological parameters $\mathcal{K}$, $\mathcal{L}$, $\mathcal{M}$, $\mathcal{N}$ to describe dissipative processes:

$$\mathcal{K} \equiv \left(\frac{\partial\ln Q^-}{\partial\ln H}\right)_\Sigma \qquad \mathcal{L} = \left(\frac{\partial\ln Q^-}{\partial\ln\Sigma}\right)_H$$

$$\mathcal{M} \equiv \left(\frac{\partial\ln Q^+}{\partial\ln H}\right)_\Sigma \qquad \mathcal{N} = \left(\frac{\partial\ln Q^+}{\partial\ln\Sigma}\right)_H . \quad (5.42)$$

In the case of the standard disk model these coefficients have well-determined values,

$$\mathcal{K}=4\frac{1+\beta}{4-3\beta};\ \mathcal{L}=\frac{-\beta}{4-3\beta},\ \mathcal{M}=2,\ \mathcal{N}=1 , \quad (5.43)$$

but it would be convenient to give general criteria here, in which these coefficients are arbitrary.

The general stability criteria found by Piran can be written in such a way that they directly correspond to the thermal stability criterion

$$\mathcal{K} - \mathcal{M} > 0 , \quad (5.44)$$

and the viscous stability criterion

$$\frac{\mathcal{N}\mathcal{K} - \mathcal{M}\mathcal{L}}{\mathcal{K} - \mathcal{M}} > 0 . \quad (5.45)$$

For stable disk models any of these conditions is necessary, but only two of them together are sufficient for stability. Because in the case of the standard thin-disk model $\mathcal{N}\mathcal{K} - \mathcal{M}\mathcal{L} = (4 + 6\beta)/(4 - 3\beta)$ is positive, the necessary and sufficient criterion for both thermal and viscous instabilities is either of the three equivalent conditions:

$$\left[\mathcal{K} - \mathcal{M} > 0\right] \Leftrightarrow \left[\beta > 2/5\right] \Leftrightarrow \left[d\ln\dot{\mathfrak{M}}/d\ln\Sigma > 0\right] . \quad (5.46)$$

This formally proves our point.

*Thermal Instabilities*. When equation (5.46) is not fulfilled, thermal perturbations grow exponentially. In the limit of wavelength of the perturbation $\Lambda$ going to infinity, $\Lambda/H \gg 1$, the rising time does not depend on the wavelength. It is equal to

$$T_{th} = t_{th}\frac{56 - 57\beta - 3\beta^2}{30(2/5 - \beta)} . \quad (5.47)$$

Here $t_{th}$ is the characteristic time for thermal processes

$$t_{th} = 2\pi/\Omega\alpha . \quad (5.48)$$

This type of instability was discovered and studied by Pringle, Rees, and Pacholczyk (1973) and Shakura and Sunyaev (1976).

*Viscous Instability*. When equation (5.44) is not also fulfilled viscous instabilities grow exponentially at the rate (for $\Lambda/H \gg 1$)

$$T_{vis} = t_{vis}\left(\frac{\Lambda}{R}\right)^2\frac{3}{10}\frac{2 - 3\beta}{2/5 - \beta} . \quad (5.49)$$

Here $t_{vis}$ is the characteristic time for viscous processes:

$$t_{vis} = \frac{t_{th}}{(H/R)^2} . \quad (5.50)$$

This type of instability was found by Lightman and Eardley (1974) and Lightman (1974).

D. *Possibility of a Limit-Cycle Behavior*

The equations describing the standard $\alpha$ disk models (with $\alpha$ = const, $k = k_{es}$ = const, $q = 0$) are all linear in their logarithmic versions with the sole exception of the equation of state [11].

We have seen that the nonlinearity causes the $\dot{\mathfrak{M}}(\Sigma)$ curves describing equilibrium sequences of models (at a given radius) to bend. At the turning points the stability properties change, as a positive slope of $\dot{\mathfrak{M}}(\Sigma)$ corresponds to stable and a negative slope to unstable disk models. Abramowicz *et al.* (1987) have recently shown that the $\dot{\mathfrak{M}}(\Sigma)$ curve bends again, to form another stable branch, at higher accretion rates $\dot{\mathfrak{M}} \sim \dot{\mathfrak{M}}_c$. This bending is also due to additional nonlinearity, connected with a new cooling mechanism operating at $\dot{\mathfrak{M}} \sim \dot{\mathfrak{M}}_c$. In this case the Shakura-Sunyaev model is not adequate and one must consider its modifications to include horizontal pressure gradients and horizontal heat transport (Maraschi, Reina, and Treves 1974). Thus the $\dot{\mathfrak{M}}(\Sigma)$ curve is characteristi-

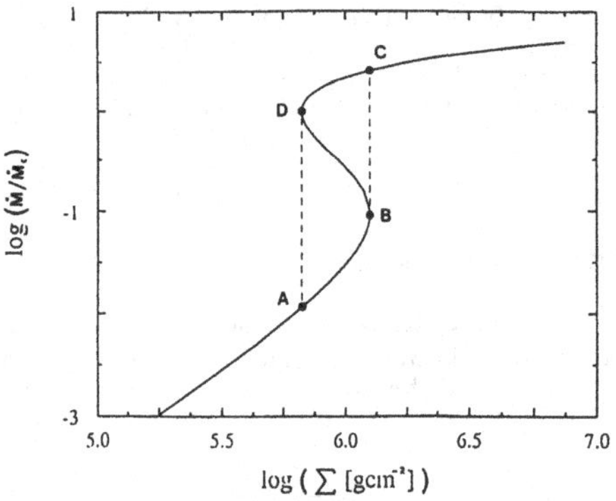


FIG. 8(b)—Limit cycle behavior $\mathfrak{M} = 10\,\mathfrak{M}_\odot$, $\alpha = 1$, $R = 5\,R_G$.

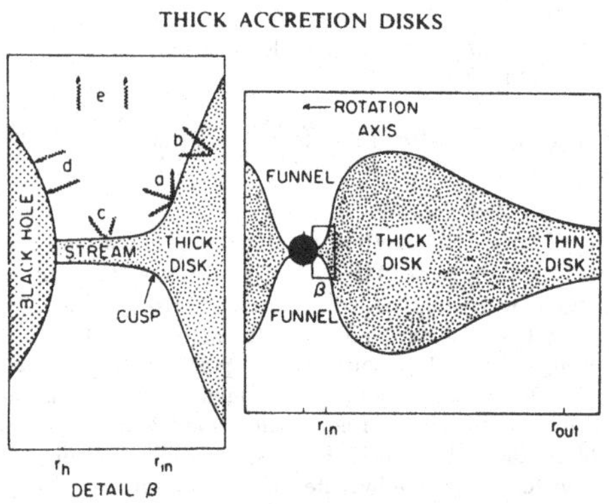


FIG. 9—A sketch of a thick accretion disk (from Abramowicz *et al.* 1980).

cally S-shaped with the upper and lower branches corresponding to stable and the middle branch to unstable disk models. Bath and Pringle (1983) were the first to notice the similarity of this situation to the one known in nonlinear dynamics, where an S-shaped phase portrait of a system indicates a limit-cycle behavior.

If the accretion rate fixed by some outside condition lies in the instability strip (region AB of Fig. 8(b)), then stationary accretion is impossible. Consider point A. The accretion rate is smaller than the supply rate causing local oversupply and an increase of surface density. The disk must evolve in the direction indicated by the arrow up to point B from which further evolution is only possible after a jump to C. Here, however, the accretion rate is higher than the supply rate. The surface density now decreases and the disk evolves down to point D, where another jump, to A, must take place. This closes the cycle.

Such limit-cycle behavior is observed in accretion disks in dwarf novae (see Section VII).

## VI. Basics of Thick Accretion Disks

As shown in the previous section (in particular eqs. (5.21) and (5.28)), the geometrical thickness of the disk $H(R)$ becomes comparable to the radius $R$ at the inner boundary of the disk, if the accretion rate is such as to produce a luminosity of the order of the Eddington one. A crucial assumption of the standard disk model, i.e., the disk thinness, becomes invalid; hence, the necessity of models for geometrically thick disks. These have been proposed originally by Paczinski and Wiita (1980) and Jaroszynski, Abramowicz, and Paczynski (1980).

As explained by Abramowicz, Calvani, and Nobili (1980), super-Eddington luminosities do not imply any dramatic consequences for the equilibrium structure of disks, in particular no strong winds as a general property. On the other hand, the high luminosity implies that the horizontal pressure gradient, neglected in the standard model, becomes dynamically important, so that the Keplerian approximation breaks down (e.g., Maraschi *et al.* 1974).

For a wide class of non-Keplerian angular momentum distributions the inner part of the disk can form a toroidal structure, in some cases resembling a sphere with two deep and narrow funnels along the rotation axis (see Fig. 9).

The physics of thick disks is more complex than that of thin disks: assuming that axially symmetric thin disks are one dimensional, we find that the horizontal and vertical structures separate, and disk models are described by ordinary differential equations, exactly integrable in the general stationary case. The stationary equilibria are therefore always given by a set of algebraic equations. As we have discussed already in the previous section, they are linear in the asymptotic cases when either the gas pressure or the radiation pressure dominates, the opacity is given by Thomson scattering or free-free absorption, and the viscosity law is the Shakura-Sunyaev form (eq. (5.4)). In these cases one has an explicit analytic solution in the deep interior of the disk, i.e., in the equatorial plane. Every physical quantity is expressed in terms of three control parameters: viscosity, accretion rate, and mass of the central object $\mathfrak{M}$ (and the distance from the center $R$). The dependence on the vertical coordinate Z follows from the physical boundary conditions on the surface $Z = H(R)$ and the fact that $H \ll R$ everywhere.

The deep interior of a thick disk is, on the contrary, an extended region. Its structure is described by rather complicated partial-differential equations. The complexity is caused by (a) a priori unknown angular momentum—the angular momentum cannot be assumed Keplerian as for the thin disk, (b) importance of the horizontal gradients of pressure and temperature, and (c) nonlocal heat balances. Note that all these effects are also important very close to the inner edge of thin disks around compact objects, but they are ignored in the standard model. In the following we will quote some of the main results of the theory of thick disks, referring to Abramowicz *et al.* (1980) for a detailed treatment.

The most complex part of a super-Eddington accretion flow is located near the central compact object. In particular, for accretion onto a black hole (or a neutron star), a self-crossing equipotential surface—the Roche lobe—which has the shape of a cusp, exists due to general relativistic gravity competing with the disk rotation (see Abramowicz *et al.* 1980). The location of the cusp $R = R_{in}$ follows from the condition that the Keplerian angular momentum at $R = R_{in}$ (given by the gravity of the hole) equals the angular momentum of the rotating matter there. The nonzero thickness of the disk in the cusp implies Roche-lobe overflow and, as in the case of close binaries, dynamical mass loss. The gas lost through the cusp goes toward the central body with roughly free-fall velocity. The cusp should be considered, therefore, as the inner edge of the disk. The accretion rate through the cusp $\dot{\mathfrak{M}}_{in}$ and the energy loss rate $L_{in}$ scale as

$$\dot{\mathfrak{M}}_{in} \sim \Sigma_{in} H_{in} \tag{6.1}$$

$$L_{in} \sim \Sigma_{in} H_{in}^3 \; . \tag{6.2}$$

The cusp is located between the marginally bound and marginally stable circular orbit and very close to the sonic point $R = R_s$. For a Schwarzschild black hole it is

$$2\,R_G < R_S < 3\,R_G \; . \tag{6.3}$$

The lower limit corresponds to the location of a circular orbit which is marginally bound in the sense that all the orbits with $R < 2R_G$ are unbound, and all those with $R > 2R_G$ are bound. The upper limit corresponds to the marginally stable circular orbit: orbits with $R < 3R_G$ are unstable and those with $R > 3R_G$ are stable. When $\dot{\mathfrak{M}} \ll \dot{\mathfrak{M}}_c$ the cusp and the sonic point coincide with the marginally stable orbit $R_S < R_{in}\, R_G$, while for $\dot{\mathfrak{M}} \gg \dot{\mathfrak{M}}_c$ the cusp goes very close to the marginally bound orbit. The energy per particle released by the process is the binding energy of the circular orbit located at the cusp. Since this goes to zero at the marginally bound orbit at which the cusp ends for $\dot{\mathfrak{M}} \gg \dot{\mathfrak{M}}_c$, also the efficiency of the process tends to zero. It can be shown, in particular, that the radiated luminosity $L$ grows only logarithmically with the accretion rate for $\dot{\mathfrak{M}} \gg \dot{\mathfrak{M}}_c$. One can therefore conclude that stationary disks with $\dot{\mathfrak{M}} \gg \dot{\mathfrak{M}}_c$ have low efficiency compared to standard disks. However, they have $L > L_E$ for $\dot{\mathfrak{M}} \gg \dot{\mathfrak{M}}_c$.

The thick-disk photosphere emits relatively soft thermal radiation with a spectrum close to a blackbody, characterized by temperatures not very different from the corresponding ones for thin disks. Deep in the funnel at $R = 5\,R_G$ one has

$$T_{max} \sim 10^7 (\dot{\mathfrak{M}}/\dot{\mathfrak{M}}_c)\,(\mathfrak{M}/\mathfrak{M}_\odot)^{-1/4}\,[\mathrm{K}] \tag{6.4}$$

$$h\nu_{max} \sim 1 \times (\dot{\mathfrak{M}}/\dot{\mathfrak{M}}_c)^{1/4}\,(\mathfrak{M}/\mathfrak{M}_\odot)^{-1/4}\,[\mathrm{keV}] \; . \tag{6.5}$$

The interior of the funnel is much hotter than the rest of the disk surface and therefore the same thick disk appears different when observed at different aspect angles.

The stability of thick accretion disks is a subject of very active research at present. Papaloizou and Pringle (1984, 1985) demonstrated that nonaccreting perfect fluid tori orbiting a Newtonian center of gravity are subject to violent global, nonaxially symmetric instability. It is not known at present whether this instability destroys the astrophysically relevant tori (thick disks) because: (1) Frank and Robinson's (1987) numerical models suggest that the importance of this instability decreases with increasing width of the torus. For astrophysically relevant, very big tori, the growth rate of the instability is too slow to be of importance. (2) Blaes (1987) showed that even a modest amount of accretion completely stabilizes the tori; and (3) Goodman and Narayan (1987) demonstrated that self-gravitation also has a stabilizing effect. More theoretical work and sophisticated 3-dimensional hydrodynamical simulations are needed before a final conclusion on the stability of thick disks can be reached.

## VII. Accretion and the Realm of Observations

Within the Galaxy, accretion plays a fundamental role in binary systems where the normal star transfers matter to a compact companion. This may happen through a wind or when the normal star overflows the critical equipotential contour connecting the two components of the binary (Roche-lobe overflow). The first instance requires an early-type star, the second a close binary and an evolved companion. Therefore, systems in which the mass transfer occurs through the wind usually contain a massive star and are relatively young, while systems accreting through Roche-lobe overflow are associated with an older stellar population. The accreting object may be a white dwarf, a neutron star, or a black hole. The combination of these possibilities gives rise to a variety of systems carrying information on properties of the collapsed objects (like the spin and the magnetic field), on the astrophysical evolution of the binary, as well as on the physics of the accretion process. Some phenomenological aspects of accreting binaries are illustrated in the following.

An entirely different astrophysical field for which accretion may be relevant is the activity observed in the nuclei of galaxies. The enormous luminosities and the small scales, implied by variability, strongly suggest that the powerhouse is gravitational. However, the connection of the observed phenomena with accretion is still under debate. Some positive indications on the relevance of the accretion process in AGNs will be presented at the end of the chapter.

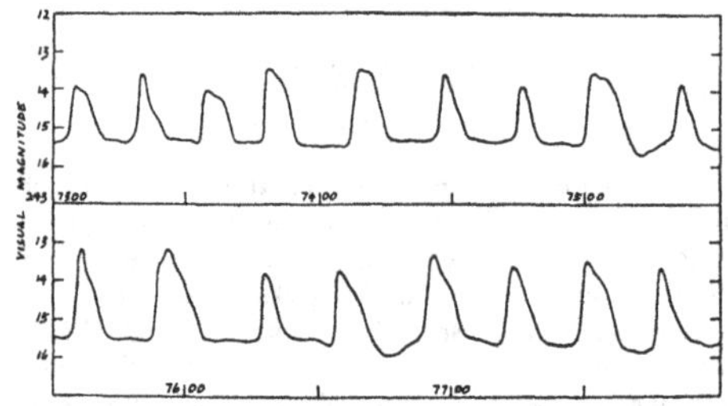

FIG. 10–Light curve of FO Aql (from Glasby 1970).

A. *Cataclysmic Variables*

Cataclysmic variables (CV) are interacting binary systems consisting of a white dwarf (primary) and a late-type star (secondary) transferring mass to the dwarf. The two stars closely orbit each other and the orbital period is always very short: for U Geminorum about four hours. The mass transfer occurs through the Roche lobe, which lies very close to the surface of the secondary.

Direct and detailed observations of disks in cataclysmic variables provide most important data on accretion. From the optical light curves of the eclipsing systems one can infer the contributions of four components: the normal star, an extended accretion disk, a hot spot on the outer rim of the disk, and a bright point source at the disk center. With new techniques of data acquisition and processing, the light distribution in the system can be reconstructed starting from general assumptions.

1. *Dwarf Novae*. One of the basic properties of this subclass of CV is the extreme variability, which is apparent at all wavelengths. In dwarf novae it has the form of recurrent eruptions (see Fig. 10) which may be interpreted as accretion-disk instabilities (e.g., Smak 1984). In these systems the optical (and most probably the total) luminosity of the hot spot is much higher than that of the disk during the quiescence periods. This suggests that during these periods matter is supplied to the outer part of the disk, but only very little actually flows through the disk. From the observation of the changes in luminosity of the dwarf, disk, and hot spot, it was estimated that the amount of matter accreted onto the spot during the quiescence equals the amount accreted onto the dwarf during eruptions. This suggests that matter is gradually accumulated in the outer parts during quiescent periods and accreted onto the dwarf during the eruptions. The eruptions are therefore high-accretion-rate events triggered by an instability in the disk. The recurrence of the eruptions is due to a limit cycle for the disk instability connected with a nonunique dependence of the surface density on the accretion rate, similar to the "S curve" discussed in Section V.F (see also Bath and Pringle 1981, 1983). The calculation of how the limit-cycle behavior at a given radius might translate into a global disk instability propagating over the whole disk structure has been made by several authors (e.g., Meyer and Meyer-Hofmeister 1981; Smak 1984; Papaloizou, Faulkner, and Lin 1983).

Contrary to the case of stationary optically thick disks, where the emitted power and temperature do not depend on the viscosity parameter $\alpha$, the nonstationary accretion models of dwarf novae strongly depend on $\alpha$. This gives a unique possibility of an observational evaluation of the viscosity parameter. The typical value needed to fit the observational data to the disk instability models is $\alpha = 0.1$.

It should be noted, however, that in these models $\alpha$, during the eruptions, is different (higher) than in quiescence. This may follow from the fact that viscosity is due to turbulence connected with a strong convection present during eruptions (Smak 1984).

2. *AM Herculis-Like Objects*. Another especially interesting subclass of cataclysmic variables is represented by AM Her-like objects. These are systems where the white dwarf is endowed with a strong magnetic field $B \sim 10^7$ G. The orbital periods are a few hours, and the separation of the components is smaller than the Alfven radius (eq. (3.7)). In these conditions magnetic torques lock the white-dwarf spin period to the orbital one, and the magnetic field prevents the formation of a disk. The mass transfer occurs directly through a magnetic funnel, where essentially radial inflow is established. As discussed in Section III one expects the formation of a shock, whose temperature is roughly given by equation (3.4). The gas heated at this temperature will emit mainly by bremsstrahlung and cyclotron radiation, because of the strong

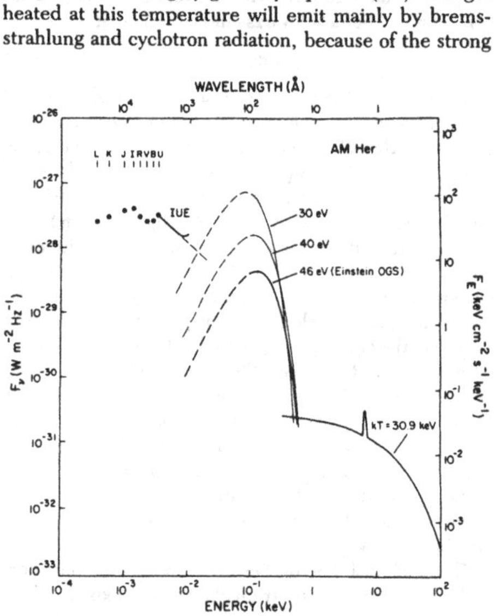


FIG. 11–Spectrum of AM Her (from Lamb 1985).

magnetic field. Part of this energy and of the kinetic energy of the shocked electrons liberated just above the star will be reprocessed at the surface and reemitted as blackbody radiation. The resulting spectrum is rather complicated, since one should take into account the details of radiative transfer. However, in a first approximation one expects three main components, one corresponding to the bremsstrahlung photons at $kT \simeq 10$ keV (eq. (3.2)), one to the cyclotron photons at IR-optical frequencies, and one to blackbody photons at the temperature given by equation (3.1). In Figure 11 we report the observed spectrum of AM Her distinguishing the three components. This agrees in general, but not in detail, with the expectations, in that the blackbody component appears too dominant. For a detailed discussion of the emission of AM Her-like objects we refer, for instance, to Liebert and Stockman (1985) and Lamb (1985).

B. *X-Ray Binaries*

Accreting binary systems containing a neutron star or a black hole are usually, though vaguely, defined as X-ray binaries. The X-ray luminosity ($10^{37}$ erg $s^{-1}$) is in fact much higher than for cataclysmic variables, due to the fact that the absolute efficiency of accretion reaches the limiting value of 10%. A distribution of the ratio of the X-ray to optical luminosities of identified galactic sources is shown in Figure 12, where the different classes are indicated.

If the neutron star has a high magnetic field ($\sim 10^{12}$ G), the accretion flow is guided through a magnetic funnel (see Section III). The radiative transfer within the funnel is very complex; the emerging radiation is expected to be beamed. The rotation of the neutron star, together with its funnel, gives rise to a periodic modulation of the observed X-ray flux, whence the name of X-ray pulsators. Examples of X-ray light curves are reported in Figure 13, where the modulation with the spin period is apparent. The X-ray spectrum is usually quite hard, and in a few cases cyclotron features around 50 keV corresponding to a field of $10^{12}$ G have been observed. X-ray pulsators are usually found in systems with a massive young companion. They have orbital periods of order of days and spin periods of order of a second. The mass transfer occurs via a wind.

If the companion of the neutron star is a low-mass (low-luminosity) star, the emission of the system is strongly dominated by the accretion power, which makes it difficult to derive the parameters of the normal star. These systems are old, and the magnetic field is supposedly small ($10^9$ G), so that it plays a minor role in the dynamics of accretion. The low-mass X-ray binaries (LMXRB) would seem to be ideal systems for the study of accretion disks. However, the actual situation is complicated by the effects on the structure of the accretion disk of the high central X-ray luminosity, released at the boundary of the disk with the neutron star. In particular it has been recognized that above the accretion disk, a corona is present which in some cases may be quite optically thick and reprocess a large fraction of the disk

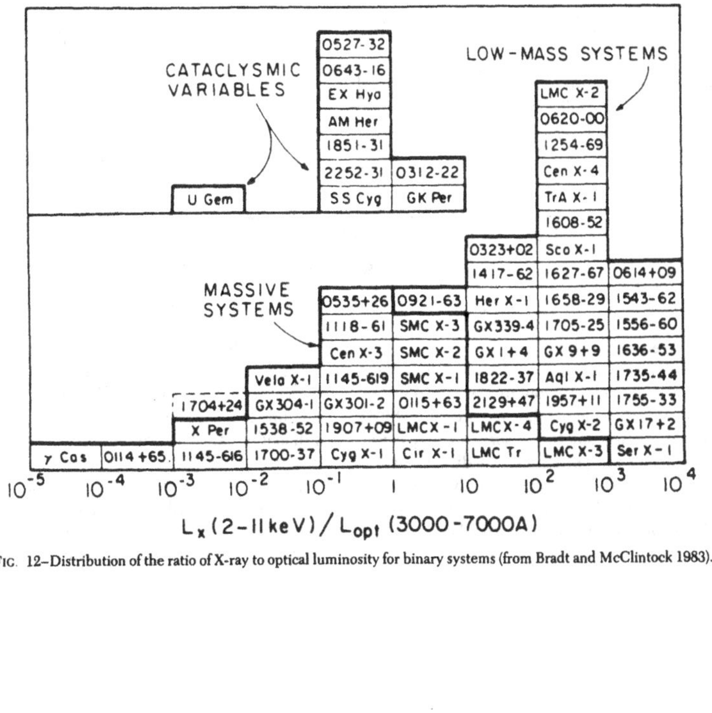


FIG. 12–Distribution of the ratio of X-ray to optical luminosity for binary systems (from Bradt and McClintock 1983).

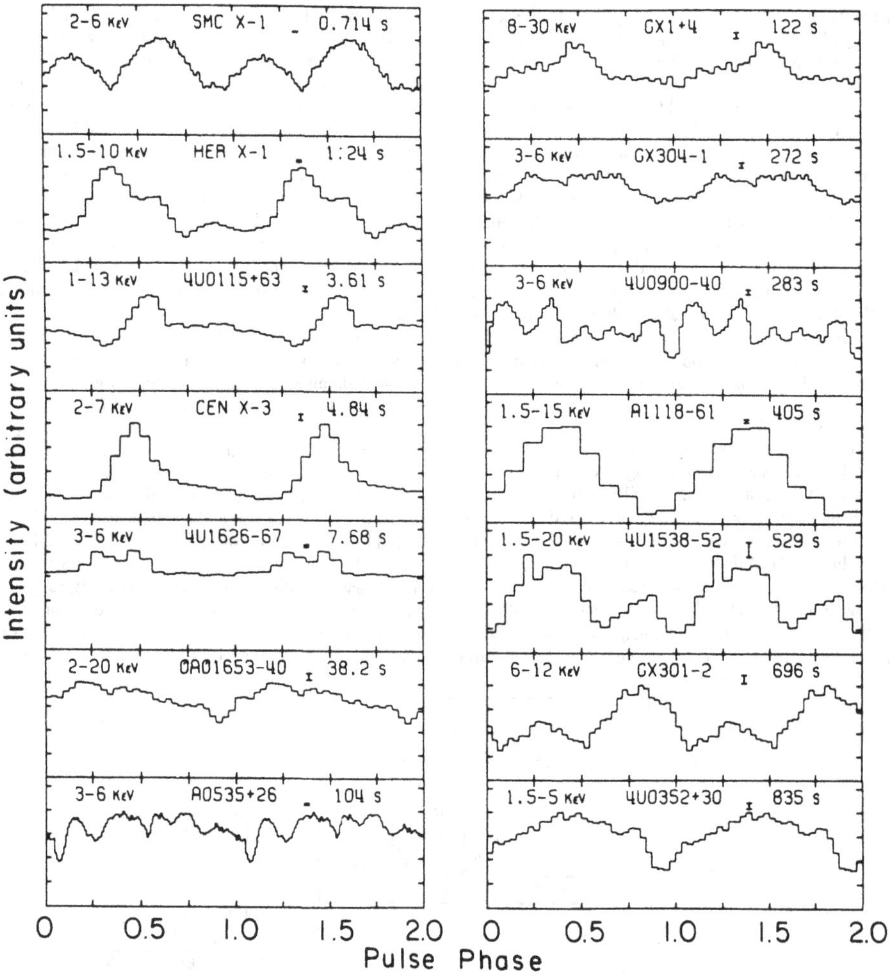


FIG. 13–Pulse profiles of some X-ray pulsators (from Joss and Rappaport 1984).

and boundary-layer luminosity.

X-ray spectra of LMXRB are generally softer than those of pulsators. Spectral models involving the superposition of a blackbody component and a thermal bremsstrahlung can satisfactorily account for the observations (see Fig. 14). While the former component may be directly associated with the neutron-star emission, as indicated by the dimension of the emitting region, the origin of the bremsstrahlung component, and in particular its relation with the emission from the disk or its corona, are still unclear. The fact that the bremsstrahlung component contains the bulk of the luminosity appears as a severe difficulty (e.g., White, Stella, and Parmer 1988). It is possible that the spectral deconvolution used up to now is still inadequate to represent the intrinsic energy distribution.

No strictly periodic component is found in the X-ray emission of most LMXRB, which is taken as evidence of the modestness of the magnetic field. However, recently it was recognized that quasi-periodic short-lived modula-

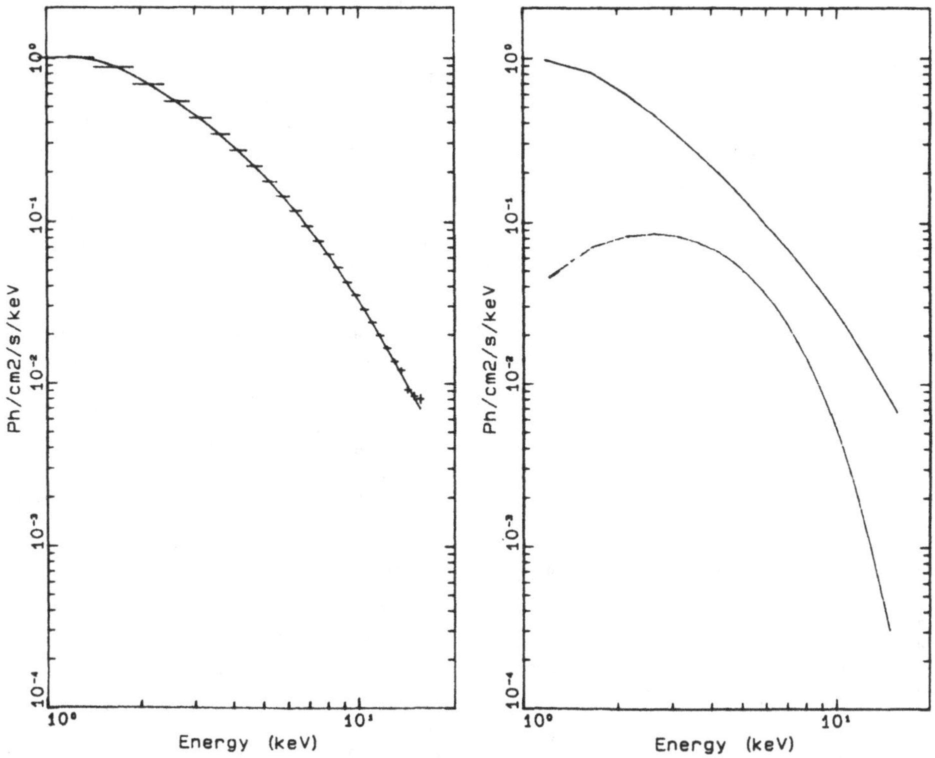


FIG. 14–X-ray spectrum of Cyg X2 and its deconvolution into a blackbody and a thermal bremsstrahlung (TB) component (courtesy of L. Chiappetti).

tion is present in various sources. Examples of power spectra are reported in Figure 15. The origin of quasi-periodic oscillations (QPO) is as yet unclear. There are proposals of a relation to accretion-disk instabilities, or to the interaction of the inner disk with an Alfven surface, with dimensions comparable to the neutron star radius. In any case it appears that QPO will become a prime instrument to gain information on the mode of accretion of LMXRB.

There are three binary X-ray sources which are thought to contain a black hole: Cygnus X-1, LMC X-3, and A 0620-00. The main evidence on the nature of the collapsed object derives from the measurement of the mass function, which yields a mass larger than the upper bound for neutron stars ($\sim 3\,\mathfrak{M}_\odot$). For two of these objects, Cyg X-1 and LMC X-3, good-quality X-ray spectra are available. They are shown in Figure 16. Both are well fitted by models in which the spectra are the product of multi-Compton scattering of a soft photon population in a hot thermal plasma. The process has been briefly described in Section IV. In such a picture the spectrum is determined by the optical depth and temperature. For both objects the derived temperatures are much lower and the optical depths larger than those expected in the proximity of the hole in a spherical model (§ IV). A general and quantitative treatment of the disk spectrum in the Comptonization regime has not been given yet. In any case the optical depths are smaller than would be obtained applying the standard model in a consistent way. In particular, the observed trend of lower optical depth with lower luminosity is opposite to that predicted by the "standard" model.

## C. *Active Galactic Nuclei*

Active Galactic Nuclei (AGN) have been the subject of several excellent reviews in recent years (e.g., Begelman, Blandford, and Rees 1984; Lawrence 1987). An enlightening discussion of models of accretion onto massive black

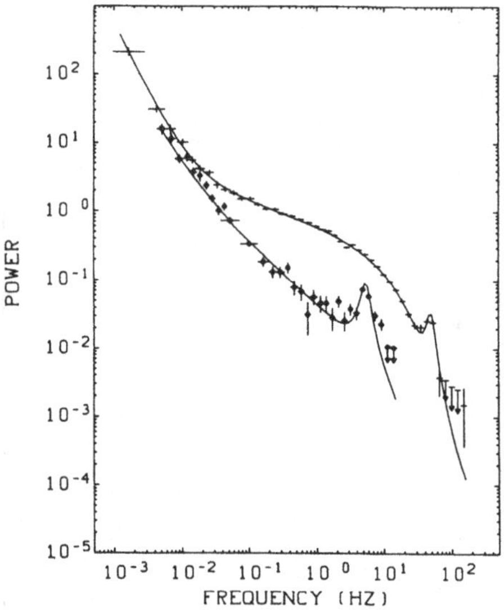


FIG. 15–Power spectrum of sources exhibiting quasi-periodic oscillations of Cyg X2 (from Hasinger 1988).

holes in relation to AGNs was given by Rees (1984). It is beyond the scope of this paper to give an overview of various possible schemes, since the activity of galactic nuclei involves many types of disparate phenomena which are interrelated. We merely wish to stress here two aspects. The first concerns the broad-band energy distribution and the second the variability of AGNs.

The continuum emitted by AGNs extends, in most cases, from the far infrared to the X-rays. This is probably due to the contribution of several components, which are not easy to distinguish on the basis of narrow-band spectra alone. Only recently, broad-band energy distributions have been determined for a large number of objects (e.g., Edelson and Malkan 1986). From this work it clearly appears that the optical-ultraviolet continuum of most quasars and Seyfert galaxies is characterized by a flat shape, energy spectral index 0.5, at variance with the overall energy distribution, which is generally steeper. This excess, which is generally referred to as the "3000 Å bump" (see Fig. 17) is such as to imply that most of the power is emitted in the far-ultraviolet band and is strongly suggestive of a thermal process, although information on the shape of the peak is still scarce (Edelson and Malkan 1986). The idea that the bump may be due to optically-thick emission from an accretion disk, originally proposed by Shields (1978) and Malkan and Sargent (1982), encounters increasing consensus, although detailed spectral modeling has only recently begun (e.g., Czerny and Elvis 1987; Madau 1988; Malkan and Sun 1988). This spectral component appears as the most promising link between AGNs and accretion models, and we expect that this area will see a major development over the next few years.

A second crucial issue is that of time variability. Many Seyfert galaxies and a few quasars have been observed to be variable in X-rays on time scales of hours. The data do not allow us to define unambiguously the characteristic time scale; nevertheless, it is interesting to evaluate the maximum observed luminosity variation and compare with the predictions of models. Since it is difficult to exceed the Eddington luminosity (see Section IV), the larger the luminosity, the larger should be the mass of the central black hole. Assuming that the minimum time scale of variability is related to the size of the gravitational radius of the black hole, one expects that the larger the luminosity is the longer the variability time scale becomes. It is interesting that the data, though still somewhat confused, show indeed some correlation of the time scale with the luminosity (see Fig. 18).

In particular, the present data seem to be in agreement with the expectations of the accretion models in the case of the X-ray emission of Seyfert galaxies and low-luminosity quasars, but not for BL Lacertae objects which could represent a different class of AGNs.

It would be a major step forward in our knowledge of AGNs if X-ray astronomy in the near future should provide a statistically firm basis for this claim.

## VIII. Conclusions

The formidable observational advances in the last two decades have assured a fundamental role to accretion theory in today's astrophysics. A comparison can be made, perhaps, with the theory of energy production by nuclear processes, which was one of the triumphs of astrophysics in the thirties. At present the basic frame of accretion theory seems well established, with the theory entering into a phase of maturity. The principal problem is now a direct connection of the theory with observations. The main difference with respect to stellar interiors is that, due to the very different opacity and time scale involved, the former are extremely steady while, in the case of accretion, variability plays a major role. One can hope to reconstruct the dynamics from the study of variability. Furthermore, one of the fascinating aspects of the accretion process is that the region of maximum energy production may in some cases be directly observable.

Substantially new input will derive in the near future from spectral and variability observations, especially at UV and X-ray frequencies. Progress in the theory is expected from numerical work for constructing more realistic models along the lines that we have tried to expose in the previous sections and, hopefully, from radically new approaches.

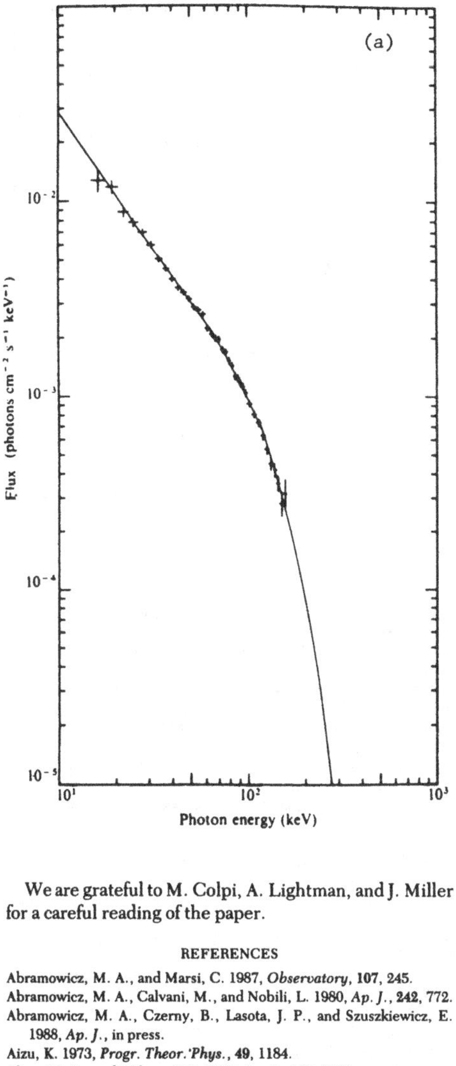


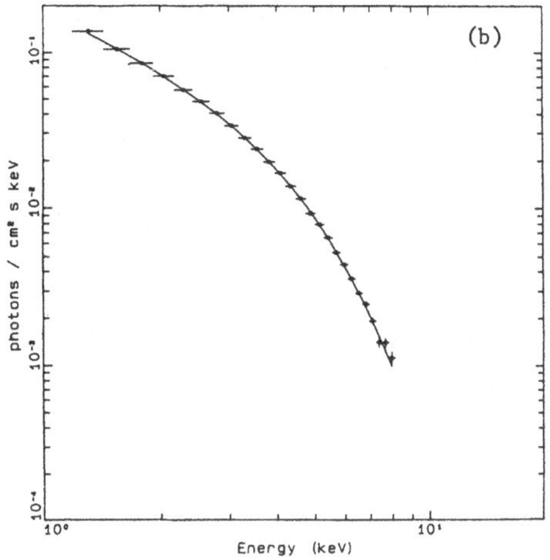


FIG. 16–X-ray spectra of two black hole candidates: Cyg X1 (a) and LMC X3 (b) (from Sunyaev and Truemper 1979 and Treves *et al.* 1988).

We are grateful to M. Colpi, A. Lightman, and J. Miller for a careful reading of the paper.

REFERENCES

Abramowicz, M. A., and Marsi, C. 1987, *Observatory*, **107**, 245.
Abramowicz, M. A., Calvani, M., and Nobili, L. 1980, *Ap. J.*, **242**, 772.
Abramowicz, M. A., Czerny, B., Lasota, J. P., and Szuszkiewicz, E. 1988, *Ap. J.*, in press.
Aizu, K. 1973, *Progr. Theor. Phys.*, **49**, 1184.
Almę, M. L., and Wilson, J. R. 1973, *Ap. J.*, **186**, 1015.
Baan, W. A., and Treves, A. 1973, *Astr. Ap.*, **22**, 421.
Barr, P. 1986, in *The Physics of Accretion onto Compact Objects*, ed. K. O. Mason, M. G. Watson, and N. E. White (Berlin: Springer).
Bath, G. T., and Pringle, J. E. 1981, *M.N.R.A.S.*, **194**, 967.
______. 1983, *M.N.R.A.S.*, **199**, 267.
Begelman, M. C. 1978, *M.N.R.A.S.*, **184**, 53.
______. 1979, *M.N.R.A.S.*, **187**, 237.
Begelman, M. C., and Chiueh, T. 1987, preprint.
Begelman, M. C., Blanford, R. D., and Rees, M. J. 1984, *Rev. Mod. Phys.*, **56**, 225.
Bisnovatyi-Kogan, G. S. 1979, *Riv. Nuovo Cimento*, **2**, 1.
Blaes, O. M. 1987, *M.N.R.A.S.*, **227**, 975.
Bondi, H. 1952, *M.N.R.A.S.*, **112**, 195.
Bradt, H. V., and McClintock, J. E. 1983, *Ann. Rev. Astr. Ap.*, **21**, 13.
Colpi, M. 1988, *Ap. J.*, **226**, 223.
Colpi, M., Maraschi, L., and Treves, A. 1984, *Ap. J.*, **280**, 319.
______. 1986, *Ap. J.*, **311**, 150.
Cowie, L. L., Ostriker, J. P., and Stark, A. A. 1978, *Ap. J.*, **226**, 1041.
Czerny, B., and Elvis, M. 1987, *Ap. J.*, **321**, 305.
Davidson, K., and Ostriker, J. P. 1973, *Ap. J.*, **179**, 585.
Edelson, P. A., and Malkan, M. A. 1986, *Ap. J.*, **308**, 59.
Elvis, M., Czerny, B., and Wilkes, B. J. 1986, in *The Physics of Accretion onto Compact Objects*, ed. K. O. Mason, M. G. Watson, and N. E. White (Berlin: Springer).
Fabian, A. C., Pringle, J. A., and Rees, M. J. 1976, *M.N.R.A.S.*, **175**, 43.
Flammang, R. A. 1982, *M.N.R.A.S.*, **199**, 833.
______. 1984, *M.N.R.A.S.*, **206**, 589.
Frank, J., and Robinson, J. 1987, preprint.
Frank, J., King, A. R., and Raine, D. J. 1985, *Accretion Power in Astrophysics* (Cambridge: Cambridge University Press).
Glasby, J. S. 1970, *The Dwarf Novae* (London: Constable and Co.).
Goodman, J., and Narayan, R. 1987, preprint.
Hasinger, G. 1988, *Astr. Ap.*, in press.
Holzer, T. E., and Axford, W. I. 1970, *Ann. Rev. Astr. Ap.*, **8**, 31.
Hoshi, R. 1973, *Progr. Theor. Phys.*, **49**, 776.
Hoyle, F., and Lyttleton, R. A. 1939, *Proc. Cam. Phil. Soc.*, **35**, 405.
Ipser, J. R., and Price, R. H. 1977, *Ap. J.*, **216**, 578.
______. 1982, *Ap. J.*, **255**, 654.
______. 1983, *Ap. J.*, **267**, 371.
Jaroszynski, M., Abramowicz, M. A., and Paczynski, B. 1980, *Acta. Astr.*, **30**, 1.
Joss, P. C., and Rappaport, S. A. 1984, *Ann. Rev. Astr. Ap.*, **22**, 537.
Klein, R. L., Stockman, H. S., and Chevalier, R. A. 1980, *Ap. J.*, **237**, 912.
Lamb, D. Q. 1985, in *Cataclysmic Variables and Low Mass X-Ray*

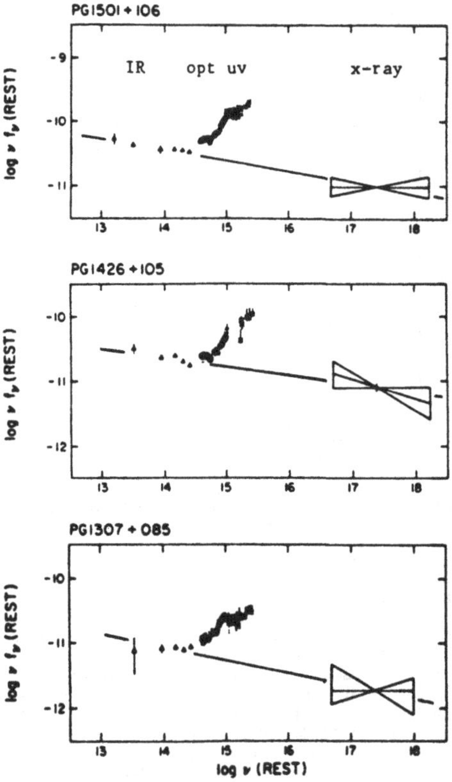


FIG. 17–Examples of blue bumps in quasars (from Elvis *et al.* 1986).

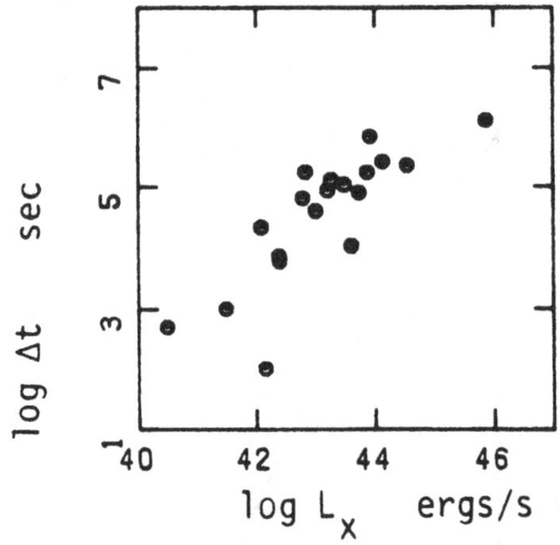


FIG. 18–Variability time scale vs. X-ray luminosity for Active Galactic Nuclei (from Barr 1986).

*Binaries*, ed. D. Q. Lamb and J. Patterson (Dordrecht: Reidel).

Lamb, F. K., Pethick, C. J., and Pines, D. 1973, *Ap. J.*, **174**, 271.

Landau, L. D., and Lifshitz, E. 1959, *Fluid Mechanics* (New York: Pergamon).

Lawrence, A. 1987, *Pub. A.S.P.*, **99**, 309.

Liebert, J., and Stockman, H. S. 1985, in *Cataclysmic Variables and Low Mass X-Ray Binaries*, ed. D. Q. Lamp and J. Patterson (Dordrecht: Reidel).

Lightman, A. P. 1974, *Ap. J.*, **194**, 429.

Lightman, A. P., and Eardley, D. M., 1974, *Ap. J. (Lettters)*, **187**, L1.

Lightman, A. P., Zdziarski, A. A., and Rees, M. J. 1987, *Ap. J. (Letters)*, **315**, L113.

Lyndell-Bell, D. 1969, *Nature*, **226**, 64.

Madau, P. 1988, *Ap. J.*, **327**, 116.

Malkan, M. A., and Sargent, W. L. 1982, *Ap. J.*, **254**, 22.

Malkan, M. A., and Sun, W. H. 1988, preprint.

Maraschi, L., Perola, G. C., Reina, C., and Treves, A., 1979, *Ap. J.*, **230**, 243.

Maraschi, L., Reina, C., and Treves, A. 1974, *Ap. J.*, **35**, 389.

———. 1978, *Astr. Ap.*, **66**, 99.

Maraschi, L., Roasio, R., and Treves, A. 1982, *Ap. J.*, **253**, 312.

Meszaros, P. 1975, *Astr. Ap.*, **44**, 59.

Meszaros, P., and Nagel, W. 1985*a*, *Ap. J.*, **298**, 147.

———. 1985*b*, *Ap. J.*, **299**, 138.

Meyer, F., and Meyer-Hofmeister, E. 1981, *Astr. Ap.*, **104**, 40.

Novikov, I. D., and Thorne, K. S. 1973, in *Black Holes*, ed. C. DeWitt and B. DeWitt (New York: Gordon and Breach).

Paczynski, B., and Wiita, P. 1980, *Astr. Ap.*, **88**, 23.

Papaloizou, J. C., and Pringle, J. E. 1984, *M.N.R.A.S.*, **208**, 721.

———. 1985, *M.N.R.A.S.*, **213**, 799.

Papaloizou, J., Faulkner, J., and Lin, D. N. 1983, *M.N.R.A.S.*, **205**, 487.

Parker, E. N. 1963, *Interplanetary Dynamical Processes* (New York: Interscience).

Piran, T. 1978, *Ap. J.*, **221**, 652.

Pozdnyakov, L. A., Sobol', I. M., and Sunyaev, R. A. 1983, *Ap. Space Sci. Rev.*, **2**, 189.

Prendergast, K. H. and Burbidge, G. R. 1968, *Ap. J. (Letters)*, **151**, L83.

Pringle, J. E., 1981, *Ann. Rev. Astr. Ap.*, **19**, 137.

Pringle, J. E., and Rees, M. J. 1972, *Astr. Ap.*, **21**, 1.

Pringle, J. E., Rees, M. J., and Pacholczyk, A. G. 1973, *Astr. Ap.*, **29**, 179.

Rees, M. J. 1978, *Phys. Scripta*, **17**, 193.

———. 1984, *Ann. Rev. Astr. Ap.*, **22**, 471.

Rees, M. J., Begelman, M. C., Blandford, R. D., and Phinney, E. S. 1982, *Nature*, **295**, 17.

Rybicki, G. B., and Lightman, A. 1979, *Radiative Processes in Astrophysics* (New York: Wiley).

Salpeter, E. E. 1964, *Ap. J.*, **140**, 796.

Shakura, N. I. 1972, *Astr. Zu.*, **49**, 921.

Shakura, N. I., and Sunyaev, R. A. 1973, *Astr. Ap.*, **24**, 337.

———. 1976, *M.N.R.A.S.*, **175**, 613.

Shapiro, S. L. 1973*a*, *Ap. J.*, **180**, 531.

———. 1973*b*, *Ap. J.*, **185**, 69.

Shapiro, S. L., and Salpeter, E. E. 1975, *Ap. J.*, **198**, 671.

Shapiro, S. L., and Teukolsky, S. A. 1983, *Black Holes, White Dwarfs and Neutron Stars* (New York: Wiley).
Shapiro, S. L., Lightman, A. P., and Eardley, D. M. 1976, *Ap. J.*, **204**, 187.
Shields, G. 1978, *Nature*, **272**, 706.
Shklovsky, I. S. 1967, *Ap. J. (Letters)*, **148**, L1.
Shvartsman, V. F. 1971, *Soviet Astr.*, **15**, 37.
Smak, J. I. 1984, *Pub. A.S.P.*, **96**, 5.
Sunyaev, R. A., and Titarchuk, L. G. 1980, *Astr. Ap.*, **86**, 121
Sunyaev, R. A., and Truemper, J. 1979, *Nature*, **279**, 508.
Takahara, F., Tsuruta, S., and Ichimura, S., 1981, *Ap. J.*, **251**, 26.
Thorne, K. S., Flammang, R. A., and Zytkow, A. N. 1981, *M.N.R.A.S.*, **194**, 475.
Treves, A., Belloni, T., Chiappetti, L., Maraschi, L., Stella, L., Tanzi, E. G., and van der Klis, M. 1988, *Ap. J.*, **325**, 119.
Turolla, R., Nobili, L., and Calvani, M. 1986, *Ap. J.*, **303**, 573.
White, N. E., Stella, L., and Parmar, A. N. 198, *Ap. J.*, **324**, 363.
Zdziarski, A. A. 1985, *Ap. J.*, **289**, 51.
Zel'dovich, Ya. B. 1964, *Soviet Phys. Doklady*, **9**, 195.
Zel'dovich, Ya. B., and Novikov, I. D. 1971, *Relativistic Astrophysics*, Vol. 1 (Chicago: University of Chicago Press).
Zel'dovich, Ya., and Shakura, N. I. 1969, *Soviet Astr.*, **13**, 175.

# Chapter 2  THE SEMINAL PAPERS

# THE EFFECT OF INTERSTELLAR MATTER ON CLIMATIC VARIATION

By F. HOYLE and R. A. LYTTLETON

*Received* 19 April 1939

## 1. Introduction

There is direct astronomical evidence for the existence of diffuse clouds of matter in interstellar space. Any section of the Milky Way containing a large number of stars usually shows regions in which no stars appear, and the extent of these patches is often large compared with the average apparent distance between the stars themselves (see, for example, Russell, Dugan, and Stewart (2), p. 820). The existence of the so-called cosmical cloud in interstellar space, sharing in the general motion of the galaxy, is now well established, and observational investigation shows that the obscuration referred to above occurs also on a galactic scale. Thus the diffuse obscuring clouds appear as irregularities in the general cosmical cloud. The dimensions of such regions are comparable with the distances between the stars, and may be very much greater. In some instances the presence of such clouds is revealed by their illumination by a star or stars lying in, or near them, so that the matter then can be directly observed. In shape the clouds are very irregular; some appear like long dark lanes, while other tracts are devoid of any particular form.

Since the existence of such clouds appears to be general in the galaxy it is of importance to consider the effects that could be produced if a star passed through one of them. The frequency of such occurrences for a particular star would clearly depend upon the distribution of the clouds in space, and the intervals between these events would accordingly be irregular. But it is to be observed at once that the intervals would in general be of the order of the periods of time occurring in galactic problems, that is, of the order of $10^7$ or $10^8$ years, the average period of revolution of a star in the galaxy being about $2{\cdot}5 \times 10^8$ years.

The density of an obscuring cloud and the velocity that a star would have relative to it are known as far as orders of magnitude are concerned from astronomical considerations, and it is shown in the sequel that these clouds may have a considerable effect upon a star's radiation during the time of passage of the star through the cloud. The importance to terrestrial climate of such an effect upon the sun is at once evident, and it is to this aspect of the process that the present paper is directed, though it would seem that encounters between stars and the diffuse clouds may also have some bearing on questions of a more general astronomical nature. If any appreciable change in the sun's radiative power were brought about, it would certainly have considerable effects upon the climate of the earth, and the investigations contained in the present paper lead to the suggestion that the passage of the sun through nebulous clouds may give an account of certain epochs in the earth's history for which no tenable theory has hitherto been given. The actual mechanism that leads to the necessary increase in the solar output of energy is simply the addition to the sun of kinetic energy by the infall of part of the cloud; the gravitational energy is transformed into kinetic energy, and subsequently into radiant energy. Before proceeding with the details of this process we propose to give in brief outline some description of the climatic changes that have taken place during geological time.

## 2. Past epochs in the earth's climatic history

### (a) *The Ice Epochs*

At least eight Ice Epochs have been surmised from geological evidence (see, for example, Holmes (1), p. 217). These are separated by periods of the order of a hundred million years. More detailed evidence is available for the last of them, this being known as the Quaternary Ice Epoch, which seems to have comprised four separate ice ages and to have lasted as a whole not less than 200,000 years. Between these four ice ages the ice retreated to something like its present position, the intervening warm periods being long compared with the duration of the ice ages themselves.

### (b) *The Carboniferous Epoch*

Much later than the last of the Pre-Cambrian Ice Epochs but before the Permian period, there occurred the Carboniferous period, during which the coal deposits were laid down. The coal originates from the plant life of the period, and judged by the thickness of the deposits this must have been prolific. From comparison with similar plant species in existence at the present time it is concluded that the climate during the Carboniferous Epoch must have been very warm and humid, these being the requirements now associated with such species. This is most remarkable when it is remembered that coal deposits are widespread over the earth and occur as far north as Spitzbergen, whose latitude is about 78° N. Since dynamical reasons make it certain that the land masses cannot have moved horizontally through appreciable distances (Jeffreys (3), p. 304), it follows that the climate over the whole earth must have been very much warmer than at the present time.

## 3. Theories of climatic variation

The theory that the hot climate of Carboniferous times was due to a period of great volcanic activity (which is known to have taken place about this time) cannot be described as indicating a satisfactory course of climatic variation, while the theories that have been put forward to explain the Ice Epochs also seem inadequate. The latter theories may be divided into two kinds: (1) terrestrial causes, and (2) astronomical causes.

It is not proposed to discuss the details of these theories apart from drawing attention once more to their defects that are overcome by the process arising from the present hypothesis.

The theories under (1) attribute variations in climate to differing terrestrial conditions at past times. Many of these, for example the change of sea-level, must have had some effect upon climate. It has been held that a small effect only is required to produce an ice age; the ice fields would begin to spread towards the equator, the process supposedly becoming cumulative, and if only small changes were required at the start, effects such as the change of sea-level might be sufficient. This theory would, however, meet difficulties in attempting to cope with the warm periods between the successive ice ages (of the Quaternary Ice Epoch) unless a suitable periodicity in the terrestrial conditions can be established.

The theories of the second type (2) based on extra-terrestrial causes, that have been proposed, have mainly invoked dynamical effects such as precession and changes in the solar eccentricity. But even if these suggestions were otherwise satisfactory, there is a fixed period associated with such dynamical motions, and this would be reflected in the effects of the climatic changes. The periods are not at all comparable with the irregular intervals at which exceptional climatic conditions have occurred as indicated by geographical and geological evidence.

## 4. Accretion of interstellar matter

The mechanism advocated in the present paper, and already adumbrated in the introduction, is of an astronomical nature and does not give rise to the defect of periodicity. It is shown that the effect of the passage of the sun through a cloud of diffuse matter is to increase the solar radiation, and that this process is able to bring about changes quantitatively in agreement with the requirements of climatic variation. The results obtained show that the percentage increase in the sun's radiation depends on the density of the cloud and on the velocity of the sun relative to it, being directly proportional to the first factor and inversely proportional to the cube of the latter factor. Thus slight changes in these factors bring about considerable ranges of variation in the solar radiation, and it may be caused to change from 0·1 to 1000 % according to the density and velocity of the cloud. Meteorological considerations appear to indicate that an ice age is unlikely to result from small decreases in the amount of radiation received at the earth, for it is essential that evaporation should be increased in order to increase the precipitation of snow in those regions normally within the snow line, and the temperature of the oceans must accordingly be raised. On our view each Ice Epoch is to be regarded as due to the passage of the sun through a diffuse cloud with relative velocity sufficient to prevent more than a small percentage increase in the sun's output of radiation. Thus the difference of time between successive Ice Epochs will depend on the intervals between successive encounters of the sun with clouds of suitable relative velocity; from considerations of motion in the galaxy such periods would be expected to be of the order of 100 million years. (It is to be noticed that intervals as low as 10 million years are by no means excluded, for in its nature the process is far from periodic.) From their very existence it could be surmised that there will be considerable variations of density within any one cloud, and this is confirmed observationally. Any such large-scale differences in a cloud would have the effect of changing the sun's radiation by different amounts during its passage through the cloud. This would be capable of causing the type of climatic variations occurring in the Quaternary Ice Epoch. For, if an ice age is regarded as the period during which the sun's radiation changes slowly from its normal value to some suitable figure, a 10 % increase say, corresponding to the amount that would bring about a warm climate, then during the passage of the sun through a cloud in which the density increased towards the inner regions, two ice ages might occur separated by a warmer period. It would be necessary to assume only two such rises and falls in temperature (and density) in order to give the climatic form of the Quaternary Ice Epoch. (That a rise and fall in the solar radiation could produce such effects appears to have been first suggested by Simpson.) The two successive changes in the sun's total emission could then be accounted for by supposing the sun to have passed in turn through two adjacent clouds.

The Carboniferous Epoch would similarly be explicable as arising in the case when the sun passes through a cloud having either high average density or, what is more probable, low relative velocity, for then a large increase in the sun's radiation would occur. In order to produce a hot and humid climate even in polar regions the radiative power may have to be more than doubled. From the astronomical evidence it can be decided that vicissitudes of this intensity would be far less frequent, for the requisite density and velocity would be less likely to occur. For it turns out that the relative velocity of the sun and cloud required to produce an ice age is of just the order of magnitude of the solar velocity relative to the stars in its neighbourhood, while the value required to produce a period of exceedingly high terrestrial temperature is much smaller.

Consider now the process of accretion by the sun of the material of the cloud lying near its path. At first sight it might appear that the only particles of the cloud to be captured would be those which actually strike the sun as they attempt to describe orbits round its centre. The number of such particles under reasonable conditions would be quite insufficient to change the sun's radiation appreciably. It is clear, however, that some kind of condensation will be formed in the cloud by the sun's gravitational attraction, and the action of collisions in this condensation can be shown to give the sun an effective capture radius much larger than its ordinary radius.

*(a) Calculation of the capture radius of the sun*

Imagine the cloud to be streaming past the sun, from right to left in the figure, and let the velocity of any element of it relative to the sun when at great distances be $v$. Consider the part of the cloud that if undeflected by the sun would pass within a distance $\sigma$ or less of its centre. It is clear that collisions will occur to the left of the sun because the attraction of the latter will produce two opposing streams of particles and the effect of such collisions is to destroy the angular

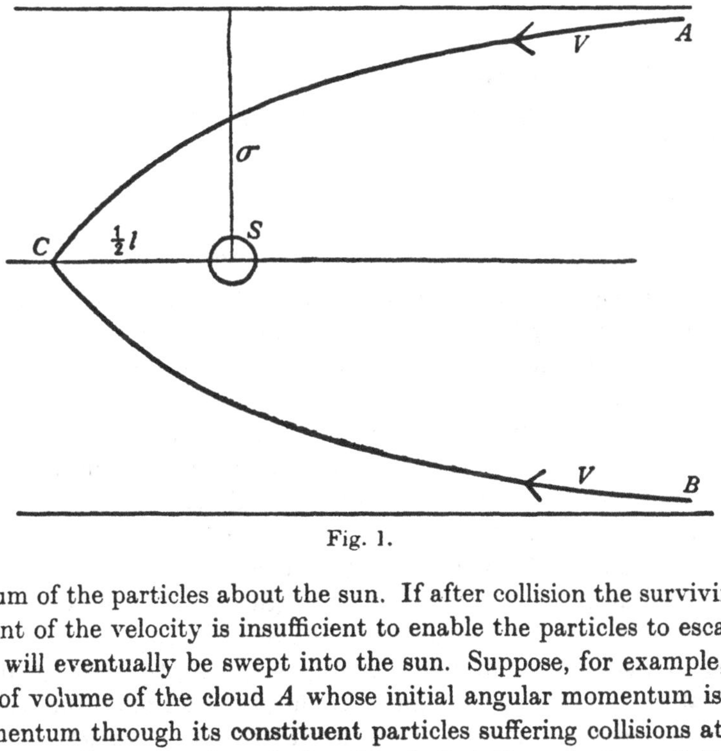


Fig. 1.

momentum of the particles about the sun. If after collision the surviving radial component of the velocity is insufficient to enable the particles to escape, such particles will eventually be swept into the sun. Suppose, for example, that an element of volume of the cloud $A$ whose initial angular momentum is $v\sigma$ loses this momentum through its constituent particles suffering collisions at $C$; then the effective radius $\sigma$ can be calculated such that the velocity radially at $C$ is less than the escape velocity at this distance. The element describes a hyperbola whose equation, with the usual notation, may be written

$$\frac{l}{r} = 1 + e\cos\theta.$$

The direction parallel to the initial asymptote corresponds to

$$e\cos\theta + 1 = 0,$$

and hence the direction $SC$ corresponds to

$$e\cos\theta = 1, \quad \text{i.e. } e\sin\theta = (e^2-1)^{\frac{1}{2}}.$$

Thus the distance $SC$ is simply $\frac{1}{2}l$. Since $r^2 d\theta/dt = h$, and is constant, the radial component of velocity is given by

$$\frac{dr}{dt} = \frac{eh}{l}\sin\theta.$$

Now $h = \sqrt{(\mu l)} = v\sigma$, where $\mu/r^2$ is the attraction of the sun at distance $r$. Hence, at $C$, $dr/dt$ has the value $v$, the velocity of the cloud at infinity. The part of the cloud will accordingly fail to escape after collision if

$$v^2 < \frac{2\mu}{SC} = \frac{4\mu}{l} = \frac{4\mu^2}{v^2\sigma^2},$$

so that

$$\sigma < 2\mu/v^2.$$

Accordingly, the effective radius corresponds to the distance at which $v$ is the parabolic velocity. If $\gamma$ denotes the constant of gravitation and $M$ is the mass of the sun then

$$\sigma = 2\gamma M/v^2.$$

For $v = 20$ km. per sec., $\sigma$ is as large as 1000 solar radii.

### (b) *Collisions in the cloud*

What this foregoing calculation has shown is that, provided collisions occur with sufficient frequency to be effective in reducing the angular momentum of the particles about the sun, and if $\sigma$ is defined as above, then nearly all particles passing within perpendicular distance $\sigma$ when at great distances will eventually be captured by the sun. It is therefore necessary to consider the question of the number of collisions which a particle would undergo in describing an orbit round the sun. Since the cloud must be regarded as diffuse, the mean free path of a particle is long, but, on the other hand, the length of the path over which collisions take place is also very long. Consider the limiting case of a particle with initial angular momentum $v\sigma$. The length of path over which collisions are effective in reducing its angular momentum is of order $\sigma$. It is necessary only to show that the number of collisions is sufficiently great to justify the previous calculation of $\sigma$ as the capture radius; in order to do this the problem may be conveniently simplified. Consider, therefore, a particle of velocity $v$ fired into a tube of length $\sigma$ and cross-section 1 cm.$^2$, through which the cloud is streaming with velocity $v$ in the opposite direction. If the density of the cloud is known, the effective number of collisions is readily obtainable. If we assume that the density is $10^{-18}$ g. per cm.$^3$ (Russell, Dugan and Stewart (2), p. 832), then if the cloud is composed of hydrogen the corresponding number $N$ of atoms per cm.$^3$ is about $6\times10^5$. (The assumption of atomic hydrogen is made in agreement with the supposition that the stars have evolved from primitive hydrogen nebulae.)

Thus, in passing along the tube, the one hydrogen atom in which interest is fixed moves past a total number of particles at least as great as $2N\sigma$. If we take the radius of the hydrogen atom to be $10^{-8}$ cm., the number of collisions is accordingly $2\pi N\sigma \times 10^{-16}$. (A simple classical kinetic model is used for the collisions. If a quantum treatment is introduced the effective scattering of the hydrogen atom is greater than in the classical picture but is of the same order of magnitude.) If we express $\sigma$ in terms of the mass of the sun and the relative velocity of the sun and cloud, the number of collisions is

$$4\pi\gamma NM/10^{16}v^2.$$

If we take
$$\gamma = \tfrac{20}{3}\times 10^{-8}, \quad N = 6\times 10^5,$$
$$M = 2\times 10^{33}\text{ g.} \quad\text{and}\quad v = 2\times 10^6 \text{ cm. per sec.,}$$
the number of collisions is about $2{\cdot}5\times 10^4$. This number is certainly sufficiently high to destroy the angular momentum of such particles about the sun. Possibly the estimate of $\sigma$ as the length of track over which collisions occur is too large, but even if it is reduced by a factor $\frac{1}{10}$ the number of collisions would nevertheless be of order $10^3$. Particles that possess initial angular momentum less than $v\sigma$ have less length of effective track, but this is compensated by an increase in the density of the cloud towards the sun. Thus it may be concluded that, for a cloud of the assumed mean density, nearly all particles crossing a circle of radius $\sigma = 2\gamma M/v^2$ are captured by the sun. Since the number of collisions has been shown to be large, this conclusion must also be valid for considerable ranges in the assumed density and is probably applicable for densities as low as $10^{-23}$ g. per cm.$^3$.

The cloud has been regarded as un-ionized in calculating the collisions. For an ionized cloud the chief process in the present connexion will be the scattering of protons by protons, and it may be verified that this would be rather more effective in destroying the angular momentum of the particles of the cloud than collisions of neutral atoms.

### (c) *The accretion of mass*

The important question is how to calculate the amount of kinetic energy that the material of the cloud captured by the sun will bring to the latter due to its fall on to the solar surface. Since the velocity of escape at the surface of the sun is so large, we can suppose that all the particles reaching the surface of the sun arrive with the escape velocity, that is, $6{\cdot}2\times 10^7$ cm. per sec.; this is permissible provided that there is no other way in which the particle can get rid of its energy. For a hydrogen atom, having mass $1{\cdot}66\times 10^{-24}$ g., the corresponding kinetic energy is slightly in excess of $3\times 10^{-9}$ ergs. Now the ionization energy of the hydrogen atom is about $4\times 10^{-11}$ ergs, hence it can be concluded that the particle cannot get rid of any appreciable portion of its energy by ionization processes before reaching the sun. Thus the obscuring effect of the particles lying between the earth and the sun would be quite negligible compared with the increase of the sun's radiation due to the kinetic energy of the particles. The sun must re-emit this energy.

The total number of particles reaching the sun per second is simply the number crossing an area bounded by a circle of radius $\sigma$ perpendicular to the direction of $v$, the velocity of the cloud. Their mass is accordingly $\pi\sigma^2\rho v$, and, if we express $\sigma$ in terms of the mass of the sun, then the mass added per second is $4\pi\gamma^2\rho M^2/v^3$, and this formula is true for all $\rho$, $M$, and $v$, so long as collisions remain important at distances comparable with $\sigma$. The kinetic energy brought to the sun per second is obtained at once, since we know the velocity of escape at the sun's surface. The energy is found to be

$$4 \times 10^{68} \rho v^{-3} \text{ ergs.}$$

For climatic effects to be appreciable, this amount of energy must be comparable with the sun's ordinary output. We see that the increase is more sensitive to changes in $v$ than in $\rho$. Suppose then that $\rho = 10^{-18}$, for example, and consider the value which $v$ would have to take in order to bring about a 10 % change in the sun's normal radiation, which means a change of about $4 \times 10^{32}$ ergs per sec. Then

$$4 \times 10^{50} \times v^{-3} = 4 \times 10^{32},$$

and $v = 10$ km. per sec. For a 1 % increase, $v$ would be approximately 20 km. per sec., while for a 1000 % increase $v$ would be about 2 km. per sec. On the other hand, if $\rho = 10^{-20}$, an increase of 1 % would occur for a velocity of about 5 km. per sec.

Stellar considerations indicate that a velocity of $2 \times 10^6$ cm. per sec. would be about the most probable value for an encounter, and the agreement of the results of the present investigation as regards orders of magnitude is manifest. The velocity required to cause a tremendous change in the solar radiation is not unacceptably low, and the value obtained shows the degree of improbability that such changes would occur frequently. In conjunction with this remark it should be remembered that the tendency of the cloud would be to increase the velocity of approach of the sun; this is now considered.

### (d) *Acceleration of the sun*

It is necessary to consider the interval for which the sun is likely to be exposed to an encounter. In general the time could be made of any desired duration by adjusting the size of the supposed cloud. But a massive cloud would increase the velocity of the sun relative to it, unless the sun was permanently inside it, and it appears that if the mass of the cloud were very much greater than that of the sun, the relative velocity might be increased by its attraction to a value much greater than the estimate of $10^6$ cm. per sec. used above. This in turn would bring about only a very slight increase in the radiation of the sun; for a relative velocity of $10^7$ cm. per sec. the increase is about 0·01 % only. Thus for an encounter to cause a suitable change in radiation, the alteration of the velocity of the sun must not exceed $10^6$ cm. per sec. The effective mass of the cloud for this purpose will depend largely on its shape, and upon what region of it the sun happens to traverse, and may vary widely from its actual total mass.

Let $m$ be of the order of the effective mass of the cloud, and let $2a$ be the length of the path which the sun travels in it. Then the gravitational potential at the surface is of the order of $\gamma m/a$, where $\gamma$ is the usual constant of gravitation. If $v$ is the velocity of the sun, the change $\Delta v$ produced by the attraction of the cloud is therefore given, so far as order of magnitude is concerned, by

$$v\Delta v = \gamma m/a,$$

provided that $\Delta v$ is not very much greater than $v$. If $v = 10^6$ cm. per sec., in accordance with the previous requirements, and if $\Delta v$ is taken as $10^5$ cm. per sec., then it is found that

$$m/a = 1{\cdot}5 \times 10^{18}.$$

Now $m$ is of order $4\rho a^3$, and if we put $\rho = 10^{-18}$ we find that

$$a = 6 \times 10^{17} \text{ cm.}$$

The time taken to travel through the cloud is therefore of order $2a/v$, i.e. about $10^{12}$ sec. (about $10^5$ years). From the way this has been computed, we see that this gives an estimate of the minimum time required in general to cross a cloud. Furthermore, if the cloud were patchy, an increase in the time of traversing it would also arise. The effective mass of the cloud is of order $4\rho a^3$ and this is about 500 times the mass of the sun. In view of the rough methods which the nature of the problem compels us to adopt, the agreement with the requirements of observation are completely in accordance with the proposed hypothesis.

### (e) *Total mass accreted by the sun and the earth*

It is important to show also that the matter added to the sun during the whole period of its passage through the cloud will not appreciably affect its mass.

To change the solar radiation by 1 % requires an energy addition of $3{\cdot}78 \times 10^{31}$ ergs per sec., and, since the velocity at the surface of the sun is about 620 km. per sec., this amount implies an increase in mass of $2 \times 10^{16}$ g. per sec. Hence in $10^{12}$ sec. the change is $2 \times 10^{28}$ g., about 3 times the mass of the earth, i.e. about $10^{-5}$ times that of the sun itself. According to the mass-luminosity relation, the usual radiation (when the sun is not passing through a cloud) can be taken as being roughly proportional to a small power of the sun's mass, say $L \propto M^3$, and hence, even if considerable allowance is made for the material added being hydrogen, the change in luminosity could not be as great as 0·01 %. Thus the sun would revert to its former luminosity to within this margin.

It is of some interest also to consider the accompanying change of mass of the earth. The velocity of the latter relative to the cloud would vary in any case, but at the earth's distance the cloud captured by the sun would presumably be mainly moving radially inwards with the velocity of escape at the earth's distance, which is over 40 km. per sec. But even if this effect is ignored and if the earth's velocity is taken as $2 \times 10^6$ cm. per sec. relative to the cloud, the addition of mass turns out to be insensibly small. For it has been seen that the amount added is proportional to the square of the mass of the central body, so that other things being equal the amount added by the sun will be $10^{11}$ times the amount added by the earth. Thus the earth would receive $2 \times 10^{17}$ g. in $10^{12}$ sec., and this is less than $10^{-4}$ the mass of the terrestrial atmosphere. If allowance is made for the orbital motion of the earth and the radial motion of the cloud towards the sun, this would be reduced by a factor of about $\frac{1}{30}$ and the amount would be about a millionth the mass of the present atmosphere. This would represent a wholly negligible change, and since in addition it would take $10^5$ years to be added, could itself produce no sensible climatic effects.

Accompanying the change in mass of the sun there will be a change in the mean distance of the earth. For a nearly circular orbit the change $\Delta a$ of the major semi-axis $a$ is given by

$$\frac{\Delta a}{a} = -\frac{\Delta M}{M}.$$

Hence the total change in $a$ will be a decrease of the order of $10^{-5}a$. The consequent change in the solar radiation received at the earth will be of the same order, and is therefore quite inappreciable. It appears, therefore, that in the course of describing its orbit in the galaxy, the sun would change its luminosity whenever it happened to pass through a nebulous cloud of any appreciable density. The density of the cosmical cloud spread throughout the galaxy is of the order of $10^{-23}$ g. per c.c., and its presence will not affect the luminosity of the sun, but the existence of huge irregular tracts of much higher density is well known. For instance, the Orion nebula may have an average density as great as $10^{-18}$ g. per c.c. (Russell, Dugan and Stewart (2), p. 832). Such clouds are prevalent in the galaxy and it may reasonably be expected that the sun has undergone several encounters with clouds of these densities, since the period of geological time and the period of motion of the sun in the galaxy are comparable.

If the conclusions of this paper are correct the solar system must have emerged from such a cloud within the last few thousand years, since an ice age took place as recently as this. Thus the cloud responsible could *not* be as remote as the brightest stars. But, even if the density were as high as $10^{-18}$ g. per c.c., starlight in traversing its breadth, which must be of order $6 \times 10^{17}$ cm., would have to pass through the equivalent of a layer of 1 cm. of gas of density 1 g. per c.c. Only one light quantum in a hundred would be absorbed in such a layer as far as visual frequencies are concerned, though higher energy quanta would be more strongly absorbed. Thus there would be no reason to suspect appreciable obscuration of the background stars to be caused by such a cloud.

We submit, therefore, that the present paper represents a discussion of a process that is of importance in considering the past history of the sun and that some, if not all, of the climatic variations occurring on earth may have resulted from the consequent changes in solar radiation.

## 5. Conclusion

The foregoing investigation relates to a process that seems hitherto to have been overlooked. The number of mechanisms that have been put forward as responsible for the Ice Epochs is manifold, but the present suggestion would seem to have the merit of giving effects quantitatively in agreement with the requirements and free from any periodicity. The present hypothesis is not advocated with a view to contradicting the opinions and results of those investigators who have confined their researches to terrestrial sources of climatic variation. The claim made is simply that an important process is brought to light and must be accorded some place in discussions of terrestrial climate.

## 6. Summary

The effect of interstellar matter on the sun's radiation is considered with a view to explaining changes in terrestrial climate. It appears that a star in passing through a nebulous cloud will capture an amount of material which by the energy of its fall to the solar surface can bring about considerable changes in the quantity of radiation emitted. The quantity of matter gathered in by the star depends directly on the density of the cloud and inversely on the cube of its velocity relative to the cloud. Thus vastly different effects on the solar radiation can be brought about under fairly narrow ranges of density and relative velocity (ranges that are in accordance with astronomical evidence). In this way the process is able to explain the small changes in the solar radiation that are necessary to produce an ice age and, under conditions less likely to have taken place frequently, the high increase in radiation required for the Carboniferous Epoch. Despite the large effects that the mechanism can bring about, it is shown that the mass of the sun does not undergo appreciable change and hence reverts to its former luminosity once the cloud has been traversed.

## REFERENCES

(1) Holmes, A. *The age of the Earth.* Nelson Classics. (1937.)
(2) Russell, H. N., Dugan, R. S. and Stewart, J. Q. *Astronomy.* Ginn and Company (1938).
(3) Jeffreys, H. *The Earth.* Cambridge (1929).

St John's College, Cambridge

# ON SPHERICALLY SYMMETRICAL ACCRETION

*H. Bondi*

(Received 1951 October 3)

*Summary*

The special accretion problem is investigated in which the motion is steady and spherically symmetrical, the gas being at rest at infinity. The pressure is taken to be proportional to a power of the density. It is found that the accretion rate is proportional to the square of the mass of the star and to the density of the gas at infinity, and varies inversely with the cube of the velocity of sound in the gas at infinity. The factor of proportionality is not determined by the steady-state equations, though it is confined within certain limits. Arguments are given suggesting that the case physically most likely to occur is that with the maximum rate of accretion.

1. The importance of the accretion of interstellar gas by stars has been recognized since the work of Hoyle and Lyttleton (**1, 2, 3**). In their work the problem, later investigated in detail by Bondi and Hoyle (**4**), was that in which the rate of accretion was limited principally by the relative motion of the star and the gas cloud, the effects of pressure being considered negligible in comparison with the dynamical effects. The result derived in these papers was that

$$dM/dt = 2\pi\alpha(GM)^2V^{-3}\rho_\infty, \tag{1}$$

where $M$ is the mass of the star, $dM/dt$ is the rate of accretion, $\rho_\infty$ the density of the gas cloud far from the star, $V$ is the relative velocity of the star and the distant (undisturbed) parts of the cloud, $G$ is the constant of gravitation, and $\alpha$ is a numerical constant which was first estimated to be equal to 2. Later work (**4**) showed that the steady-state equations did not determine $\alpha$, although it seemed likely that it should always be between 1 and 2. J˙ was also shown that if the star entered a cloud of uniform density with a plane ˙oundary, $\alpha$ settled down to a value near 1·25.

/ In all this work pressure effects were neglected, the argument being that any heat generated would be radiated away rapidly, so that the temperature of the gas was always very low. Considerable mathematical simplification is introduced by this assumption, and it was shown that it was likely to be satisfied in most cases of astrophysical interest (**3**). The mathematical difficulties of the more general problem, in which both dynamical and pressure effects are considered, seem insuperable at present. However, the extreme case of negligible dynamical effects is again far simpler, and will be discussed in this paper. It may reasonably be expected that the case discussed here together with the case discussed previously bracket the complete problem.

2. The problem to be discussed may be defined as follows:

A star of mass $M$ is at rest in an infinite cloud of gas, which at infinity is also at rest and of uniform density $\rho_\infty$ and pressure $p_\infty$. The motion of the gas is spherically symmetrical and steady, the increase in mass of the star being ignored so that the field of force is unchanging. The pressure $p$ and density $\rho$ are related everywhere by

$$p/p_\infty = (\rho/\rho_\infty)^\gamma, \tag{2}$$

where $\gamma$ is a constant satisfying $1 \leqslant \gamma \leqslant \frac{5}{3}$.

With a suitable choice of $\gamma$, equation (2) is equivalent to the physical condition that no heat is radiated or conducted away. Hence the solution should provide the most complete contrast possible with the problem previously investigated. The equations governing the problem are easily set up. If we take $r$ to be the radial coordinate and $v$ the *inward* velocity of the gas, the equation of continuity is

$$4\pi r^2 \rho v = \text{constant} = A \text{ (say)}, \tag{3}$$

where $A$ is the accretion rate.

Bernoulli's equation is

$$\frac{v^2}{2} + \int_{p_\infty}^{p} \frac{dp}{\rho} - \frac{GM}{r} = \text{constant } (=0). \tag{4}$$

The constant is readily seen to vanish by virtue of the boundary conditions at infinity. Combining (2) and (4) we have

$$\frac{v^2}{2} + \frac{\gamma}{\gamma-1}\frac{p_\infty}{\rho_\infty}\left[\left(\frac{\rho}{\rho_\infty}\right)^{\gamma-1} - 1\right] = \frac{GM}{r}. \tag{5}$$

Equations (3) and (5) are two equations for the two variables $v$ and $\rho$ in terms of $r$, the distance from the centre of the star.

The equations may be made non-dimensional by the appropriate use of the velocity of sound in the gas at infinity, which as usual we denote by $c$. By the well-known formula

$$c^2 = \gamma p_\infty/\rho_\infty. \tag{6}$$

Let us introduce non-dimensional variables, $x$, $y$, $z$, to replace $r$, $v$, $\rho$, respectively, as follows:

$$\begin{aligned} r &= xGM/c^2, \\ v &= yc, \\ \rho &= z\rho_\infty. \end{aligned} \tag{7}$$

Then (3) and (5) take the non-dimensional form

$$x^2 yz = \lambda, \tag{8}$$

$$\tfrac{1}{2}y^2 + (z^{\gamma-1} - 1)/(\gamma - 1) = 1/x, \tag{9}$$

where $\lambda$ is given by

$$A = 4\pi\lambda(GM)^2 c^{-3}\rho_\infty. \tag{10}$$

Accordingly $\lambda$ is the non-dimensional parameter determining the accretion rate. It plays the same role as $\alpha$ in equation (1). It will also be observed that the relative velocity $V$ of equation (1) has been replaced by $c$ in (10).

3. The explicit solution of equations (8) and (9) for general $\gamma$ is possible not in terms of the variables $y$ and $z$ but only if an auxiliary variable depending only on $y^2/z^{\gamma-1}$ is introduced. It is particularly interesting that mathematical requirements lead to the introduction of this variable, since

$$u = yz^{-(\gamma-1)/2} \tag{11}$$

has immediate physical significance as the ratio of the local bulk velocity $v$ of the gas to the local velocity of sound $(\gamma p/\rho)^{1/2}$. Substituting (11) into (8) and solving for $y$ and $z$ we have

$$y = u^{2/(\gamma-1)}(\lambda/x^2)^{(\gamma-1)/(\gamma+1)}, \tag{12}$$

$$z = (\lambda/x^2u)^{2/(\gamma-1)}. \tag{13}$$

Then (9) becomes

$$\tfrac{1}{2}u^{4/(\gamma+1)}\left(\frac{\lambda}{x^2}\right)^{2(\gamma-1)/(\gamma+1)} + \frac{1}{\gamma-1}\left(\frac{\lambda}{x^2u}\right)^{2(\gamma-1)/(\gamma+1)} = \frac{1}{x} + \frac{1}{\gamma-1}. \tag{14}$$

Rearranging the terms and multiplying by $(x^2/\lambda)^{2(\gamma-1)/(\gamma+1)}$ we find that (14) takes the form

$$f(u) = \lambda^{-2(\gamma-1)/(\gamma+1)}g(x), \tag{15}$$

where

$$f(u) = \tfrac{1}{2}u^{4/(\gamma+1)} + \frac{1}{\gamma-1}u^{-2(\gamma-1)/(\gamma+1)} = u^{4/(\gamma+1)}\left(\frac{1}{2} + \frac{1}{\gamma-1}\frac{1}{u^2}\right), \tag{16}$$

$$g(x) = x^{4(\gamma-1)/(\gamma+1)}\left[\frac{1}{x} + \frac{1}{\gamma-1}\right] = \frac{x^{4(\gamma-1)/(\gamma+1)}}{\gamma-1} + x^{-(5-3\gamma)/(\gamma+1)}. \tag{17}$$

A study of the functions $f$ and $g$ serves to determine $u$ as a function of $\lambda$ and $x$. The variables $y$ and $z$ are then readily found by (12) and (13).

4. We shall first assume that $1 < \gamma < \frac{5}{3}$. The two limiting cases $\gamma = 1$, $\gamma = \frac{5}{3}$ will be examined later. With this assumption both $f$ and $g$ are each the sum of a positive and negative power of the respective variables and hence each of them has a minimum. The minimum of $f(u)$ occurs for $u = u_m = 1$ and is of value

$$f = \tfrac{1}{2}(\gamma+1)/(\gamma-1) = f_m \text{ (say)}.$$

The minimum of $g(x)$ occurs for $x = x_m = \frac{1}{4}(5-3\gamma)$ and is of value

$$g = \frac{1}{4}\frac{\gamma+1}{\gamma-1}[\tfrac{1}{4}(5-3\gamma)]^{-(5-3\gamma)/(\gamma+1)} = g_m \text{ (say)}.$$

In our problem $x$ varies between infinity and the value corresponding to the surface of the star. This last value is very small indeed. As an example, if the star is taken to be the Sun, and $c = 1$ km/s corresponding to a gas temperature at infinity of nearly 3 000 deg. K, the surface value of $x$ is only $5 \times 10^{-6}$. Even for a red giant the surface value of $x$ would be less than $10^{-2}$ unless the temperature of the gas at infinity were quite improbably high (more than, say, $5 \times 10^4$ deg. K). Accordingly $x$ will attain the value $x_m$ in the physically significant interval. Hence, somewhere in that interval, the right-hand side of (15) will reach a value as low as $\lambda^{-2(\gamma-1)/(\gamma+1)}g_m$. But the lowest value $f$ can reach is $f_m$. Hence $\lambda$ cannot exceed $\lambda_c$, where

$$\lambda_c = \left(\frac{g_m}{f_m}\right)^{(\gamma+1)/2(\gamma-1)} = \left(\frac{1}{2}\right)^{(\gamma+1)/2(\gamma-1)}\left(\frac{5-3\gamma}{4}\right)^{-(5-3\gamma)/2(\gamma-1)}. \tag{18}$$

**Our first result is therefore that the accretion rate $A$ cannot exceed the value**

$$4\pi\lambda_c(GM)^2c^{-3}\rho_\infty. \tag{19}$$

**Table I gives the value of $\lambda_c$ for a few values of $\gamma$.**

TABLE I

| | | | | | |
|---|---|---|---|---|---|
| $\gamma$ | 1 | 1·2 | $1·4 = \frac{7}{5}$ | 1·5 | $\frac{5}{3}$ |
| $\lambda_c$ | $\frac{1}{4}e^{3/2} \doteqdot 1·12\ldots$ | $\frac{1}{4}(0·7)^{-3·5} \doteqdot 0·872\ldots$ | 0·625 | 0·500 | 0·250 |

5. In order to obtain a more detailed picture it is necessary to discuss $f(u)$ and $g(x)$ more fully, and to take account of the boundary conditions of the problem. Since, at infinity, $v$ vanishes but $p$ and $\rho$ tend to finite limits, it follows that $u$ tends to zero there.

It is now easy to consider the problem graphically. Fig. 1 shows $f$ and $g$ drawn for the typical case $\gamma=\frac{7}{5}$. The resulting variation of $u$ as a function of $x$ is shown in Fig. 2 for the cases (i) $\lambda=\frac{1}{4}\lambda_c$, (ii) $\lambda=\lambda_c$ and (iii) $\lambda=4\lambda_c$. It will be seen that the boundary condition at infinity implies that $u$ is very small near infinity (points beyond A on the graphs). For $\lambda<\lambda_c$, as $x$ diminishes, $u$ rises gradually to a maximum (B) and diminishes to zero (C) as $x$ tends to zero. The closer $\lambda$ approaches $\lambda_c$ from below, the sharper the maximum B. No part of the curve A′B′C′ (on which $u$ is very large both at infinity and near $x=0$) is of physical significance, since $u$ given by this curve does not satisfy the boundary condition at infinity. No jump from the curve ABC to A′B′C′ is possible, since this would imply an infinite acceleration. Along ABC the variable $u$ is always less than unity, so that the motion is subsonic. Along A′B′C′ the value of $u$ exceeds unity.

The case $\lambda=\lambda_c$ is quite different. For in this the curves have contact at B = B′. Coming from A, the physically significant curve can continue either to C or to C′. In the first alternative, the curve is the limiting form of the curves for $\lambda<\lambda_c$. The curve has a discontinuous tangent at B and there is hence a finite jump in the acceleration. Although this is perhaps physically not very plausible, there does not seem to be any argument disallowing this motion altogether. It may be significant that at B the value of $u$ is unity, so that the bulk velocity equals the velocity of sound.

The curve ABC′ is perfectly smooth and monotonic. For $x>x_B$ the motion is subsonic, while for $x<x_B$ the motion is supersonic. The system is in a state quite different from any state possible for $\lambda<\lambda_c$.

If $\lambda$ exceeds $\lambda_c$ then the pattern of the curves changes as indicated, and no solution is possible.

We see hence that there are two quite different types of motion. Type I exists for $\lambda\leqslant\lambda_c$. The motion is everywhere subsonic (except at $x=x_m$ if $\lambda=\lambda_c$), and $u$ has a single maximum which is less than or equal to unity. The bulk velocity $v$ has a maximum if $\gamma<\frac{3}{2}$ but not if $\gamma\geqslant\frac{3}{2}$. (This follows from a simple consideration of $y'$ in terms of $u$.) For $\gamma<\frac{3}{2}$ the velocity $v$ tends to zero as $r$ tends to zero, for $\gamma=\frac{3}{2}$ it tends to a limit (equal to $\frac{1}{4}c\lambda$), and for $\gamma>\frac{3}{2}$ it tends to infinity. The density is always a monotonic function of the radius. Type II exists only if $\lambda=\lambda_c$. In this case $u, v, \rho$ are all monotonic functions of the radius.

The special case $\lambda=0$ may be briefly referred to here. In this case the gas is at rest, forming a tenuous continuation of the star. Since $y=0$ we have by (9)

$$z^{\gamma-1}=1+(\gamma-1)/x. \tag{20}$$

Figs. 3, 4 and 5 show $y$ and $z$ as functions of $x$ for $\gamma=1$, $\frac{7}{5}$ and $\frac{5}{3}$ respectively. Three values of $\lambda$ are taken in each case, namely $\lambda=\lambda_c$, $\lambda=\frac{1}{4}\lambda_c$ and the case $\lambda=0$ just referred to. It will be seen from the figures that $z$ does not depend very critically on $\lambda$, varying only slightly between the extreme cases $\lambda=0$ and $\lambda=\lambda_c$, especially in the case of the higher $\gamma$ values.

6. It remains to discuss the two limiting cases $\gamma=1$ and $\gamma=\frac{5}{3}$ respectively.

If $\gamma=1$ equation (9) becomes

$$\tfrac{1}{2}y^2+\ln z=1/x. \tag{9'}$$

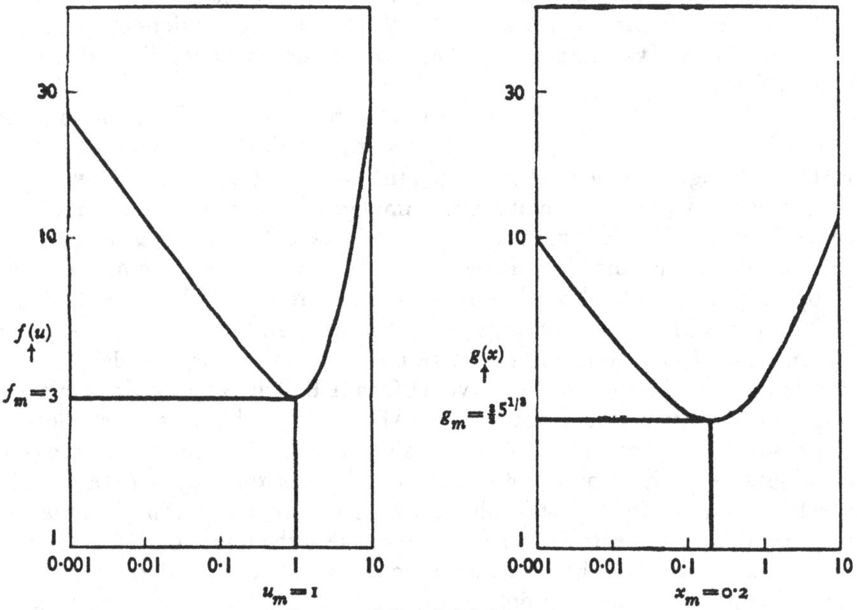


FIG. 1.—$f(u)$ and $g(x)$ for $\gamma = \frac{7}{5}$.

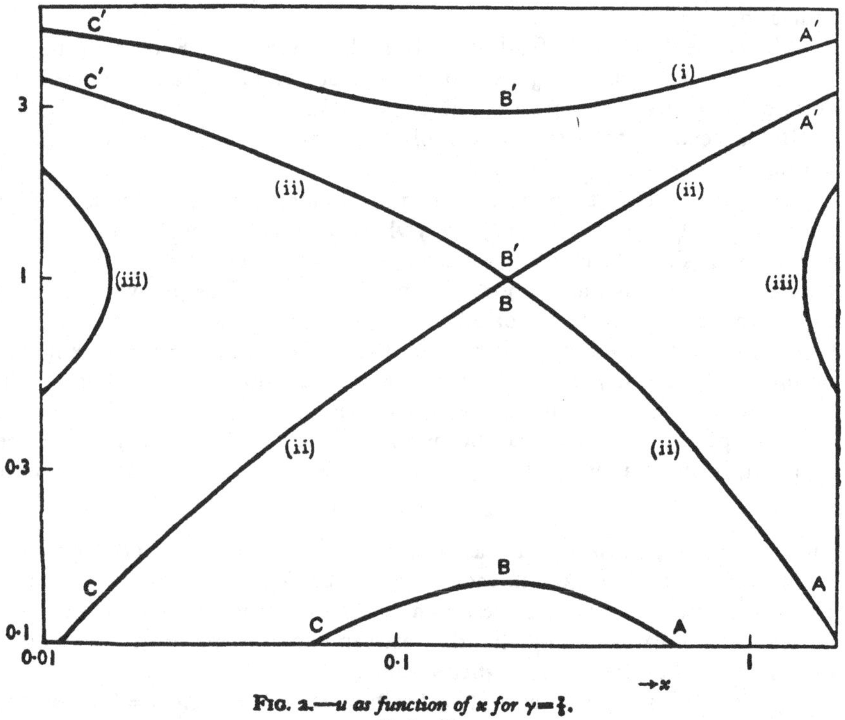


FIG. 2.—$u$ as function of $x$ for $\gamma = \frac{7}{5}$.
(i) $\lambda = \frac{1}{2}\lambda_c$;
(ii) $\lambda = \lambda_c$;
(iii) $\lambda = 4\lambda_c$.

It is easy to eliminate $z$ between (8) and (9′) since, by (11), $y=u$ in the present case. Accordingly

$$\tfrac{1}{2}y^2-\ln y=-\ln\lambda+(1/x+2\ln x). \tag{14′}$$

The minimum of the left-hand side occurs for $y=1$ and equals $\frac{1}{2}$, while the minimum of the bracket on the right-hand side occurs for $x=\frac{1}{2}$ and equals $2-2\ln 2$. Accordingly

$$\lambda_c=\tfrac{1}{4}e^{3/2}=1\cdot120\ldots,\quad \ln\lambda_c=0\cdot1138\ldots. \tag{18′}$$

This is also the limit of expression (18) as $\gamma\to1$, so that $\lambda_c$ is continuous at $\gamma=1$. Fig. 4 represents this case.

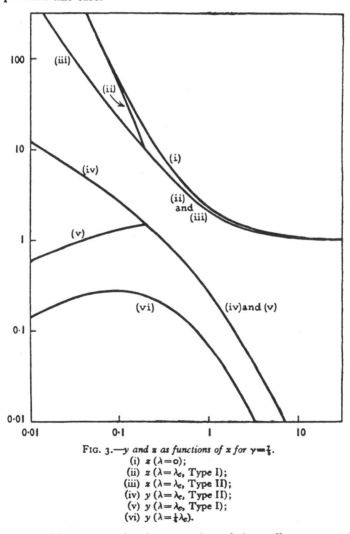

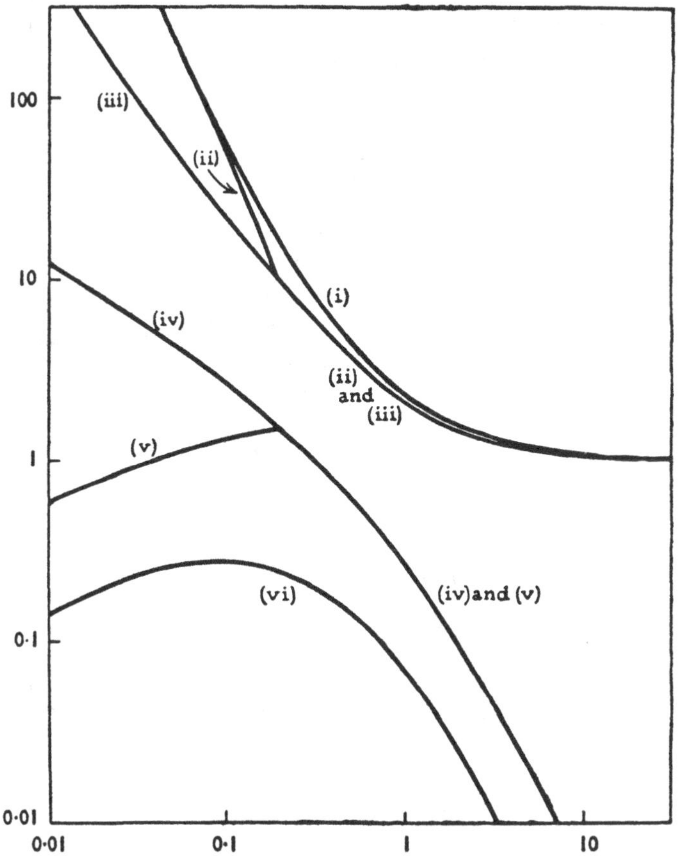


FIG. 3.—*$y$ and $z$ as functions of $x$ for $\gamma=\frac{4}{3}$.*
(i) $z$ ($\lambda=0$);
(ii) $z$ ($\lambda=\lambda_c$, Type I);
(iii) $z$ ($\lambda=\lambda_c$, Type II);
(iv) $y$ ($\lambda=\lambda_c$, Type II);
(v) $y$ ($\lambda=\lambda_c$, Type I);
(vi) $y$ ($\lambda=\frac{1}{2}\lambda_c$).

The case $\gamma=\frac{5}{3}$ is an even simpler extension of the ordinary case. The only real change is that $x_m$ (which decreases monotonically as $\gamma$ increases) now equals zero. Accordingly there is now no difference between Type I and Type II motions for $\lambda=\lambda_c=\frac{1}{4}$ (Fig. 5).

[It can be seen that $u$, $y$ and $z$, considered as functions of $x$ and $\gamma$, are continuous in $1 \leqslant \gamma \leqslant \frac{5}{3}$.

7. The final question that must be considered is what determines the value of $\lambda$ in any actual case. It has been seen that the steady-state equations possess a solution whenever $0 \leqslant \lambda \leqslant \lambda_c$. The particular value of $\lambda$ actually occurring must therefore be determined by other considerations. This is analogous to the

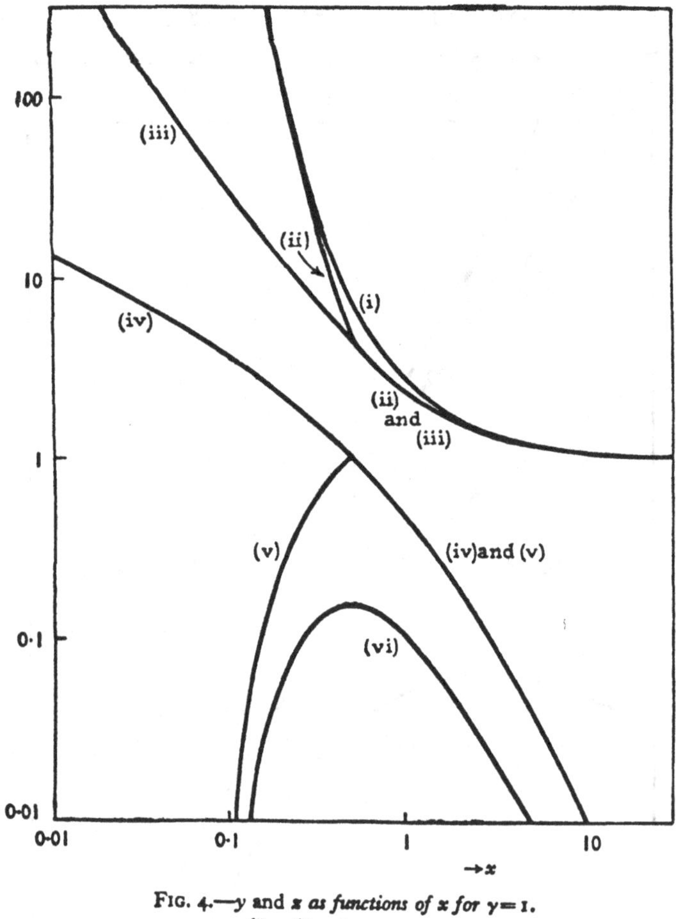


FIG. 4.—$y$ and $z$ *as functions of* $x$ *for* $\gamma = 1$.
(i) $z$ ($\lambda = 0$);
(ii) $z$ ($\lambda = \lambda_c$, Type I);
(iii) $z$ ($\lambda = \lambda_c$, Type II);
(iv) $y$ ($\lambda = \lambda_c$, Type II);
(v) $y$ ($\lambda = \lambda_c$, Type I);
(vi) $y$ ($\lambda = \frac{1}{2}\lambda_c$).

velocity-limited case of accretion where $\alpha$ is not determined by the steady-state equations (4). The method used there to determine a specific value of $\alpha$ was to consider the case in which the star entered a cloud of gas with a plane boundary. The use of such a model would lead to very great mathematical difficulties in the present case, since the conditions of time-independence and spherical symmetry would have to be dropped simultaneously.

The boundary conditions at the surface of the star do not seem to be of help in the problem, since the star will swallow up any material falling into it without imposing any real conditions on the velocity, density, or pressure of the incoming materials.

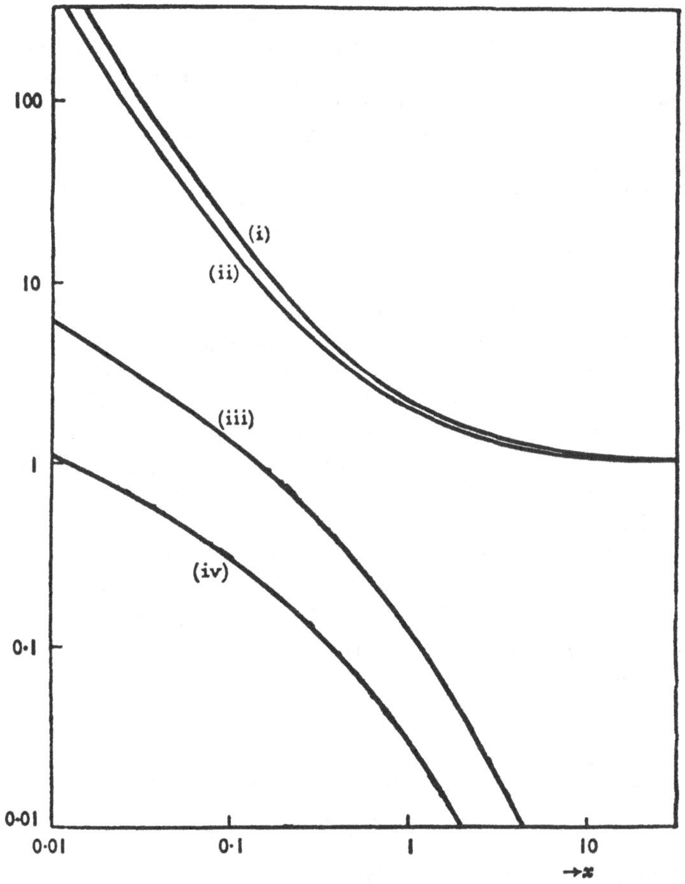


FIG. 5.—*y and z as functions of x for* $\gamma=\frac{4}{3}$.
(i) $z\ (\lambda=0)$;
(ii) $z\ (\lambda=\lambda_c)$;
(iii) $y\ (\lambda=\lambda_c)$;
(iv) $y\ (\lambda=\frac{1}{2}\lambda_c)$.

There remains the possibility of investigating the stability of the system with respect to small disturbances. Even if only spherically symmetrical perturbations are admitted, a partial differential equation of considerable complexity results. It is easily seen from it that disturbances are in part propagated with the velocity of sound relatively to the material, but the nature of the part that remains behind is not easily found. However, this may be a possible method of approach to the problem.

There is yet another possibility of investigating the stability of the system, and that is by comparing the energy of the system in its various states. The state with the lowest energy would then be expected to be the only stable state. Owing to the fact that our system is not isolated (the star itself not being considered part of the system) the validity of this approach is not quite assured, and similarly doubts may arise owing to the infinite extent of the system. Nevertheless, since the comparison of the energy of the system in every spherical shell leads to the same result, it seems very likely that the method gives the correct answer.

The energy of the gas per unit mass is constant by virtue of our assumptions (*cf.* equation (4)). Accordingly a comparison of the densities is all that is involved. Consider now $z$ as function of $\lambda$ for fixed $x$. Then it may be seen from equations (13), (15) and (16) that for $u$ less than unity $z$ decreases as $\lambda$ increases. Accordingly the energy of the system in every spherical shell is lower for the Type I state with $\lambda=\lambda_c$ than for any other Type I state. Comparing now, for $\lambda=\lambda_c$, the Type I and the Type II states, it is immediately seen that, for $x \geqslant x_m$, the densities are the same, but that for $x<x_m$ the density (and hence the energy) is lower in the Type II state.

Accordingly the system has, in the sense described, the lowest energy in the Type II state, and we may expect to find a natural system in this state with $\lambda=\lambda_c$. If $\gamma=\frac{5}{3}$, the difference between the Type II state and the Type I state with $\lambda=\lambda_c$ disappears, and we would expect to find the system in this joint state. The result that the Type II state is the one most likely to be realized is very satisfactory, since the behaviour of all the functions is most uniform and smooth in this state. The result is also in agreement with the intuitive idea that, since there is nothing to stop the process of accretion, it takes place at the greatest possible rate, i.e. with $\lambda=\lambda_c$.

8. The two cases of accretion that have been examined so far may be called velocity-limited and temperature-limited respectively. The intermediate range of cases presents far greater difficulties. However, it may be possible to conjecture what the result is in the following way.

In the velocity-limited case of accretion the accretion rate $A$ is given (4) by

$$A \doteqdot 2{\cdot}5\pi(GM)^2V^{-3}\rho_\infty, \tag{21}$$

while for, say, $\gamma=\frac{3}{2}$ in the temperature-limited case the result is

$$A=2\pi(GM)^2c^{-3}\rho_\infty. \tag{22}$$

If we therefore write down the formula

$$A \doteqdot 2\pi(GM)^2(V^2+c^2)^{-3/2}\rho_\infty,$$

it seems likely that it represents the order of magnitude of the accretion rate in the intermediate case, in which a star of mass $M$ moves with relative velocity $V$ in a uniform cloud of gas, in which the undisturbed density and the velocity of sound have the values $\rho_\infty$ and $c$ respectively. This formula, in agreement with intuitive ideas, suggests that if $c$ exceeds $V$, temperature (pressure) imposes the chief limitation on the rate of accretion; whereas if $V$ exceeds $c$, dynamical limitations are of greater importance.

The limitations due to pressure have probably been somewhat overestimated in this work. For if the cloud is able to radiate away some of the heat of compression then the adiabatic law will not apply, the pressure near the star will be diminished, and the accretion rate somewhat increased. How large this effect will be depends on the composition of the cloud. If there is a high proportion of constituents (such as hydrogen molecules (2)) that easily radiate at moderate temperatures, then the effect will be appreciable. In this case the effective value of $\gamma$ will be closer to unity than to the standard value for the gas in question. If $\gamma$ equals unity the process is isothermal.

The work of the present paper, together with previous work, is likely to give a fair estimate of the order of magnitude of accretion in all cases of physical interest. Further progress in this field will probably require the consideration of non-steady states.

*Trinity College,*
*Cambridge :*
1951 *October* 2.

*References*

(1) F. Hoyle and R. A. Lyttleton, *Proc. Camb. Phil. Soc.*, **35**, 405, 1939.
(2) F. Hoyle and R. A. Lyttleton, *Proc. Camb. Phil. Soc.*, **36**, 325, 1940.
(3) F. Hoyle and R. A. Lyttleton, *Proc. Camb. Phil. Soc.*, **36**, 424, 1940.
(4) H. Bondi and F. Hoyle, *M.N.*, **104**, 273, 1944.

# NOTES

## ACCRETION OF INTERSTELLAR MATTER BY MASSIVE OBJECTS

Observations of quasi-stellar radio sources have indicated the existence in the Universe of extremely massive objects of relatively small size. The present note discusses the possible further growth in mass of a relatively massive object, by means of accretion of interstellar gas onto it, and the accompanying energy release. Although there is no evidence for (and possibly some evidence *against*) quasi-stellar radio sources occurring inside ordinary galaxies, for the sake of concreteness we consider the fate of an object of mass $M > 10^6$ (masses in solar units throughout) in an ordinary spiral galaxy somewhat like ours.

We first re-examine the hypothetical problem of an object of mass $M$ moving with velocity $U$ (in km/sec) relative to a completely uniform gas medium of density $n$ (expressed as H-atoms per cm³) and thermal speed $U_{th}$. We define (Hoyle and Lyttleton 1939) a characteristic length $s_0$ and express the rate of accretion in terms of a dimensionless parameter $\alpha$ to be determined,

$$s_0 = GM/U^2 = (M/U^2) \times 4.3 \times 10^{-3}\ \text{pc}\,,$$

$$dM/dt = 2\pi\alpha s_0^2\, nU \equiv \alpha M/t_0\,, \tag{1}$$

$$t_0 = (U^3/Mn) \times 3.3 \times 10^{11}\ \text{years}\,.$$

We assume that the size of the object, as well as the collision mean free path in the gas, is very much smaller than $s_0$. For cases with large Mach number, $U \gg U_{th}$, a standing (relative to the object) shock front develops behind the object, with a large pressure increase across the front. The shape of the shock front is conical about the accretion axis, as shown schematically in Figure 1, but the cone angles depend on the specific heat ratio $\gamma$ of (and on any irreversible energy loss from) the gas. If $\gamma - 1 \ll 1$ (almost isothermal compression) then the density increases greatly across the shock front, the cone angles are everywhere small, and pressure gradients parallel to the accretion axis are small compared with gravitational forces. For such cases Bondi and Hoyle (1944) found that $1 < \alpha < 2$, no matter what the degree of mixing in the shocked region is. For the special case of purely laminar flow, the results are as though molecules moved without collisions except for giving up their transverse momentum when they reach the accretion axis. In this case all the gas with impact parameters less than $\sqrt{(2)}\, s_0$ is accreted and $\alpha = 2$.

If $\gamma - 1$ is appreciable, then the cone angles of the shock front are not very small, and the gas molecules lose less of their transverse momentum when crossing the shock front and also regain some longitudinal momentum during the re-expansion in the shocked region (see Ruderman and Spiegel [1964]). These effects result in $\alpha$ decreasing with increasing $\gamma$. Explicit calculations for $\alpha(\gamma)$ have not yet been carried out but plausibility arguments lead to the conjecture that $\alpha$ is non-zero for $\gamma < \frac{5}{3}$ and not much less than unity for $\gamma < \frac{4}{3}$. The $\alpha$ is also reduced slightly if $U$ is only slightly larger than $U_{th}$ (see Bondi [1952] for cases with $U < n\, U_{th}$).

McCrea (1953) has discussed the slowing down of an object during its passage through a gas cloud with velocity $U \ll U_{th}$. This retardation, analogous to that experienced by a charge moving through a plasma, can be expressed as

$$dU/dt = -\,\beta U/t_0\,, \qquad \beta \approx 1 + \ln(b_{max}/s_0)\,, \tag{2}$$



Reprinted courtesy of the author(s) and *The Astrophysical Journal*, published by the University of Chicago Press.

where $t_0$ and $s_0$ are defined in equation (1). For a uniform medium of infinite extent the limiting impact parameter would be $b_{max} \sim U/\sqrt{(4\pi\rho G)}$ as in the plasma problem. In practice the limit in the gravitational problem is the extent of the uniform medium (e.g., thickness of the galactic disk). Consider a massive object in the "halo population" with $U \sim 100$ or 200 (relative to the galactic disk) which spends a fraction of about $(10/U)$ of its time in the disk. According to equations (1) and (2) the time required for slowing down is $(U/10)\ t_0/4\beta$ (if mass increase is neglected and $\beta$ assumed constant). Unlike accretion of mass, the retardation involves only distant gravitational encounters and does not require collisions by the "gas molecules," so that the stars in the disk also contribute to the retardation. We assume an average gas density $n \sim 1$ in the disk and $n_{tot} \sim 10n$ (including stars) and $\beta \approx 5$. A halo object with initial velocity $U$ (relative to the disk) will slow down in a time less than $10^{10}$ years, roughly the age of our Galaxy, if its mass $M$ exceeds a critical value of

$$M_{cr,\, sl.} \approx (U/100)^4 \times 1.5 \times 10^8 . \tag{3}$$

For a highly supersonic object retardation is more rapid than accretion ($4\beta n_{tot} \gg an$), and we need to consider accretion only after the object has slowed down to a velocity

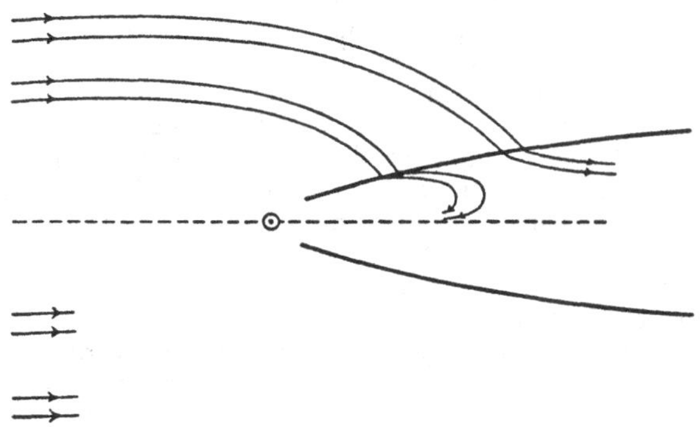

FIG. 1.—A schematic view of the flow. The dotted line is the axis of accretion. The heavy line is the standing shock front; the thin lines are (laminar) flow lines.

$U$ not much bigger than $U_{th}$. For our large mass $M$ the characteristic impact parameter $s_0$ exceeds the "spacing between interstellar gas clouds" (or "turbulence scale length") $d$, and we have to include the random cloud velocities in the thermal velocity. We assume $U_{th} \sim 10$ or 15, $U \sim 25$, $n \sim 1$. We thus cannot make use of the much lower thermal velocity and higher density available in individual H I region, as attempted in previous work for smaller masses. On the other hand, the difficulties encountered previously (Schatzman 1955) due to the heating and expansion of H I regions are absent when $s_0 > d$. The energies released by the compression at distances near $s_0$ can easily be radiated away at typical H II temperatures, so the compression is almost isothermal ($\gamma - 1$ small). The presence of turbulent magnetic fields in the gas makes the compression more adiabatic and raises the effective $\gamma$ toward $\frac{4}{3}$. The accretion time $(t_0/a)$ is less than $10^{10}$ years if the mass $M$ exceeds

$$M_{cr,\, acc.} \approx (U/25)^3\ (0.25/an) \times 2 \times 10^6 , \tag{4}$$

and $a$ probably lies between 0.1 and 1. The characteristic impact parameter is $s = \sqrt{(a)}\ s_0$ (with $s \approx 7$ pc for $M = M_{cr,\, acc.}$) and the size of the object must be less than $s$. The mass of typical halo globular clusters in our own Galaxy is too small for retardation and accretion to become catastrophic by one order of magnitude or so (and their diameter would be sufficiently small).

So far we have only discussed the gravitational capture of gas atoms from the stream at radial distances $r$ from the central object with $r \sim s$, where the energy released per H atom ($\sim GMH/S \sim \frac{1}{2}HU^2$) is only a few eV. If the original gas stream possessed no angular momentum about the accretion axis at all (and $\gamma < \frac{5}{3}$) the captured gas would immediately collapse toward the center, somewhat as in star formation but with a continuously increasing mass in the condensation. The evolution of massive stars takes less than $10^7$ years and no zero-temperature spherical equilibrium models exist exceeding two or three solar masses. After a relatively short time, then, most of the accreted matter would have collapsed to the "Schwarzschild singularity" ($r = 2GM/c^2$). Although the gravitational redshift for radiation from "collapsed matter" is large, note that its "active gravitational" (as well as "passive inertial") mass is essentially undiminished. If the original star cluster (or other non-condensed object) exceeded both critical masses in equations (3) and (4), then the accreted collapsed matter eventually dominates as the cause for further accretion.

Theories for quasi-stellar (or other condensed) objects which invoke the gravitational collapse of an *isolated* mass or cluster usually encounter some difficulties in shedding original angular momentum. In our problem the incident "turbulent" gas stream will also carry some small angular momentum at any instant of time, and a captured gas "turbule" will not be able to penetrate to very small radial distances by itself. For our continuous gas stream, however, the angular momentum is only a statistical quantity, provided that the accretion impact parameter $s$ is much smaller than characteristic dimensions of the galaxy. The angular momentum then changes sign over times of the order of $s/U$ and in about that time a gas "turbule" can shed its angular momentum by collisions with other turbules and spiral in toward the Schwarzschild radius. In this turbulent manner gravitational energy is continually converted into bulk kinetic energy, compressional heating, and compression of turbulent magnetic fields. Densities $n(r)$ at radial distance $r \ll s$ must lie in the range between $(s/r)^{1.5}n$ and $(s/r)^3n$ and radiating away energy (bremsstrahlung or synchrotron radiation via energetic electrons) at these high densities presents no great problems.

Each accreted gas atom eventually approaches the Schwarzschild radius $r = 2r_0$ (with $r_0 = GM/c^2 = (M/10^8) \times 4.7 \times 10^{-6}$ pc), and we have to estimate the fraction $f_{tot}$ of its rest-mass energy which is radiated away altogether. As a simplified model for the spiraling in via many collisions, consider a particle in circular orbits with diminishing radii caused by a slow drain of angular momentum and energy. Due to general relativistic effects (see the Appendix) the orbits become unstable to *spontaneous* spiraling in at $r = 6r_0$. By this time a fraction $f = 0.057$ of the rest-mass energy has escaped to infinity. If, at $r = 6r_0$, the particle could be brought to rest (by the unlikely collision with a matched particle of opposite momentum) an additional fraction $\Delta f = 0.126$ could in principle escape. After free fall to $r = 3r_0$ the additional fraction is 0.22, and so on. However, the chances of arresting the free fall with appreciable efficiency seem small. In addition, even in the absence of any optical absorption, 50 per cent of all radiation (or highly relativistic particles) emitted isotropically at $r = 3r_0$ would be prevented from escaping by general relativistic effects (but only 14.5 per cent at $r = 6r_0$). We estimate $f_{tot}$ (net energy escaping to infinity in all forms) to lie in the range of 0.05–0.20. Note that the gravitational mass of the collapsed source is increased by a fraction $1 - f > 0.8$ of the accreted mass, and we omit a corresponding correction factor which should be applied to equation (1).

The optical luminosity $L_{opt}$ can be written in the form

$$L_{opt} = \left(\frac{25}{U}\right)^3 n \, \frac{a}{0.25} \, \frac{f_{opt}}{0.03} \left(\frac{M}{10^7}\right)^2 \times 2 \times 10^9 L_\odot . \qquad (5)$$

If $a$ (and $n$) remained constant, then $M$ (and $L$) would increase to infinity after a finite time ($t_0/a$), according to equation (1). However, the effects of radiation pressure decrease the effective value of the gravitational constant appreciably as $(L_{opt}/M)$ approaches a limiting value which depends on the average opacity coefficient. $(L_{opt}/M)_{lim}$ can be as low as 100 (in solar units) in H I-regions in our Galaxy, due to the presence of dust grains (van de Hulst 1955) and as large as $3 \times 10^4$ in pure ionized hydrogen if only Thomson scattering is important. Once the mass $M$ has grown sufficiently for equation (5) to give a value of $(L_{opt}/M)$ close to this limit, radiation pressure automatically lowers $\alpha$ in such a way as to keep $L_{opt}/M$ almost constant. Subsequently, $M$ (and hence $L_{opt}$) increases with time only exponentially as $e^{t/\tau}$, where

$$\tau = (f_{opt}/0.03)(10^3\, M/L_{opt}) \times 4 \times 10^8 \text{ years}, \tag{6}$$

and the characteristic accretion impact parameter $s = \sqrt{(\alpha)}\, s_0$ increases only as $\sqrt{M}$. For instance, under typical conditions for our galactic disk with $(L_{opt}/M)_{lim} \sim 200$, the exponential growth would start at $M \sim 10^7$ and $s$ would reach about 350 pc when $M \sim 10^9$ and $L_{opt} \sim 2 \times 10^{11}$.

To summarize the situation for a highly evolved spiral galaxy like ours (with only a few per cent of the mass remaining in the form of gas): Objects of the order of $10^6\, M_\odot$, more than typical globular cluster masses, are required to initiate a catastrophic accretion process. Unlike theories involving the gravitational collapse of *isolated* objects, the time scale of our process never becomes very short due to the self-limiting effects of radiation pressure. In our Galaxy, in fact, dust grains would keep the time scale well above $10^8$ years, which is much longer than required for quasi-stellar objects, and the luminosity too low ($< 10^{12}\, L_\odot$). The situation is likely to be more favorable in systems less evolved than our Galaxy where (*a*) a larger fraction of the mass is still in the form of gas and (*b*) the relative abundance of heavier elements, and hence of dust grains, is lower. This leads to a shorter time scale and a larger limiting value for the luminosity-mass ratio for the accreting condensation.

I am indebted to Drs. R. P. Feynman, M. Ruderman, M. Schwarzschild, E. Spiegel, L. Spitzer, and L. Woltjer for helpful criticism and suggestions. In fact, they have contributed most of the positive ideas in this note without being responsible for any of the unwarranted conjectures. I am also grateful to the National Academy of Sciences for a senior postdoctoral fellowship.

E. E. SALPETER*†

Received May 7, 1964

NEWMAN LABORATORY OF NUCLEAR STUDIES AND
CENTER FOR RADIOPHYSICS AND SPACE RESEARCH
CORNELL UNIVERSITY
AND
GODDARD INSTITUTE FOR SPACE STUDIES
NEW YORK, NEW YORK

## APPENDIX

We use the usual Schwarzschild metric outside of a mass $M$ (and put $c = 1$),

$$ds^2 = (1 - 2u)\,dt^2 - (1 - 2u)^{-1} dr^2 - r^2 d\theta^2, \tag{A1}$$

where $u = GM/r$, and we consider the motion of a single test particle of unit mass in this field. As discussed by Feynman in unpublished lecture notes, the following quantities are constants of the motion

$$k \equiv (1 - 2u)\,t^{(1)}, \qquad l \equiv r^2 \theta^{(1)}, \tag{A2}$$

* Permanent address: Cornell University, Ithaca, New York.

† This work was supported in part by contract AFOSR-321-63 with Cornell University.

where a superscript (1) denotes differentiation with respect to proper time. The terms $l$ and $k$ are generalizations of angular momentum and energy, respectively, and are also conserved in any *local* collision processes (any number of particles at the same $u$). The radial distance $r$ satisfies the equation

$$[r^{(1)}]^2 = k^2 - 1 + 2u - (1 - 2u)l^2u^2 . \tag{A3}$$

The circular orbit with the smallest radius which is stable is then at $u = \frac{1}{6}$ with $l = 2\sqrt{3}$ and $k = \sqrt{\frac{8}{9}}$. A photon (or an extremely relativistic particle) can have an unstable circular orbit at $u = \frac{1}{3}$. A particle at rest at distance $r(r^{[1]} = l = 0)$ has $k = \sqrt{(1 - 2u)}$. Consider a particle which started at rest far away ($k = 1, u = 0$) and ends up at some finite distance from the source with some value of $k$. If the energy released during this change all escapes "to infinity" then the escaped energy (*after* all redshift corrections) is simply $(1 - k)c^2$.

## REFERENCES


Bondi, H. 1952, *M.N.*, **112**, 195.
Bondi, H., and Hoyle, F. 1944, *M.N.*, **104**, 273.
Hoyle, F., and Lyttleton, R. A. 1939, *Proc. Cambridge Phil. Soc.*, **35**, 405.
Hulst, H. C. van de. 1955, *Mém. Soc. R. Sci. Liège*, **15**, 393.
McCrea, W. H. 1953, *M.N.*, **113**, 162.
Ruderman, M., and Spiegel, E. 1964, unpublished.
Schatzman, E. 1955, *Gas Dynamics of Cosmic Clouds* (I.A.U. Symposium No. 2 [Amsterdam: North-Holland Publishing Co.]), p. 193.

SOVIET PHYSICS–DOKLADY VOL. 9, NO. 3 SEPTEMBER, 1964

ASTRONOMY

# THE FATE OF A STAR AND THE EVOLUTION OF GRAVITATIONAL ENERGY UPON ACCRETION

Ya. B. Zel'dovich

Translated from Doklady Akademii Nauk SSSR, Vol. 155, No. 1, pp. 67-69, March, 1964
Original article submitted December 6, 1963

It is well known that cold matter cannot resist the compressive action of gravitational attraction if its mass is greater than the mass of the sun. This result was obtained by Oppenheimer and Volkoff [1] in 1938 in a study of the degenerate neutron gas in the Einstein theory of gravitation (the "general theory of relativity") and is to be found in textbooks [2]. This result is not affected qualitatively by any assumptions concerning the interactions of elementary particles in the presence of high matter density.

The general theory of relativity makes a radical modification of the picture of the dynamics of condensation. In the classical theory an infinite density is attained after a finite time; one may suppose that an expansion results from this, or a shock wave arises, traveling outward from the center and ejecting a portion of the matter.

As was shown by Oppenheimer and Snyder [3], the general theory of relativity leads to the conclusion that an infinite density is indeed reached after a finite proper time (as measured by an observer moving with any particle of the star). However, one must take the change in the time scale into account if signals are exchanged between the star and an external observer located outside the star's gravitational field. The red shift of lines emitted from the surface of a star is a special case of this change in the time variation. It turns out that for an external observer, the external surface of a star only attains the so-called gravitational (Schwarzschild) radius $r_g = 2GM/c^2$ asymptotically at $t \to \infty$. For every particle (for example, the central one) it is possible to determine the moment at which the particle must emit a signal which will arrive at the external observer as $t \to \infty$. One may speak in this sense of the gravitational self-collapse of a star [4].

At the moment of emission of the signal, the density for each particle is less than the characteristic value

$$\rho_g = \frac{3M}{4\pi r_g^3} = \frac{3}{32\pi}\frac{c^6}{M^2G^3} = 1.8\cdot 10^{16}\left(\frac{M}{M_\odot}\right)^{-2}\frac{g}{cm^3}.$$

As a result, the attainment of an infinite density during the condensation is not observable, and the acquisition of information is accomplished long before this moment. Hence, the question of what happens after $\rho = \infty$ has even less meaning.

All of these conclusions remain valid when pressure is taken into account [5], even for the case of burning matter. One should regard the entropy of the matter as constant in studying the problem of the existence and stability of mechanical equilibrium in the presence of pressure and gravitational attraction. Equilibrium corresponds to a minimum of the total energy for a given entropy (and also, of course, for fixed number of conserved particles – the baryons). The entropy may be considered as constant during a rapid condensation.

For each value of entropy S there is a number of equilibrium configurations of a burning gas (the star) of different masses M; such configurations exist, however, only for masses less than the critical mass. The value of the critical mass $M_c$ increases with increasing S; $M_c = M_c(S)$. For $S = 0$ $M_c = M_c(0) \cong M_\odot$. Stars with $M > M_\odot$ may be found in a state of mechanical equilibrium to the extent that they are burning and to the extent that, the densities being equal, the pressure of the burning matter exceeds that of the cold matter [6].

As a result, the final stage in the evolution of every nonrotating star whose mass is much greater than that of the sun consists of an uncontrollable condensation, i.e., a collapse.

The brightness of a burning star drops off exponentially quite rapidly during gravitational self-collapse, with a time constant of the order $r_g/c$, i.e., $10^{-4}$ sec for $M \sim 10M_\odot$. As a result, condensed stars must be dark bodies which interact with the surrounding medium only via their gravitational field. Since the star's radiation, and therefore its loss in mass, during the collapse is small [4, 7], the gravitational field of a condensing star at large distances is no different from the field of the same star prior to the condensation. The question has been previously raised [8] as to what portion of all nucleons in the universe exist at a given moment in dark condensed stars. From considerations relating to the problem of the growth of the universe, one obtains only an inequality for the over-all density $\bar{\rho} < 2 \cdot 10^{-28}$ g/cm$^3$, whereas the density for normal, visible stars $\rho \cong (0.3\text{-}1) \cdot 10^{-30}$ g/cm$^3$. Hoyle, Fowler, and Burbidge have decisively raised the question of whether a large number of dark stars is present. According to their estimate the mass of such stars is several times larger than the mass of the luminescent stars, which puts the mean density close to the upper limit [8]. The interest in the catastrophic condensation of stars arose in connection with the discovery of optically intense distant radio sources [9, 10], and Hoyle's hypothesis [11] that these sources are superstars with masses of the order of $10^8$ $M_\odot$. Until recently, the source of the energy of the powerful radio galaxies was not understood; ideas such as the annihilation of matter and antimatter, collisions between galaxies, and the simultaneous explosion of many supernova turned out to be untenable. How can the condensation of superstars lead to the emission of necessarily gigantic quantities of energy? According to Hoyle et al. [7] the condensation is accompanied by fluctuations, whose density reaches $10^{30}$ g/cm$^3$; ultrarelativistic particles are emitted when the density reaches its maximum. It is impossible to agree with this point of view, merely because the break at $\rho_m = 10^{30}$ g/cm$^3$ depends on a hypothetical C-field with very strange properties [7, 12]. For an external observer the growth in the density as a result of gravitational self-collapse stays asymptotically at a quite modest value of the order of 2-200 g/cm$^3$ (for $M = 10^8\text{-}10^7$ $M_\odot$).

An alternative mechanism of energy emission is examined, in the present note, which is associated with a decrease in the surface mass in the gravitational field of a condensing star.

The velocity of a freely falling particle approaches the gravitational radius.

The relative velocity is of the order c in the collision of two particles at a distance of the order (no larger!) of the gravitational radius. The energy radiated by relativistic particles during a collision can therefore be $\alpha mc^2$; the energy carried out to infinity is less because a part of the radiation and the particles falls to the star as a result of the red shift and the geometrical factor. The net amount of energy which escapes is $\alpha\beta mc^2$, where m is the mass of the particle, $\alpha < 1$, $\beta < 1$. This product attains a maximum which is of the order $0.1\ mc^2$ for a collision at $r = 1.5\ r_g$. It is essential that the colliding particles have different angular momenta relative to the star; the numbers given refer to the case $\vec{M}_1 = -\vec{M}_2$. If the collisions are infrequent, the energy emission is proportional to the collision cross section for a given velocity distribution of the particles far away from the star. If the cross section is large, making the mean free path small compared to the dimensions of the star, the motion of the particles is governed by collisions, making it necessary to use a hydrodynamic description of the motion. At the same time it appears that for spherically symmetric motion the gravitational energy is transformed principally into kinetic energy of motion in the radial direction; the amount of energy which can be radiated amounts to a negligibly small fraction of the rest mass of the falling matter.

However, if a flow of matter is directed against the star with a supersonic velocity at a great distance from the star, the picture of the motion is radically changed. A stationary shock wave develops on the side of the star opposite the direction of the incident matter flow. Near the star the velocity change at the front of the wave is of the order of c, and an appreciable portion of rest mass of the matter being compressed by the wave is radiated. According to Bernoulli's theorem, not even a small part of the matter can be ejected to infinity with a velocity greater than the initial velocity far from the star. But when the cloud of matter arrives at the star, surrounds it and strikes the rear side, the motion ceases to be stationary and the cumulative ejection of part of the matter with a velocity of the order of c becomes possible.

In conclusion we note that the "particles" of which we have been speaking are not necessarily atoms and molecules, but may be localized plasma concentrations with a frozen magnetic field; the production of relativistic electrons by collisions then becomes especially probable. When examining the flow of matter from a distance one should evidently not visualize it as inter-star gas and flame with $\bar{\rho} = 10^{-25}$ g/cm$^3$, which would give a small output.

During the collapse of one member of a closely bound pair of stars, one may regard the matter of the other member as the falling material. This can be that part of the shell of the collapsing star itself which was ejected before the instant of gravitational self-collapse; in addition to the matter which acquires a hyperbolic velocity, part of the ejected matter can be stored in distant but closed orbits.

The idea of collapse in a powerful gravitational field as a source of the radiated energy of radiosources was advanced in its most general form by I. S. Shklovskii [13].

We take this opportunity to express our sincere gratitude to I. D. Novikov and I. S. Shklovskii for numerous discussions.

## LITERATURE CITED


1. J. Oppenheimer and G. Volkoff, Phys. Rev., 55, 374 (1938).
2. L. D. Landau and E. M. Lifschitz, Statistical Physics [in Russian] (1962).
3. J. Oppenheimer and H. Snyder, Phys. Rev., 56, 455 (1939).
4. Ya. B. Zel'dovich, Astr. Tsikulyar, No. 250, July 1 (1963).
5. M. A. Podurets, Doklady Akad. Nauk SSSR, 154, No. 2 (1964).
6. Ya. B. Zel'dovich, Vopr. Kosmogonii, 9, 80 (1963).
7. F. Hoyle, W. Fowler, G. Burbidge, and M. Burbidge, Relativistic Astrophysics, Preprint (1963).
8. Ya. B. Zel'dovich and Ya. A. Smorodinskii, ZhÉTF, 41, 907 (1961)[Soviet Physics—JETP, Vol. 14, p. 647].
9. T. Matthews and A. Sandage, Publ. Astron. Soc. Pacific, 74, 406 (1962).
10. M. Schmidt, Astrophys. J., 136, 684 (1962); Nature, 197, 1040 (1963).
11. F. Hoyle and W. Fowler, Nature, 197, 533 (1963).
12. F. Hoyle and J. V. Narlikar, Proc. Roy. Soc., 273, No.1352, 1 (1963).
13. I. S. Shklovskii, Astron. Zhurn., 39, 591(1962) [Soviet Astronomy—AJ, Vol. 6, p. 465].


All abbreviations of periodicals in the above bibliography are letter-by-letter transliterations of the abbreviations as given in the original Russian journal. *Some or all of this periodical literature may well be available in English translation.* A complete list of the cover-to-cover English translations appears at the back of this issue.

# BINARY STARS AMONG CATACLYSMIC VARIABLES
# I. U GEMINORUM STARS (DWARF NOVAE)

ROBERT P. KRAFT
Mount Wilson and Palomar Observatories
Carnegie Institution of Washington, California Institute of Technology

*Received October 9, 1961*

## ABSTRACT

A spectroscopic test for binary motion among several U Gem variables is presented. Five stars are shown to be binaries with $P < 9$ hours (SS Cyg, U Gem, RX And, RU Peg, SS Aur), one spectrum is composite (EY Cyg), and one other (Z Cam) shows evidence of binary motion with an as yet undetermined period. There is no evidence contradicting the hypothesis that all members of this group are spectroscopic binaries of short period.

The peculiar radial velocities are small, and the corresponding statistical parallax leads to $\langle M_v \rangle = +9.5 \pm 1$; the blue stars in these systems are probably white dwarfs. The masses of the red components and their spectra (when photographed) seem consistent with a star of mass $\sim 1 \mathfrak{M}_\odot$. Thus the red components of U Gem variables are seriously underluminous for their masses. Evidence is presented which indicates that the red stars overflow their lobes of the inner Lagrangian surface; the ejected material forms, in part, a ring, or disk, surrounding the blue star. Several lines of argument suggest that the U Gem variables are descendants of the W UMa stars.

## I. INTRODUCTION

The cataclysmic variables of type U Geminorum (dwarf novae) are characterized by a small range (2–6 mag.) and by the existence of an "induction time" between outbursts. The latter has a time scale of the order of days or weeks, which can be described as "periodic" when averaged for any one variable over a long enough time interval. Of the roughly 100 known stars of this type, two subgroups are recognized: SS Cygni and Z Camelopardalis stars. The former have a more or less well-defined minimum brightness; the latter do not always descend to the minimum after each outburst but rather maintain some magnitude intermediate between minimum and maximum. The properties of U Gem itself are similar to those of SS Cyg, and this has led to the designation of the whole group, by some authors, as SS Cyg stars; this will not be done here. A few of these variables have been shown to vary irregularly in light at minimum (Walker 1957); this "flickering" has a time scale of the order of 1–10 minutes and a range of a few hundredths to a tenth of a magnitude or more. The variations are similar to those found in various old novae and nova-like objects (Walker 1957).

In 1956 Joy announced that SS Cyg was a spectroscopic binary with $P = 6^h 38^m$ and with components of spectral types dG5 and sdBe. This result, coupled with Walker's (1954, 1956) discovery that Nova (DQ) Her is an eclipsing binary of period $4^h 39^m$, led to the speculation that all cataclysmic variables might be binary systems. The present paper describes a spectroscopic test for binary motion in several U Gem variables; remaining papers of the series will deal with other stars of the group and with certain of the old novae. The results for U Gem stars can be summarized as follows: five are definitely binaries with $P < 9$ hours; one is composite, and one other shows evidence of velocity variation but with an as yet undetermined period. There is at present no evidence in direct contradiction to the hypothesis that all stars of this group are, indeed, spectroscopic binaries of short period.

It should be stated at the outset that U Gem stars are very faint at minimum light; thus observational material is gathered rather slowly even with the 200-inch telescope. The data reported here were obtained over a 30-month interval. Certain fairly definite conclusions can be reached from this limited material, and it was thought best to make a somewhat preliminary report at this time. The possibility remains, of course, that new observations may radically alter some of the conclusions drawn.



Reprinted courtesy of the author(s) and *The Astrophysical Journal*, published by the University of Chicago Press.

## II. A SURVEY OF SPECTRA AT MINIMUM LIGHT

A survey of spectra of U Gem variables and related objects was carried out by Elvey and Babcock (1943) at the McDonald Observatory. Since the brightest of these stars is twelfth magnitude at minimum, low dispersion was employed (~340 A/mm) except during the moments of outburst. In most cases the minimal spectra contained rather broad, bright Balmer lines, feebler emission of the triplet He I lines, and bright lines of Ca II (H and K) superimposed on an apparently continuous background having an intensity distribution similar to spectral type G or K. At, or near, maximum light the spectra were found to have a continuous distribution corresponding to type B or A, occasionally crossed by broad, very shallow, absorption lines of H; these sometimes contained faint, narrow emission features. At this dispersion, however, a satisfactory exposure required 2–3 hours (Babcock, private communication)—rather too long to provide sufficient time resolution, as we shall see.

Somewhat earlier, Joy (1940) had indicated that the spectrum of RU Peg at minimum had the absorption lines of a star of type dG3 as well as the usual emission features. Later

TABLE 1

SUMMARY OF OBSERVATIONAL DATA FOR FIGURE 1

| Star | Plate No.* | Date (U.T.) (Mid-exp.) | Length of Exposure (min.) | Approx. Dispersion (A/mm) | Remarks |
|---|---|---|---|---|---|
| U Gem...... | N1324a | 1961 Feb. 10.162 | 31 | 180 | |
| EX Hya..... | B1607a | 1960 Feb. 23.443 | 101 | 180 } | Sharp emission lines arise from Hg vapor in Los Angeles city lights |
| Z Cam....... | β 1758b | 1960 Dec. 19.444 | 122 | 180 } | |
| T Leo........ | B 925a | 1956 Mar. 5.371 | 150 | 180 } | |
| RX And..... | N1156b | 1960 Aug. 28.320 | 16 | 180 | |
| SS Aur....... | N1164 | 1960 Aug. 30.494 | 29 | 180 | |
| RU Peg...... | N1149a | 1960 Aug. 26.329 | 30 | 90 | |
| | b | Aug. 26.352 | 30 | 90 | |
| | c | Aug. 26.375 | 31 | 90 | |
| EY Cyg...... | N1162a | 1960 Aug. 30.312 | 40 | 180 | |

* N = 200-inch prime-focus spectrograph; B = Mount Wilson Newtonian focus spectrograph operating at the 100-inch; β = same, operating at the 60-inch.

Joy (1956) showed that the spectrum of SS Cyg at minimum was composite and that the star was a binary of short period, as already mentioned. Since the present survey has as its purpose the detection of binary motion, it differs from that of Elvey and Babcock in three important respects: (1) almost the entire effort was put on obtaining spectrograms at minimum light; (2) no dispersion less than 180 A/mm was employed; (3) the program was, to a considerable extent, planned to permit sufficient time resolution.

Several spectrograms have been obtained with the grating nebular spectrograph at Mount Wilson, using both the 60-inch and the 100-inch telescopes; the dispersion was 180 A/mm. Because of the faintness of the stars in question, these spectrograms do not, in general, satisfy the last criterion mentioned above; they must be regarded as exploratory in character. The larger bulk of spectrograms was obtained with the nebular spectrograph at Palomar; these have dispersion of 180 A/mm, except for the plates of RU Peg, which were obtained at a dispersion of 90 A/mm.

Some representative spectra of U Gem and suspected U Gem variables at minimum light are shown in Figure 1. Not all stars observed are included. The spectra are arranged roughly in order of decreasing emission-line width; observational details are summarized in Table 1. All spectrograms were obtained with the 200-inch, except for EX Hya and T Leo (100-inch) and Z Cam (60-inch). It is not certain, therefore, to what extent the wide lines of these three stars may result from integration over significant velocity changes.

The spectra reproduced in Figure 1 confirm the results obtained by Elvey and Babcock and also give some new information. He II (λ 4686) is present in a few spectrograms of U Gem (see Fig. 6); there may also be lines of Fe II (see appendix). A faint trace of the Balmer jump in emission is present in SS Aur and T Leo; however, in all the stars, the Balmer lines converge long before the series limit is reached, and thus they effectively merge with the Balmer continuum. The emission lines of U Gem, EX Hya, and possibly Z Cam are doubled—in the case of U Gem, very conspicuously so. We are not yet certain whether the emitting region is optically thick or thin, so it is not clear whether self-absorption plays a role. The absorption lines of a late-type dwarf star are present in the spectra of RU Peg and EY Cyg; some indication of these lines is found in the spectrum of SS Aur, but completely satisfactory agreement with the features of a G- or K-type dwarf star has not been found.

TABLE 2

ORBITAL ELEMENTS FOR U GEMINORUM VARIABLES

| | RX And | SS Aur | U Gem | EY Cyg† | SS Cyg | RU Peg |
|---|---|---|---|---|---|---|
| $P$ (hr., min.) | 5 05 | 3 30(?) | 4 10.5 | ... | 6 38 | 8 54 |
| $P$ (days) | 0.21173 | 0.15(?) | 0.1739825 | ... | 0.276244 | 0.3708 |
| $K_1$* (km/sec) | 77.5 | ∼85 | 265 | (?) | 122 | 137 |
| $K_2$ (km/sec) | ... | ... | ... | ... | 115 | 112 |
| $\gamma$ (km/sec) | −18 | ∼+45 | +42 | ∼−10 | −9 | −7 |
| Sp. (1) | sdBe | sdBe | sdBe | sdBe | sdBe | sdBe |
| Sp. (2) | ... | ... | ... | K0 V | dG5 | G8 IVn‡ |
| $a_1 \sin i \times 10^{-10}$ (cm) | 2.09 | ∼0.88 | 6.31 | ... | 4.63 | 7.02 |
| $a_2 \sin i \times 10^{-10}$ (cm) | ... | ... | ... | ... | 4.37 | 5.73 |
| $e$ | 0.40 | (?) | 0.05 | ... | ∼0 | ∼0 |
| $\omega$ | 220° | (?) | 160° | ... | ... | ... |
| $\mathfrak{M}_1 \sin^3 i$ (⊙) | ... | ... | ... | ... | 0.18 | 0.27 |
| $\mathfrak{M}_2 \sin^3 i$ (⊙) | ... | ... | ... | ... | 0.20 | 0.32 |
| $\mathfrak{M}_1/\mathfrak{M}_2$ | ... | ... | ... | ... | 0.90 | 0.85 |

* Subscript 1 refers to the blue star, regardless of whether or not it is the more massive.

† Probably viewed nearly pole-on.

‡ The luminosity class varies during the cycle, but averages MK class ∼ IV.

In spite of the considerable variation in width from star to star, the emission-line spectrum of these variables is very characteristic. We have not been able so far to confirm the finding by Elvey and Babcock that AY Lyr has no emission lines at minimum; the star is very faint and is in a crowded field—misidentification is a possibility. However, even if this result should be confirmed, it is reasonable to state that a sufficient condition for class membership is the presence of an emission spectrum similar to those of Figure 1. On this basis we include EX Hya and T Leo in the group (cf. Brun and Petit 1952, 1959; Kukarkin, Parenago, Efremov, and Kholopov 1958; Petit 1959).

### III. RADIAL VELOCITIES AND ORBITAL ELEMENTS

The journal of observations and velocity-curves for RX And, U Gem, and RU Peg are given in the appendix. In Table 2 we summarize the orbital elements for these stars, including also the results from Joy's (1956) study of SS Cyg. In addition, preliminary re-

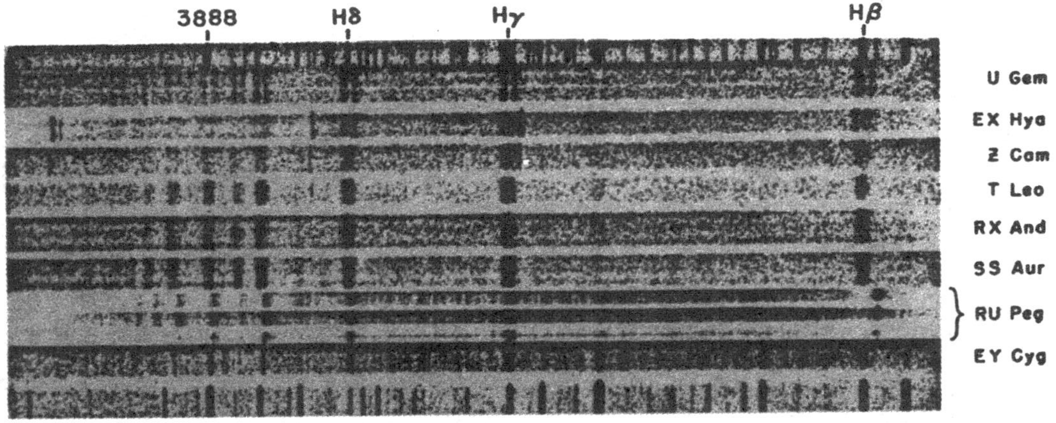


FIG. 1.—Representative sample of spectra of U Gem (and suspected U Gem) variables at minimum light. Observational details are summarized in Table 1. Sharp emission lines of mercury vapor are present in the spectra of EX Hya, Z Cam, and T Leo.

sults for SS Aur and EY Cyg are given; we expect to obtain more spectrograms of the former. EY Cyg may have a small velocity variation, but it is close to the limit of detection; the star may be a system viewed nearly "pole-on." Our assurance of binary characteristics is based on the fact that the spectrum is composite. The spectrum of Z Cam shows evidence of velocity variation, but the period cannot be determined from the available data. A short description of some of the spectral details of these stars is also given in the appendix.

The main conclusions summarizing Table 2 are the following:

1. No U Gem variable so far studied in detail has failed to show evidence of binary characteristics either from velocity variations or from the presence of a composite spectrum.
2. The mean period of the four best-studied stars is $\langle P \rangle = 0.25$ day.
3. In those cases in which a late-type spectrum is detected, the spectral type is dG or dK; the mass ratio is near unity, with the red star slightly more massive than the blue.
4. From the material at hand, it appears that the late-type absorption spectrum is most easily detected in stars having the narrowest emission lines.
5. It is possibly, though not necessarily, significant that the only system of high eccentricity is also the only star belonging to the Z Cam subgroup.

In the cases of SS Cyg and RU Peg, a comparison of the probable radius of the late-type star and the size of the corresponding lobe of the inner Lagrangian surface (given as soon as $\mathfrak{M}_1/\mathfrak{M}_2$ and $a_2 \sin i$ are known) indicates that the star, even if a dwarf, may fill that surface.[1] Following Kuiper and Johnson (1956), if $\mu = \mathfrak{M}_2/(\mathfrak{M}_1 + \mathfrak{M}_2)$, the effective radius $r_2$ of the larger inner lobe is given by

$$\log \frac{r_2}{a} = -0.335 \log \mu - 0.511\,, \qquad (1)$$

where $a$ is the separation of the centers of mass. Thus we have $\log r_2/a = -0.396$ and $-0.400$ for SS Cyg and RU Peg, respectively. The minimum values of $a$ for each star are $9.00 \times 10^{10}$ cm and $12.75 \times 10^{10}$ cm; hence the minimum values of $r_2$ for SS Cyg and RU Peg are $3.6 \times 10^{10}$ cm and $5.1 \times 10^{10}$ cm, respectively. But for stars of types dG5 and dG8, we would have $R_{SS} = 6.2 \times 10^{10}$ cm and $R_{RU} = 5.9 \times 10^{10}$ cm. If we set $r_2 = R$, we can estimate the inclination of the orbit. Thus $i = 35°$ and $59°$, respectively, for SS Cyg and RU Peg. Considering the crudity of the argument, these inclinations can scarcely be regarded as more than rough estimates. However, Grant (1955) found that SS Cyg does not eclipse. Since the blue star is quite small (the results of Sec. V show that it is a white dwarf) in terms of the dimensions of the system, this would mean that $i < 76°$; the limiting value would be the same if RU Peg, as well, did not eclipse; this is not known, however. The values of $i$ found above certainly satisfy this limit.

However, there is more compelling evidence that, at least in the case of RU Peg, the late-type component acts as if it filled one lobe of the inner Lagrangian surface. Classification on the MK system is possible, since the dispersion is high enough to resolve the ratio λ 4254/λ 4260. Less satisfactory as a basis for classification is the absolute intensity of λ 4226; it is filled in to some extent by the blue continuum of the hot component. The hydrogen lines obviously cannot be used. On this basis, spectral types are listed in the appendix. The lines are always somewhat diffuse (broadened by rotation?), and the spectral type is roughly constant at G8. However, the luminosity class runs from V to III, apparently depending on aspect, as judged from the strengths of λ 4077 and λ 4215 of Sr II. The highest luminosity corresponds to phases when the stars are in conjunction, red star behind. A star of dwarf dimensions which, at the same time, fills a lobe of the inner Lagrangian surface might be expected to imitate a star of higher luminosity,

[1] Throughout this paper, the subscript 1 refers to the blue star, whether or not it is the more massive component.

especially at superior conjunction, because of the reduction in effective surface gravity. (The effect would be more pronounced, of course, if $i$ were nearer to 90° than is probably the case in RU Peg.) If the star were, in fact, a normal star of type G8 IV, it would have a radius of $1.5 \times 10^{11}$ cm, the orbital inclination would therefore be about 18°, and the mass of the red star would be 12☉! This is completely unreasonable, since, as we shall see in Section V, the luminosity is probably near $M_V = +9.5$. Thus it seems likely that the spuriously high spectroscopic luminosity results from the reduction in surface gravity caused by the centrifugal acceleration.

### IV. MASSES FOR THE COMPONENTS

The preceding discussion suggests that the model advanced for AE Aqr (Crawford and Kraft 1956), T CrB (Kraft 1958), and DQ Her (Kraft 1959) is probably also applicable to the U Gem stars. We suppose that the late-type component fills its lobe of the inner Lagrangian surface. Material spills out from the inner Lagrangian point and forms a ring or disk around the blue companion. If the emitting region is truly flattened and not spherical, the shape of the emission line will depend on the orbital inclination. Regardless of its optical thickness, the disk will produce a doubled emission if seen in the plane of the orbit, and a single emission if $i$ deviates from 90° to any appreciable extent. We may tentatively suppose, therefore, that U Gem and possibly EX Hya and Z Cam are viewed near the plane of the orbit—they may even be eclipsing binaries. Support for this proposal is found in the presence of conspicuously doubled emission lines arising near the hot component of DQ Her (Greenstein and Kraft 1959), a well-known eclipsing binary.

If the preceding model is correct, we can make a rough estimate of the masses to be expected for U Gem variables. The argument of Section III, coupled with the "single" character of the emission lines of SS Cyg and RU Peg, suggests that the minimum masses listed in Table 2 are rather smaller than the actual masses. Though $i$ is probably near 90° for U Gem, unfortunately the spectrum of the red star has not yet been photographed. Consider, then, the following indirect argument for the determination of masses.

Kuiper (1941) has shown, on the basis of Jacobi's integral, that a particle, in falling from the inner Lagrangian point to the ring around the blue star, approximately conserves its angular momentum with respect to the blue star. In that case, we can derive a relation between the width of the emission lines, the period in the orbit, and the masses of components (Kraft 1959); viz.,

$$\frac{G\mathfrak{M}_1 \sin^3 i}{v \sin i} = \lambda (a_1 \sin i)^2 \left(1.39 \frac{\mathfrak{M}_1}{\mathfrak{M}_2} - 0.39\right)^2 \frac{2\pi}{P}, \tag{2}$$

where $v \sin i$ is the projected half-half-width of the emission line, $\mathfrak{M}_1$ and $\mathfrak{M}_2$ are the masses of the blue and red stars, respectively, and $\lambda$ is a factor of proportionality describing the fractional change of angular momentum. For both SS Cyg and RU Peg, $v \sin i = 500$ km/sec, as derived from microphotometer tracings. For SS Cyg, the left- and right-hand sides of equation (2) are $5.3 \times 10^{17}$ and $4.2 \times 10^{17}\,\lambda$, respectively; thus $\lambda = 1.26$. Corresponding values for RU Peg are $8.0 \times 10^{17}$, $5.9 \times 10^{17}\,\lambda$, and $\lambda = 1.36$. Thus the values of $\lambda$ are very similar, and $\langle\lambda\rangle = 1.31$.

If the same value of $\lambda$ applies to U Gem, we have

$$\frac{G\mathfrak{M}_1}{v} = 1.31\, a_1^2 \left(1.39 \frac{\mathfrak{M}_1}{\mathfrak{M}_2} - 0.39\right)^2 \frac{2\pi}{P}, \tag{3}$$

where $a_2 = 6.31 \times 10^{11}$ cm and $P = 1.50 \times 10^4$ sec. The value of $v$, however, depends on whether the emitting region is optically thick or thin—the latter leads to $v \sin i = v = 670$ km/sec. Equation (3) can then be balanced reasonably well for $\mathfrak{M}_1/\mathfrak{M}_2$ lying between 0.8 and 1.9, implying $1.9\ \mathfrak{M}_\odot < \mathfrak{M}_1 < 5.5\ \mathfrak{M}_\odot$. Since, as already mentioned, $M_V \sim +9.5$ for these stars, a mass greater than $1.2\ \mathfrak{M}_\odot$ is unlikely because the blue star must be essentially a white dwarf (Schwarzschild 1958). On the other hand, equation (3) cannot be balanced within the error of measuring the width of the emission line if $\mathfrak{M}_1 < 1.0$. Moreover, if the emitting region is optically thick, $v < 670$ km/sec, and the value of $\mathfrak{M}_1$ required to balance the equation is driven up still further. It seems most likely that $\mathfrak{M}_1$ lies in the 0.9–1.2$\odot$ range and that the emitting region is optically thin.

If the masses of the red components of SS Cyg and RU Peg are $\sim 1\ \mathfrak{M}_\odot$ as well, the orbital inclinations would be in quite reasonable agreement with those derived by the argument of Section III.

## V. STATISTICAL PARALLAX

The systemic velocities and corresponding peculiar radial velocities (i.e., radial velocities corrected for solar motion) are listed in Table 3, along with the proper motions. A rough statistical parallax can be obtained, following Smart (1938), from the radial velocities and proper motions of four stars: U Gem, SS Cyg, RU Peg, and EY Cyg. If

TABLE 3

MAGNITUDES, PROPER MOTIONS, PECULIAR RADIAL VELOCITIES, ETC., FOR U GEMINORUM VARIABLES

| Star | $V$* (km/sec) | $\mu$† (sec of arc/yr) | $m_V$ (min) | Kind | Outburst Period (days) |
|---|---|---|---|---|---|
| RX And | −12 | ........... | 13.6 | Z | 14 |
| SS Aur | +35 | ........... | 14.8 | SS | 54 |
| U Gem | +37 | 0.078 | 14.0 | SS | 103 |
| EY Cyg | ∼+13 | .062 | 15.0: | SS | ∼1000 |
| SS Cyg | +16 | .116 | 12.1 | SS | 52 |
| RU Peg | +11 | 0.070 | 13.1 | SS | 70 |

* Radial velocity with solar motion removed.

† Mean of values given by Mannino and Rosino (1950) and by Miczaika and Becker (1948).

$r$ is the distance of any star, then, under the assumption that $\langle \log r \rangle \cong \log \langle r \rangle$, we find $\langle M_V \rangle = +9.5 \pm 0.6$ (m.e.) at minimum light. The quoted mean error is internal; a value of $\pm 1$ mag. would be more realistic.

There are two direct checks on this value of $\langle M_V \rangle$. Strand's (1948) well-determined parallax for SS Cyg gives $M_V = +9.5$ at minimum light. From a study of the color of UZ Ser, Herbig (1944) concluded that the star is in front of an absorbing cloud not farther away than 200 pc. From this he found $M_V \sim +10.1$ at minimum. Both these values agree well with our statistical parallax. It must be admitted, however, that U Gem stars need not all have the same absolute magnitude. We have no way, at present, of determining the dispersion about the mean, if any.

A further indication that $\langle M_V \rangle = +9.5$ is approximately correct is found from a consideration of the colors at maximum light. At a mean distance of only 66 pc, little or no reddening would be expected. At maximum, the spectra of U Gem and SS Cyg are very nearly continuous (Elvey and Babcock 1943). (Occasionally, feeble emission or absorption lines of H and He II are seen [cf. also Adams and Joy 1922].) Thus we make the assumption that SS Cyg and U Gem radiate as black bodies at maximum light. The $U-B$, $B-V$ colors of these stars (Grant and Abt 1959; Wallerstein 1959) may be compared with black-body radiators (Arp 1961). This was done earlier by Wallerstein (1959), but a slightly incorrect black-body trajectory in the color-color plot was used. With Arp's new curve, we have the comparison shown in Figure 2. We conclude that, if

SS Cyg and U Gem radiate as black bodies at maximum light with temperatures of 12000° and 15000° K, respectively, there is no interstellar reddening; this is consistent with $\langle M_v \rangle_{min} = +9.5$.

This result is not, however, compatible with a spectral type of G5–K0 and $\mathfrak{M} \sim 1.0 \odot$ for the red components of these systems. It would appear that these stars are significantly underluminous for their masses. In this respect and a number of others, the U Gem variables are similar to the stars of type W UMa.

### VI. EVOLUTIONARY CONSIDERATIONS

The small peculiar velocities of U Gem stars indicate that they belong to a moderately flattened stellar population—the galactic disk. They are not members of the galactic "halo"; neither can they be described as belonging to Baade's classical population II.

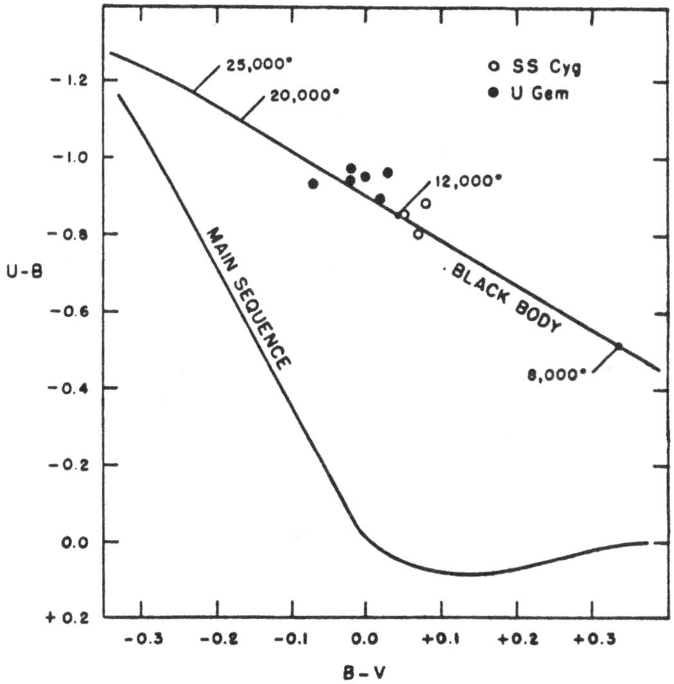


FIG. 2.—The $U - B$, $B - V$ colors of U Gem and SS Cyg at maximum light. The black-body locus computed by Arp (1961) is shown.

The only other binaries of comparably short period which also belong to the galactic disk are the W UMa stars (cf. Struve 1950; Kitamura 1959). The remainder of the paper is concerned with evidence supporting the proposition that a *U Gem variable represents a later stage in the evolution of a W UMa star.* This view has been advanced by Sahade (1959) and also independently by the writer for some years.

The physical properties of the U Gem and W UMa systems are summarized in Table 4; data for the latter are taken from the compilation by Kitamura (1959).

In Figure 3 we plot the $z = r \sin b$ distribution for each group. The U Gem stars have been taken to the limit $m_{pg}$ (min.) $= +17.5$; the corresponding limit for the W UMa stars must therefore be $m_{pg} = +12.5$. For each group, 82 and 123 stars, respectively, were found in the *Variable Star Catalogue* (Kukarkin *et al.* 1958). Twenty-five of the former did not have known magnitudes at minimum; the average range of 4 mag. was therefore applied to the magnitudes at maximum. Considering the various uncertainties and particularly the rather different basis for discovery of stars in the two groups, one finds that the $z$-distributions are extraordinarily similar. This result, coupled with our finding that the total masses and mean peculiar radial velocities are closely alike, strongly suggests a generic relation between the two kinds of stars.

There is, however, also a physical point of similarity not covered in the table. Kitamura (1959) has shown that in the W UMa stars is it likely that the primary overflows its lobe of the inner Lagrangian surface. The star therefore loses mass, which may be collected by the secondary or lost to the system. The main point, however, is that the primary is seriously underluminous for its mass (~3 or 4 mag.). The same is true for the red components of U Gem systems, for, if $M_V \sim +10$ but $\mathfrak{M} \sim 1\odot$, these stars are underluminous by perhaps 5 mag. But if our model is correct, these also lose mass through the inner Lagrangian surface.

Having established the similarity in kinematics and space distribution for the two kinds of variables, we now turn to the question: Are the periods, mass ratios, and total

TABLE 4

PHYSICAL PROPERTIES OF U GEM AND W UMA STARS COMPARED

| Property | U Gem | W UMa |
|---|---|---|
| $\langle P\rangle$ | 0ᵈ25 | 0ᵈ37 |
| $\mathfrak{M}_1/\langle\mathfrak{M}_1+\mathfrak{M}_2\rangle$ | ~1/2 | ~1/3 |
| $\langle\mathfrak{M}_1+\mathfrak{M}_2\rangle$ | ~1.5–2.0$\odot$ | ~1.2–2.5$\odot$ |
| $\langle M_V\rangle$ | +9.5 | +4.5 |
| $\langle z\rangle$* = $\langle r \sin b\rangle$ | 37 pc | 40 pc |
| $\langle V\rangle$ | +16 km/sec | −8 km/sec |
| $\sigma_V$ | 18 km/sec | 26 km/sec |
| No. stars† | 6 | 10 |

* For U Gem and W UMa stars to apparent magnitude limit, $m_{pg}$ = 17.5 and $m_{pg}$ = 12.5, respectively.

† Used in determination of $\langle V\rangle$ and $\sigma_V$ only.

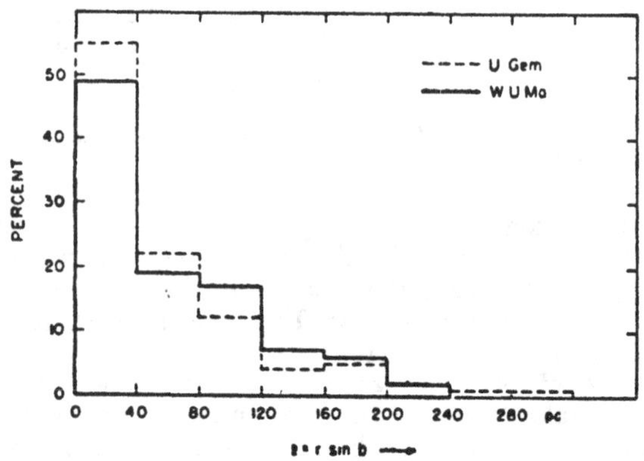


FIG. 3.—The $z$-distributions of U Gem and W UMa stars compared. The limits of the survey are $m_{pg}$ (min.) = +17.5 for U Gem stars and $m_{pg}$ = +12.5 for W UMa systems.

masses compatible with the proposed evolutionary pattern? If so, we must simultaneously satisfy the following conditions: (1) the total mass remains nearly constant or perhaps slightly decreases; (2) the mass ratio decreases from about 2 to 1; (3) the period decreases from an average value of 0.37 day to 0.25 day. The most important clue suggesting that these conditions are compatible, qualitatively at least, comes from the work of Huang (1956), who showed that if a component of a binary star loses mass to its companion, the period will decrease, but if it loses mass into space, the period will increase.

In the present picture, we imagine that the W UMa primary fills its lobe of the inner Lagrangian surface and loses mass both to the secondary and to outer space. Its evolution is speeded up to the point that it rapidly becomes a white dwarf. Later, the secondary begins to overflow its lobe of the Lagrangian surface, giving the conditions now observed for U Gem variables. If the ejection velocity is small compared with the relative velocity of the components in the orbit and if $e \sim 0$, we can write (cf. Huang 1956)

$$\frac{\delta a}{a} = \frac{1-\alpha}{\alpha}\,\frac{\delta \mathfrak{M}_2}{\mathfrak{M}_1+\mathfrak{M}_2} - 2\,\frac{\delta \mathfrak{M}_2}{\mathfrak{M}_2}, \qquad \frac{\delta P}{P} = \frac{2(1-\alpha)}{\alpha}\,\frac{\delta \mathfrak{M}_2}{\mathfrak{M}_1+\mathfrak{M}_2} - 3\,\frac{\delta \mathfrak{M}_2}{\mathfrak{M}_2}, \tag{4}$$

where $\alpha$ is the fraction of mass lost by the primary that is collected by the secondary. Integration leads to

$$\frac{a}{a_0} = \left[\frac{(\mathfrak{M}_1+\mathfrak{M}_2)_0}{(\mathfrak{M}_1+\mathfrak{M}_2)}\right]\left[\frac{(\mathfrak{M}_2)_0}{\mathfrak{M}_2}\right]^2, \qquad \frac{P}{P_0} = \left[\frac{(\mathfrak{M}_1+\mathfrak{M}_2)_0}{(\mathfrak{M}_1+\mathfrak{M}_2)}\right]^2\left[\frac{(\mathfrak{M}_2)_0}{\mathfrak{M}_2}\right]^3, \tag{5}$$

where the zero subscripts indicate initial values. The first terms in equation (4) correspond to loss of mass from the system, and the second to the transfer of mass from primary to secondary.

Let us begin with a W UMa star having $(\mathfrak{M}_1)_0 = 1.43\odot$, $(\mathfrak{M}_2)_0 = 0.76\odot$, $P_0 = 0.37$ day, and $a_0 = 1.98 \times 10^{11}$ cm. Then, if we lose $\mathfrak{M} = 0.41\odot$ from the system and transfer $\mathfrak{M} = 0.20\odot$ to the secondary, we wind up with $\mathfrak{M}_1 = 0.82\odot$, $\mathfrak{M}_2 = 0.96\odot$, $P = 0.28$ day, and $a = 1.93 \times 10^{11}$ cm. These values are quite reasonable for SS Cyg. On the other hand, if no mass were lost from the system and we transferred $\mathfrak{M} = 0.3\odot$ from primary to secondary, the final values would be $\mathfrak{M}_1 = 1.13\odot$, $\mathfrak{M}_2 = 1.06\odot$, $P = 0.14$ day, and $a = 1.03 \times 10^{11}$ cm. These values give a reasonable fit to U Gem. The numbers all become invalid, however, if the velocity of ejection from the primary becomes large; observational evidence on this point is not readily obtainable, however.

Though the preceding discussion seems plausible enough, it leaves unanswered a number of perhaps more serious questions, among them, the following:

1. How can a star of mass $\mathfrak{M} \sim 1.4\odot$ lose one-fourth to one-third of its own mass and become a white dwarf in the process?

2. Why is the U Gem phenomenon observed only when the mass ratio achieves a value near unity? In other words, why do we not observe intermediate cases between W UMa stars and U Gem variables?

3. What is the relation between the binary characteristic and the presence of outbursts in U Gem variables?

4. How can a star losing mass have a spectrum corresponding to a mass of the order of $1\odot$, yet have a luminosity 5 mag. fainter than expected for this mass?

With regard to the last, we might suppose that the outflow of energy is taken up mostly by increasing the potential energy of the ejected particles, leaving little left over for the luminous flux. For a W UMa star, we would then have

$$\frac{1}{2}\left(\frac{dm}{dt}\right) v^2 = L \cong 10^{35} \text{ erg/sec} . \qquad (6)$$

With an ejection velocity of (say) 10 km/sec, the equation can be balanced if $dm/dt = 2 \times 10^{23}$ gm/sec $= 6 \times 10^{30}$ gm/yr. This would mean that W UMa stars last for only 500 years—a result unacceptable not only because of the excessively large number of W UMa stars it would imply but also because the change in period per unit time would greatly exceed that observed unless the amounts lost to the system and collected by the secondary were very closely related.

Evidence of the time scale that might be expected comes from the work of Morton (1960), who showed that the mass loss must proceed on the Kelvin time scale. In this case, $t_K = 1.5 \times 10^6$ years. Thus, even if the ejection velocity were 100 km/sec, we could not satisfy the Kelvin time scale and at the same time maintain the validity of equation (6).

However, the Kelvin time is quite short compared with the age of Praesepe, viz., 1–5 $\times 10^8$ years; there is strong evidence that Praesepe contains the W UMa star TX Cnc (Haffner 1937; Eggen 1961). This presumably means that the components of TX Cnc have only recently begun the activity presently identified with the W UMa characteristic. The mass of the primary is about 1.6☉ (Kitamura 1959). However, the masses of the stars now breaking off the main sequence and evolving to the right in the HR diagram are about 2.0☉. If the lifetime of TX Cnc is, indeed, significantly less than the age of the cluster, one could imagine that the original system consisted of an A or F primary of mass ~2☉, together with a much fainter and less massive secondary. At first, neither star filled its lobe of the inner Lagrangian surface. As evolution proceeded, the primary expanded to fill its lobe in a time of the order of $10^8$ years; loss of mass has brought it rapidly to its present position in the cluster HR diagram. If this scheme is to be maintained, it will be necessary to explain how the primary can evolve along a line nearly parallel to the main sequence itself.

## APPENDIX

### Journal of Observations, Velocity-Curves, and Discussion of Radial Velocities for Certain U Geminorum Variables

I. *RX And.*—The emission spectrum of hydrogen is strong and shows the Balmer jump in emission. He I (λ 4771, λ 4026) and Ca II are present in emission; there is a faint indication of bright He II (λ 4686). No trace of an absorption-line spectrum has been found.

The journal of observations is given in Table 5, and the velocity-curve plotted in Figure 4. All spectrograms have dispersion ~180 A/mm. The star is unusual in having an orbit of high eccentricity. The distortion of a sine-wave velocity-curve by streams of gas cannot, of course, be ruled out.

II. *SS Aur.*—Only nine spectrograms (dispersion ~180 A/mm) are available. The emission lines of hydrogen are strong and have relatively sharp edges; the Balmer jump is in emission. Ca II (K) is prominent in emission, but He I is weak. The underlying continuum is irregular and may correspond to the spectrum of a late-type star, but the match is not completely free from ambiguity.

The journal of observations is found in Table 6. A period of about 3½ hours fits the velocities, but more observations are needed.

III. *U Gem.*—The hydrogen emission lines and Ca II (K) are distinctly double. λ 4471 and λ 4026 of He I are feebly present, and there is a bright trace of λ 4686 of He II. The difference in velocity at opposite elongations is illustrated in Figure 7. As was the case in the spectrum of DQ Her (Greenstein and Kraft 1959), the $V/R$ ratio reverses at opposite elongations. When the spectra are lined up, as in Figure 8, some faint emission features at λλ 4172, 4232, 4301, 4512(?), and 4582 emerge which may correspond to the Fe II lines λλ 4173, 4233, 4303, 4508 and 4522 (blended), and 4584.

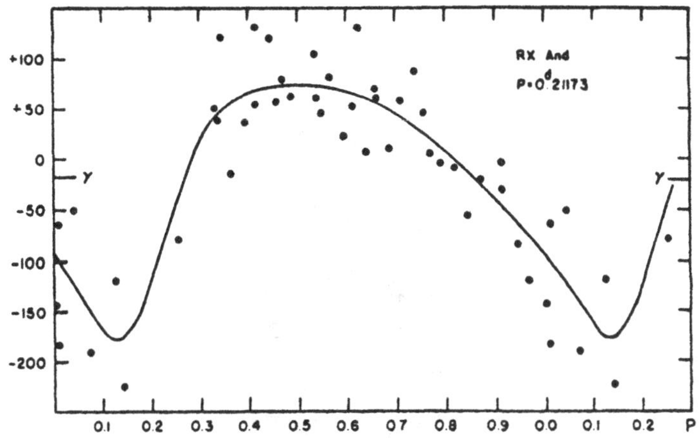


FIG. 4.—The radial-velocity curve (emission lines) of RX And

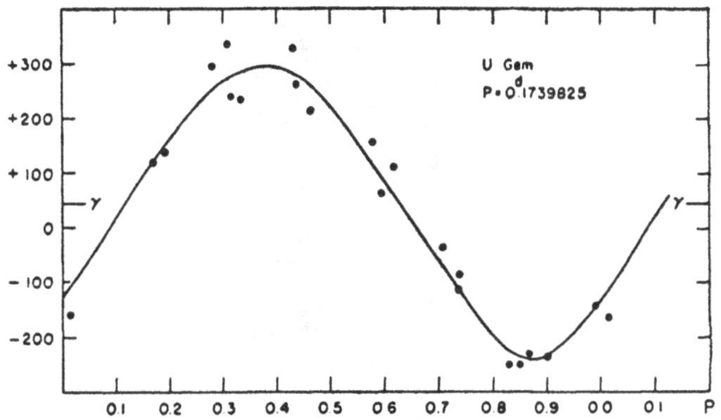


FIG. 5.—The radial-velocity curve (emission lines) of U Gem

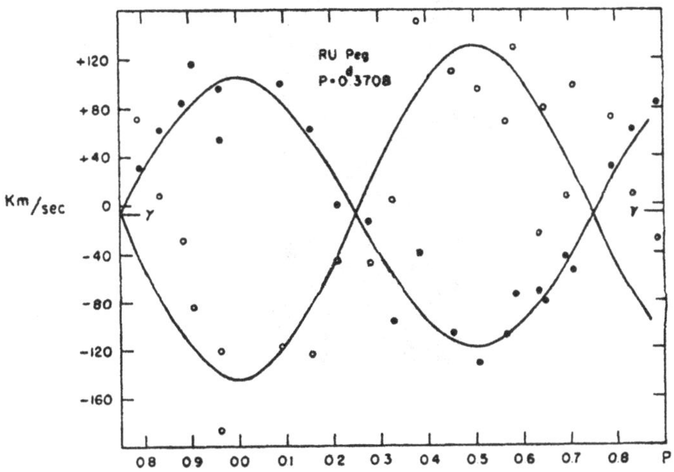


FIG. 6.—The radial velocities of RU Peg. Filled and open circles refer to absorption and emission, respectively.

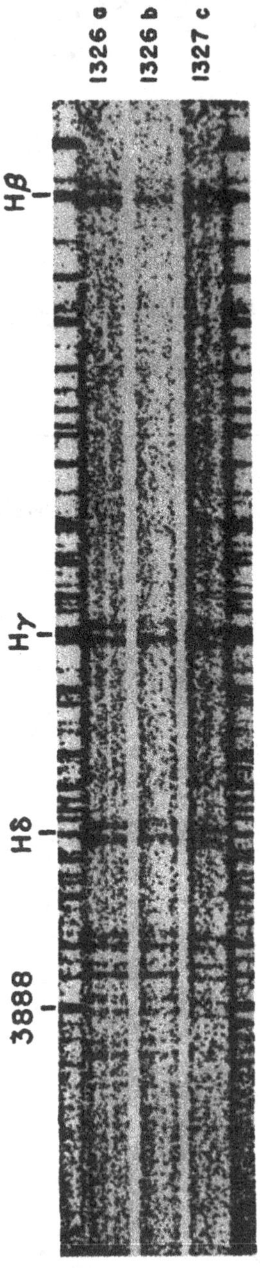


FIG. 7.—The difference in radial velocity of U Gem at opposite elongations

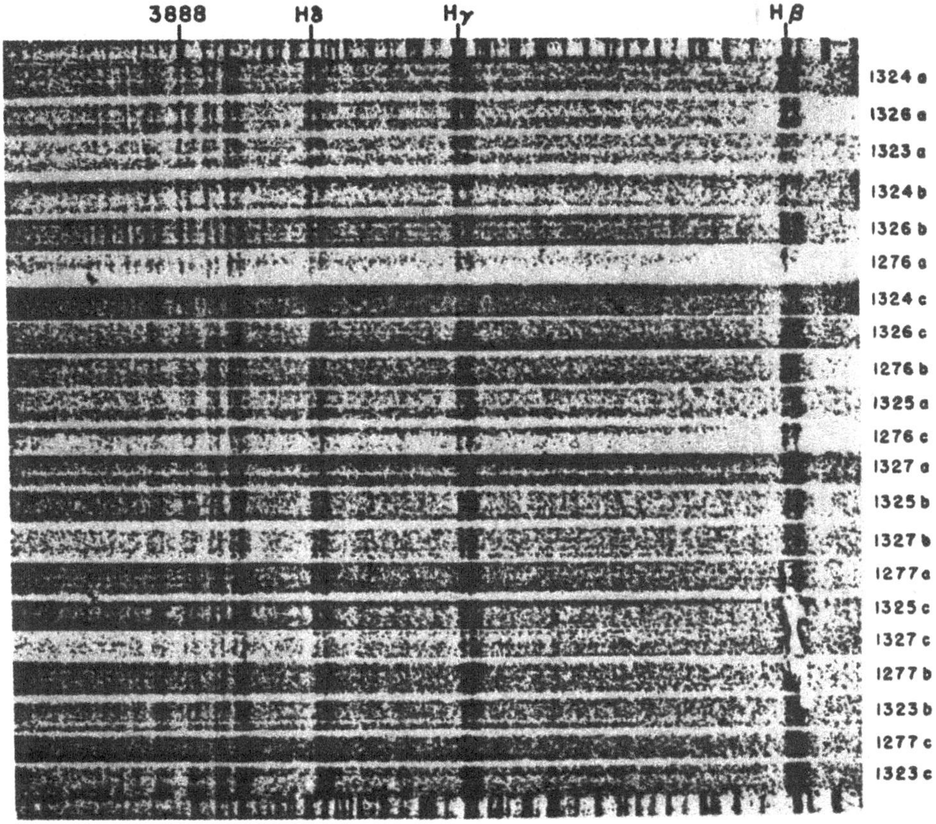


FIG. 8.—Spectra of U Gem around the $4^h10^m.5$ cycle. No attempt has been made to show the radial velocity variation. Notice the change in visibility of the higher members of the Balmer series.

TABLE 5

JOURNAL OF OBSERVATIONS FOR RX AND

| Plate No. | J. D. (Helioc.) (mid-exp) | V (km/sec) | Phase* (P) | Plate No. | J. D. (Helioc.) (mid-exp) | V (km/sec) | Phase* (P) |
|---|---|---|---|---|---|---|---|
| | 2437000+ | | | | 2437000+ | | |
| N 1151 a | 173.8284 | − 5 | 0.912 | N 1205 a | 221.6256 | + 68 | 0.658 |
| b | 173.8520 | − 65 | .011 | b | 221.6423 | + 87 | .737 |
| c | 173.8735 | −139 | .126 | c | 221.6534 | − 6 | .790 |
| N 1152 a | 173.9235 | − 16 | .362 | N 1206 a | 221.6909 | −120 | .967 |
| b | 173.9437 | + 57 | .457 | b | 221.6999 | −184 | .009 |
| c | 173.9610 | + 61 | .539 | | | | |
| | | | | N 1207 a | 222.6163 | + 38 | .338 |
| N 1153 a | 173.9819 | + 6 | .638 | b | 222.6326 | + 54 | .415 |
| b | 173.9971 | + 57 | .709 | c | 222.6479 | + 62 | .487 |
| N 1154 c | 174.7479 | − 79 | .256 | N 1208 a | 222.6645 | + 80 | .565 |
| | | | | b | 222.6770 | +130 | .624 |
| N 1155 a | 174.7667 | +121 | .344 | c | 222.6902 | + 8 | .687 |
| b | 174.7820 | +131 | .417 | | | | |
| c | 174.7931 | + 78 | .469 | N 1209 a | 222.7076 | + 4 | .769 |
| | | | | b | 222.7236 | − 57 | .844 |
| N 1156 a | 174.8097 | + 45 | .547 | c | 222.7381 | − 31 | .913 |
| b | 174.8229 | + 51 | .610 | | | | |
| c | 174.8334 | + 58 | .659 | N 1210 a | 222.7562 | − 52 | .041 |
| | | | | b | 222.7722 | −192 | .074 |
| N 1157 a | 174.8535 | + 45 | .754 | c | 222.7867 | −225 | .142 |
| b | 174.8671 | − 11 | .818 | | | | |
| c | 174.8778 | − 22 | .869 | N 1211 a | 222.8270 | + 51 | .332 |
| | | | | b | 222.8402 | + 37 | .395 |
| N 1158 a | 174.8938 | + 84 | .945 | c | 222.8506 | +120 | .444 |
| b | 174.9063 | −144 | 0.003 | | | | |
| | | | | N 1212 a | 222.8701 | +104 | .536 |
| | | | | b | 222.8826 | + 22 | 0.594 |

*Zero phase taken arbitrarily at J. D.$_\odot$ 2437173.0000

TABLE 6

JOURNAL OF OBSERVATIONS FOR SS AUR

| Plate No. | Date (U.T.) (Mid-exp) | V (km/sec) |
|---|---|---|
| N 1164 | 1960 Aug 30.494 | + 108 |
| N 1165 a | Aug 31.381 | + 77 |
| b | 31.409 | + 99 |
| N 1217 a | Oct 17.389 | + 15 |
| b | 17.420 | + 101 |
| c | 17.448 | + 130 |
| N 1218 a | 17.479 | - 21 |
| b | 17.502 | - 45 |
| c | 17.520 | - 9 |

## TABLE 7

## JOURNAL OF OBSERVATIONS FOR U GEM

| Plate No. | J. D. (Helioc.) (mid-exp) | V (km/sec) | Phase* (P) |
|---|---|---|---|
| | 2437000+ | | |
| N 1276 a | 266.9270 | + 235 | 0.328 |
| b | 266.9499 | + 213 | .460 |
| c | 266.9722 | + 58 | .588 |
| N 1277 a | 266.9978 | − 86 | .735 |
| b | 267.0201 | − 228 | .863 |
| c | 267.0416 | − 143 | .987 |
| N 1323 a | 339.6428 | + 290 | .277 |
| b | 340.6206 | − 237 | .897 |
| c | 340.6407 | − 160 | .013 |
| N 1324 a | 340.6675 | + 117 | .167 |
| b | 340.6914 | + 332 | .304 |
| c | 340.7129 | + 325 | .428 |
| N 1325 a | 340.7383 | + 153 | .574 |
| b | 340.7609 | − 39 | .704 |
| c | 340.7820 | − 252 | .825 |
| N 1326 a | 340.8456 | + 135 | .190 |
| b | 340.8668 | + 238 | .312 |
| c | 340.8876 | + 259 | .432 |
| N 1327 a | 340.9188 | + 108 | .611 |
| b | 340.9397 | − 116 | .731 |
| c | 340.9595 | − 250 | 0.845 |

*Zero phase taken arbitrarily at J. D.$_{\odot}$ = 2437266.0000

## TABLE 8

## JOURNAL OF OBSERVATIONS FOR RU PEG

| Plate No. | J.D. (Helioc.) (mid-exp) | V (km/sec) em. | V (km/sec) abs. | Phase* (P) | Sp. Type |
|---|---|---|---|---|---|
| | 2437000+ | | | | |
| N 1143 a | 171.7137 | + 68 | −108 | 0.566 | G8 IV n |
| b | 171.7380 | − 24 | − 71 | .633 | G8 IV n |
| c | 171.7613 | + 6 | − 43 | .691 | G8 IV n |
| N 1144 a | 171.8123 | + 7 | + 62 | .836 | G8 V n |
| b | 171.8360 | − 84 | +116 | .906 | G8 V n |
| c | 171.8585 | −187 | + 96 | .961 | K0 V n |
| N 1145 a | 171.9085 | −118 | +100 | .092 | G8 IV n |
| b | 171.9311 | −124 | + 62 | .157 | G8 IV n |
| c | 171.9533 | − 46 | 0 | .210 | G8 III n |
| N 1146 a | 171.9818 | − 48 | − 14 | .277 | G8 IV n |
| b | 171.9995 | + 3 | − 97 | .328 | K0 IV n |
| N 1148 a | 172.7596 | +150 | − 40 | .382 | G8 V n |
| b | 172.7842 | +109 | −107 | .453 | G8 V n |
| c | 172.8061 | + 95 | −132 | .506 | G8 V n |
| N 1149 a | 172.8346 | +129 | − 74 | .584 | G8 V n |
| b | 172.8582 | + 80 | − 80 | .648 | G8 IV n |
| c | 172.8807 | + 98 | − 54 | .708 | G8 IV, V n |
| N 1150 a | 172.9099 | + 72 | + 31 | .790 | G8 IV n |
| b | 172.9419 | − 29 | + 84 | .885 | G8 IV n |
| c | 172.9748 | −121 | + 54 | 0.963 | G8 IV n |

*Zero phase taken arbitrarily at J.D. 2437171.873, the moment of max. positive absorption-line velocity.

The cyclical behavior of the spectrum is illustrated also in Figure 8, and the velocity-curve is shown in Figure 5. Elongation (1) with the blue star receding corresponds to plates near N 1324 b and c; superior conjunction (blue star behind) to N 1276 c; elongation (2) with blue star approaching to N 1325 c and N 1327 c. The visibility of the higher members of the Balmer series varies in the cycle; they are strongest near elongation (1) and virtually disappear at superior conjunction. There seems to be a tendency for the appearance of duplicity (among the higher members) to be more pronounced, the stronger the emission. The large changes in continuum intensity from plate to plate, though enhanced in the process of reproduction, are at least partially due to changes in stellar brightness.

The journal of observations is given in Table 7. The radial velocities have been corrected (slightly) for the effect of exposure time, which is a small, but significant, fraction of the period (cf. Herbig 1960).

IV. *EY Cyg.*—From five spectrograms, there is no significant velocity variation. For N 1162 a, b, c, the mean velocities of emission ($H\gamma + H\delta$) and absorption ($\lambda$ 4045 and $\lambda$ 4226) are +26 and −81 km/sec, respectively. For N 1163 a, b, we have +18 and −8 km/sec, respectively. The mean errors of measurement are large enough that the difference in absorption-line velocity between the two plates is probably not significant.

The emission lines are quite narrow for a U Gem variable. The quoted velocity is a mean between absorption and emission velocities.

V. *RU Peg.*—With the exception of SS Cyg, this is the only other star showing the absorption spectrum of a late-type star in the photographic region (II*a*-O plates). MK spectral types as well as radial velocities are listed in Table 8. The emission lines have amorphous edges at 90 A/mm and are difficult to measure. Ca II (K) sometimes appears doubled; He I is rather weak in emission. The higher Balmer lines converge, and the Balmer continuum is probably in emission. The velocity-curve is illustrated in Figure 6.

## REFERENCES


Adams, W. S., and Joy, A. H. 1922, *Pop. Astr.*, **30**, 103.
Arp, H. C. 1961, *Ap. J.*, **133**, 874.
Brun, A., and Petit, M. 1952, *Bull. Assoc. française Obs. étoiles var.*, **12**, 1.
———. 1959, *Peremennye Zvezdy*, **12**, 18.
Crawford, J. A., and Kraft, R. P. 1956, *Ap. J.*, **123**, 44.
Eggen, O. J. 1961, *Royal Obs. Bull.*, No. 31.
Elvey, C. T., and Babcock, H. W. 1943, *Ap. J.*, **97**, 412.
Grant, G. 1955, *Ap. J.*, **122**, 566.
Grant, G., and Abt, H. A. 1959, *Ap. J.*, **129**, 323.
Greenstein, J. L., and Kraft, R. P. 1959, *Ap. J.*, **130**, 99.
Haffner, H. 1937, *Zs. f. Ap.*, **14**, 285.
Herbig, G. H. 1944, *Pub. A.S.P.*, **56**, 230.
———. 1960, *Ap. J.*, **132**, 76.
Huang, S.-S. 1956, *A.J.*, **61**, 49.
Joy, A. H. 1940, *Pub. A.S.P.*, **52**, 324.
———. 1956, *Ap. J.*, **124**, 317.
Kitamura, M. 1959, *Pub. Astr. Soc. Japan*, **11**, 216.
Kraft, R. P. 1958, *Ap. J.*, **127**, 625.
———. 1959, *ibid.*, **130**, 110.
Kuiper, G. P. 1941, *Ap. J.*, **93**, 133.
Kuiper, G. P., and Johnson, J. R. 1956, *Ap. J.*, **123**, 90.
Kukarkin, B., Parenago, P., Efremov, Y., and Kholopov, P. 1958, *General Catalogue of Variable Stars*, Vol. 1 (Moscow: Akademiia Nauk U.S.S.R.).
Mannino, G., and Rosino, L. 1950, *Padua Pub.*, No. 14.
Miczaika, G., and Becker, W. 1948, *Heidelberg Veröff.*, Vol. 15, No. 8.
Morton, D. C. 1960, *Ap. J.*, **132**, 146.
Petit, M. 1959, *Peremennye Zvezdy*, **12**, 4.
Sahade, J. 1959, *Liege Symposium: Modèles d'étoiles et évolution stellaire*, p. 76.
Schwarzschild, M. 1958, *Structure and Evolution of the Stars* (Princeton: Princeton University Press), p. 233.
Smart, W. M. 1938, *Stellar Dynamics* (Cambridge: Cambridge University Press), p. 208.
Strand, K. Aa. 1948, *Ap. J.*, **107**, 106.
Struve, O. 1950, *Stellar Evolution* (Princeton: Princeton University Press), p. 183.
Walker, M. F. 1954, *Pub. A.S.P.*, **66**, 230.
———. 1956, *Ap. J.*, **123**, 68.
———. 1957, in *I.A.U. Symposium*, No. 3, ed. G. H. Herbig (Cambridge: Cambridge University Press).
Wallerstein, G. 1959, *Pub. A.S.P.*, **71**, 316.

THE ASTROPHYSICAL JOURNAL
LETTERS TO THE EDITOR

VOLUME 148 APRIL 1967 NUMBER 1, PART 2

# ON THE NATURE OF THE SOURCE OF X-RAY EMISSION OF SCO XR-1

Recently the brightest source of X-ray emission Sco XR-1 was identified with an optical object of 13 mag (Sandage, Osmer, Giacconi, Gorenstein, Gursky, Waters, Bradt, Garmire, Sreekantan, Oda, Osawa, and Jugaku 1966). This object is by all its characteristic features reminiscent of an old Nova. At the present time most investigators believe that the mechanism of the X-ray emission of the source may be explained by the bremsstrahlung of an optically thin layer of very hot plasma ($T \sim 5 \times 10^7$ ° K). According to this interpretation the main part of the optical emission is a low-frequency continuation of the bremsstrahlung.

However, the latter conclusion in our opinion does not seem to be right in view of the following considerations.

1. The recently discovered soft X-ray emission from Sco XR-1 in the region 60 Å > $\lambda_2$ > 44 Å (Byram, Chubb and Friedman 1966) has a rather high spectral flux density; $F_{\nu_2} \sim 5 \times 10^{-24}$ ergs cm$^{-2}$ sec$^{-1}$. If the emitting hot plasma were transparent at optical frequencies, the optical intensity can be extrapolated from $F_{\nu_2}$ according to the bremsstrahlung spectral law. The apparent magnitude of the optical object, which is identified with Sco XR-1 would then be brighter than 9 mag, which is contradictory to observation.

2. The optical object shows changes of color. The observed variations in intensity of the emission lines of H, He II, C III, and N III are insufficient to cause these color changes, because of the small equivalent widths of the lines. In the case of bremsstrahlung from a very hot plasma, such variations in color are impossible. The absence of a sufficiently bright optical object in the proximity of the source may be explained if we assume that the hot plasma is opaque at optical frequencies

$$\frac{2\pi kT}{\lambda_1^2}\frac{4\pi R_1^2}{4\pi r^2} < F_{\nu_1}, \qquad (1)$$

where $F_{\nu_1} \sim 3 \times 10^{-25}$ ergs cm$^{-2}$ sec$^{-1}$ is the flux density from the optical object with apparent magnitude $m = 12.6$, which is identified with Sco XR-1, $r$ is the distance to the source, and $R$ is its radius. The temperature of the plasma responsible for the emission in the spectral region 44 Å < $\lambda$ < 60 Å must be higher than $2 \times 10^5$ ° K, since otherwise the flux of the X-ray emission for a given $F_{\nu_1}$ would be too low. The upper limit of the temperature is about $2 \times 10^6$ ° K, as computed by comparing the flux at $\lambda_2 = 50$ Å with the previously measured flux at $\lambda_3 \sim 10$ Å (Hayakawa, Matsuoka, and Yamashita 1966).

From equation (1) we may obtain the upper limit of the linear dimension $R$ of the source (using the upper limit on temperature)

$$R_1 < 10^9 \, (r/200) \text{ cm} , \tag{2}$$

where r is expressed in parsecs.

In addition there must be some much hotter plasma with temperature $T \sim 5 \times 10^7 \, ^\circ$ K in the source Sco XR-1. This conclusion is derived from its spectrum in the region of $\lambda < 10$ Å (Hayakawa *et al.* 1966). The ratio of optical emission in the visible to emission at $\lambda < 10$ Å shows that this plasma must also be optically thick in the visible. Equation (1) applied to this region at $T \sim 5 \times 10^7 \, ^\circ$ K gives an upper limit for its linear dimension,

$$R_2 < 1.5 \times 10^8 \left(\frac{r}{200 \text{ pc}}\right) \text{cm} . \tag{3}$$

The conception that two small, very dense plasma clouds exist side by side seems to be rather artificial. It is much more natural to suppose that there is one source; that is, one hot ball whose temperature and density are increasing toward its center. According to such a model the emission of Sco XR-1 in the region of $\lambda < 12$ Å arises in the inner, hotter part of the source, while the emission in the region $\lambda > 44$ Å arises in the outer colder part.

Using this model we can estimate the lower limit of the dimensions of the source from the requirement that emission from the inner part with the $\lambda < 12$ Å is not absorbed in the outer part of the source for which $T < 2 \times 10^6 \, ^\circ$ K. The requirement is

$$\tau_2 = \phi_2 R_1 = \frac{1.5 \times 10^{-2}}{\nu^2 T^{3/2}} N_e N_i g R_1 < 1 , \tag{4}$$

where $g$ is the gaunt factor (which for this spectral region is close to unity), $\phi_2$ is the X-ra[y] absorption coefficient, and $\tau_2$ is the optical depth of the outer part of the source.

On the other hand, considering that the emission in the region 44 Å $< \lambda <$ 60 [Å] arises in an optically thin layer of plasma with the temperature of about several hundre[d] thousand degrees, we may derive from the $F_{\nu_2}$ the "volume emission measure" on th[e] basis of the well-known formula

$$E_2 = \frac{4\pi}{3} R_1^3 N_e N_i \sim 10^{61} \left(\frac{r}{200 \text{ pc}}\right) e^{\tau_X} , \tag{5}$$

where $\tau_X$ is the optical depth of the interstellar medium for X-ray emission. If th[e] plasma in the region of 44 Å $< \lambda <$ 60 Å is not yet transparent, then $E_2$ will be highe[r]. From equations (4) and (5) it follows that

$$R_1 > 3 \times 10^8 \left(\frac{r}{200 \text{ pc}}\right) e^{1/2 \tau_X} .$$

Comparing equations (2) and (6) we come to the conclusion that the dimensions of t[he] source Sco XR-1 must be close to $5 \times 10^8$ cm if the distance $r \sim 200$ pc, which seems [to] be most probable. With such dimensions the mean electron concentration in the out[er] layers of the source (where $T$ is less than $3 \times 10^6 \, ^\circ$ K) is $N_e \sim 2 \times 10^{17}$ cm$^{-3}$. T[he] total mass of the plasma is $3 \times 10^{20}$ gm. and its heat content is $\frac{4}{3}\pi R_1^3 \times 3kT \sim 3 \times 1$[...] ergs. This store of heat energy will be sufficient to maintain an emission of the observ[ed] power only for one tenth of a second. Consequently a powerful and highly efficient mechanism for continual heating of the plasma is needed.

The volume emission measure corresponding to an emission in the spectral region of $\lambda < 12$ Å is $E_3 \sim 2 \times 10^{59}$ cm$^{-5}$. It is quite natural to assume that for a hotter plasma with the temperature $T \sim 5 \times 10^7$ ° K in the inner part of the source the electron concentration is higher than in the outer part. Thus it follows that the dimensions of this inner hotter part are less than $10^8$ cm. The lower limit on the dimensions of this region, determined from the condition that for $\lambda \sim 10$ Å the plasma remains optically thin, is equal to $2 \times 10^6$ cm. It is possible that the dimension of the region with $T \sim 5 \times 10^7$ ° K is close to $10^7$ cm. In this case $N_e \sim 10^{19}$ cm$^{-3}$, and the total mass is very small, $10^{17}$ gm. One more circumstance must be taken into account. As we may see from the spectrum of Sco XR-1 there is a very flat secondary maximum in the region of $35 \pm h\nu < 50$ keV (Peterson and Jacobson 1966). In the frame of our model of the source this spectral feature may be naturally explained as the thermal emission of a still hotter and denser plasma in the central part of the source. The temperature of this plasma is about $5 \times 10^8$ ° K. The volume emission measure of this innermost source may be estimated as $\sim 10^{58}$ cm$^{-5}$ and the dimensions as $\sim 10^6$ cm.

The "three-layer" model of the source Sco XR-1, described above, is certainly a very rough approximation to reality. In fact it must be expected that the physical properties of the source change more or less continually while the plasma remains transparent for sufficiently energetic X-ray quanta.

By all its characteristics this model, obtained only from the analysis of the data of observations without any a priori hypothesis about the nature of the source, corresponds to a neutron star in a state of accretion. If the identification of the optical object similar to an old nova with the X-ray source is correct, then the natural and very efficient supply of gas for such a accretion is a stream of gas, which flows from a secondary component of a close binary system toward the primary component which is a neutron star. In this case it may be that we observe a binary system similar to WZ Sagittae in which one of the components is a neutron star.

From the analysis of the emission lines we may first of all conclude that the regions of emission for He II, H, and C III–N III do not coincide. The emission in the lines of H and He II is caused by recombinations while the emission in the lines of C III–N III is due to electron collisions. The ionization equilibrium of different elements in the stream is governed by the intensity of X-ray radiation which is very high. From the observed intensities of the lines the volume emission measure of the comparatively cold plasma in the stream may be estimated as $10^{55}$ cm$^{-5}$. From this we may infer that the electron concentration in the stream is $10^{13}$ cm$^{-3}$, the latter being opaque for X-ray radiation. The emission in H-lines arises in the remotest part of the stream from the neutron star where $T \sim 10^4$ ° K. The emission in the lines of C III–N III arises in the part of the stream which is nearer to the neutron star and where $T \sim 5 \times 10^4$ ° K. Finally the region of emission in the lines of He II (partly overlapping the preceding region) is placed still nearer to the neutron star.

The flux of gas in the stream is estimated as $10^{16}$–$10^{17}$ gm/sec ($\sim 10^{-9}$ $M_\odot$/year). When this gas falls on the neutron star the production of energy per unit mass may amount to $\sim 10^{20}$ ergs/gm. Thus it follows that the suggested modification of the mechanism of the accretion of gas on the neutron star gives the possibility of explaining the power of X-ray emission of the source Sco XR-1.

I. S. SHKLOVSKY

February 3, 1967

STERNBERG ASTRONOMICAL INSTITUTE
MOSCOW STATE UNIVERSITY
MOSCOW, U.S.S.R.

## REFERENCES

Byram, E. T., Chubb, T. A., and Friedman, H. 1966, *Science*, **153**, 1527.
Hayakawa, S., Matsuoka, M., and Yamashita, K. 1966, Nagoya University preprint.
Peterson, L. E., and Jacobson, A. J. 1966, *Ap. J.*, **145**, 962.
Sandage, A., Osmer, P., Giacconi, R., Gorenstein, P., Gursky, H., Waters, J., Bradt, H., Garmire G., Sreekantan, B. V., Oda, M., Osawa, K., and Jugaku, J. 1966, *Ap. J.*, **146**, 316.

THE ASTROPHYSICAL JOURNAL, Vol. 151, February 1968

# ON THE NATURE OF SOME GALACTIC X-RAY SOURCES

K. H. PRENDERGAST
Department of Astronomy, Columbia University
AND
G. R. BURBIDGE
Department of Physics, University of California, San Diego
*Received December 8, 1967; revised December 15, 1967*

ABSTRACT

A model to explain the generation of X-ray sources in close binary systems is described. It is concluded that models of this type may be able to account for the X-ray flux emitted by Sco X-1 and Cyg X-2.

## I. INTRODUCTION: POINT-SOURCE IDENTIFICATIONS AND OPTICAL DATA AVAILABLE

The majority of the X-ray sources so far detected are thought to be galactic objects. The identification of the Crab Nebula has led to the conclusion that at least some of these sources are supernova remnants. However, following the optical identification of Sco X-1 with a comparatively bright ultraviolet star (Sandage *et al.* 1966), it was realized that other classes of objects are strong X-ray emitters. This early identification was with an optical object which has some of the outward characteristics of an ex-nova. Since it is believed that ex-novae are highly evolved close binary systems, the first proposals made by a group of theoreticians at the Noordwijk conference in August, 1966 (cf. G. Burbidge 1967), and followed in more detail by Shklovsky (1967) and Cameron and Mock (1967), included the idea that one of the stars was a neutron star or a white dwarf and that the X-ray emission was produced by matter ejected from one star falling into the highly evolved star. Prendergast (1967) explored the possibility that the X-ray emission might arise in interaction between gas streams in close binary systems. A suggestion that X-ray emission might be produced by gas streams in early-type binaries had been made earlier by Hayakawa and Matsuoka (1964). Manley (1966) and Tucker (1967) have made alternative suggestions.

With the identification of a second X-ray source with a starlike object, Cyg X-2 (Giacconi *et al.* 1967), new information is available on the optical properties of the object. Following the first spectrogram obtained by Lynds (1967), investigations by Burbidge, Lynds, and Stockton (1967), Kristian, Sandage, and Westphal (1967), and Kraft and Demoulin (1967) have shown that this object is indeed a binary system. In this Letter we outline the kind of theoretical model which is compatible with the observations, both for Cyg X-2 and Sco X-1. The observations of Cyg X-2 can be summarized as follows:

*a*) The major part of the optical radiation may come from a G-type subdwarf. However, there is a complication associated with this conclusion. When Cyg X-2 was first identified, it was estimated that the object should have an apparent magnitude approximately equal to 16.1 mag, with $B - V \approx 0.4$, $U - B \approx -0.7$, when the effect of galactic obscuration is taken into account. But the accurate observations by Kristian *et al.* (1967) give $m_v = 14.46$–$14.89$, with $B - V$ ranging from 0.5 to 0.4 while $U - B$ varies between 0 and $-0.3$, and, as they have pointed out, these are not the color and magnitude expected from extrapolated thermal bremsstrahlung. The color is, however, according to Kraft and Demoulin (1967), that of a G-type subdwarf with variations due to a companion or gas streams; the absolute magnitude would in this case be $+5.5$, giving a distance of about 600 pc. The total luminosity in the X-ray flux is then found from the measures of Gorenstein, Giacconi, and Gursky (1967) to be about $10^{35}$ ergs $sec^{-1}$. If this is thermal bremsstrahlung flux, the extrapolation into the optical region shows that the optical bremsstrahlung is only a few per cent of the stellar radiation ($4 \times 10^{33}$ ergs $sec^{-1}$), and it is thus not surprising that the color is not compatible with thermal bremsstrahlung. At the same time this result casts doubt on the identification insofar as it depends on the idea that the color and magnitude of the optical object must be compatible with the extrapolation of a bremsstrahlung spectrum, although this method worked for Sco X-1.

The basis for the identification lies now in the fact that the optical object is variable and shows peculiarities. However, the optical radiation may not be coming from a normal stellar atmosphere; this will be discussed later.

*b*) Spectrograms obtained at comparatively long intervals suggested that the object was a double-line binary with a period of $\sim M/M_{\odot}$ days. However, spectrograms taken at much shorter intervals indicate that the object is a single-line binary or a more complex system with a period of 5–7 hours. Closely associated with the G-type subdwarf there is a faint companion which contributes none of the spectral features. Also, more rapid light variations have been detected (Kristian *et al.* 1967).

*c*) Some of the anomalous features in the spectrum may be due to gas streams.

Sco X-1 undergoes rapid, erratic light variations. However, no radial-velocity variations have been established for it, and its spectrum shows significant differences from those of ex-novae. Thus, although it has been shown that Cyg X-2 is a binary or multiple star, the preliminary conclusion that can be drawn from the data is that galactic X-ray sources of this type are not necessarily ex-novae. They are likely to be very close binaries in which gas streams play an important role. Before describing the model, we briefly relate the optical features seen in Cyg X-2 and Sco X-1 with those detected in well-studied binary systems.

Many of the features which are seen in Sco X-1 and Cyg X-2 have been seen in binary systems which are not known to be X-ray sources. For example, consider the binary UX Mon (Struve 1950), which consists of a G-type star and an A-type star with a period of 5.9 days. The spectroscopic features due to the G-type star are very similar to those seen in Cyg X-2; the large negative velocity obtained from the Ca II absorption lines in UX Mon is attributed to gas streams. The large negative velocity measured in Ca II in Cyg X-2 may have the same origin. In this case it does not represent the radial velocity of the system relative to the Sun, so that one of the arguments which has led to the conclusion that Cyg X-2 is a Population II object is removed. Rapid variations in light with time scales of hours have also been detected in UX Mon (Struve 1950), and these are similar to the variations detected in Sco X-1 and Cyg X-2.

The difference between the systems UX Mon and Cyg X-2 may be that in Cyg X-2 the two stars are much closer together and a comparatively small, dense star is present rather than the A-type star in UX Mon. The observations (Struve 1950) of many binary systems show that there is nothing in the optical observations of Sco X-1 and Cyg X-2 which has not been seen before in complicated binary systems. What is new and must be presumed a rare phenomenon, since X-ray sources are few, whereas a considerable fraction of stars are in binary systems, is that X-ray emission can be produced in great intensity.

## II. MODEL FOR X-RAY EMISSION FROM A GASEOUS DISK

Since the optical energy emitted by Sco X-1 is well accounted for by the optically thin bremsstrahlung at a temperature of about $50 \times 10^6$ ° K, we assume that both stars in that system are comparatively faint. The line spectrum may arise from the gaseous envelope. However, in the case of Cyg X-2 the optical energy which is emitted comes from the G-type star, or it might conceivably come from a gaseous envelope. Thus, in any case, it dominates as compared with the optical component of the radiation from the hot gas at a temperature near $40 \times 10^6$ ° K (Gorenstein *et al.* 1967). It should be emphasized that, since we have discarded the ex-nova hypothesis, we have no direct evidence that either component of Sco X-1 or the faint star in Cyg X-2 is a highly evolved star. However, the condition that gas streams arising on one star are able to generate X-rays by interacting with the other really determines a value for $M/R$, where $M$ is the mass of the star, henceforth designated the "primary," and $R$ is its radius. The maximum temperature $T$ that can be reached by gas falling onto the star is given approximately by

$$T \simeq 10^7 \left(\frac{M}{R}\right)\left(\frac{R_\odot}{M_\odot}\right). \qquad (1)$$

Thus X-ray sources with $T \sim 10^7$–$10^8$ ° K will be detected from binary systems with $M/R$ in solar units lying between 1 and 10. For a neutron star $M/R \approx 10^4$, and, as Cameron and Mock have pointed out, the temperature generated will be too high; but for a white dwarf $M/R$ has approximately the correct value. However, any other stellar configuration with $M/R$ in the appropriate range can in principle give rise to hard radiation by this process. How much of the radiation will be emitted as X-rays from a gas close to the maximum temperature depends on the model, and in the following discussion we shall be using parameters appropriate for Cyg X-2.

As we have already pointed out, gas streams are common in close binaries and originate when the secondary overflows its lobe of the critical Lagrangian surface. Any element of gas leaving the surface of the secondary can go to one of three places: it can return to the secondary, it can leave the system entirely, or it can coalesce with the primary.

Consider the third possibility. What radius must the primary have if the infall of material is to amount to about $10^{35}$ ergs sec$^{-1}$ emitted as bremsstrahlung from a gas at a temperature near $40 \times 10^6$ ° K? For a star of mass 0.5 $M_\odot$ the radius will be $\sim$0.1 $R_\odot$. Similar figures were arrived at by Cameron and Mock (1967). Since each gram of gas acquires energy $GM/R$ in its fall, some $10^{19}$ gm sec$^{-1}$, or 1 $M_\odot$ in $10^6$ years, is required. This is not unreasonably high, but it does mean that this phase of energy dissipation is comparatively short-lived.

It is clear that the matter cannot fall directly onto the primary, as it has a large angular momentum when it leaves the secondary. We must therefore consider what mechanism will decrease the angular momentum of the gas near the primary. Consider the case of a gaseous disk surrounding the primary, with each element moving in a circle at the local Keplerian velocity $V = (GM/r)^{1/2}$. Since any real gas has a viscosity which tends to make the angular velocity constant, the inner, rapidly rotating parts of such a disk will be slowed down and the outer parts speeded up. The matter near the primary gradually loses its centrifugal support and must move inward. This inward drift can end only when the gas becomes attached to the primary and is supported by a nearly hydrostatic pressure gradient. The angular momentum transferred to the outer parts of the stream will accelerate the gas until it can pass the Lagrangian point or even leave the system. A crude estimate indicates that roughly half the gas leaving the secondary will end on the primary.

Having shown that the mass transfer will take place into a disk around the primary, it is necessary to show that the energy is likely to be released largely in the form of X-rays in the kilovolt range. In order to find out whether this is the case, a number of detailed models of a gaseous disk have been constructed. A full account of these calculations will be published elsewhere. Here only a summary of the results will be given.

To avoid geometrical complications arising from the presence of the secondary, we suppose that the flow near the primary can be treated as axisymmetric, and we also suppose that a steady state exists. We take the gravitational field to be that of the primary only and ignore the rotation of the binary system. It is also assumed that the disk is mainly supported against gravity by its own rotation. Velocities perpendicular to the disk are neglected, and it is assumed that radial and tangential velocities are almost independent of the distance from the central plane. We then set up the hydrodynamic and radiative-transfer equations for the disk.

Solution of the hydrodynamic equations will tell us whether the assumption of Keplerian velocity is adequate. We write the equation for the conservation of momentum in the radial direction, neglecting only compressive viscosity, and the equation for the conservation of momentum in the azimuthal direction assuming a mixing-length theory for the tangential stress tensor (a viscosity).

The mixing length is taken to be about half the thickness of the disk. In the $z$-direction it is assumed that the disk is in hydrostatic equilibrium. For the energy equation there is a gain due to the work done by the tangential component of the stress tensor, and there is a loss due to the radiative flux from the transfer equation for the atmosphere. The transfer equation which is coupled to the hydrodynamic equations through this term is set up and solved, using the Eddington approximation, to obtain the energy and flux. It is assumed that both scattering and absorption occur, that the absorption is due to free-free processes, and that electron scattering is the dominant scattering mechanism.

## III. RESULTS AND DISCUSSION

The coupled hydrodynamic and radiative-transfer equations reduce to a system of third-order differential equations, with the initial values of temperature, projected density, and tangential velocity to be prescribed, together with the values of the parameters which specify the flux of mass and of angular momentum.

The mass and radius of the primary were put equal to $10^{33}$ gm and $1.5 \times 10^{10}$ cm, and the flux of mass was put equal to $2 \times 10^{19}$ gm sec$^{-1}$. The flux of angular momentum was calculated by assuming an initial density of about $10^{-14}$ gm cm$^{-3}$ and initial temperature near 30000° K from spectroscopic studies of binaries. From these we obtained a scale height and surface density $\sigma_i = 30$ gm cm$^{-2}$. The flux of angular momentum was then found to be $2\pi \times 10^{38}$ cm$^2$ sec$^{-2}$. The ratio of mixing length to scale height was put equal to 0.5. The equations were then integrated, and the dependence of the physical parameters of the disk on distance $x$ from the center of the star is as follows ($x$ is measured in units of $1.5 \times 10^{10}$ cm): $T = 1.4 \times 10^6\, x^{-1.05}$ ° K, $\sigma = 10^3\, x^{-0.9}$ gm cm$^{-2}$, $V_{\tan} = 6.7 \times 10^7\, x^{-0.5}$ cm sec$^{-1}$, radiative flux (top + bottom) $= 3.56 \times 10^{15}\, x^{-3.45}$ ergs cm$^{-2}$ sec$^{-1}$, total flux $= 3.47 \times 10^{36}(x_1^{-1.45} - x_2^{-1.45})$ ergs sec$^{-1}$. In this last expression $x_1$ is the inner radius of the disk and $x_2$ is the outer radius. These results are all given as approximate power-law dependences based on exact numerical integrations. By varying the initial conditions, it has been shown that they are not important, since the starting values are quickly forgotten in the integration.

It is clear from these results that the optical radiation will be emitted from the outer parts of the disk and the X-rays from the inner part, where the gas is optically thin. There is a continuous transition between an atmosphere dominated by absorption and one dominated by electron scattering, with the transition point roughly at the radius where the product of the two relevant optical depths is about unity.

In this model the tangential velocity, and hence the angular momentum at the inner radius, is very high, and the question of the connection of the disk to the star is not discussed. In fact the problem is very difficult. The magnitude and even the sign of the flux of angular momentum cannot be determined in the present model, since their determination requires the solution of a two-point boundary-value problem, the important parameter being the equatorial velocity of the primary.

These representative calculations still give us only a qualitative idea of what is taking place in Cyg X-2 and Sco X-1. For Cyg X-2 two possible situations can be envisaged. The bulk of the optical radiation can come from the G-type star, or it can come from the outer part of the disk. If the first possibility is the correct one, it is easily shown that the X-rays from the disk will not significantly change the structure of the companion G-type star but will only produce limited hot regions on it. If both stars are emitting little in the way of optical flux, as must be the case for the Sco X-1 source, the only optical flux, which is small compared with the bremsstrahlung component, will come from the outer parts of the disk and from the gas streams before they interact.

It is not clear from these calculations under what conditions the color and the optical spectrum seen in Cyg X-2 can be reproduced from the disk. However, the structure of the disk depends sensitively on the value of $M/R$ and the flux of angular momentum, and it is entirely possible that the right conditions can be obtained. As far as the X-ray flux is concerned, it is clear that for the value of $M/R$ chosen the temperature does not reach values greater than about $10^7$ ° K before a disk radius comparable with the radius of the primary is attained. However, the flux of energy contained in the rapidly rotating inner parts of the disk is still exceedingly high and remains to be dissipated. It is this dissipation which gives rise to the highest-temperature gas. The details of this dissipation are tied to the problem of the interaction of the disk with the star's surface, which has not yet been solved. The maximum temperature given by equation (1) is obtained by assuming that all the gravitational energy has been dissipated.

Our conclusions can be stated as follows: If we take it that the binary nature of X-ray sources has been established by the recent investigations of the optical object originally identified with Cyg X-2, it is plausible that the X-rays are produced by the mass transfer taking place in close binary systems of the type associated with Cyg X-2. The optical spectrum can arise either in the atmosphere of a comparatively normal star, together with the cool parts of the gas streams, or in a pseudo-atmosphere forming part of the disk of gas surrounding the small component of the binary system. The X-ray flux originates in the disk, and the models outlined above suggest that the observed energetic and spectroscopic characteristics may be satisfied.

For this mechanism to work, it is necessary that the separation between the stars be small, that mass exchange be able to occur, and that the primary have an appropriate value of $M/R$. For stars with masses $\sim 1\ M_{\odot}$ and separations $\sim 10^{11}$ cm, as is probably the case for Cyg X-2, X-ray emission is the major mechanism of energy dissipation in the binary. The total energy in Cyg X-2 is about $10^{48}$ ergs. The only other mechanism of energy dissipation in such a system is the radiation of gravitational waves. It is easily shown, using the expressions for gravitational radiation given by Landau and Lifshitz (1951), and the parameters assumed above, that the gravitational radiation amounts to about $10^{32}$ ergs sec$^{-1}$ if both stars have masses of $1\ M_{\odot}$. In this model the rarity of X-ray sources as compared with binary stars is due to the fact that only in a small fraction of binary systems will the conditions given above be fulfilled.

Variability of X-ray sources can then be due to changes in the rate of, or to the cessation of, mass transfer. The short-period variations of light or of X-ray flux may be explained in this way. A larger variation in the rate of mass transfer might explain the discovery as a strong source, and the subsequent fading, of Cen X-2 (Francey, Fenton, Harries, and McCracken 1967).

There still remains a basic uncertainty in all this discussion, centered on the question of whether or not the optical object originally identified with Cyg X-2 is the true X-ray source. If it is, our arguments may have some validity. It is then clear, however, that it will not be possible in future to identify X-ray sources with optical objects by the method used in the case of Sco X-1 and Cyg X-2—that is, by looking for an object which has the color and magnitude to be expected from an extrapolation of the thermal bremsstrahlung spectrum. Much more accurate positions will be required, so that most of the optical objects in the field can be eliminated.

If the optical identification of Cyg X-2 is incorrect, there is still no certainty that X-ray sources are associated with close binary systems. Moreover, the distance determination is no longer valid, and the intrinsic power of the X-ray source is not known.

We are indebted to Margaret Burbidge, Robert Kraft, and Allan Sandage for informing us of the results obtained by them and their colleagues prior to publication. This work has been supported in part by grants from the National Science Foundation and in part by NASA through Grant NsG-357. We are indebted to Dr. R. Jastrow for affording us computing facilities at the Institute for Space Studies.

## REFERENCES

Burbidge, E. M., Lynds, C. R., and Stockton, A. N. 1967, *Ap. J. (Letters)*, **150**, L95.
Burbidge, G. 1967, "Radio Astronomy and the Galactic System," *I.A.U. Symp. No. 31* (London and New York: Academic Press), p. 463.
Cameron, A. G. W., and Mock, M. 1967, *Nature*, **215**, 464.
Francey, R. J., Fenton, A. G., Harries, J. R., and McCracken, K. G. 1967, *Nature*, **216**, 773.
Giacconi, R., Gorenstein, P., Gursky, H., Usher, P. D., Waters, J. R., Sandage, A., Osmer, P., and Peach, J. V. 1967, *Ap. J. (Letters)*, **148**, L129.
Gorenstein, P., Giacconi, R., and Gursky, H. 1967, *Ap. J. (Letters)*, **150**, L85.
Hayakawa, S., and Matsuoka, M. 1964, *Progr. Theoret. Phys. Suppl.*, No. 30, p. 204.
Kraft, R. P., and Demoulin, M. H. 1967, *Ap. J. (Letters)*, **150**, L183.
Kristian, J., Sandage, A. R., and Westphal, J. 1967, *Ap. J. (Letters)*, **150**, L99.
Landau, L., and Lifshitz, E. 1951, *Classical Theory of Fields* (Cambridge, Mass.: Addison-Wesley Press), chap. xi.
Lynds, C. R. 1967, *Ap. J. (Letters)*, **149**, L41.
Manley, O. 1966, *Ap. J.*, **144**, 1253.
Prendergast, K. H. 1967 (unpublished).
Sandage, A. R., Osmer, P., Giacconi, R., Gorenstein, P., Gursky, H., Waters, J., Bradt, H., Garmire, G., Sreekantan, B. V., Oda, M., Osawa, K., and Jugaku, J. 1966, *Ap. J.*, **146**, 316.
Shklovsky, I. S. 1967, *Ap. J. (Letters)*, **148**, L1.
Struve, O. 1950, *Stellar Evolution* (Princeton, N.J.: Princeton University Press), chap. iii.
Tucker, W. H. 1967, *Ap. J. (Letters)*, **149**, L105.

SOVIET ASTRONOMY – AJ VOL. 13, NO. 2 SEPTEMBER-OCTOBER, 1969

# X-RAY EMISSION ACCOMPANYING THE ACCRETION OF GAS BY A NEUTRON STAR

Ya. B. Zel'dovich and N. I. Shakura

Institute of Applied Mathematics, Academy of Sciences of the USSR
Physics Department, Moscow University
Translated from Astronomicheskii Zhurnal, Vol. 46, No. 2,
pp. 225-236, March-April, 1969
Original article submitted August 19, 1968

A discussion is given of the accretion of gas by a neutron star and the spectrum of the radiation emitted in this process. The entire treatment applies to the spherically symmetric case. A systematic and internally consistent method is developed for calculating the physical parameters of the atmosphere of a neutron star and the electromagnetic radiation spectrum. Calculations are performed under two assumptions regarding the mean free path of the incident protons: 1) that the protons are decelerated by pair collisions with the particles in the atmosphere, and 2) that collective plasma oscillations would serve to reduce the mean free path. The computed spectrum differs considerably from a Planck curve, particularly in the second case.

## INTRODUCTION

Neutron stars were born at the tip of the theoretician's pen over 30 years ago, but a convincing identification of such a star has yet to be made. The same is also true of stars in a state of relativistic collapse.

Calculations indicate that neutron stars would develop as a result of catastrophic contraction in a state of intensive oscillations and with a high temperature. The initial state at the time when stability is lost (prior to the catastrophe) has been discussed by Bisnovatyi-Kogan [1], and calculations of the catastrophic stage itself have been performed by Arnett [2]. However, the heat would rapidly be dissipated by neutrino radiation [3], and the oscillations would also readily die out because of gravitational radiation [3].

Accretion of gas from the surrounding space could ensure a very prolonged maintenance of x-ray emission by a neutron star.

The lifetime of a neutron star, if measured as the period during which its mass increases, say, from 1 to 1.5 $M_{\odot}$, would be $3 \cdot 10^{8}$ yr for a luminosity $L = 10^{37}$ ergs/sec. The energy released per unit mass of infalling material, equal to the gravitational potential at the surface, would be roughly $(0.1\text{-}0.2)c^2$, a value 10-20 times the energy released in nuclear reactions. These considerations were advanced in 1964 and at that time a distinctive, self-regulating accretion mechanism was described: the pressure of the radiant flux would restrict the accretion rate to a value of the order of $1.5 \cdot 10^{-8}$ $M_{\odot}$ per year, corresponding to a "critical" luminosity $L_c \approx 6 \cdot 10^4 L_{\odot}$ for neutron stars with $M \approx M_{\odot}$.

It has been pointed out [5-8] that in addition to interstellar gas, some of the gas ejected at the time when a neutron star was formed may be subject to accretion, or gas flowing away from the secondary component if the neutron star belongs to a binary system.

Shklovskii [9] has analyzed the radiation of Sco X-1, representing it as the bremsstrahlung of optically thin gas layers at different temperatures. He claims to have proved the gas-accretion mechanism in a binary star directly from observational data, and thereby independently of prior theoretical hypotheses.

Cameron and Mok [10] have considered accretion by white dwarfs and have pointed out that soft x rays might be emitted in this case as well. One can show without difficulty that the lifetime should then be substantially shorter, of order $3 \cdot 10^5$ yr for $L \approx 10^{37}$ ergs/sec, since the gravitational potential is smaller at the surface. Possibly nuclear-reaction bursts and the discharge of gas [11] would lengthen this period.

Melrose and Cameron [12] have recently considered the generation of fast particles and plasma oscillations when gas is accreted by a neutron star in conjunction with the instability for $L \approx L_c$. This mechanism has been applied to explain the compact radio source in the Crab Nebula.

To conclude this survey we may mention the distinctive accretion by a collapsing ("solidifying") star [7]; in this instance spherically symmetric accretion in general would not, in the hydrodynamic approximation, be accompanied by the release of energy.

We would also point out that it is meaningful to consider accretion only for stars that are in the closing stage of their evolution, since in the case of stars whose energy source is nuclear burning a phenomenon opposite to accretion occurs—the "stellar wind."

The great diversity of accretion effects will be evident from this brief survey. Yet until recently no accurate solutions have been obtained for even the simplest and most idealized problems.

In this paper we shall consider the accretion of gas by a neutron star and the radiation arising in this process. The entire treatment will refer to the simplest idealized case of spherically symmetric gas motion. Adopting a law for the deceleration of the incoming particles, we shall find the distribution of temperature, density, and pressure in the layer where particle deceleration occurs. We shall then find the spectrum of the emergent radiation. The calculation will be performed in two approximations: 1) for particle deceleration by collisions, and 2) with allowance for plasma instability, serving to reduce the mean free path.

We shall not take into account the influence of the magnetic field, either that frozen into the incoming gas or the field of the star. We assume that $L < L_c$, so that we may neglect the influence of radiation on the incident flux and the inverse action of the rarefied incoming gas on the emission spectrum. A preliminary report of the results was presented at the 13th General Assembly of the International Astronomical Union [13].

## 1. EQUATIONS FOR THE ENERGY BALANCE, TEMPERATURE DISTRIBUTION, AND DENSITY

All the numerical computations are very approximate in character and have been made for a neutron star with $M \approx M_\odot$, $R \approx 10^6$ cm, and a luminosity $L_1 = 1.3 \cdot 10^{37}$ ergs/sec for $L_2 = 1.3 \cdot 10^{36}$ ergs/sec, equal to 0.1 or 0.01 of the critical luminosity respectively. A stream of ionized hydrogen impinges on the surface of the neutron star; according to the free-fall law the velocity $v = (2GM/R)^{1/2}$ of the stream reaches $\approx 0.5c$. The mass flux $dM/dt = 4\pi\rho v R^2 = 10^{17}$ g/sec or $10^{16}$ g/sec, corresponding to a density in the incident stream of $5 \cdot 10^{-7}$ g/cm³ or $5 \cdot 10^{-8}$ g/cm³ at the surface.

As it decelerates in the atmosphere of the neutron star the stream relinquishes its kinetic energy, which is transformed into radiation. Extreme limits on the temperature can easily be given. Under laboratory conditions, when a flow encounters an obstacle a shock wave is formed, kinetic energy is transformed into thermal energy, and radiation is slowly released. In this approximation we find a temperature $T = 10^{12}$ °K along a Hugoniot adiabat. We can obtain another estimate from the energy balance, assuming that the neutron star radiates like a blackbody:

$$L = \frac{GM}{R}\frac{dM}{dt} = 4\pi R^2 b T_{\rm eff}^4,$$

$$b = 5.75 \cdot 10^{-5} \quad \text{erg/cm}^2 \cdot \text{deg}^4.$$

We thereby have $T_{\rm eff} = 1.15 \cdot 10^7$°K for $L = 0.1\,L_c$ or $T_{\rm eff} = 6.5 \cdot 10^6$ °K for $L = 0.01\,L_c$.

The enormous disparity between these two estimates shows how important a task the detailed study of the phenomenon is. The essential aspect of the problem is that the protons in the incident stream are not decelerated instantaneously. They relinquish their energy to a comparatively extended layer of the atmosphere. Let us consider the thermal balance of this layer, or more accurately, of the electrons located in the layer. They receive their energy from the incoming protons (either directly or through protons in the layer which receive their energy from the incoming protons); the electrons release their energy by bremsstrahlung or through the inverse Compton effect.

We introduce the quantity $y = \int_{x_0}^{\infty} \rho(x)\,dx$; y has the meaning of the amount of material lying above a given point. This integral contains the density $\rho(x)$ of the material in the neutron-star atmosphere, which is essentially at rest.

Without making the mean free path of the incident particles more specific at this point, we shall denote it by $y_0$.

Then in the first approximation the energy released per gram of atmospheric material is

$W' = Q/y_0$ for $y < y_0$ and $W = 0$ for $y > y_0$, where $Q = L/4\pi R^2$ is the energy flux per unit surface area of the star. The thermal balance of the electrons will be determined by the arriving energy W and the expenditure to bremsstrahlung,

$$W' = 5 \cdot 10^{20} \sqrt{T_e} \rho \,, \qquad (1.1)$$

as well as the energy lost to Comptonization, that is, the change in the energy of the photons because of scattering by Maxwellian electrons [14, 15]:

$$W'' = \frac{4 \varepsilon c \sigma_c}{m_p} \frac{kT_e}{m_e c^2}; \qquad (1.2)$$

here $\varepsilon$ is the radiant energy density and $\sigma_c = 6.65 \cdot 10^{-25}$ cm$^2$.

These expressions do not take the inverse processes into consideration. Evidently in an equilibrium radiation field with $T_{ph} = T_e$, $W' = W'' = 0$. We shall allow for the inverse processes by introducing effective radiation temperatures T' and T"; the energy-balance equation will then take the form

$$\frac{Q}{y_0} = 5 \cdot 10^{20} \sqrt{T_e} \rho \left(1 - \frac{T'}{T_e}\right) + 6{,}5 \varepsilon T_e \left(1 - \frac{T''}{T}\right). \qquad (1.3)$$

The quantities T' and T" depend on the spectrum of the radiation, but below, in the first approximation, we shall adopt $T' = T'' = T_{ph} = (\varepsilon/a)^{1/4}$; $a = 7.6 \cdot 10^{-15}$ erg/cm$^3$ · deg$^4$.

The radiant energy density is determined by the diffusion equation:

$$q = Q \frac{y - y_0}{y_0} = -\frac{c}{3\sigma} \frac{d\varepsilon}{dy}, \quad y < y_0, \qquad (1.4)$$

with $\sigma = 0.38$ cm$^2$/g the opacity to fully ionized plasma and q the radiant energy flux. For $y > y_0$, $q = 0$ and $\varepsilon$ = const. For fully ionized hydrogen plasma this equation is valid independently of the emission spectrum.

If we take $\varepsilon = (3Q)^{1/2}/c$ at the boundary of the atmosphere, where $y = 0$, we have

$$\varepsilon = \frac{Q}{c}\left\{\sqrt{3} + 3\sigma y_0\left[\frac{y}{y_0} - \frac{1}{2}\left(\frac{y}{y_0}\right)^2\right]\right\}, \quad 0 < y < y_0,$$
$$\varepsilon = \frac{Q}{c}\left\{\sqrt{3} + \frac{3}{2}\sigma y_0\right\}, \quad y > y_0. \qquad (1.5)$$

In the interior, where $y \gg y_0$, complete thermodynamic equilibrium will be established, with a temperature which we can find from the condition $\varepsilon = aT^4$. Using Eqs. (1.5), we find that for $y \gg y_0$, $T \approx L^{1/4} y_0^{1/4}$. At the surface, where $\rho \to 0$ and $W' \to 0$, the temperature of the electrons will be determined by Comptonization and will not depend on the luminosity, since $\varepsilon \approx Q \approx L$.

We can find the density distribution of the material trivially by using the equation of hydrostatic equilibrium:

$$P = \frac{2\rho kT}{m_p} = \left(\frac{GM}{R^2} + \frac{\rho_0 v^2}{y_0}\right) y, \quad 0 < y < y_0,$$
$$P = \frac{2\rho kT}{m_p} = \frac{GM}{R^2} y + \rho_0 v^2, \quad y > y_0. \qquad (1.6)$$

The first term in the sum represents the weight of the atmospheric material; the second, the momentum imparted by the incident particles to the layer $y_0$.

If we now adopt a flux $Q_1 = 10^{24}$ ergs/cm$^2$ · sec or $Q_2 = 10^{23}$ ergs/cm$^2$ · sec at the surface and a mean free path $y_0$, we can obtain the temperature and density distribution in the atmosphere of the neutron star. The determination of the value of $y_0$ remains the fundamental difficulty in the problem.

## 1a. Deceleration of protons by collisions with individual particles.

The kinetic energy of the incident protons can vary within wide limits, depending on the gravitational potential of the neutron star. For a mass $M \approx M_\odot$, $E_p \approx 100$–$300$ MeV. The mean free path of such protons in fully ionized plasma will be determined by Coulomb collisions, and we obtain $y_0 \approx 5$–$30$ g/cm$^2$. For our numerical computations we shall adopt $y_0 \approx 20$ g/cm$^2$. Since in the interior $T \sim y_0^{1/4}$, a different value for $y_0$ will not materially alter the physical conditions for $y \gg y_0$, while at the surface the electron temperature will not depend on $y_0$. To be sure, for small $y_0$ the contribution of Comptonization to the energy exchange will increase, and this will affect the emergent spectrum, but we shall take the influence of Comptonization into account in another limiting case (Sec. 1b). For $y_0 \approx 20$ g/cm$^2$ at depths where $y \gg y_0$ we have $T_1 = 1.5 \cdot 10^7$°K or $T_2 = 8.6 \cdot 10^6$°K for the luminosities adopted. The electron temperature at the surface is $\approx 10^8$°K in both cases. We shall obtain the temperature distribution in the $y_0$ layer by solving Eqs. (1.3), (1.5), and (1.6) numerically. The results of the computations are displayed in Fig. 1, but here we merely wish to call attention to the effective temperatures obtained from general energy estimates.

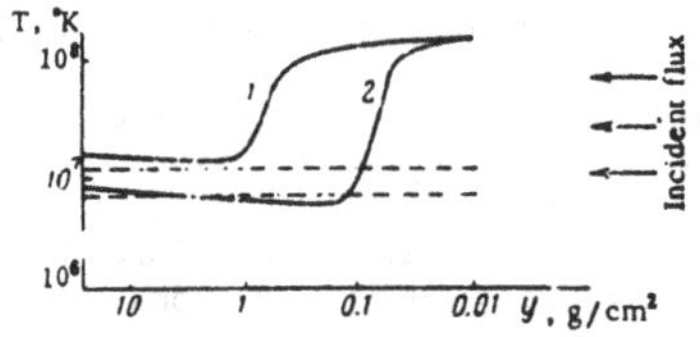


Fig. 1. Distribution of electron temperature in the atmosphere of a neutron star. 1) Luminosity $L_1 = 1.3 \cdot 10^{37}$ ergs/sec; 2) luminosity $L_2 = 1.3 \cdot 10^{36}$ ergs/sec. The dashed lines designate the effective temperatures computed from the overall energy balance. The value $y_0 = 20$ g/cm² has been adopted.

We introduce the quantity

$$\eta = \frac{\int_0^{y_0} W'' dy}{Q}, \qquad (1.7)$$

which represents the fraction of energy released by electrons to Comptonization. For $y_0 \approx 20$ g/cm², $\eta < 0.05$ if $L_1 = 0.1\, L_C$ or $\eta < 0.01$ if $L_2 = 0.01\, L_C$, so that the contribution of Comptonization to the overall heat exchange is not large; accordingly, Comptonization will have a small effect on the emergent radiation.

1b. The influence of plasma oscillations. It is well established that when a beam of charged particles passes through a plasma, oscillations will appear in the plasma, primarily at the Langmuir frequency $\nu_{pl} = (\pi e^2 n/4m_e)^{1/2}$. The plasma oscillations will interact efficiently with the ions in the beam, decelerating the beam.

The reduction in the mean free path due to plasma instabilities will lead to a rise in the electron temperature. We may regard the rise in temperature as restricted by the circumstance that for $v_{t,e} \geq v = (2GM/R)^{1/2}$ the rate of generation of plasma oscillations will decline sharply, and the plasma oscillations will be subject to Landau absorption [16]. We shall assume below that a regime of plasma oscillations and a value of $y_0$ are established such that $v_{t,e} = v = (2GM/R)^{1/2}$. The electron temperature determined by this condition will be $\theta \approx 10^9$ °K.

The mean free path for deceleration will be found from the condition that the heating and losses to radiation and Comptonization occur at a given temperature $\theta$. Using Eqs. (1.3, 1.5, and 1.6), and setting $\theta = 10^9$ °K throughout the $y_0$ layer, we obtain $y_0 \approx 2$ g/cm².

We can determine the temperature of the material in the interior, for $y \gg y_0$, in the same manner as previously: $T = (\varepsilon/a)^{1/4}$. This yields $T_1 = 1.10 \cdot 10^7$ °K or $T_2 = 6 \cdot 10^6$ °K for the different luminosities.

In this case, then, the temperature in the surface layer increases by almost an order of magnitude; the fraction of Comptonization rises ($\eta = 0.96$ for $L_1$ or $\eta \approx 0.7$ for $L_2$), and hard x-ray photons of energy 50-100 keV appear (bremsstrahlung from the hot layer). The emergent spectrum will definitely be non-Planckian.

## 2. SPECTRUM OF THE RADIATION

In a real atmosphere the emergent radiation flux is determined by solving the integrodifferential transfer equation with allowance for the change in frequency through scattering. However, to a perfectly satisfactory approximation we may divide the atmosphere of the star into a relatively "cool" half-space with $T \approx 10^7$ °K and a thin hot layer with $\theta \approx 10^8$ °K or $10^9$ °K, depending on the mean free path $y_0$. The flux emitted by the isothermal half-space with the relatively low radiation temperature will be Comptonized by the hot electrons in the thin layer; the radiation of the hot layer itself will be added to this flux.

It is readily seen that for a large part of the x-ray spectrum the coefficient $\sigma = 0.38$ cm²/g for scattering by free electrons in the atmosphere of a neutron star will be many times larger than the true absorption coefficient, which is here determined by free-free transitions:

$$\varkappa_\nu = \frac{1.14 \cdot 10^{56}}{T^{1/2} \nu^3} (1 - e^{-h\nu/kT}) \rho \text{ cm}^2/\text{g}. \qquad (2.1)$$

If scattering is included, the radiation flux from the isothermal homogeneous half-space is

$$F_\nu \approx \pi B_\nu(T) \sqrt{\frac{\varkappa_\nu}{\sigma + \varkappa_\nu}} \qquad (2.2)$$

and the Wien region, where $h\nu > kT$ (in this region the inequality $\sigma/\varkappa_\nu \gg 1$ will be satisfied with large margin), $F_\nu \sim \nu^{3/2} \exp(-h\nu/kT)$. For comparison, the radiant flux from a blackbody ($\sigma = 0$) will have $F_\nu \sim \nu^3 \exp(-h\nu/kT)$. However, the atmosphere of the neutron star is not homogeneous; its density increases very rapidly with depth. This circumstance somewhat mitigates the effects of scattering, and $F_\nu \sim \nu^2 \exp(-h\nu/kT)$. A calculation of the spectrum in the Eddington approximation, for $\sigma > \varkappa_\nu$, is given in Appendix 1. For $\sigma < \varkappa_\nu$ the half-space radiates like a blackbody. In our case this condition is satisfied in the frequency $h\nu < kT$.

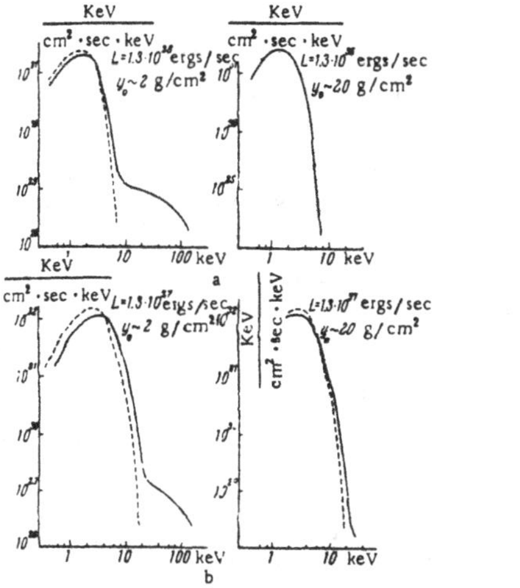

Fig. 2. Energy distribution in the spectrum of a neutron star, computed for selected values of the parameters L and $y_0$. Dashed curves, energy distribution in the spectrum of a black body of equal total luminosity.

It is very difficult to arrive at an exact solution to the problem of the spectrum with allowance for the change in frequency because of scattering. The problem is formulated as an integral equation for a function of the coordinates (x) and the wave vector of a photon; in the plane case the function takes the form $F(x, p, \mu)$, where p is the modulus of the wave vector and $\mu = p_x/p = \cos(\mathbf{p}, x)$.

If there is no change in frequency the function $F_p(x, \mu)$ is found independently for each p, and very efficient methods are known.

We present below an approximate method in which we use the solution to the problem of scattering with a change of frequency in the nonstationary but spatially homogeneous case, that is, for $F(p, t)$.

The exact solution includes a determination of the time $\tau$ that the photons remain in the high-temperature zone; in the geometry adopted here we may speak of the distribution of photons with respect to the quantity $\tau$. We shall replace this distribution below by a single mean value $\bar{\tau}$, and shall use the solution $F(p, \bar{\tau})$ to the homogeneous problem. We shall find the value of $\bar{\tau}$ from energy considerations. It was shown above that even if we do not know the spectrum we can find the energy released by the electrons through Comptonization.

The problem of the Comptonization of radiation in a homogeneous medium for scattering by electrons whose mean energy is many times larger than the mean energy of the photons has an analytic solution [17]. The change in the spectrum depends on a unique quantity $t = \bar{\tau}(kT_e/m_ec^2)$, or in the general case, a variable temperature $t = \frac{k}{m_ec^2}\int_{t_1}^{t_2} T\,dt$.

We can also express in terms of this quantity the change in the total radiant energy per unit volume: $F_2(t_2) = e^{4t}F_1(t_1)$.

In the problem of the radiation of an atmosphere we identify the quantity $F_2$ with the total radiant flux emerging outward from the hot layer, and the quantity $F_1$ with the radiant flux generated in the cool half-space and penetrating into the hot layer. We thereby find the effective value of t, which we use to transform the radiation spectrum of the cool layer; following the Comptonization process we finally obtain for the emergent spectrum

$$F_2(\nu) = \int_0^\infty K(\nu, \nu', t) F_1(\nu')\,d\nu', \qquad (2.3)$$

where

$$K = \left(\frac{\nu}{\nu'}\right)^3 \frac{1}{\nu'} \exp\left\{-\frac{\left(\ln\frac{\nu}{\nu'} + 3t\right)^2}{4t}\right\}.$$

The mathematical details are all given in Appendix 2. Qualitatively, the influence of the Comptonization on the hot electrons consists in redistributing the photons: the number of soft photons decreases and the number of hard photons increases at their expense.

However, in the region of still harder photons the bremsstrahlung of the hot layer plays the leading part, as is evident from Fig. 2.

## 3. PLASMA RADIATION

In the optical, infrared, and radio regions the radiation of heated ionized gas is weak simply because it cannot exceed the equilibrium radiation in accordance with the Rayleigh-Jeans equation: $h\nu \ll kT$. Despite the high temperature, the total flux of low-frequency radiation is small, since the surface of the neutron star is small; that a neutron star may be visible in the x-ray region but not optically has been pointed out some time ago (Ambartsumyan and Saakyan [18]).

On the other hand, evidently it is in the low-frequency range that coherent radiation mechanisms may play a decisive role. In this connection the question of plasma instabilities should be considered in the phenomenon as pictured here. It is well recognized that such instabilities will appear for practically any deviations from statistical equilibrium. In the spherical problem there is a compelling factor in the incident stream that serves to induce a departure from a Maxwellian distribution: in free fall a plasma volume element will become elongated along the radius and will contract in the perpendicular direction. In the collisionless problem the distribution in momentum space will become anisotropic (this remark is due to Sagdeev, who predicted it with reference to the "solar wind").

In our case, however, collisions of electrons with photons, the Compton effect, will not only maintain the electron temperature in the incident stream at a constant level, of order 1 keV, but will moreover ensure an isotropic Maxwellian distribution for the electrons. We may assume that no strong plasma oscillations are present in the incident stream.

Under the influence of the incident stream, strong plasma oscillations would probably be generated in the atmosphere, materially reducing the mean free path of the protons (see Sec. 1b). These oscillations should be transformed into electromagnetic waves with considerable efficiency. In the zone where the protons are decelerated the density changes from $5 \cdot 10^{-7}$ to $3 \cdot 10^{-3}$ g/cm$^3$, corresponding to $y_0 \approx 2$ g/cm$^2$. The plasma frequency changes from $4.4 \cdot 10^{12}$ to $3.4 \cdot 10^{14}$ Hz; $\lambda$, from $7 \cdot 10^5$ to 8000 Å.

The optical depth in the incident stream will be greater than unity for frequencies $\nu \lessdot 10^{14}$ Hz, so that the shortest-wave region will be of interest; rough estimates indicate that some of the optical radiation may perhaps be of plasma origin, with this portion being confined to a negligible solid angle.

It remains unclear whether the plasma oscillations in the atmosphere will generate fast electrons, and what the subsequent rate of these electrons would be. We may presume that the oscillations serving to reduce the mean free path of the protons would simultaneously reduce the path of the electrons as well, impeding the escape of fast electrons toward the stream and reducing the electron heat conductivity.

In this investigation we shall assume that the luminosity is significantly lower than the critical value, so that the influence of the outgoing photon flux on the incident particle stream will be small. Thus we shall not consider here the question of the possible instability of the stream for $L \approx L_c$ as formulated previously [6]. Cameron and Melrose [12] regard this instability as the probable cause of the generation of particles and radio emission in the compact Crab Nebula source.

## 4. INTERACTION BETWEEN THE EMERGENT RADIATION AND THE INCIDENT STREAM

Let us estimate the optical depth $\tau$ in the incident stream relative to Compton scattering. (For photons of energy $E \geq 1$ keV the absorption is negligible.)

Using the free-fall law $v \approx r^{-1/2}$ and the conservation law $\rho v r^2 = \text{const}$ for the stream, we obtain the density distribution $\rho \sim r^{-3/2}$ in the incident stream.

We now have for $\tau$

$$\tau = \sigma_0 \int_R^\infty n(R) \left(\frac{R}{r}\right)^{3/2} dr = 2\sigma_c n(R) R. \tag{4.1}$$

Thus $\tau = 0.4$ for $L = 1.3 \cdot 10^{37}$ ergs/sec, and is determined primarily by the layers that are located in the immediate vicinity of the stellar surface. Clearly the spectral energy distribution will remain practically unchanged.

However, this is fully adequate for the Compton effect to hold the temperature in the incident stream to $T \approx T_{ph} \approx 10^7$°K out to distances $r \approx 10^{10}$ cm.

The temperature $T_{ph}$ will be determined not by the integrated radiation density at a given point [see Eq. (1.3)], which is greatly weakened by dilution, but by the spectral composition of the incoming radiation. Let us now set up the energy-balance equation. For 1 g of incident material we have

$$R^* \frac{dT}{dt} = 6{,}5\varepsilon (T_{ph} - T) - 5.10^{20} \sqrt{T} \rho + R^* \frac{T}{\rho} u \frac{d\rho}{dr}, \tag{4.2}$$

$R^* = 8.3 \cdot 10^7$ ergs/mole · deg.

Here in addition to Compton scattering we have included bremsstrahlung and the change in energy due to isothermal compression of the material during the fall. We may rewrite Eq. (4.2) in the following way:

$$\frac{dT}{dt} = \frac{10^{19}}{r^2} \frac{L}{L_c} (T_{ph} - T) - 10^{12} \sqrt{T} \rho + \frac{T}{r} u. \tag{4.3}$$

By direct substitution one can readily show that out to distances $r \approx 10^{10}$ cm the term for the Compton effect will be decisive. Supersonic flow will not be destroyed in the process. We can estimate the bremsstrahlung $L' = 10^{-27} T^{1/2} \mathrm{EM}$ of this layer:

$$\mathrm{EM} = 4\pi \int_R n^2(R) \left(\frac{R}{r}\right)^3 r^2 dr = 4\pi n^2(R) R^3 \ln \frac{r}{R} \tag{4.4}$$

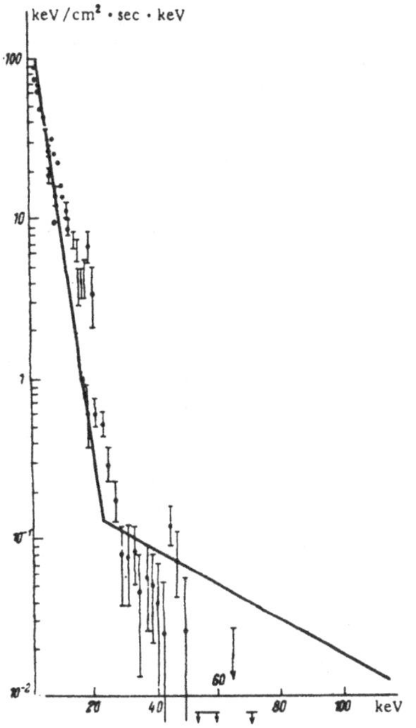


Fig. 3. Energy distribution in the spectrum of Sco XR-1 [19]. The solid line in the radiative flux from a neutron star with $L \sim 10^{37}$ ergs/sec, $y_0 \sim 2$ g/cm², and a distance of $\sim 320$ pc.

for $r \approx 10^{10}$ cm; EM $\approx 10^{54}$ cm$^{-5}$, and $L' \approx 3 \cdot 10^{30}$ ergs/sec, amounting to $\approx 10^{-7}$ of the total luminosity of the neutron star.

For $r > 10^{10}$ cm the energy balance and radiation of the rarefied incident gas will depend appreciably on the assumptions regarding the distant regions. Are we concerned with accretion from the interstellar gas, with gas flowing out of the secondary component of a binary star, or with gas that has previously been ejected in a supernova explosion? We may mention the possibility that emission lines of H, He II, N III, and C III may appear in this region, as observed in the spectrum of the optical object identified with Sco X-1. We hope to undertake a detailed analysis of these assumptions in forthcoming papers.

## 5. DISCUSSION OF THE RESULTS

The x-ray spectra obtained above depend on two parameters, the luminosity L (or the mass flux dM/dt associated with it) and the distance $y_0$ for deceleration of the stream. The general form of the spectrum is such that by assigning parameters we can obtain a satisfactory agreement with experiment (for example, for Sco X-1, $L \approx 10^{37}$ ergs/sec and $y_0 \approx 2$ g/cm² at a distance $\approx 320$ pc; see Fig. 3). A summary of the data has been given in Morrison's survey [19].

However, an agreement of this kind cannot be considered a proof that the model is correct, because of the idealized formulation of the problem and the simplifying assumptions. We may recall the most important aspects: spherical symmetry, the absence of a magnetic field, the shortening of the path by plasma oscillations, and the neglect of electron heat conductivity because of the same plasma oscillations. It is these simplifications, not the rough character of the mathematical calculations, that limit the reliability and accuracy of the conclusions.

The main result of this investigation is that for an idealized model it has been possible to find a systematic, internally consistent solution in which the temperature distribution and the distribution and spectrum of the electromagnetic radiation are taken into account. A realistic theory of accretion, including also the remote region responsible for the amount of mass flow, would utilize the results described above, although such a theory might yield different predictions regarding the spectrum.

Only at the end of such a route might one be able to ascertain whether the observed point x-ray sources are in fact neutron stars, and whether neutron stars are entitled to pass from the realm of hypothetical objects into the class of reliably identified kinds of stars.

Far more detailed study ought to be accorded accretion by white dwarfs and by collapsing stars, as well as nonthermal mechanisms for accelerating electrons which might lead to the emission of x rays.

We would like in conclusion to point out the possibility that $\gamma$ photons might be emitted in an accretion process, in principle enabling neutron stars to be distinguished from other point x-ray sources. As first indicated in previous papers [7], if a neutron star has the maximum possible potential the free-fall energy will be sufficient for $\pi$ mesons to be generated, and their decay can yield $\gamma$ photons of energy 20-60 MeV. It would also be possible for $\gamma$ photons of energy $\approx 2$ MeV to be emitted through the reaction of proton capture by a neutron, which in turn would knock an incident proton out of a $He^4$ nucleus.

Calculations show [19] that the maximum possible ratio of the energy radiated in the form of

$\gamma$-photons (20–60 MeV) to the x-ray luminosity is $\approx 10^{-4}$. For Sco X-1 the flux of $\gamma$ photons at the earth would be nearly two orders of magnitude below the sensitivity of currently available equipment.

## APPENDIX 1

### RADIATION OF AN ISOTHERMAL HALF-SPACE FOR $\sigma \gg \varkappa_\nu$

We shall use the transfer equation

$$\mu \frac{dI_\nu}{dy} = (\varkappa_\nu + \sigma) I_\nu - \sigma J_\nu - \varkappa_\nu B_\nu. \quad \text{(A1.1)}$$

Here $J_\nu = \frac{1}{4\pi}\int\limits_{4\pi} I_\nu d\Omega$ is the mean intensity at a given point and $B_\nu$ is the Planck function.

The transfer equation may evidently be used in this form if the scattering is isotropic and occurs without change of frequency. Moreover, operating in the standard manner for a calculation in the Eddington approximation, we introduce $H_\nu = \frac{1}{4\pi}\int I_\nu \mu d\Omega$ and $K_\nu = \frac{1}{4\pi}\int I_\nu \mu^2 d\Omega$, obtaining from Eq. (A1.1)

$$\frac{dH_\nu}{dy} = \varkappa_\nu (J_\nu - B_\nu),$$
$$\frac{dK_\nu}{dy} = (\sigma + \varkappa_\nu) H_\nu \approx \sigma H_\nu. \quad \text{(A1.2)}$$

We shall assume that the relation $K_\nu \approx J_\nu/3$ is satisfied for the whole atmosphere; the second equation may then be written as $dJ_\nu/dy = 3\sigma H_\nu$, and from the system of two first-order equations we obtain the equation

$$\frac{d^2 J_\nu}{dy^2} = 3\sigma\varkappa_\nu (J_\nu - B_\nu). \quad \text{(A1.3)}$$

An isothermal atmosphere in a gravity field is distributed according to the barometric law

$$\rho = \rho_1 + y/h, \quad \text{(A1.4)}$$

where $h = kT/\mu_e m_p g$ is the scale height of the atmosphere and $\rho_1$ is the density at the boundary of the isothermal half space. If we introduce in Eq. (A1.3) the density $\rho$ in place of y as the variable, we have

$$\frac{d^2 J_\nu}{d\rho^2} = 3\sigma\varkappa_\nu h^2 (J_\nu - B_\nu). \quad \text{(A1.5)}$$

We may rewrite Eq. (A1.5) as follows:

$$\frac{d^2 J_\nu}{d\rho^2} = \frac{3\sigma\varkappa_\nu h^2}{\rho}\rho (J_\nu - B_\nu) \quad \text{(A1.6)}$$

setting $k_\nu = 3\sigma\varkappa_\nu h^2/\rho$. Evidently $k_\nu$ will be constant for a given frequency and will not depend on the density.

Introducing now the variable $\xi = k_\nu^{1/3}\rho$, we obtain the Airy equation

$$\frac{d^2 J_\nu}{d\xi^2} = \xi (J_\nu - B_\nu), \quad \text{(A1.7)}$$

whose solution may be written in terms of the Airy function:

$$J_\nu = B_\nu + CA_i(\zeta) + DBi(\zeta). \quad \text{(A1.8)}$$

D = 0 because of the condition that the solution be bounded as $y \to \infty$.

Using the condition $H_\nu(0) = J_\nu(0)/2$, we can determine C and the emergent radiant flux

$$F_\nu = \pi B_\nu \frac{4k_\nu^{1/3}}{2k_\nu^{1/3} - 3\sigma h \dfrac{A_i(k_\nu^{1/3}\rho_1)}{A_i'(k_\nu^{1/3}\rho_1)}}. \quad \text{(A1.9)}$$

As $\rho_1 \to 0$, $A_i/A_i' \approx -1.4$; for large frequencies $2k_\nu^{1/3}$ will be small, and $F_\nu \approx (3\varkappa\nu/\rho\sigma^2 h)^{1/3}\pi B_\nu$.

## APPENDIX 2

### CHANGE IN FREQUENCY THROUGH COMPTONIZATION

Following Kompaneets [14] and Weymann [15], we may write the interaction of photons with electrons for $kT_e \gg h\nu$ as a kind of diffusion equation in frequency space, for a population number $n = I_\nu c^2/2h\nu^3$:

$$\frac{\partial n(x,t)}{\partial t} = \frac{1}{x^2}\frac{\partial}{\partial x} x^4 \frac{\partial n}{\partial x}, \quad x = \frac{h\nu}{kT_e}. \quad \text{(A2.1)}$$

The total number of photons will be $N = a\int\limits_0^\infty n x^2 dx$. Multiplying both members of Eq. (A2.1) by $x^2$ and integrating with respect to x, we obtain $dN/dt = 0$, so that the total number of photons will be conserved in the scattering process.

The expression for the total energy is $E = a\int\limits_0^\infty n x^3 dx$. Multiplying both sides of the equation by $x^3$ and integrating with respect to x, we obtain $E_2 = E_1 e^{4t}$, the change in the total energy through Comptonization. Knowing $\Delta E = E_2 - E_1$, we can determine the effective value of t from the total energy balance (see Sec. 2). We then proceed in the manner described elsewhere [17].

In order to obtain an analytic expression for the change in energy at each frequency of the spec-

...rum, we introduce in place of x the variable y = ...x, and then transform to the variable z = 3t + y.

After this replacement Eq. (A2.1) reduces to the heat-conductivity equation

$$\frac{\partial n}{\partial t}=\frac{\partial^2 n}{\partial z^2}. \quad (A2.2)$$

The solution of Eqs. (A2.1, A2.2) has the form

$$n(z,t)=\frac{1}{\sqrt{4\pi t}}\int_{-\infty}^{\infty} n(y)\exp-\left\{\frac{(\ln x+3t-y)^2}{4t}\right\}dy. \quad (A2.3)$$

We finally obtain for the intensity

$$I(\nu)=\frac{1}{\sqrt{4\pi t}}\int_0 I(\nu')\left(\frac{\nu}{\nu'}\right)^3\exp-\left\{\frac{\left(\ln\frac{\nu}{\nu'}+3t\right)^2}{4t}\right\}\frac{d\nu'}{\nu'}. \quad (A2.4)$$

## LITERATURE CITED

1. G. S. Bisnovatyi-Kogan, Astron. Zh., 43, 89 (1966) [Sov. Astron. – AJ, 10, 69 (1966)].
2. W. D. Arnett, Canad. J. Phys., 45, 1621 (1967); 44, 2553 (1966).
3. H. Y. Chiu, Ann. Phys., 26, 364 (1964).
4. Chao Wai-yin, Astrophys. J., 147, 664 (1967).
5. Ya. B. Zel'dovich, Dokl. Akad. Nauk SSSR, 155, 67 (1964) [Sov. Phys. – Dokl., 9, 195 (1964)].
6. Ya. B. Zel'dovich and I. D. Novikov, Dokl. Akad. Nauk SSSR, 158, 811 (1964) [Sov. Phys. – Dokl., 9, 834 (1965)].
7. Ya. B. Zel'dovich and I. D. Novikov, Usp. Fiz. Nauk, 86, 447 (1965) [Sov. Phys. – Usp., 8, 522 (1966)]; Relativistic Astrophysics [in Russian], Nauka, Moscow (1967).
8. I. D. Novikov and Ya. B. Zel'dovich, Suppl. Nuovo Cimento, 40, 810, 19 (1966).
9. I. S. Shklovskii, Astron. Zh., 44, 930 (1967) [Sov. Astron. – AJ, 11, 749 (1968)]; Astrophys. J., 148, L1 (1967).
10. A. G. W. Cameron and M. Mok, Nature, 215, 464 (1967).
11. W. C. Saslaw, Mon. Not. Roy. Astron. Soc., 138, 337 (1968).
12. D. B. Melrose and A. G. W. Cameron, Preprint (1968).
13. Ya. B. Zel'dovich, Report to 13th Gen. Assembly, IAU, Prague (1967).
14. A. S. Kompaneets, Zh. Eksp. Teor. Fiz., 31, 876 (1956) [Sov. Phys. – JETP, 4, 730 (1957)].
15. R. Weymann, Phys. Fluids, 8, 2112 (1965).
16. A. A. Vedenov, E. P. Velikhov, and R. Z. Sagdeev, Usp. Fiz. Nauk, 73, 701 (1961) [Sov. Phys. – Usp., 4, 332 (1961)].
17. Ya. B. Zel'dovich and R. A. Syunyaev, Usp. Fiz. Nauk (in press).
18. V. A. Ambartsumyan and G. S. Saakyan, Voprosy Kosmog., 9, 91 (1963); Astron. Zh., 37, 193 (1960).
19. P. Morrison, Ann. Rev. Astron. Astrophys., 5, 325 (1967).
20. V. F. Shvartsman, Astrofizika (in press).

690 NATURE. VOL. 223. AUGUST 16. 1969

# Galactic Nuclei as Collapsed Old Quasars

by

D. LYNDEN-BELL

Royal Greenwich Observatory,
Herstmonceux Castle, Sussex

**Powerful emissions from the centres of nearby galaxies may represent dead quasars.**

RYLE gives good evidence[1] that quasars evolve into powerful radio sources with two well separated radio components, one on each side of the dead or dying quasar. The energies involved in the total radio outbursts are calculated to be of the order of $10^{61}$ erg, and the optical variability of some quasars indicates that the outbursts probably originate in a volume no larger than the solar system. Now $10^{61}$ erg have a mass of $10^{40}$ g or nearly $10^7$ Suns. If this were to come from the conversion of hydrogen into helium, it can only represent the nuclear binding energy, which is 3/400 of the mass of hydrogen involved. Hence $10^9$ solar masses would be needed within a volume the size of the solar system, which we take to be $10^{15}$ cm (10 light h). But the gravitational binding energy of $10^9$ solar masses within $10^{15}$ cm is $GM^2/r$ which is $10^{62}$ erg. Thus we are wrong to neglect gravity as an equal if not a dominant source of energy. This was suggested by Fowler and Hoyle[2], who at once asked whether the red-shifts can also have a gravitational origin. Greenstein and Schmidt[3], however, earlier showed that this is unlikely because the differential red-shift would wash out the lines. Attempts to avoid this difficulty have looked unconvincing, so I shall adopt the cosmological origin for quasar red-shifts. Even with this hypothesis the numbers of quasar-like objects are very large, or rather they were so in the past. I shall assume that the quasars were common for an initial epoch lasting $10^9$ yr, but that each one only remained bright for $10^6$ yr, and take Sandage's estimate (quoted in ref. 4) of $10^7$ quasar-like objects in the sky down to magnitude 22. This must represent a snapshot of the quasar era, so only one in a thousand would be bright. If these represent all the quasar-like objects that there are, then the density of dead ones should be $10^7 \times 10^3 = 10^{10}$ per Hubble volume. The distance between neighbouring dead ones is then an average of $10^{-3}$ Hubble distances ($10^{10}$ light yr) or 3 Mpc. From these statistics it seems probable that a dead quasar-like object inhabits the local group of galaxies and we must expect many nearer than the Virgo cluster and $M87$. If we restrict ourselves to old quasars bright at radio wavelengths, then Sandage reduces his estimate by a factor of 200 so the average distance between dead quasars is around 20 Mpc. This is typical of the distance between clusters of galaxies.

If some $10^7$–$10^8$ solar masses were involved in the quasar, releasing $10^7$ solar masses as energy, then the dead quasar is likely still to be in the range $10^7$–$10^8$ solar masses and to be bound still within a radius of the size of the solar system. Such an object is unlikely to exist for $10^{10}$ yr without burning out its nuclear fuel. There are no equilibria for burnt-out bodies of masses considerably in excess of a solar mass, however. Even uniform rotation hardly increases Chandrasekhar's critical mass of about 1·4 $M_\odot$ and non-uniform rotation always leads to the generation of magnetic fields and to angular momentum transport. For masses already of the size of the solar system such periods of angular momentum transport will not be very long. In a few thousand years the outer parts acquire a large fraction of the angular momentum and slow down in circular orbit while the more massive inner portion contracts and spins faster. This central portion will collapse and finally fall within its Schwarzschild radius and be lost from view. Nothing can ever pass outwards through the Schwarzschild sphere of radius $r = 2\ GM/c^2$, which we shall call the Schwarzschild throat. We would be wrong to conclude that such massive objects in space-time should be unobservable, however. It is my thesis that we have been observing them indirectly for many years.

## Effects of Collapsed Masses

As Schwarzschild throats are considerable centres of gravitation, we expect to find matter concentrated toward them. We therefore expect that the throats are to be found at the centres of massive aggregates of stars, and the centres of the nuclei of galaxies are the obvious choice. My first prediction is that when the light from the nucleus of a galaxy is predominantly starlight, the mass-to-light ratio of the nucleus should be anomalously large.

We may expect the collapsed bodies to have a broad spectrum of masses. True dead quasars may have $10^{10}$ or $10^{11}$ $M_\odot$ while normal galaxies like ours may have only $10^7$–$10^8$ $M_\odot$ down their throats. A simple calculation shows that the last stable circular orbit has a diameter of $12\ GM/c^2 = 12m$ so we shall call the sphere of this diameter the Schwarzschild mouth. Simple calculations on circular orbits yield the following results, where $M_7$ is the mass of the collapsed body in units of $10^7$ $M_\odot$, so that $M_7$ ranges from 1 to $10^4$.

Circular velocity

$$V_c = [GM/(r-2m)]^{1/2} \text{ where } r > 3m \qquad (1)$$

Binding energy of a mass $m^*$ in circular orbit

$$m^*\varepsilon = m^*c^2\{1-(r-2m)[r(r-3m)]^{-1/2}\} \qquad (2)$$

Angular momentum of circular orbit per unit mass

$$h = [mc^2 r^2/(r-3m)]^{1/2} \qquad (3)$$

The maximum binding energy in circular orbit is

$$m^*c^2(1-2\sqrt{2}/3) = 0{\cdot}057\ m^*c^2 \sim m^*c^2/18 \qquad (4)$$

which occurs at $r = 6m$ and $h = \sqrt{12}\ mc$ (equation (5)). This orbit is also the circular orbit of least angular momentum. The period of the circular orbit as seen from infinity is $(2\pi r/V_c)(1-[2m/r])^{-1/2} = 2\pi(r^3/GM)^{1/2}$ (equation (6)). The maximum wavelength change toward the blue visible from infinity is $\lambda_e/\lambda = 2^{1/2}$ for a stable circular orbit while for a stable parabolic orbit it is $\lambda_e/\lambda = 1+2^{-1/2}$. For parabolic orbits that will disappear down the Schwarzschild throat the greatest blueward change is seen from $r = 27m/8$ when $\lambda_e/\lambda = 1{\cdot}77$. There is no possibility of synchrotron-like blue-shifts.

Numerical values are: $V_c = 200\ (M_7/r)^{1/2}$ km s$^{-1}$ at $r$ pc. $12m$ diameter of Schwarzschild mouth $= 1{\cdot}6 \times 10^{13}\ M_7$ cm $\simeq M_7$, AU. Roche limit for a star of density $\rho^*$ g cm$^{-3}$ $(R) = 4{\cdot}1 \times 10^{13}(M_7/\rho^*)^{1/3}$ cm $= 2{\cdot}8\ (M_7/\rho^*)^{1/3}$ AU. Greatest swallowable angular momentum per unit mass $h_s = 4mc = 0{\cdot}5\ M_7$ pc km s$^{-1}$. Once the initial very low angular momentum stars have been swallowed (those with $h < h_s$) the star swallowing rate rapidly declines to a negligible trickle of about $10^{-8}$ stars yr$^{-1}$. Because the Roche limit is near the Schwarzschild mouth we may likewise neglect the tearing apart of stars by tides. Spectroscopic velocity dispersion of the same energy as the circular orbit $\sigma^2 = \frac{1}{3}V_c^2$, that is, $\sigma = 116\ (M_7/r)^{1/2}$ km s$^{-1}$. Period of circular orbit at $r$ pc $= 3 \times 10^4\ (r^3/M_7)^{1/2}$ yr.

It is thus by no means impossible that there are collapsed masses in galactic nuclei. Considerable support for the notion comes from a detailed consideration of what happens when a cloud of gas collects in the galactic nucleus. (This was first considered by Salpeter, who also derived the 0·057 $Fc^2$ power output[5] which is now described.)

## Gas Swallowing

The total mass loss from all stars in a galaxy will be roughly 1 $M_\odot$ per year. A fraction of this accumulates in galactic nuclei, which are the centres of gravitational attraction. There is dissipation when gas clouds collide, due to shock waves that radiate the energy of collision. For a given angular momentum the orbit of least energy is circular, so we must expect gas to form a flat disk held out from the centre by circular motion. Such a differentially rotating system will evolve due to "friction" just as described earlier for a dying quasar. Nothing happens in the absence of friction so the energy is liberated via the friction. In the cosmic situation molecular viscosity is negligible and it is most probable that magnetic transport of angular momentum dominates over turbulent transport just as it does in Alfvén's theory of the primaeval solar nebula. To give a sensible model of the swallowing rate we must estimate the magnetic friction and investigate what happens to the energy acquired by the magnetic field. Before doing this let us assume a conservative swallowing rate of $10^{-3}$ $M_\odot$ per year and work out the power available through magnetic friction. We assume that this mass flux is processed down through the circular orbits until it reaches the unstable orbit at $r=6m$. We shall assume that the flux is swallowed by the throat without further energy loss. The power for a mass flux $F$ is then 0·057 $Fc^2 = 3\cdot5\times10^{42}$ erg s$^{-1} \simeq 10^9$ $L_\odot$ where the values are for $F=10^{-3}$ $M_\odot$ yr$^{-1}$ and $L_\odot$ is the power of the energy output of the Sun. This sort of power emitted as light could just noticeably brighten a nucleus. If a fraction were emitted in the radio region the nucleus would be a radio source. Clearly it only requires a mass flux of 1 $M_\odot$ yr$^{-1}$ and a conversion into light at 10 per cent efficiency for the nucleus to equal the stellar light output from the whole galaxy. Can this be the explanation of the Seyfert galaxies?

We now return to the model for the magnetic transfer of angular momentum in a disk. As a magnetic field $B$ is sheared by an initially perpendicular displacement. The component across the shear is left unchanged but the component down the shear is progressively amplified. We may therefore expect the magnetic field in a shearing medium to be progressively amplified until it somehow changes either the motion or its own configuration. The magnetic field first has a significant effect when it can bow upwards and downwards out of the differentially rotating disk leaving the material to flow down the field lines and so collect into clouds in the disk[6]. For significant cloudiness to result, the magnetic field pressure $B^2/8\pi$ must equal the turbulent and gas pressures $\rho c_s^2$. Here $\rho$ is the density and $c_s^2$ the combined velocity dispersion of microscopic and molecular motion. The thickness of the disk $2b$ is determined by the balance between the gravity of the central mass and the non-magnetic pressure, $GMr^{-3}z\rho = -c_s^2\,\partial\rho/\partial z$, which gives

$$\rho = \rho_0 \exp[-z^2/2b^2] \text{ where } b^2 = r^2c_s^2/GM \qquad (7)$$

The local shear rate is given by Oort's constant $A$

$$A = -\tfrac{1}{2} r\,\mathrm{d}(v/r)/\mathrm{d}r = \tfrac{3}{4}(GM/r^3)^{1/2} \qquad (8)$$

$$\text{Hence } \frac{B^2}{8\pi} \simeq \rho_0 c_s^2 = \frac{16}{9}A^2b^2\rho_0 = \left(\frac{16}{9\sqrt{2\pi}}\right)A^2b\Sigma \qquad (9)$$

Here $\Sigma$ is the surface density of matter in the disk, $\Sigma = \sqrt{2\pi}\, b\, \rho_0$.

On multiplication by $b^3$, equation (9) tells us that the magnetic field energy in a volume $b^3$ is about equal to the kinetic energy of the shearing in the same region. Equation (9) may be obtained approximately from a dimensional analysis of a general shearing sheet of highly conducting material. As such, it claims general validity, so it is interesting to apply it to the gas in the neighbourhood of the Sun. The very reasonable result, $B=4\times10^{-6}$ G, restores confidence in this rather inadequate treatment.

We are now in a position to estimate the magnetic frictional force per unit length in a shearing disk. The Maxwell stress $B_x.B_y/(4\pi)$ acts over a thickness of about $2b$. Once the medium has broken up into clouds the magnetic field will be rather chaotic but the shearing will still give it some systematic tendency to oppose the shear. The greatest possible value of $B_xB_y$ for given $|\mathbf{B}|$ is $B^2/2$, so we estimate $B^2/4$ as a typical value in a sense directed to oppose the shear. The force per unit length is then

$$bB^2/8\pi = [16/(9\sqrt{2\pi})]\,A^2b^2\Sigma \qquad (10)$$

Unlike a normal viscous drag, the force here depends on the square of $A$, the rate of shear. This is reasonable because the shear itself is needed to build up the magnetic field which eventually opposes it. We shall use this estimate of the friction to make a model of the disk, but it is important to consider first where the energy goes. We must consider why the field is not further amplified by further shearing once it has reached the value estimated. Once the medium has split into clouds each cloud acts like a magnet. As the medium is sheared these magnets try to re-align themselves to take up a configuration of minimum energy. The intercloud medium, although dominated by the magnetic field, is still a highly conducting medium, however. In free space re-alignment of the magnets involves reconnexion of magnetic field lines through neutral points. This cannot happen in a force-free, strictly perfect, conductor. Rather one may show that neutral sheets develop with large sheet currents flowing through them. This situation does not really happen; any small resistivity causes a return to the reconnecting case and it is of the greatest interest to see how a plasma of particles behaves near a neutral point at which reconnexion takes place. (In the frame in which the neutral point is fixed this involves an electric field $E$. The mechanism of acceleration is basically that of Syrovatskii[7].) Tritton at this observatory has been studying with me the details of an exact model. To reconnect a finite flux in a finite time through a point at which $B=0$, it is clear that the lines of force must move infinitely fast at the neutral point. They are not material lines and there is nothing wrong with this; even the normal formula for their velocity $v=c\,(\mathbf{E}\times\mathbf{B}/B^2)$ gives superluminous velocities when $|\mathbf{E}\times\mathbf{B}|>B^2$ (that is, for $|E|>|B|$ in gaussian units assuming they are perpendicular). The particles gyrate about their field lines until the lines move with the velocity of light; thereafter the electric field predominates, the magnetic field is too weak to change anything significantly and the particles are electrostatically accelerated. The potential drop follows directly from Faraday's law

$$\Phi = -\frac{1}{c}\,\mathrm{d}N/\mathrm{d}t$$

where $\mathrm{d}N/\mathrm{d}t$ is the rate of flux reconnexion. To calculate this e.m.f. we estimate the reconnexion rate as a flux of $2b^2B$ in a time of $(2A)^{-1}$. This gives an e.m.f. of $4b^2AB/c$ (equation (11)). We shall see that these e.m.f.s are of the order of $10^{11}$ V. Because I believe that this can be the primary source of dissipation for the highly conducting disk I deduce that the energy of dissipation can be converted directly into cosmic rays in the GeV range. We have now good reason to believe that galactic nuclei ought to be radio sources because such cosmic rays will clearly radiate by the synchrotron mechanism in the magnetic field. In summary, we expect from a shear $2A$ the generation of a magnetic field given by equation (2), a shearing force given by equation (3), and a power $p$ per unit area dissipated into cosmic rays of energy up to $4b^2AB/c$ given by $p=2AbB^2/8\pi=(32/(9\sqrt{2\pi}))\;A^2b^2\Sigma$ (equation (12)). These formulae have very general application in the astrophysics of the Galaxy, in the early history of the solar nebula and in the origin of peculiar A stars from binaries. There is, however, one caveat: the reconnexion energy only goes into cosmic rays if the density is low enough in the reconnecting region. If the particle being accelerated has a collision before it reaches the r.m.s. velocity, then it cannot run away to high energy; instead the reconnexion energy is dissipated by ohmic heating.

If $m_A$, $m_p$ are the masses of the accelerated particle and the proton, then the condition for acceleration is

$$e\, m_A^{-1} E\, t_s > \left(\frac{3}{2} kT m_p^{-1}\right)^{1/2}$$

where $t_s$ is the time between scatterings of the particles A. This time may be taken from Spitzer's book[8] to be approximately

$$t_s = m_A\, \rho^{-1} T^{3/2} m_A/(m_A+m_p) \text{ seconds}$$

$T$ is the temperature and $\rho$ is the density at the acceleration region. Putting in our expressions for the electric field $\Phi/2b$ we obtain for our disk in c.g.s. units the acceleration condition

$$\rho < 10^{-24}\left(\frac{2m_A}{m_A+m_p}\right) bABT$$

Notice the density for proton acceleration can be 918 times that for electron acceleration.

## The Steady Model Disk

We look for a steady state of gas swallowing with a mass flux $F$. We shall assume that a small fraction of the rotational energy is converted into random motions so that $c_s \propto V_c$. Sensible values would be $c_s = 10$ km s$^{-1}$ when $V_c = 200$ km s$^{-1}$ so we write $c_s = (1/20)\, x\, V_c$, where $x$ is of order unity but might be a weak function of $r$. The couple on the material inside $r$ due to magnetic friction is from equations (10), (8) and (7)

$$g = 2\pi r^2 bB^2/8\pi = \sqrt{2}\pi c_s^2 r^2 \Sigma \qquad (13)$$

Apart from the trickle of angular momentum $\sqrt{12}\, mcF$ into the singularity $g$ must be balanced in a steady state by the inward flux of angular momentum carried by the material. Thus, using equations (3) and (5)

$$g = \sqrt{12}\, mcF + F\,[mc^2r^2(r-3m)^{-1}]^{1/2} \text{ for } r > 6m \qquad (14)$$

that is

$$g \simeq F\,(GMr)^{1/2} \qquad (15)$$

Because $g \propto r^{1/2}$ we see from equation (13) that $\Sigma \propto r^{-3/2} c_s^{-2} \propto r^{-3/2} V_c^{-2} \propto r^{-1/2}$ and that the radial velocity

$$V_r = F\,(2\pi r\Sigma)^{-1} \qquad (16)$$

The power $2\pi r p(r)\delta r$ liberated into heat or cosmic rays in the region between $r$ and $r+\delta r$ is $-F d\epsilon/dr$ where $-d\epsilon/dr$ is the derivative of the binding energy in circular orbit given by equation (2).

$$p(r) = \frac{Fc^2}{4\pi r}\,\frac{m\,(r-6m)}{[r(r-3m)]^{3/2}} \quad \text{for } r \geq 6m \qquad (17)$$

$$p(r) \simeq FGM/(4\pi r^3) \quad \text{for } r \gg m \qquad (18)$$

For a given total mass and radius of the disk, a given Schwarzschild mass and choice of the parameter $x \simeq 1$, we have now a unique model of the disk including the cosmic ray power, the heating per unit area, the magnetic field and the mass flux. Rather than give the total mass and radius of the disk, we shall determine everything in terms of the mass flux. Any chosen maximum radius then determines the total mass within. We shall calculate two temperatures $T_1$ and $T_2$ as follows. $T_1$ is the temperature that the disk would have if it radiated as a black body just the power per unit area that is locally generated. Thus $T_1(r) = (P(r)/(2\sigma))^{1/4}$ where $\sigma$ is Stefan's constant and the factor 2 arises because the disk has two sides. $T_2(r)$ is the ambient black body temperature at a distance $r$ from a source the total power of which is the total power of one disk. Thus

$$T_2(r) = [P/(16\pi\sigma r^2)]^{1/4}$$

We now give the numerical values in terms of the following variables: $F_{-3}$ the mass flux in units of $10^{-3} M_\odot$ per year, $M_7$ the Schwarzschild mass in units of $10^7 M_\odot$, $x$ the ratio of the turbulent velocity to $1/20\, V_c$, $r_0$, $m_0$ the running variable $r$ in units of 1 pc and $GM/c^2$ respectively. From equation (15) couple $g = 4\times10^{48}\, r_0^{1/2} F_{-3} M_7^{1/2} = 2.6\times10^{45}\, m_0^{1/2}\, F_{-3} M_7$ g cm$^2$ s$^{-1}$ ($m_0 \gg 6$). From equation (13) surface density $\Sigma = 0.16\, r_0^{-1/2} x^{-2} F_{-3} M_7^{-1/2} = 240\, m_0^{-1/2}\, x^{-2}\, F_{-3} M_7^{-1}$ g cm$^{-2}$ ($m_0 \gg 6$). $\int^r 2\pi r\Sigma dr = 6.7\times10^{36}\, r_0^{3/2}\, x^{-2}\, F_{-3} M_7^{-1/2}$ g $= 3.4\times10^3\, r_0^{3/2}$ solar masses. From equation (7) $b/r = 0.05x$. $\rho_0 = (2\pi)^{-1/2}\Sigma b = 4.6\times10^{-19} r_0^{-3/2} x^{-3} F_{-3} M_7^{-1/2} = 1.4\times10^{-9} m_0^{-3/2} x^{-3} F_{-3} M_7^{-2}$ g cm$^{-3}$ ($m_0 \gg 6$). From equation (13), $B = 3\times10^{-3}\, r_0^{-5/4}\, x^{-1/2}\, F_{-3}^{1/2} M_7^{1/4} = 2.7\times10^5\, m_0^{-5/4}\, x^{-1/2}\, F_{-3}^{1/2} M_7^{-1}$ G ($m_0 \gg 6$). From equation (18), $p(r) \simeq 0.22\, r_0^{-3} F_{-3} M_7$ ergs cm$^{-2}$ s$^{-1}$ ($r \gg 6m$). Notice that the power is strongly concentrated towards the centre, where we need the accurate relativistic formula (17)

$$p(r) = 0.22\, r_0^{-3}\left[\frac{1-2.6\times10^{-6} M_7 r_0^{-1}}{(1-1.3\times10^{-6} M_7 r_0^{-1})^{3/2}}\right] F_{-3} M_7$$

$$= 2.6\times10^{19}\left[\frac{1-6\, m_0^{-1}}{(1-3\, m_0^{-1})^{3/2}}\right] m_0^{-3} F_{-3} M_7^{-2} \text{ erg cm}^{-2}\text{ s}^{-1}$$

Total power $P = 0.057\, Fc^2 = 3.2\times10^{43} F_{-3}$ erg s$^{-1}$ $\sim 10^9 L_\odot$

$$T_1(r) = 6.7\, r_0^{-3/4}\left[\frac{1-2.6\times10^{-6} M_7 r_0^{-1}}{(1-1.3\times10^{-6} M_7 r_0^{-1})^{3/2}}\right]^{1/4} F_{-3}^{1/4} M_7^{1/4}$$

$$= 3.7\times10^5\, m_0^{-3/4}\left[\frac{1-6\, m_0^{-1}}{(1-3\, m_0^{-1})^{3/2}}\right]^{1/4} F_{-3}^{1/4} M_7^{-1/2} \text{ K} \quad (m_0 \geq 6)$$

$$T_2(r) = 100\, r_0^{-1/2} F_{-3}^{1/4} = 1.6\times10^5\, m_0^{-1/2} F_{-3}^{1/4} M_7^{-1/2} \text{ K}$$

Maximum cosmic ray energy $e\Phi = 1.5\times10^{13}\, r_0^{-3/4}\, x^{3/2}\, F_{-3}^{1/2} M_7^{3/4} = 10^{18}\, m_0^{-3/4}\, x^{3/2}\, F_{-3}^{1/2}$ eV. Period of circular orbit (seen from infinity) $3\times10^4\, r_0^{3/2} M_7^{-1/2}$ yr $= 9.8\times10^{-6}\, m_0^{3/2} M_7$ yr. Circular velocity $V_c = 200\, r_0^{-1/2}(1-9.7\times10^{-7} M_7 r_0^{-1})^{-1/2} = 3\times10^5\, m_0^{-1/2}(1-2\, m_0^{-1})^{-1/2}$ km s$^{-1}$ ($m_0 > 6$). Radial velocity $V_r = 0.2\, r_0^{-1/2}\, x^2\, M_7^{1/2} = 3\times10^2\, m_0^{-1/2}\, x^2$ km s$^{-1}$ ($m_0 \gg 6$). Condition for electron acceleration is $\rho < 3\times10^{-20}\, T_4\, r_0^{-7/4} x^{1/2} F_{-3}^{1/2} M_7^{3/4}$ and for proton acceleration $\rho < 3\times10^{-17}\, T_4$, etc., where $T_4$ is the temperature in the acceleration region in units of $10^4$ K and $\rho$ is the density. Acceleration will be between the clouds so at any region of the disk we should probably take $\rho_0$ 10 for $\rho$ (and $T_4 \sim 1$ unless the ambient temperature $T_1$ is greater.)

Diameter of Schwarzschild mouth $12m = 5.7\times10^{-6} M_7$ pc, that is, $m_0 = 12$. Diameter of region producing half the power $44m = 2.1\times10^{-5} M_7$ pc, that is, $m_0 = 44$. Rotational period at that radius $9\times10^{-4} M_7$ yr $= 0.32\, M_7$ days. Inward movement time $r/V_r$ at that radius $1.6\times10^{-1} M_7\, x^{-2}$ yr $\sim 59\, M_7$ days.

It should by now be clear that with different values of the parameters $M_7$ and $F_{-3}$ these disks are capable of providing an explanation for a large fraction of the incredible phenomena of high energy astrophysics, including galactic nuclei, Seyfert galaxies, quasars and cosmic rays. The next section is therefore devoted to predicting the spectra.

## Spectrum

The maximum temperature is at $m_0 = 7.05$ and is $T_1 = 6.6\times10^4\, F_{-3}^{1/4} M_7^{-1/2}$ K. The medium will be optically thick for $\Sigma = 90\, x^{-2}\, F_{-3}\, M_7^{-1}$ g cm$^{-2}$. The disk is in danger of becoming optically thin around $\Sigma = 1$, but there the temperature has fallen to $T_2 \sim 700$ K so dust will take over as a source of opacity (this may not happen for large $M_7$). Our standard model with the parameters all at unity will provide opacity out to about a parsec or so. Because all but the centre of our disk obeys a law $T = Ar^{-2/a}$ with $a = 4$ in the outer parts where $T_2$ is relevant and $a = 8/3$ in the inner parts where $T_1$ is relevant, we study the radiation from disks with such power law tem-

NATURE, VOL. 223, AUGUST 16, 1969 

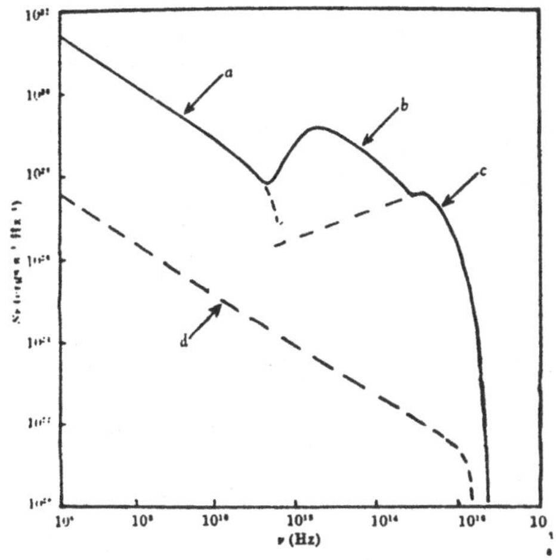


Fig. 1. The emitted spectrum of disk and synchrotron radiation for the standard model. The flux from Sagittarius *A* is weaker by a factor $10^4$, indicating only 1 per cent efficiency of the proton synchrotron. *a*, Proton synchrotron; *b*, outer disk; *c*, central disk; *d*, electron synchrotron.

perature distributions. The total emission at frequency ν is given by

$$S_\nu = \int_0^\infty \frac{c}{4} u_\nu\,(T(r))\,4\pi r \mathrm{d}r = \frac{8\pi^2 h}{c^2}\int_0^\infty \frac{\nu^3\, r\mathrm{d}r}{\exp(h\nu/kT)-1}$$

Writing $x = h\nu/kT = h\nu r^{2a}(kA)$ we find

$$S_\nu = \frac{4\pi^2 h}{c^2}\left(\frac{kA}{h}\right)^a \int_0^\infty \frac{a\,x^{a-1}}{e^x - 1}\,\mathrm{d}x\;\nu^{3-a}$$

where

$$\int_0^\infty \frac{a\,x^{a-1}}{e^x-1}\,\mathrm{d}x = a\,\Gamma(a)\zeta(a)$$

Thus for $a = 8/3$ we have $S_\nu \propto \nu^{1/3}$ while for $a = 4$ we have $S_\nu \propto \nu^{-1}$. Before trying to use these formulae it is important to find out at what radius the main contributions to $S_\nu$ arise. For $a < 1$ the main contributions come from radii close to those for which $h\nu \sim kT$. We may therefore deduce that for our standard model $S_\nu \propto \nu^{-1}$ when 100 K $< h\nu/k <$ 3,000 K and $S_\nu \propto \nu^{1/3}$ when $3\times10^4$ K $< h\nu/k < 10^5$ K, and that for frequencies corresponding to temperatures of $10^5$ K or greater the system shines like a black body of $10^5$ K. In practice it is known that at least for Seyfert galaxies the reddening by dust takes a large fraction of the energy out of the ultraviolet and replaces it in the infrared. Because I have no theory for the amount of dust at each radius I cannot predict the final optical spectrum in detail. But because dust evaporates at a thousand degrees or so it would seem likely that the radiation should be peaked on the red side of the corresponding frequency. Fig. 1 shows the details of the emitted "black body" radiation. It is clear that fluorescence from the ultraviolet will mean that the optical spectrum should be full of emission lines. Those arising from where the disk has ambient temperatures near $2\times10^3$ K come from regions $r_9 \simeq 10^{-3}$ where the circular velocities are $V_c \simeq 6\times10^3$ km s$^{-1}$. These emission lines should therefore be very broad. We may expect that the real disk is not steady, although exact periodicity due to a source in orbit is unlikely. Rather, we should expect variations on a time scale given by $2r/V_r$ at $m = 22$ (120 days) because that is the time scale in which the material flux can vary over the region in which most of the flux is emitted. Using our theory in the most straightforward way it is clear that more power is used in accelerating protons than is used in accelerating electrons. Protons can be accelerated in a density 918 times as great as that in which the electron acceleration can operate. Density in our model behaves like $r^{-3/2}$ while total power between $3/2\,r$ and $1/2\,r$ behaves like $r^{-1}$. We deduce that the proton power is about $10^2$ times the electron power. The steady state spectrum is easily determined. From our accelerator we expect energy proportional to potential drop. Because particles start from all points the energy spectrum ejected by the accelerator is uniform up to the maximum energy, $e\Phi \sim 10^{13}$ eV.

Each particle of energy $E$ radiates at a rate proportional to $E^2$. If there were constant monoenergetic injection into the medium, the flux of particles downwards in energy would be constant. Thus the number per unit area at any $E$ less than the injection energy would follow the law $N(E) = KE^{-2}$ for $E < E_{max} \equiv E_m$. This law is only slightly modified by our uniform injection at all energies up to $E_{max}$. It is

$$N(E) = K\,(1 - E/E_m)\,E^{-2} \quad E < E_m$$

where $K$ is related to the total power of injection power per unit area $p$ by

$$K = \frac{2p}{E_m}\,\frac{9\,m_A^2\,c^7}{4e^4}\,B^{-2}$$

Because the power law is near $E^{-2}$ except close to $E_{max}$, we expect the γ of synchrotron radiation theory to be close to two, and the corresponding spectrum to be close to $S_\nu \propto \nu^{-\alpha}$ with $\alpha = \frac{1}{2}$. It is possible to work out a better approximation using the δ function approximation to the frequency spectrum of a single electron[9]. For our disk model the flux is

$$S_\nu = \int_{r_A}^{\infty}\int_0^{E_m} \frac{2p}{E_m}\,\delta\,(\nu - \nu_m)\left(1 - \frac{E}{E_m}\right)\mathrm{d}E\,2\pi r\,\mathrm{d}r$$

where $r_A$ is the least radius at which the particles can be accelerated and

$$\nu_m = 0{\cdot}07\left(\frac{2}{3}\right)^{1/2}\frac{e}{m_A^3 c^5}BE^2$$

Using our power laws for $p(r)$, $E_m(r) \equiv e\Phi$, $B(r)$, we obtain

$$S_\nu = 3{\cdot}8\times10^{26}\left(\frac{m_A}{m_p}\right)^{12/11}\nu_9^{-7/11}$$

$$\left[1 - \frac{1}{11}\left(14 - 3\left(\frac{\nu_9}{\nu_A}\right)^{1/2}\right)\left(\frac{\nu_9}{\nu_A}\right)^{3/22}\right] x^{-10/11}\,F_{-3}^{5/11}\,M_7^{4/11}$$

where this formula holds for $\nu < \nu_A \equiv \nu_m(r_A)$ and $\nu_9$ is ν in units of GHz. This formula has assumed that all the power dissipated goes into fast protons or electrons as the case may be. In regions where both electrons and protons are accelerated the power should obviously be divided by two. In practice it is probable that $S_\nu$ should be reduced by some efficiency factor because probably only a fraction of the total reconnexion energy really gets into fast particles. The radius of the source at frequency ν is about

$$r_9 = 0{\cdot}7\,\nu_9^{-4/11}\left(\frac{m_A}{m_p}\right)^{-12/11} x^{10/11}\,F_{-3}^{6/11}\,M_7^{7/11},\ \nu < \nu_A$$

Notice that the fast electrons can only be produced much further out than the protons but that they nevertheless produce radiation to much higher frequencies.

## Comparison with Observations

In the Galaxy it is not clear that the circular velocity near the centre falls below 200 km s$^{-1}$, but the OH observations do suggest velocities as low as 100 km s$^{-1}$ within 70 pc of the centre. This indicates a nuclear mass of $M_7 \sim 3$ for the central singularity. The size and flux from Sagittarius *A* are in rough accord with our estimate of the synchrotron spectrum. An infrared flux found at

100 μm could be due to dust from an ultraviolet source radiating $10^9 L_\odot$, so the flux of mass into the throat must be around $F_{-3} = 1$. The general level of activity observed at radio wavelengths close to the nucleus indicates that high energy phenomena are involved[10].

The Magellanic clouds have no nucleus. *M*31 has a strong radio source rather larger but weaker than that found in the Galaxy[11]. Code has discovered strong ultraviolet emission from the nucleus, which Kinman finds to have a large mass-to-light ratio and to contain some $10^8$ solar masses. This suggests a small mass flux into the centre of *M*31 and only a very small ultraviolet disk about the Schwarzschild mouth. *M*32 has a nucleus which is not a radio source but the system is very deficient in gas. We suggest that this system has a Schwarzschild mass but the Galaxy has run out of gas and left it hungry. *M*82 has had a recent violent radio explosion and an infrared nucleus with a small bright radio source in the centre. I suspect $M_7 \sim 3$ and $F_{-3} \sim 10$ but that $F_{-3}$ was larger in the recent past. *M*81 has a very small flux but an intense radio source at its nucleus. I suggest that it is intermediate between the Galaxy and *M*31. *M*87 is the nearest really bright radio galaxy. Luckily the velocity dispersion in its nucleus has been measured[12]. We can therefore measure $M_7$ with some pretence of accuracy to be about $4 \times 10^{10} M_\odot$, that is, $M_7 = 4 \times 10^3$. Over $10^{10}$ yr it would take an $F_{-3}$ of $4 \times 10^3$ to build such an object. This is probably the nearest old dead radio-bright quasar. It is only a shadow of its former self, as $F_{-3}$ has declined severely as the gas has run out. Its electron synchrotron still produces copious X-rays[13], however. *NGC* 4151 is a Seyfert galaxy, and $M_7$ need not be greater than 3, or more likely 30, but $F_{-3}$ is high because there is still much gas in the central regions. Seyfert galaxies that are active have $F_{-3} = 10^3$ but probably are only active at this flux level for one-hundredth of the time. The breadths of the wings of the Balmer lines are 6,000 km $s^{-1}$—I suggest that these are Doppler widths[14]. *NGC* 4151 is a strong infrared source.

## Quasars

When $F_{-3}$ achieves large values $\sim 10^6$ or $10^7$ (that is, $10^{3-4}$ $M_\odot$ $yr^{-1}$) the mass of the Schwarzschild throat rapidly builds up to $10^9$–$10^{10}$ $M_\odot$. When galaxies first formed there was this amount of gaseous material in them. Large proto-galaxies rapidly achieved large Schwarzschild throats and greedily swallowed gas. It is clear that the right energy is available and by making *M* close to $10^{10}$ $M_\odot$ we lower the densities close to the Schwarzschild throat. This allows the radio phenomena to occur closer to the singularity where more of the power is.

*Note added in proof*. Low's recent observations of the galactic centre at 100 μm (reported at the Cambridge conference on infrared astronomy, 1969) suggest that $F_{-3}$ for the galaxy is nearer $10^{-2}$ than 1. A dust model by Rees can explain the infrared observations assuming a single central source of visible or ultraviolet light. The light pressure from such sources will expel dusty material from the nuclei of Seyfert galaxies causing the observed outflow as suggested by Weymann. Such a mechanism could cut off the flux $F_{-3}$ and therefore produce the changes in the emitted flux. The light pressure may drive the dust out of the nucleus so that no dust is ever swallowed. This could leave a great enhancement of dust in the surroundings of the nucleus corresponding to the dust content of all material swallowed in the past. A violently active outburst of such a nucleus would then be associated with the expulsion of great swathes of dust such as those seen across several radio galaxies.

The proton synchrotron radiation discussed here is probably replaced in practice by synchrotron radiation from electron secondaries and *X* and γ-ray bremsstrahlung corresponding to a sizable fraction of the power input into fast protons.

I thank the Astronomer Royal for discussions, and Drs Pagel, Bingham, Tritton, Rowan-Robinson, Weymann and Osterbrock for further help and encouragement.

Received July 8, 1969.

[1] Ryle, M., *Highlights of Astronomy* (edit. by Perek, L.) (D. Reidel, 1968).
[2] Hoyle, F., Fowler, W. A., Burbidge, G., and Burbidge, E. M., *Astrophys. J.*, **139**, 909 (1964).
[3] Greenstein, J. L., and Schmidt, M., *Astrophys. J.*, **140**, 1 (1964).
[4] Schmidt, M., *Texas Conf. Relativistic Astrophys.* (edit. by Maran, S. P., and Cameron, A. G. W.) (1968).
[5] Salpeter, E. E., *Astrophys. J.*, **140**, 796 (1964).
[6] Parker, E. N., *Astrophys. J.*, **149**, 517 (1967).
[7] Syrovatskii, S. I., *IAU Symp. No. 31, Radio Astronomy and the Galactic System* (edit. by Van Woerden, H.), 133 (Academic Press, 1967).
[8] Spitzer, L., *Physics of Fully Ionised Gases* (Interscience, 1955).
[9] Ginzburg, V. L., and Syrovatskii, S. I., *Ann. Rev. Astron. Astrophys.*, **3**, 297 (1965).
[10] Lequeux, J., *Astrophys. J.*, **149**, 393 (1967).
[11] Poolley, G. G., *Mon. Not. Roy. Astron. Soc.*, **144**, 101 (1969).
[12] Brandt, J. C., and Rosen, R. G., *Astrophys. J. Lett.*, **156**, L59 (1969).
[13] Byram, E. T., Chubb, T. A., and Freidman, H., *Science*, **152**, 66 (1966).
[14] Woltjer, L., *Astrophys. J.*, **130**, 38 (1959).

Publ. Astron. Soc. Japan **26**, 429-436 (1974)

# An Accretion Model for the Outbursts of U Geminorum Stars

Yoji OSAKI

*Department of Astronomy, University of Tokyo, Tokyo*

(Received 1974 March 7; revised 1974 April 24)

**Abstract**

A working model is proposed in which the outbursts of U Geminorum stars (or dwarf novae) may be caused by sudden gravitational energy release due to *intermittent* accretion of material onto the white dwarf component of the close binary from the surrounding disk. The intermittent accretion has been hypothesized to be triggered by some kind of instability within the disk. It is demonstrated that this accretion model naturally explains amplitudes of outbursts and the period-amplitude relation for the outbursts. The standstills observed in Z Cam stars are then interpreted as a more or less continuous accretion of matter from the disk.

Key words: Accretion; Close binary; U Gem stars.

## 1. *Introduction*

Dwarf novae (or U Geminorum stars) are stars exhibiting repeated outbursts of the amplitudes 2-6 mag with mean interval 10-200 day. It now seems well-established that probably all dwarf novae are close binaries, in which the cool component fills its critical Roche lobe, while the hot component (presumably a white dwarf) is surrounded by a rotating ring or disk (KRAFT 1962). There had been some controversy as to which component erupts in U Gem stars. KRZEMINSKI (1965) suggested from his observations of U Gem itself that the cool star is the seat of outbursts. Based on this, some attempts were made to explain the outbursts by instabilities of mass loss from the cool components (PACZYNSKI 1965; BATH 1969, 1972; OSAKI 1970). However, recent observations (WARNER and ROBINSON 1972; WARNER 1973b) and reinterpretations of KRZEMINSKI's (1965) observations (WARNER and NATHER 1971; SMAK 1971) strongly indicate the hot star as the seat of outbursts.

The mechanism of outbursts of dwarf novae based on the hot star has not been worked out yet. A popular conjecture is that the U Gem star is a miniature nova and that outbursts are caused by the thermal runaway of the shell-hydrogen burning in the envelope of the white dwarf. The time scale for the thermal runaway is determined essentially by the Kelvin time of the hydrogen-rich envelope. We may get in principle a Kelvin time as short as we desire, by reducing the envelope mass. However, in order to have a thermal runaway, we are constrained by the condition that the hydrogen-rich envelope lies deep enough to burn the hydrogen at its bottom. It seems, therefore, highly unlikely that this model can explain the recurrence periods of U Gem stars as short as 10 day, although it may be the most promising theory for the outbursts of ordinary nova (see, e.g., STARRFIELD et al. 1972). In this paper, we propose an alternative model for the outbursts of U Gem stars.

## 2. *Accretion Model for the Outbursts of U Gem Stars*

### a) *Importance of Accretion onto the Compact Components of Close Binaries*

The recent discovery of several binary X-ray sources throws light on the importance of the accretion of material onto the compact component of a close binary. A widely accepted model for the pulsating binary X-ray sources such as Her X-1 and Cen X-3 is that the material is lost from the larger component and is accreted onto a magnetic neutron star, giving rise to the observed pulsed X-ray (PRINGLE and REES 1972; DAVIDSON and OSTRIKER 1973; LAMB, PETHICK, and PINES 1973). In this model, the X-ray luminosity of the order of $10^{37}$ erg s$^{-1}$ is naturally explained by the mass accretion of $10^{-9} M_{\odot}$ yr$^{-1}$. One of the outstanding features in binary X-ray sources is the X-ray "on" and "off" transition, in particular the 35-day periodicity of Her X-1 in which X-ray intensity is "on" for 11 day and "off" for the remaining 24 day (GIACCONI et al. 1973). Although there is no generally-accepted theory for 35-day periodicity, one of the explanations for the sudden onset of X-rays may be that the sudden accretion of material, which piles up at the Alfvén surface of the neutron star's magnetosphere, is triggered by the precession motion (PINES et al. [1972] quoted by GIACCONI et al. [1973]). The commonness of on-off transitions among binary X-ray sources may indicate that the mass accretion onto a neutron star is not continuous, but more or less intermittent.

Let us now consider what happens if the accreting compact component is a white dwarf, instead of a neutron star. Let us take the mass and the radius of a white dwarf as $1M_{\odot}=2\times10^{33}$ g and $5\times10^{8}$ cm, respectively. If material is accreted onto the white dwarf at rates of $10^{-9}$-$10^{-8} M_{\odot}$ yr$^{-1}$, the luminosity due to accretion is 4-40 $L_{\odot}$, which far exceeds the luminosities of typical white dwarfs. According to KRAFT's (1963) close binary model, the U Gem star is exactly this kind of close binary system. Since the amplitudes of outbursts of U Gem stars are 2-6 mag and their absolute visual magnitude at minimum is between 5 and 10 mag with the mean $\langle M_v \rangle \sim +7.5$ mag (KRAFT and LUYTEN 1965), the luminosity during outbursts is supposed to be 1-$10^{2} L_{\odot}$. Since this agrees quite well with the accretion luminosity estimated above, and since the mass and the accretion rates used above seem to be appropriate for U Gem stars (WARNER 1973a; ROBINSON 1973a) we shall consider more details of this accretion model in the next sub-section. An accretion model for outbursts of dwarf novae is not new. In fact, CRAWFORD and KRAFT (1956) have already suggested that outbursts of nova-like binary AE Aqr are caused by accretion of matter onto the blue component. The importance of accretion onto a white dwarf in the old nova DQ Her was also pointed out by STARRFIELD (1970).

### b) *Intermittent Accretion Model for U Gem Outbursts*

The binary model for the U Gem stars proposed by WARNER and NATHER (1971) and SMAK (1971) is now summarized as follows. The cool component overflows the Roche limit and loses material toward the white dwarf through the inner Lagrangian point. Because of high angular momentum, this material does not fall directly onto the white dwarf but forms a rotating disk or ring around it. The streaming material which flows out of the cool component, collides with material in the disk and forms the "hot spot" (or bright spot). During the quiet stage of the outburst cycle, the hot spot and the rotating disk contribute appreciably to the total light, and possibly most of the luminosity of the system comes from them.

We now advance the following working hypothesis for the outbursts of U Gem stars. "The material is lost from the secondary *at a constant rate* and collected in the rotating disk around the white dwarf. If the material piled up in the disk exceeds some critical amount, it is then hypothesized that some kind of instability in the disk sets in, which involves the exchange of angular momentum within the disk's material, and a considerable amount of mass falls onto the white dwarf with a shorter time scale than that of piling up, resulting in sudden release of gravitational energy. Some of the remaining material may leave the system, carrying away the excees angular momentum. This process then repeats and some sort of steady state is established for material in the disk."

In this paper, we shall not discuss the instability mechanism, but rather study how well this working hypothesis explains observed outbursts of U Gem stars. However, it may be noted that an instability of accretion disk has been discussed recently by LIGHTMAN and EARDLEY (1974). Let us first consider the amplitude of outbursts. In this model, the luminosity during an outburst is determined by the mass accretion rate onto the white dwarf, and it is given by

$$L_{max}=\frac{GM_1}{R_1}Q_1\,, \tag{1}$$

where $Q_1$ is the accretion rate upon the white dwarf during the outburst and $M_1$ and $R_1$ are the mass and the radius of the white dwarf. The luminosity due to the hot spot in minimum is also supposed to come from the gravitational energy released (ROBINSON 1973a), and it is written as

$$L_{spot}\sim\frac{GM_1}{r_{spot}}Q_2\,, \tag{2}$$

where $Q_2$ denotes a mass accumulation rate on the disk from the secondary (assumed to be constant in time) and $r_{spot}$ is the distance of the hot spot from the white dwarf. From the mass continuity, we have a relation

$$Q_1/Q_2\sim\tau_R/\tau_{outburst}\,, \tag{3}$$

where $\tau_R$ and $\tau_{outburst}$ denote the recurrence period of outbursts and the duration of an outburst, respectively. Here we have neglected any small amount of mass leaving the system (see section 4). From equations (1), (2), and (3), we obtain

$$L_{max}/L_{spot}\sim\frac{r_{spot}}{R_1}\frac{\tau_R}{\tau_{outburst}}\,. \tag{4}$$

As already noted, the hot spot contributes appreciably to the total light. For instance, the height of the shoulder of U Gem itself amounts to 0.5 mag so that the hot spot contributes one half of the total luminosity of the system at minimum. In what follows, we set $L_{spot}\sim L_{min}/2$ for simplicity, where $L_{min}$ is the luminosity of the system at minimum. The position of the hot spot is approximately estimated by assuming that material goes into the Keplerian circular orbit around the white dwarf, conserving the angular momentum it has at the inner Lagrangian point. This gives an expression for the radial distance of the spot from the primary:

$$r_{spot}/A \simeq \frac{M_1+M_2}{M_1}\left(\frac{r_L}{A}\right)^4 , \tag{5}$$

where $A$ and $r_L$ denote the separation of centers of the two stars and the distance of the inner Lagrangian point from the primary, respectively, and $M_2$ is the mass of the secondary. From equation (5) we obtain $r_{spot}/A \simeq 1/8$ for $M_1/M_2 \sim 1$. Since we get $A \sim 1.3 \times 10^{11}$ cm with parameters appropriate for U Gem binaries (KRAFT 1962), we estimate $r_{spot}$ to be $\sim 1.6 \times 10^{10}$ cm. If we put $R_1 \sim 5 \times 10^8$ cm and $\tau_R/\tau_{outburst} \sim 10$ for a typical U Gem star (for instance, $\tau_R \sim 100$ day and $\tau_{outburst} \sim 10$ day for U Gem itself), equation (4) gives then

$$L_{max}/L_{min} \sim 160 . \tag{6}$$

This is just the right order of magnitude for observed amplitudes of U Gem stars. The more detailed comparisons of equation (4) with observations of individual stars do not, however, seem to be fruitful, because the contribution of the red star to the total light at minimum varies very much with individual stars.

It is well known that there exists a correlation between the scale of outbursts and the recurrence cycle among U Gem stars. KUKARKIN and PARENAGO (1934, see also, e.g., LEDOUX and WALRAVEN [1958]) proposed the relation

$$\text{Amp.} = 0.63 + 1.667 \log_{10} \tau_R , \tag{7}$$

where amplitudes of outbursts and $\tau_R$ are expressed in magnitudes and days, respectively. This kind of relation naturally follows in our model. Since the recurrence cycle among individual stars varies much more than the durations of outbusts, we simply assume, for instance, $\tau_{outburst} \sim 10$ day. We then obtain with parameters used above

$$L_{max}/L_{min} \sim 1.6\tau_R \quad \text{for} \quad 20 < \tau_R < 200 . \tag{8}$$

This is certainly in satisfactory agreement with equation (7) in view of our over-simplified assumptions and of deviations of individual stars from equation (7). It is, however, more appropriate to consider the total energy output during one outburst instead of the amplitude of the outburst. The relation between the cycle length and the scale of outburst is then written as

$$k = L_{min}\tau_R/E , \tag{9}$$

where $E$ stands for the total energy emitted during one outburst and $k$ should be a constant. The observational value of $k$ has been obtained by ZUCKERMANN (1954) for several U Gem stars, and it is $k_{obs} = 0.1\text{-}0.2$. The quantity $k$ is estimated in our model:

$$k \simeq \frac{R_1}{r_{spot}} \frac{L_{min}}{L_{spot}} \sim 0.06 . \tag{10}$$

By comparing the above two values, we may conclude that our model is not unsatisfactory in explaining the cycle-range relationship in U Gem stars, if we take into account possible uncertanties involved both in observations and in theories.

Let us now consider the "standstills" of Z Cam stars. The light curves of Z Cam stars show, from time to time, curious brightness halts (called standstills) usually on the descending branch at a level between maximum to minimum, and they last from several multiples to several tens of multiples of the regular cycle time of outbursts. The scale of outbursts of Z Cam stars is small as compared with other U Gem stars, and their periods of regular cycle are some 20 day. In our model, Z Cam stars may be regarded as less unstable stars, and the standstill is interpreted as a more or less continuous accretion of matter from the disk onto the white dwarf. The luminosity at the standstill is then given by

$$\left.\begin{aligned} L_{\rm standstill}/L_{\rm min} &\sim \frac{r_{\rm spot}}{R_1}\frac{L_{\rm spot}}{L_{\rm min}} \simeq k^{-1} \sim 16\,, \\ L_{\rm max}/L_{\rm standstill} &\sim \tau_R/\tau_{\rm outburst} \sim 2\text{-}3\,. \end{aligned}\right\} \qquad (11)$$

This is again in reasonable agreement with observations, because observed amplitudes of outbursts for Z Cam stars are two to three magnitudes, and the standstills occur at 0.5-1 mag below the maximum.

## 3. *Interpretations of Other Observational Characteristics of U Gem Stars*

In this section, we shall discuss how other observational characteristics of U Gem stars are interpreted in our accretion model.

a) *Relation of U Gem Stars to Old Novae and Nova-Like Binaries*

In our model, the U Gem phenomenon is hypothesized to be caused by intermittent accretion of matter onto the white dwarf. Obviously, this model cannot be extended as an explanation for the outburst of ordinary novae, in which mass ejection has been observed instead of mass accretion. The outburst of the ordinary nova may rather be caused by a thermal runaway of shell-hydrogen burning in the envelope of the white dwarf.

Old novae and nova-like binaries do not show U Gem-like activities (i.e., small-scale outbursts with recurrence periods of some 50 day), although they share most of the binary characteristics with U Gem binaries. Since the white dwarf components in old novae and nova-like binaries are also supposed to accrete matter, the above-mentioned fact can be understood in our model as indicating that the white dwarf components in old novae and nova-like binaries may *continuously* accrete material from the disk, and that they may correspond in effect to the standstill phase of the Z Cam stars. If so, there must be a qualitative difference in radiation between the quiet phase of U Gem stars and that of old novae (or nova-like binaries). In our model most of the radiation at minimum of an outburst cycle in a U Gem star is considered to come from the hot spot, which is formed by the collision between the disk material and the stream flowing directly out of the secondary, while most of the radiation in an old nova (or a nova-like binary) comes from the gravitational energy release of the disk material which infalls gradually onto the white dwarf.

Obervationally, SMAK (1972) classifies eruptive binaries into two types. According to his classification a Type I system is a system in which the luminosity of the spot is high, and Type II is one in which the luminosity of the spot is low as compared with that of the disk. It is interesting that all of the U Gem stars are classified as Type I, while most of the old novae and the nova-like binaries are classified as Type II, and this is in perfect agreement with the above picture.

In our model there exists a possibility that the material which ends up accreting onto the white dwarf may produce an ordinary nova outburst after some time. Let us give a rough estimate of how many U Gem stars may become nova per year. At the present time about 100 stars have been classified as U Gem stars. If we assume that a close binary of this sort erupts as a common nova once every $\tau_{nova}$ yr, the number of U Gem stars which become nova per year will be $N \sim 100/\tau_{nova}$. Since details of nova outbursts have not been worked out, a reliable value for $\tau_{nova}$ cannot be given. However, since the time-averaged accretion rate onto the white dwarf is $10^{-9} M_{\odot} yr^{-1}$ in our model, and since the hydrogen-rich material ejected by a nova is considered to be about $10^{-4} M_{\odot}$ from observations, we estimate $\tau_{nova} \gtrsim 10^{5}$ yr. With this value we find $N \lesssim 10^{-3}$, which does not contradict the observational fact that no U Gem stars are known to be novae.

b) *Continuum Radiation during Outbursts of U Gem Stars*

Let us consider the nature of radiation during outbursts. In our model an outburst is caused by a sudden accretion of the disk material onto the white dwarf. Since the instability mechanism itself is not now known, details of the accretion process are not clear. We are then forced to adopt one of the simplest models, in which all of the released gravitational energy is assumed to be radiated as a black body at the surface of the white dwarf. If we take $L_{max} \sim 10 L_{\odot}$ and $R_1 \sim 5 \times 10^{8}$ cm as a typical luminosity during outbursts and the radius of the white dwarf respectively, the effective temperature of the white dwarf at the maximum will be $(T_{eff})_{max} \sim 120{,}000$ K, and most of the radiation will therefore be emitted in the far ultraviolet. Although the assumption of black body radiation may not be good, the conclusion that most of the radiation is emitted in the far ultraviolet and possibly in the soft X-ray region is inevitable. However, since the white dwarf component is embeded in the optically thick disk of material, most of this UV radiation will be absorbed in the disk and transformed ultimately into visible radiation when it emerges from the disk.

Photometric colors during outbursts of U Gem stars have been discussed by WALLERSTEIN (1961) and SMAK (1972). It is noted there that the colors of the U Gem stars at light-maximum are very similar to those of supergiants with $T_{eff} \sim 12{,}000$ K in the $B-V$, $U-B$ plane. As SMAK (1972) argued, it may not be unlikely that the optically thick disk mimics a supergiant star in photometeric colors, when it is heated by the strong ultraviolet radiation of the central source. There exists other observational evidence which suggests that most of the visible radiation during outbusts comes from the disk, and not directly from the central white dwarf. In WARNER's (1973b) observations of Z Cha, the eclipse is found to be partial near the maximum of the outburst while it is total at light-minimum. This indicates that the radiation at light maximum comes from a much extended source, and it is unlikely that the white dwarf component expands to that size because there is no spectroscopic evidence for any expansion during the outbursts of U Gem stars.

## 4. *Discussion*

We have proposed a working model for the outbursts of U Gem stars in which the gravitational energy is suddenly released by intermittent accretion of material onto the white dwarf component from the disk. This model naturally explains amplitudes of outbursts and the cycle length-amplitude relation of U Gem outbursts. Obviously, the most serious defect of this theory is that we have failed to present the definite instability mechanism within the disk, which is required to explain intermittent accretion of matter. Because of the same reason we fail to predict the cycle length (or the induction time) of outbursts. Except for this difficulty this working model looks attractive, and we feel it worthwhile to investigate possible instabilities within the disk.

In section 2b, we have neglected mass leaving the system, which is supposed to carry away excess angular momentum of the rotating disk. Recently, ROBINSON (1973b) has discovered spectroscopically expanding circum-system shells in Z Cam itself, and the rate of mass loss has been estimated to be comparable to the mass transfer rate from the secondary to the rotating ring. He suggests that nearly all of the transferred mass is lost from the system and only a small fraction (less than 10 percent) is accreted onto the white dwarf. His suggestion is based on the assumption that the energy of outbursts comes from the nuclear burning of the accreted hydrogen. However, we shall show from an argument of angular momentum that the mass loss from the system must be less than half of the transferred mass. To do so, we adopt a simplified model for the rotating disk in which material is orbiting around the central star with the Keplerian velocity and the gravitational effect of the secondary is neglected. The specific angular momentum of material within the disk is then given by $(GM_1r)^{1/2}$ where $r$ denotes the distance of material from the central star, and the angular momentum flux transferred to the disk from the secondary is $(GM_1r_{\rm spot})^{1/2}(dM/dt)_{\rm transfer}$. We suppose that some amount of mass leaves the system carrying away excess angular momentum, and its rate is written as $(GM_1r_{\rm boundary})^{1/2}|dM/dt|_{\rm loss}$. Here $|dM/dt|_{\rm loss}$ is the mass loss rate from the system and $r_{\rm boundary}$ represents the radius of the outer boundary of the disk from which the material escapes the system. Since the angular momentum flux carried away by the escaping mass must be smaller than that transferred, we may write

$$(GM_1r_{\rm boundary})^{1/2}|dM/dt|_{\rm loss}\lesssim(GM_1r_{\rm spot})^{1/2}(dM/dt)_{\rm transfer}\,. \tag{12}$$

If we reasonably assume $r_{\rm boundary}\sim r_L$, $r_{\rm spot}/A\sim1/8$, and $r_L/A\sim1/2$, we finally obtain

$$|dM/dt|_{\rm loss}\lesssim(r_{\rm spot}/r_{\rm boundary})^{1/2}(dM/dt)_{\rm transfer}\sim(1/2)(dM/dt)_{\rm transfer}\,. \tag{13}$$

Thus we may conclude that the mass loss rate from the system is smaller than one half of that transferred, unless some other mechanism for the ejection of matter is involved.

In almost completing this work, I have noticed a paper by BATH (1973), who has suggested a somewhat similar mechanism for the outbursts of U Gem stars. However, there exists an essential difference between our model and that of BATH (1973). In his model, the outbursts are assumed to be caused by the *unsteady* mass overflow from the secondary, while a steady mass loss from the secondary is assumed in our model.

RAPPAPORT et al. (1974) have recently detected a new source in soft X-rays, and they have identified it with SS Cygni. Their observation was made when SS Cyg was in its optical outburst. If their identification is correct, this may support strongly accretion as the cause of outbursts in U Gem stars.

## *References*

BATH, G. T. 1969, *Astrophys. J.*, **158**, 571.
BATH, G. T. 1972, *Astrophys. J.*, **173**, 121.
BATH, G. T. 1973, *Nature Phys. Sci.*, **246**, 84.
CRAWFORD, J. A., and KRAFT, R. P. 1956, *Astrophys. J.*, **123**, 44.
DAVIDSON, K., and OSTRIKER, J. P. 1973, *Astrophys. J.*, **179**, 585.
GIACCONI, R., GURSKY, H., KELLOGG, E., LEVINSON, R., SCHREIER, E., and TANANBAUM, H. 1973, *Astrophys. J.*, **184**, 227.
KRAFT, R. P. 1962, *Astrophys. J.*, **135**, 408.
KRAFT, R. P. 1963, *Adv. Astron. Astrophys.*, **2**, 43.
KRAFT, R. P., and LUYTEN, W. J. 1965, *Astrophys. J.*, **142**, 1041.
KURARKIN, B. V., and PARENAGO, P. P., 1934, *Perem. Zvezdy Gorki*, **4**, 251.
KRZEMINSKI, W. 1965, *Astrophys. J.*, **142**, 1051.
LAMB, F. K., PETHICK, C. J., and PINES, D. 1973, *Astrophys. J.*, **184**, 271.
LIGHTMAN, A. P., and EARDLEY, D. M. 1974, *Astrophys. J. Letters*, **187**, L1.
LEDOUX, P., and WALRAVEN, Th. 1958, *Handbuch der Physik*, Bd. 51 (Springer-Verlag, Berlin), p. 419.
OSAKI, Y. 1970, *Astrophys. J.*, **162**, 621.
PACZYNSKI, B. 1965, *Acta Astron.*, **15**, 89.
PINES, D., PETHICK, C. J., and LAMB, F. K. 1972, paper presented at the Sixth Texas Symposium on Relativistic Astrophysics.
PRINGLE, J. E., and REES, M. J. 1972, *Astron. Astrophys.*, **21**, 1.
RAPPAPORT, S., CASH, W., DOXSEY, R., McCLINTOCK, J., and MOORE, G. 1974, *Astrophys. J. Letters*, **187**, L5.
ROBINSON, E. L. 1973a, *Astrophys. J.*, **180**, 121.
ROBINSON, E. L. 1973b, *Astrophys. J.*, **186**, 347.
SMAK, J. 1971, *Acta Astron.*, **21**, 15.
SMAK, J. 1972, *Veröff. Reimeis-Sternw. Bamberg*, **9**, 248.
STARRFIELD, S. G. 1970, *Astrophys. J.*, **161**, 361.
STARRFIELD, S., TRURAN, J. W., SPARKS, W. M., and KUTTER, G. S. 1972, *Astrophys. J.*, **176**, 169.
WALLERSTEIN, G. 1961, *Astrophys. J.*, **134**, 1020.
WARNER, B. 1973a, *Monthly Notices Roy. Astron. Soc.*, **162**, 189.
WARNER, B. 1973b, *Sky Telesc.*, **46**, 298.
WARNER, B., and NATHER, R. E. 1971, *Monthly Notices Roy. Astron. Soc.*, **152**, 219.
WARNER, B., and ROBINSON, E. L. 1972, *Nature Phys. Sci.*, **239**, 2.
ZUCKERMANN, M. C. 1954, *Ann. d'Ap.*, **17**, 243.

# Chapter 3 THE STANDARD ACCRETION THEORY

SOVIET ASTRONOMY – AJ VOL. 15, NO. 3 NOVEMBER-DECEMBER, 1971

# HALOS AROUND "BLACK HOLES"

V. F. Shvartsman

Shternberg Astronomical Institute, Moscow
Institute of Applied Mathematics, Academy of Sciences of the USSR
Translated from Astronomicheskii Zhurnal, Vol. 48, No. 3,
pp. 479-488, May-June, 1971
Original article submitted July 20, 1970

"Black holes" – bodies confined within their gravitational radius – will necessarily attract interstellar gas. If $M_{hole} > 0.03\ M_{\odot}$, at least 0.01 $mc^2$ of the infalling material should be converted into radiation. The corresponding luminosity would be of order $10^{32}\,(M/10\,M_{\odot})^{3/2}\,(\rho_{gas}/10^{-24}\,g\cdot cm^{-3})^{1/2}$ erg/sec, and would result from synchrotron radiation by magnetized plasma that would be heated to $T \approx 10^{12}$°K during the infall process. The spectrum would have a very mild slope extending from optical to radio wavelengths. Black holes might be observable as faint optical stars with no lines; they could be distinguished by intensity fluctuations on a time scale $\Delta t \approx 10^{-5}$–$10^{-2}$ sec, with no periodic component whatever. In many cases accretion by massive holes ($10 \leqslant M \leqslant 10^4\ M_{\odot}$) should engender hard radiation (x and/or $\gamma$ rays) exhibiting a flare behavior on a time scale ranging from a few months to tens of years; the peak intensity should be 10-100 times the synchrotron intensity. This phenomenon is associated with the turbulence of the interstellar medium: the angular momentum of the gas would halt the infall near the gravitational radius of the hole, and the momentum would be "annihilated" when the object moves into the adjacent turbulence cell. Only if $M > 10^6\,M_{\odot}$ would the angular momentum of the gas be capable of diminishing the mass falling into an isolated hole (that is, its luminosity). During infall toward a hole of $M \approx 10^5\,M_{\odot}$, the gas will initially remain cool ($T \approx 5000$°K) and will display an emission spectrum similar to the optical spectra of quasars. Because of accretion, a hole in a binary-star system might be observable as a visible secondary component. Accretion by a hole can be distinguished observationally from accretion by a neutron star. Possible candidates for black holes that may actually have been detected include certain type-Dc white dwarfs, the $\gamma$-ray star Sgr $\gamma$-1, the x-ray flare stars Cen X-2 and Cen X-4, and such objects as Sco X-1 and Cyg X-2.

## 1. INTRODUCTION

Perhaps the most interesting implication of general relativity theory is the prediction that the universe may contain masses that are confined within their gravitational radius $r_g = 2GM/c^2$. To an external observer such objects should appear rather like "black holes," drawing matter and radiation into themselves. We will recall that although a collapsed body will of itself radiate nothing at all – neither light, neutrinos, nor gravitational waves – it will nevertheless possess a static gravitational field which will influence its surroundings. Matter that is drawn in will reach $r_g$ only asymptotically, after an infinite time; the region $r < r_g$ could not be observed at all by an external observer, and would thereby "drop out" of our space [1].

How might "black holes" be formed? There are at least five ways:

1) holes may have been present in the universe "from the beginning," that is, left over from the epoch of singularity;

2) holes may have developed from density fluctuations during the prestellar stage;

3) holes may have formed from supermassive first-generation stars;

4) holes may have resulted from the relativistic evolution of close clusters, galaxies, and the like;

5) finally, holes may have developed from ordinary but sufficiently massive stars.

It is highly probable that this last possibility has been realized. The discovery of pulsars, as is now well recognized, has confirmed the old theoretical prediction that massive stars are unstable and should suffer a catastrophic collapse at the end of their evolution. But theory predicts two final states after collapse: a neutron star in the event that the mass of the object is less than 1.5 $M_\odot$; and an uninterrupted infall, or a "collapsed" star, if M > 1.5 $M_\odot$.

Neutron stars have been discovered because of their activity: the generation of relativistic particles and radio waves. Collapsed stars, however, are passive by their very nature. Consequently, it is usually considered that "black holes" might most likely be discovered through their influence on radiating matter: on the motion of the normal component in a binary star [2-4], on stars in globular clusters [5] or galaxies [6], on the motion of the galaxies themselves [7] and so on.

However, the method of the "excluded third" is risky in astronomy. And furthermore, when configurations with an invisible component are selected the objects sought might be left out of the list altogether. As we shall demonstrate in this paper, holes whose mass exceeds the solar mass ought to be surrounded by highly luminous halos.

The question of the energy released through accretion by collapsed bodies was first posed by Zel'dovich [8] and Salpeter [9]. The accretion of gas by "superholes" ($M > 10^7 M_\odot$) located at the centers of galaxies has been discussed by Lynden-Bell [10]. We shall be interested primarily in "stellar" masses, with $10^{-1} M_\odot \leqslant M \leqslant 10^6 M_\odot$. Their luminosity will turn out to be given by the single equation $L \approx 0.1\, c^2\, dM/dt$, while their spectra can be highly diversified.

## 2. INTENSITY OF THE ACCRETION AND FLOW REGIMES

Let us imagine a massive object moving at a velocity u through a gas possessing no angular momentum. On the "back" side of such an object a conically shaped shock wave will be formed in which the gas will lose the component of its velocity perpendicular to the shock front. After compression in the shock wave, particles having sufficiently small impact parameters will fall into the star [1]. One can show that the infall of gas will begin at the characteristic distance

$$r_c = \alpha GM/v_c^2 = \alpha 10^{14} M v_{c(10)}^{-2}\ \mathrm{cm}, \tag{1}$$

the distance at which the potential energy of the particles becomes comparable to the kinetic energy; and the flux of mass at the star will be

$$\frac{dM}{dt} = \beta 10^{11} M^2 v_{c(10)}^{-3} n_c\ \mathrm{g/sec}. \tag{2}$$

In Eqs. (1) and (2) the mass M is expressed in solar masses, the velocity $v_c = (u^2 + a_c^2)^{1/2}$ is expressed in units of 10 km/sec, and the sound velocity $a_c$ in the gas and the density $n_c$ [atoms/cm$^3$] of the material drawn in refer to distances $r > r_c$. The dimensionless coefficients $\alpha$ and $\beta$ are functions of the adiabatic index of the gas and the ratio $u/a$. We may take the rough approximations $\alpha \approx 1$, $\beta \approx 1$. Numerical values for various $\gamma$ and $u/a$ have been given by Salpeter [9]. Historically, Eq. (2) was first obtained by Hoyle, Lyttleton, and Bondi [11-13].

Note that for any reasonable value of the adiabatic index $\gamma$ the diameter of the tail will be $d \approx r_c \gg r_g$, and the pressure will be finite. It therefore seems to us beyond doubt that a symmetrization of the flow should take place during infall, so that for $r \ll r_c$ the motion of the plasma may be considered radial (see Fig. 1).

In the spherically symmetric case, with back pressure absent, a supersonic flow of the gas will be established at $r \ll r_c$, that is, practically free fall [1]. To find the temperature variation, we shall write the second law of thermodynamics, $dE = -p\,dV + dQ$, in the form

$$\frac{3}{2} R^* \frac{dT}{dt} = R^* \frac{T}{\rho} \frac{d\rho}{dt} - 5 \cdot 10^{20} T^{1/2} \rho \varkappa + \frac{dQ'}{dt}. \tag{3}$$

Here R* is the gas constant; the second term on the right-hand side describes the losses of 1 g of plasma to radiation ($\varkappa = 1$ corresponds to bremsstrahlung from a fully ionized plasma), and the third term represents the energy variation due to other nonadiabatic processes. Since $\rho \propto r^{-3/2}$, we have

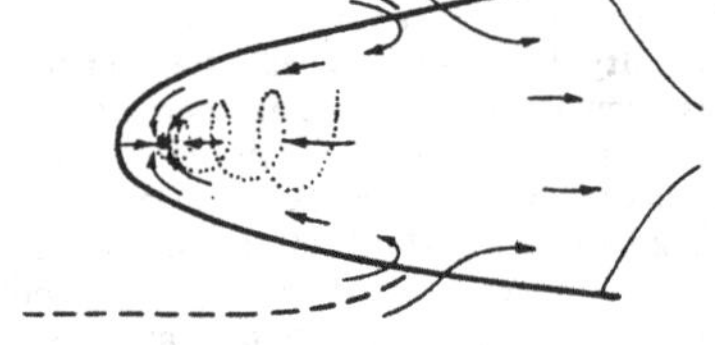


Fig. 1. The pattern of hydrodynamic accretion in the case where the star's own velocity u is much greater than the sound velocity $a$ in the gas. The dashed curves correspond to the critical impact parameter. The dotted curve denotes the trajectory (helical) that the particles describe if an angular momentum relative to the star is present. In the figure the momentum vector is oriented parallel to the direction of motion of the object.

$$\frac{dT}{dr} = -\frac{T}{r} + [2\cdot 10^{-4} M v_{c(10)}^{-3/2} n_c]\frac{\sqrt{T}}{r}\varkappa + \frac{dQ'/dt}{v\cdot \frac{3}{2} R^*}. \quad (4)$$

As it falls in toward a massive object the gas will be efficiently cooled through radiation ($T = [A \ln (r/r_0) + T_0^{1/2}]^2$); when $T \approx 5000°K$ is reached, recombination will begin and a temperature $T \approx$ const will be established. In the case of infall to an object of low mass, radiation will play a minor role; if the last term is neglected, the temperature variation will approach a $T \propto r^{-1}$ law, and the adiabatic index of the gas $\gamma \to 5/3$. The physical explanation of this difference is clear: the smaller the object, the smaller its gravitational radius, that is, its characteristic scale and the time for infall, and the radiation processes will become slow compared to the contraction process. Taking $dQ'/dt = 0$ (see the Appendix), we obtain for the critical mass, by Eq. (4),

$$M_{cr} \sim 10^4 M_\odot [T'_{(4)}]^{1/2} v_{c(10)}^{3} n_c^{-1} \varkappa_{(2)}^{-1}. \quad (5)$$

Here $T'_{(4)}$ denotes the temperature at the distance of interest to us in units of $10^4$°K, and $\varkappa_{(2)} \equiv \varkappa/100$.

## 3. THE GAS-DYNAMICS APPROXIMATION

The high temperatures developed during infall toward a mass $M < M_{cr}$ cast some doubt on the applicability of the gas-dynamical approach. At $r < r_c$, the mean free path with respect to Coulomb collisions,

$$l = kT^2 / ne^4 \cdot L_{\text{Coul}} \approx 10^{12} T_{(4)} n^{-1}\ \text{cm}, \quad (6)$$

will in this case far exceed the characteristic size $r$ of the region. For $M \approx M_\odot$, the free path $l \approx r$ even at the critical radius. We recall that the mass flux at the star in the approximation of noninteracting particles is smaller than the gas-dynamical flux (2) by a factor $(c/a_{\text{sound}})^2 \approx 10^9$ [1].

However, the material drawn in will contain magnetic fields. The Larmor radius of protons moving at thermal velocities will be smaller than the dimensions of the region of motion even at field strengths $H > 1.3\, T^{1/2} r^{-1}$ gauss. Inasmuch as $T \leq T_c (r_c/r)$, we have the condition for capture

$$H(r) \gg 10^{-12} T_{c(4)}^{3/2} M^{-1} (r_c/r)^{3/2}\ \text{gauss}. \quad (7)$$

The field at infinity is of order $10^{-6}$ gauss, so that in order for the condition (7) to be satisfied out to $r = r_g$ it is necessary that H increase with depth at least as $r^{-0.8}$. We shall see in Sec. 5 that there are grounds for expecting a far steeper rise in H(r), so that the gas-dynamics approach would be applicable.

The rise in the magnetic field would prevent any heat exchange between the different layers (see the Appendix). It would also affect the radiation and motion of the plasma. But first let us digress to consider one other topic.

## 4. LUMINOSITY AND EFFICIENCY IN THE IDEALIZED PROBLEM

For $M > M_{cr}$ the gas will be practically isothermal, so that the luminosity

$$L = \frac{dM}{dt} R^* \int_{r_1}^{r_2} T \frac{dn}{n} \approx 10^{34} M_{(5)}^{2} v_{c(10)}^{-3} n_c\ \text{erg/sec}. \quad (8)$$

Here $M_{(5)} \equiv M/10^5 M_\odot$, $T \approx 5000°K$, and $r_1$ is the radius below which radiation will rapidly be quenched by general-relativity effects (gravitational plus the Doppler red shift) [1]. Roughly speaking, $r_1 \approx 2r_g$. The luminosity (8) will fall mainly in the optical range and will result from line and recombination emission. If $M < M_{cr}$, the plasma temperature will increase inward along an adiabatic curve, and the luminosity of the spherical layer bounded by $r$ and $r/2$ will be

$$L(r) \approx r^3 \times 10^{-27} \cdot n^2 \cdot T^{1/2}(r)$$

$$\approx 10^{21} M^3 v_{c(10)}^{-6} n_c^2 [T_{12}(r)]^{1/2}\ \text{erg/sec}. \quad (9)$$

The value $T(r) \approx 10^{12}$°K may be regarded as an upper limit. The optical depth

$$\tau(r) = \int_r^{r_c} \sigma n_c (r_c/r)^{3/2} dr \approx 10^{-6} (\sigma/\sigma_k) M (r_g/r)^{1/2} n_c T_{c(4)}^{-3/2} \quad (10)$$

(where $\sigma_k$ is the Compton cross section) will always be much less than unity if $M < M_{cr}$, but if $M > M_{cr}$ the interior regions may become opaque, despite the Doppler shift.

In the approximation considered here, the efficiency of black holes would be extremely low. If accretion by an object with $M > M_{cr}$ takes place, only $\approx 10^{-8}$ of the mass of the infalling material will be converted into radiation; if $M < M_{cr}$ the efficiency will be even lower. The actual situation, however, is more favorable than this idealized one.

## 5. THE INFLUENCE OF MAGNETIC FIELDS[1]

Any plasma that is drawn in will necessarily contain magnetic fields. The observable effects will depend in an essential way on the law for the growth of the fields during the infall process. In our opinion there is only one real possibility: the

[1] See the note added in proof.

magnetic energy $\varepsilon_m$ and the gravitational energy $\varepsilon_{gr}$ per unit volume should be of the same order not only at the critical radius (where all four energies – gravitational, kinetic, thermal, and magnetic – would be of the same order) but also in the zone $r < r_c$. We can prove this statement by assuming the contrary. For suppose that in some region $\varepsilon_m \ll \varepsilon_{gr}$; then the field would not affect the infall, and because of strict freezing-in (it is readily seen that during infall the magnetic viscosity will always be far smaller than the kinematic viscosity) its radial component $H_r$ would increase according to the law $H_r \propto r^{-2}$, so that $\varepsilon_m = H^2/8\pi \propto r^{-4}$. However, $\varepsilon_{gr} \propto r^{-5/2}$, and an equality $\varepsilon_m \approx \varepsilon_{gr}$ would therefore be established very rapidly. On the other hand, the inequality $\varepsilon_m \gg \varepsilon_{gr}$ also would not be possible, because the field energy (due to freezing-in) would be derived from the energy of contraction, that is, from the kinetic energy, which would be smaller than $\varepsilon_{gr}$. We are therefore left with the condition $\varepsilon_m \approx \varepsilon_{gr}$. The behavior of the infall will be highly complicated in this situation: the field will have a predominantly radial character, and will be annihilated at the boundaries of sectors in which it is oppositely directed; the plasma will be inhomogeneous; the rate of infall will be nonuniform; very strong gutter instabilities will develop periodically in regions where $H \perp r$; and so on.

We shall nevertheless assume that despite the magnetic field, on the average the infall of gas does not depart seriously from free fall; that is, we shall suppose that $\bar{v} \approx (2GM/r)^{1/2}$ and $\bar{n} \propto r^{-3/2}$. Admittedly, both our idealizations may seem far-reaching, but it is natural to make them in our first steps toward a solution of the problem. We shall, then, let $H^2/8\pi = anGMm_p/r$, with $a \approx 1$.

In the case of infall toward a mass $M < M_{cr} \approx 10^4 M_\odot$, the plasma will be strongly heated, and the synchrotron losses will rapidly become decisive. In the range $T > 10^{10}$°K, Eq. (4) will take the form ($dQ'/dt = 0$)

$$\frac{dT}{dr} = -\frac{T}{r} + [3\cdot 10^{-8} M^2 v_{c(10)}^{-3/2} n_c]\frac{T^4}{r^2}a. \qquad (11)$$

Its solution is $T = (Cr + A/2r)^{-1}$; after the value

$$T_{max} = 3\cdot 10^{12} M^{-1/3} v_{c(10)}^{3/4} n_c^{-1/3} a^{-1/3}\ {}^\circ\mathrm{K} \qquad (12)$$

is reached the "temperature" will begin to fall. The efficiency of radiant energy release will be

$$\eta = 0.2\, mc^2 T_{max(12.5)}, \quad \text{if} \quad T_{max} < 3\cdot 10^{12}, \qquad (13a)$$

$$\eta = 0.2 mc^2 T_{max(12.5)}^{-3}, \quad \text{if} \quad T_{max} > 3\cdot 10^{12}. \qquad (13b)$$

Of course, since equilibrium will not have been established during the accretion process [see Eq. (6)], the "temperature" T should be interpreted not as the Maxwellian parameter but as the mean energy of motion of the electrons across the magnetic field. Equations (11-13) as well as (17-19) below are, to some extent, merely illustrative in character. (In a separate paper [17] we have given a more detailed discussion of the circumstances of magnetic accretion, a theorem regarding the "equipartition of energy," and a demonstration that in the case of accretion by "black holes" the synchrotron mechanism should convert about $0.1\ mc^2$ of the infalling material into radiation.) We shall defer to Sec. 8 a description of the corresponding spectrum; we first wish to point out one possible factor that might raise the efficiency.

## 6. THE INFLUENCE OF ANGULAR MOMENTUM

Interstellar gas, generally speaking, would possess a certain angular momentum K relative to the line of motion of a hole. It is well recognized [1] that, in general relativity, particle capture is possible if $K < K_g = 2mcr_g$. After it approaches a gravitating center an isolated particle with $K > K_g$ will again recede to infinity. According to Sec. 3, however, plasma at any distance from a collapsed object may be regarded as a continuous medium. Its compression at the tail of the flow, in the shock wave, will be accompanied by radiant energy release; if this energy becomes negative, then matter can never leave the neighborhood of the hole. The corresponding criterion is evident: it is necessary that the momentum K be much smaller than the quantity $K_c = mr_c v_c$. Thus if the momentum of the material drawn in satisfied the inequality

$$2mr_g c < K \ll mr_c v_c, \qquad (14)$$

then before they fall into the hole the particles will go into a stationary orbit. The minimum radius of a stable stationary orbit would be $r_{min} = 3r_c$ [18]. In order for a particle in Keplerian motion to reach $r_{min}$, it should emit radiation of $0.07\ mc^2$; if the role of viscosity is appreciable and nearly solid-body rotation prevails, the efficiency would be twice as great.

Might we expect the condition (14) to be satisfied under actual conditions? Denoting the tangential velocity component by $v_t$, let us write the inequality in the form

$$1.2\cdot 10^{-4} v_{c(10)}^{3} < v_{t(10)}(r = r_c) \ll v_{c(10)}. \qquad (14a)$$

First let us consider the left-hand inequality. Estimates indicate that the internal scale of interstellar turbulence, $l_0 = \mathrm{Re}_0 \nu'/v_t$, will always be less than or of the order of the radius $r_c$ ($\mathrm{Re}_0$ is the critical Reynolds number, and $\nu'$ is the magnetic viscosity). Hence $v_t(r_c) = v_t(L_t)\cdot(r_c/L_t)^x$, where $L_t \approx 100$ pc and $v_t(L_t) \approx 10$ km/sec [19]. In the interstellar medium, because of suppression of turbulence by the magnetic field and the formation of shock waves, the spectrum $v(l)$ should be steeper than a Kolmogorov spectrum ($x = 1/3$); evidently $1/2 \leqslant x < 1$ [19, 20]. Let $x \equiv 2/(3 + y)$. It is then easily seen than the first of the inequalities (14) will be satisfied for masses

$$M > M_t^{I} \approx 3.5 M_\odot v_{c(10)}^{3}[1.2\cdot 10^{-2} v_{c(10)}]^{y}. \qquad (15)$$

Thus for objects with stellar masses, spherically symmetric accretion will evidently be realized in many cases, while in some cases the plasma will be halted not far from $r_g$.

In the case of accretion of gas by very massive objects, rotation will undoubtedly play a role, and the radiant efficiency $\eta \approx 0.1\ mc^2$. It is now appropriate to inquire as to the influence of the momentum on dM/dt, that is, the right-hand inequality (14). For $L_t = 100$ pc and $v_t(L_t) = 10$ km/sec this inequality will become

$$M \ll M_t^{II} \simeq 3\cdot 10^{6} M_\odot v_{c(10)}^{3}. \qquad (16a)$$

The criterion (16a) refers to the case where $v_c \approx v_t \approx 10$ km/sec;[2] if $v_c \gg 10$ km/sec, only differential galactic rotation would be capable of preventing accretion. For a body 10 kpc away from the galactic center we would have in place of the criterion (16a):

$$M \ll M_t^{II} \simeq 3\cdot 10^{9} M_\odot v_{c(10)}^{3}. \qquad (16b)$$

The role of the momentum of the gas in accretion by objects with $M > 10^6 M_\odot$ has been pointed out qualitatively by Salpeter [9]. He has also indicated a fundamental mechanism for momentum loss: the gas in adjacent turbulence cells would have opposite directions of rotation. We shall return to this topic presently, but we first with to call the reader's attention to an important case in which the sign of the momentum might not change.

## 7. ACCRETION IN BINARY SYSTEMS

Our motive in discussing this situation is clear: in addition to a high efficiency, binary systems would ensure an enormous loss rate dM/dt. Thus, variables of the $\beta$ Lyrae type lose as much as $10^{-5}$ $M_\odot$/yr; under favorable conditions about one-half of this mass could fall into the second component. Hence the luminosity of holes in binary systems would in principle be capable of reaching $10^{38}$–$10^{39}$ erg/sec.

Around the dense component gas streams should form a disk with an approximately Keplerian velocity distribution. The mechanism for momentum loss would be viscosity, which would tend to produce solid-body rotation. The inner regions of the disk would be retarded, and the outer regions accelerated. Some of the gas would leave the binary system, while the rest would sink in toward the dense component. Very high temperatures would evidently be attained during this settling process (see the model calculation by Prendergast and Burbidge [22]).

Accretion by holes, neutron stars, and white dwarfs in close binary systems has been proposed on several occasions [8, 22–24] as a source of energy. An important question arises here: would it be possible to distinguish accretion by a hole from accretion by a star? The answer will be clear from the considerations above. In the first case the radiation would all come from a hot, thin disk; in the second case half the radiation would be due to the disk and half to emission from the surface of the star, in which an appreciable contribution should be present from equilibrium radiation [25] and/or plasma oscillations [26]. Furthermore, in the case of accretion by a neutron star the radiation ought to contain a strictly periodic term (with $p \approx 1$ sec) arising from the rotation of the star. In the case of accretion by a hole, only random luminosity variations would be expected, with $\Delta t \approx 10^{-5}$–$10^{-2}$ sec (see Sec. 9 below). Also, the emission spectrum of the disk itself probably would, in general, be quite flat. Thus a search for a hole in a binary system, based on the absence of an optically visible component [2–4], might actually exclude some of the objects sought.

## 8. GAMMA-RAY, X-RAY, AND OPTICAL STARS?

We shall now subdivide accretion into spherical, helical, and disk types, depending on the character of particle motion. The first two types would be realized around isolated objects; the last type, in binary systems. In the case of disk accretion, most of the energy should be released in the form of x-ray photons. Helical accretion would lead

[2]Note that under the conditions prevailing in the Galaxy, supermassive objects necessarily would rapidly diminish their velocity u [but not $v = (u^2 + a^2_{sound})^{1/2}$] to about 10 km/sec [21].

to the appearance of $\gamma$-ray stars as well as x-ray stars. Indeed, as we have been in Sec. 6, the spectrum of interstellar turbulence is such that in many instances the plasma would be halted (transferred from a helical to a circular orbit) in the vicinity of $r_g$. When a hole moves into the adjacent turbulence cell, where the gas is rotating in the opposite direction, particle collisions will begin to take place near $r_g$, and unlike the case of settling onto a disk, the process of momentum loss would here be very rapid.

It is worth noting that in the case of accretion by an isolated neutron star the role of the momentum of the gas would probably be small because of the small value of $r_c$ (see Sec. 6). On the other hand, even if a disk were to develop, it would be located near the surface of the neutron star and would rapidly sink into the star. The integrated luminosity of the objects in this situation would be very low, of order $10^{30}$ ergs/sec [see Eq. (2)]; and most of the radiation would fall in the ultraviolet [27]. If the neutron star has a magnetic field, the infall of gas would be stopped, in general, far beyond $r_g$, and would be accompanied by Langmuir oscillations at radio frequencies [26].[3] Evidently, then, only "black holes" could be "pure" $\gamma$-stars.

In this connection it is interesting to note the recent discovery [28] of a discrete source of $\gamma$ rays ($E_\gamma > 50$ MeV) which has not yet been identified with an x-ray object. Yet, the sensitivity of the x-ray equipment was one or two orders higher than the sensitivity of the $\gamma$-ray counters.

The intensity of disk-type accretion should remain unchanged over tens of thousands of years, apart from fluctuations associated with the inhomogeneity of the gas flow. (Possible eclipses of the disk by the normal component would be of special interest.) The radiation of isolated objects due to spiral-type accretion would undoubtedly exhibit a flare behavior. The minimum characteristic time for a flare would be of order $r_c/v_c$, that is, a few months. Curiously enough, among the known sources of hard x rays there are definitely two classes of objects: those whole luminosity has remained constant, on the average, throughout the entire period of observation (for example, Sco X-1 and Cyg X-2); and those whose luminosity has varied by tens of times within a year, either appearing or disappearing from the field of view (such as Cen X-2 and Cen X-4 [29, 30]). Observers have often suggested that Sco X-1 and Cyg X-2 might belong to binary systems [31-34].

The accretion of gas by an isolated hole of mass $M \approx 1$-$100\ M_\odot$ should, as mentioned in previous sections, be primarily spherical. Magnetic fields should play a definite role in the radiation. According to Eqs. (2) and (13a), the corresponding luminosity should be

$$L \sim 0.1 T_{\max(12.3)} c^2 dM/dt$$
$$\approx 2\cdot 10^{33} M_{100}^{9/2} v_{10}^{-9/4} n^{1/2} a^{-1/2}\ \text{erg/sec}; \qquad (17)$$

here $M_{100} = M/100\ M_\odot$. The spectrum of the radiation should have a distinctive form: the intensity should remain nearly constant over a wide frequency range near

$$\nu_{\max} \sim 10^{14}\cdot M_{100}^{-3/8} v_{10}^{-9/16} n^{-9/4} a^{-13/8}\ \text{Hz}, \qquad (18)$$

an exponential decline should set in at $\nu \gg \nu_{\max}$, and at $\nu \ll \nu_{\max}$ there should be an extremely slow decline to frequencies at which absorption of the radiation becomes appreciable.

$$L(\nu) = \int_{\nu/2}^{\nu} (dL/d\nu)\, d\nu \propto \nu^{4/13}, \qquad dL/d\nu \propto \nu^{-9/13}, \qquad (19)$$

Coherent mechanisms and negative self-absorption might be operative.

In cases where $\gamma$ rays are generated, they would be in addition to synchrotron radiation. However, because of the "accumulation" of material in circular orbits, the "peak" $\gamma$-ray luminosity should be one or two orders higher than given by Eq. (17).

The radiation of a "black hole" would of course heat the gas at $r > r_c$, thereby influencing the intensity of the accretion [see Eq. (2)]. One can show that for most cases of interest this effect would be insignificant, but in certain cases it could be decisive. A more thorough discussion of this topic, together with a solution of the self-consistent problem as exemplified by a neutron star, has been given elsewhere [27].

Conceivably, then, individual "black holes" might already be observable today, as faint optical stars with no lines but with a nonthermal spectrum extending far into the low-frequency range (as far as radio frequencies). Perhaps some of the objects heretofore regarded as type DC white dwarfs are actually "black holes."

## 9. A CRITICAL EXPERIMENT

What properties of the radiation would allow "black holes" of stellar mass to be distinguished r e l i a b l y from other objects? In our view, such properties would be the exceptionally small size

[3] From the observational standpoint such objects might appear as "second generation" pulsars. Further details have been given elsewhere [26].

of a hole together with the absence of any rotation. In other words, because of the development of instability in a magnetized plasma the luminosity of a hole would fluctuate on a time scale $\Delta t \approx 10^{-5}$–$10^{-2}$ sec, but a periodic term should be entirely absent.

## 10. QUASAR-LIKE STARS?

Equations (5) and (16) imply that in the case of accretion by a hole with $M \approx 10^5 M_\odot$ the infalling gas will be cooled up to the onset of the "spiral" mode (somewhere in the zone $r \approx 10^{-4}$–$10^{-3}$ $r_c$); $T \approx 5000°K$. The corresponding luminosities are given by Eq. (8), and are small compared to the integrated luminosity, but nevertheless they fall wholly in the optical range. The spectrum of the radiation – broadened emission lines, well-developed recombination bands, the presence of a nonthermal continuum and absorption lines – should to a large extent resemble the optical spectrum of quasars. On the other hand, holes with masses of order $10^5 M_\odot$ are interesting in that they might have come from "first generation" stars that developed from entropy perturbations in the pregalactic medium [35] (the remainder of these perturbations would presumably have served as the origin of the globular clusters [36]).

In the case of accretion by "superholes" with $M \gg 10^5 M_\odot$, the momentum of the gas would be expected to play a role from the very outset; but here too it would seem that in many cases a regime of "cold accretion" would develop, accompanied by a spectrum similar to that observed for quasars. We intend to examine this possibility in a separate paper.

## APPENDIX

### THE ABSENCE OF HEAT CONDUCTIVITY BETWEEN DIFFERENT LAYERS

Why have we consistently neglected the term $dQ'/dt$, representing the heat conductivity? For radiative conductivity, we have done so because the optical depth of the plasma is negligible [see Eq. (10)]; for ion conductivity, because the infall of the gas is supersonic; and finally, for electron conductivity, because the magnetic fields grow rapidly during the infall process. Let us consider this last point more carefully. According to Eq. (6), the time required for exchange of energy between different electrons will far exceed the time required for infall toward the hole. Thus heat conductivity could only arise from the migration of electrons. However, such migration will be severely limited by the small Larmor radius of the electrons ($l_L/r < 10^{-6}$), by the strictness with which the freezing-in conduction is satisfied, and by the tangling of the lines of force during infall (see Sec. 5). In the interior, where the temperatures and fields are large, there will in addition be a small "mean free path" relative to synchrotron losses, as compared to the characteristic scale of the motion.

The author is indebted to Ya. B. Zel'dovich for suggesting the topic and for much valuable counsel. He is also grateful to G. S. Bisnovatyi-Kogan and L. M. Ozernoi for their comments.

## NOTE ADDED IN PROOF

Kardashev [37] was the first to point out the circumstance that in accretion by a "black hole" a substantial fraction of the rest mass of the infalling material might be liberated because of the presence of a magnetic field; in this connection see also the remark on page 360 of the Russian edition of Zel'dovich and Novikov's book [1]. The synchrotron radiation emitted upon accretion by the magnetosphere of a neutron star has been considered in recent papers [14,15]. Bisnovatyi-Kogan and Syunyaev, in discussing the problem of infrared sources [16], consider that in accretion by collapsed stars the annihilation of the magnetic field may lead to the formation near $r_g$ of a shock wave at whose front a substantial part of the energy will be transformed into plasma oscillations with the emission of radiation. Our views concerning magnetic accretion have been set forth more fully in a separate paper [17].

## LITERATURE CITED

1. Ya. B. Zel'dovich and I. D. Novikov, Relativistic Astrophysics [in Russian], Nauka, Moscow (1967); English translation, Vol. 1, Univ. Chicago Press (1971).
2. O. Kh. Guseinov and Ya. B. Zel'dovich, Astron. Zh., 43, 313 (1966) [Sov. Astron.–AJ, 10, 251 (1966)]; Ya. B. Zel'dovich and O. Kh. Guseinov, Astrophys. J., 144, 840 (1965).
3. V. L. Trimble and K. S. Thorne, Astrophys. J., 156, 1013 (1969).
4. O. Kh. Guseinov and Kh. I. Novruzova, Astron. Tsirk., No. 560 (1970).
5. A. A. Wyller, Astrophys. J., 160, 443 (1970).
6. A. M. Wolfe and G. R. Burbidge, Astrophys. J., 161, 419 (1970).
7. S. van den Bergh, Nature, 224, 891 (1969).
8. Ya. B. Zel'dovich, Dokl. Akad. Nauk SSSR, 155,

67 (1964) [Sov. Phys.-Dokl., 9, 195 (1964)].
9. E. E. Salpeter, Astrophys. J., 140, 796 (1964)].
10. D. Lynden-Bell, Nature, 223, 690 (1969).
11. F. Hoyle and R. A. Lyttleton, Proc. Cambridge Phil. Soc., 35, 405 (1939).
12. H. Bondi and F. Hoyle, Mon. Not. Roy. Astron. Soc., 104, 273 (1944).
13. H. Bondi, Mon. Not. Roy. Astron. Soc., 112, 195 (1952).
14. P. R. Amnuél' and O. Kh. Guseinov, Izv. Akad. Nauk Azerbaidzhan SSR, Ser. Fiz. Tekh. Mat., No. 3, 70 (1968).
15. G. S. Bisnovatyi-Koyan and A. M. Fridman, Astron. Zh., 46, 721 (1969) [Sov. Astron. -AJ, 13, 566 (1970)].
16. G. S. Bisnovatyi-Kogan and R. A. Syunyaev, Preprinty Inst. Priklad. Matem. Akad. Nauk SSSR, No. 31 (1970); Astron. Zh. [Sov. Astron. -AJ (in press)].
17. V. F. Shvartsman, Preprinty Inst. Priklad. Matem. Akad. Nauk SSSR, No. 42 (1970).
18. S. A. Kaplan, Zh. Éksp. Teor. Fiz., 19, 951 (1949).
19. S. A. Kaplan and S. B. Pikel'ner, The Interstellar Medium, Harvard Univ. Press (1970).
20. S. A. Kaplan, Dokl. Akad. Nauk SSSR, 94, 33 (1954).
21. W. H. McCrea, Mon. Not. Roy. Astron. Soc., 113, 162 (1953).
22. K. H. Prendergast and G. R. Burbidge, Astrophys. J., 151, L83 (1968).
23. I. S. Shklovskii, Astron. Zh., 44, 930 (1967) [Sov. Astron.-AJ, 11, 749 (1968)]; Astrophys. J., 148, L1 (1967).
24. A. G. W. Cameron and M. Mock, Nature, 215, 464 (1967).
25. Ya. B. Zel'dovich and N. I. Shakura, Astron. Zh., 46, 225 (1969) [Sov. Astron.-AJ, 13, 175 (1969)].
26. V. F. Shvartsman, Radiofizika, 13, 1852 (1970).
27. V. F. Shvartsman, Preprinty Inst. Priklad. Matem. Akad. Nauk SSSR, No. 57 (1969); Astron. Zh., 47, 824 (1970) [Sov. Astron.-AJ, 14, 662 (1971)].
28. G. M. Frye, J. A. Staib, A. D. Zych, V. D. Hopper, W. R. Rawlinson, and J. A. Thomas, Nature, 223, 1320 (1969).
29. W. H. G. Lewin, J. E. McClintock, and W. B. Smith, Astrophys. J., 159, L193 (1970).
30. W. D. Evans, R. D. Belian, and J. P. Conner, Astrophys. J., 159, L57 (1970).
31. Yu. N. Efremov, Astron. Tsirk., No. 401 (1967).
32. E. M. Burbidge, C. R. Lynds, and A. N. Stockton, Astrophys. J., 150, L95 (1967).
33. R. P. Kraft and M.-H. Demoulin, Astrophys. J., 150, L183 (1967).
34. J. Kristian, A. R. Sandage, and J. A. Westphal, Astrophys. J., 150, L99 (1967).
35. A. G. Doroshkevich, Ya. B. Zel'dovich, and I. D. Novikov, Astron. Zh., 44, 295 (1967) [Sov. Astron.-AJ, 11, 233 (1967)].
36. P. J. E. Peebles and R. H. Dicke, Astrophys. J., 154, 891 (1968).
37. N. S. Kardashev, Astron. Zh., 41, 807 (1964) [Sov. Astron.-AJ, 8, 643 (1965)].

Astron. & Astrophys. 21, 1–9 (1972)

# Accretion Disc Models for Compact X-Ray Sources

J. E. Pringle and M. J. Rees
Institute of Theoretical Astronomy, Cambridge

Received April 4, revised May 9, 1972

**Summary.** The accretion process is considered, in cases when the infalling matter possesses angular momentum and forms a disc spinning around a central compact mass. This situation occurs when gas falls onto a black hole or neutron star from a binary companion, and may be relevant to galactic X-ray sources. The spectrum of the radiation emitted by gas spiralling down into a black hole is calculated and compared with the data on Cygnus X-1. Rapid irregular variability is expected in this model. If the compact object is a neutron star, the gas dynamics near the star may be controlled by the stellar magnetic field. The main X-ray emission then comes not from the disc, but from regions on the stellar surface near the magnetic poles. An interpretation of Centaurus X-3 involving a rotating magnetized neutron star is proposed. It is emphasized that accretion discs can display a wide range of properties (depending on the accretion rate, viscosity, etc.). Tentative interpretation of some other phenomena are also proposed.

**Key words:** accretion – binary stars – X-ray sources

## 1. Introduction

A number of galactic X-ray sources are now known to be components of close binary systems. Their rapid ($\lesssim 1$ s) variability suggests that a compact object – either a neutron star or a black hole – is involved. Both these facts strongly support earlier suggestions (see, e.g. Shlovskii, 1967; Prendergast and Burbidge, 1968) that X-ray sources are an accretion phenomenon: capture of material from the binary companion could supply an abundant mass flux; and each gram of matter, falling into the deep gravitational well associated with a compact object, could yield $\sim 10^{20}$ erg of radiation.

Material transferred from the companion star is likely to have so much angular momentum that it cannot fall directly onto the compact object. The fairly extensive literature on spherically symmetric accretion (reviewed by, for example, Zeldovich and Novikov, 1971) is thus not directly relevant. The matter will, instead, form a differentially rotating disc – the circular velocity at any point being approximately Keplerian – composed of material which gradually spirals inward as viscosity transports its angular momentum outward. Prendergast and Burbidge (1968) have already considered some aspects of this kind of model. These authors estimated that as much as half the material transferred from the large star could fall onto the compact object, and discussed the spectrum of the radiation from the disc under certain assumptions. Schwartzman (1971b) has also speculated along the same lines as in the present discussion. A somewhat similar situation, but on a much larger scale, was discussed by Lynden-Bell (1969) and by Lynden-Bell and Rees (1971), with application to active galactic nuclei.

In the present paper we explore some further properties of accretion models for X-ray sources. For plausible parameters the energy released by the gradually infalling matter is radiated by bremsstrahlung. Because of the small size of the effective emitting region, the temperature $\gtrsim 10^7$ °K, and most of the energy is radiated as X-rays (see also footnote 2). We consider, in particular, the spectrum of the radiation from the disc, and the effects of a strong magnetic field attached to the compact central star (which may be rotating). Our results suggest that one may envisage six types of X-ray source, which fall into two main categories, depending on the nature of the compact object.

*I. When the Central Object is a "Black Hole"*

(a) If the mass flux lies within a certain range, the observed radiation comes almost solely from the disc. We show that in this case a power law X-ray spectrum may result.

(b) If the mass flux is *either* so *large* that radiation pressure prevents steady accretion, *or* so *small* that the gravitational energy released cannot be radiated efficiently, we expect some kind of flaring, with the black hole alternately accreting matter and expelling it from its environment.

*II. When the Central Object is a Neutron Star*

(a) If the mass flux is again too large or too small, the situation resembles I(b).

(b) If the star has a negligible magnetic field, the disc extends down to the surface. We then expect comparable

amounts of X-ray power to come from the star and the disc. The spectrum would then be thermal, with perhaps an apparently non-thermal high energy "tail".
(c) If the star has a magnetic field and is spinning slowly, accretion takes place along field lines. The X-rays would then be emitted thermally from regions near the magnetic poles of the neutron star. Unless the field is axisymmetric, the received radiation would be pulsed, with a period equal to the star's rotation period.
(d) If the star rotates rapidly, or has a very strong magnetic field, accretion cannot take place, and energy would instead be supplied to the system from the stellar rotational energy. It would seem likely that, in this case, the rotation would give rise to some rapid but regular periodic effect.
The gas flow in binary systems is a complex subject which we do not even attempt to discuss here. However almost all the gravitational energy of the accreted material is liberated at distances from the compact object very small compared to the binary separation. We are therefore justified in ignoring the gravitational effect of the other star, and simply considering a circular disc surrounding a compact object. This we do in § 2. We then (§ 3) calculate the influence on the accretion process of a dipole-type field attached to the central star. In § 4 we tentatively apply our results to some particular observed sources.

## 2. Accretion Discs

The detailed properties of disc models depend on what assumptions are made about turbulence, viscosity, etc. (see, for example, Prendergast and Burbidge (1968) or Lynden-Bell (1969). We shall here, in general, follow the notation of the latter paper). For the sake of argument we shall take a specific – and, we hope, reasonable – model for the disc, at the same time allowing ourselves a few variable parameters.
The disc is held out by centrifugal force, and so the circular velocity, $V_c$, is Keplerian:

$$\text{i.e.}\quad V_c \approx 1.15\times10^{10}\,(M/M_\odot)^{1/2}\,R_6^{-1/2}\,\text{cm s}^{-1} \tag{1}$$

where $R_6$ is the radius of the orbit in units of $10^6$ cm, and $M$ is the mass of the central object. In all the situations we consider, the gravitational effects of the disc itself will be negligible. We also neglect relativistic effects. This is a good approximation when $R_6 \gg (M/M_\odot)$ and inclusion of the relativistic corrections would not significantly alter any of our conclusions.
The thickness of the disc is determined by the balance between the pressure in the disc (comprising turbulent and magnetic pressure, as well as ordinary gas and radiation pressure) and the relevant component of the gravitational pull of the central object. We shall take the disc semithickness $b$ as

$$b=\frac{x}{20}\times R=5\times10^4\,x R_6\,\text{cm}, \tag{2}$$

where $x$ is a dimensionless parameter which we expect to be of order unity, and which may depend on the radius $R$. (Strictly speaking, $b$ is the scale height of the disc, but throughout this paper we neglect any gradual dependence of density and velocity on distance from the plane of symmetry.)
The material in the disc will drift inward at a rate $V_r$ which depends on how fast angular momentum can be transported outwards, i.e. on the viscosity. We assume

$$V_r = y\,\frac{V_c}{100} \tag{3}$$

where $y$ is a parameter (which we again expect, following Lynden-Bell (1969), to be $\sim 1$) depending on the viscosity. In a steady state when the flux of accreted matter is $F$, the density $\varrho(R)$ is given by

$$F=2\pi R\cdot 2b\cdot\varrho V_r\,. \tag{4}$$

Hence

$$\varrho=1.4\times10^{-4}\,y^{-1}x^{-1}(M/M_\odot)^{-1/2}F_{16}R_6^{-3/2}\,\text{g cm}^{-3} \tag{5}$$

where $F_{16}$ is the mass flux in the convenient units of $10^{16}$ gm s$^{-1} \approx 1.5\times10^{-10}\,M_\odot$ year$^{-1}$.
To radiate the energy dissipated by viscosity, the disc must emit a power per unit area of

$$p(R)=\frac{3FGM}{4\pi R^3} = 1.1\times10^{23}F_{16}(M/M_\odot)\,R_6^{-3}\,\text{erg cm}^{-2}\,\text{s}^{-1}\,. \tag{6}$$

We note in passing that, if the central object is a neutron star of radius $R_{ns}$, and the disc extends inwards to its surface, the total luminosity of the disc is

$$L\approx 7\times10^{35}\,F_{16}\,(M/M_\odot)\left(\frac{R_{ns}}{10^6\,\text{cm}}\right)^{-1}\text{erg s}^{-1}\,. \tag{7}$$

For a disc surrounding a Kerr black hole with the maximum allowable angular momentum to mass ratio, the material continues to spiral gradually inward until it reaches the most tightly bound stable orbit, whose binding energy is $0.42c^2$ per unit mass (Bardeen, 1970). We then have

$$L\approx 4\times10^{36}\,F_{16}\,\text{erg s}^{-1}\,. \tag{8}$$

If the disc radiated like a black body, the temperature would be

$$T_{bb}(R)=\left(\frac{p(R)}{2\sigma}\right)^{1/4} = 7.3\times10^6 F_{16}^{1/4}(M/M_\odot)^{1/4}R_6^{-3/4}\,{}^\circ\text{K} \tag{9}$$

where $\sigma$ is Stefan's constant.
This is, of course, a *lower limit* to the temperature: the disc would become substantially *hotter* than $T_{bb}$ if *either*
(i) cooling processes were not efficient enough to radiate the power $p(R)$ at a temperature $\sim T_{bb}$;
or (ii) the main contribution to the opacity comes from *electron scattering*, in which case the surface brightness

cannot attain the full black body intensity (Felten and Rees, 1972).

We consider first the inner part of the disc where $T_{bb} \gtrsim 10^4$ °K. Since the actual temperature cannot be less than $T_{bb}$, these regions must be almost completely ionized, the electron density $n_e$ being comparable with the total particle density $n$.

For discs with temperatures high enough to be relevant to X-ray sources, the dominant emission mechanism is free-free radiation (bremsstrahlung [1]) which yields a cooling rate $\propto n_e^2 T^{1/2}$ per unit volume (if we neglect the $T$-dependence of the Gaunt factor). The temperature needed to give the required output by this process is

$$T_{ff}(R) \simeq 1.2 \times 10^{10} F_{16}^{-2} (M/M_\odot)^4 x^2 y^4 R_6^{-2} \text{ °K} . \quad (10)$$

If the effects of electron scattering ((ii) above) were negligible, the actual disc temperature would be

$$T \approx \max[T_{bb}, T_{ff}] . \quad (11)$$

Since $T_{ff}$ decreases more sharply than $T_{bb}$ with increasing $R$, (if $x$ and $y$ are taken to be constants) we expect the outer parts of the disc to have essentially the black body temperature given by (9); the inner parts, however may be unable to radiate the locally-generated power ($\propto R^{-3}$ per unit area) unless they are much hotter than $T_{bb}$. The critical radius at which $T_{bb} \approx T_{ff}$ – which is also the place where the disc has a free-free optical depth $\tau_{ff}(\nu)$ of order unity for photons near the black body peak ($h\nu \approx 3kT$) – is

$$\left(\frac{R_{ff}}{10^6 \text{ cm}}\right) \approx 3.4 \times 10^2 F_{16}^{-9/5} (M/M_\odot)^3 x^{8/5} y^{16/5} . \quad (12)$$

We note that this is sensitive to the uncertain parameters $x$ and $y$.

The optical depth of the disc to electron scattering is $2\tau_{es}$, where

$$\tau_{es} \approx 1.9 (M/M_\odot)^{-1/2} F_{16} R_6^{-1/2} y^{-1} . \quad (13)$$

If $\tau_{es} > 1$, *and* $\tau_{ff}(h\nu \approx 3kT) < \tau_{es}$, the peak power that can be radiated by a disc at temperature $T$ is reduced by a factor $\mathfrak{R}^{-1/2}$ below the black body intensity, where $\mathfrak{R} = \tau_{es}/\tau_{ff}(\nu = 3kT/h)$.

As a corollary, the disc may have to be hotter than (11) would imply; and we then find that the disc temperature is

$$T = \max[T_{bb}, T_{mod}, T_{ff}] , \quad (14)$$

where $T_{mod}$ ($> T_{bb}$ when $\mathfrak{R} > 1$ and electron scattering opacity is dominant) is

$$T_{mod}(R) = 4.4 \times 10^7 F_{16}^{2/9} (M/M_\odot)^{5/9} y^{2/9} x^{2/9} R_6^{-1} \text{ °K} . \quad (15)$$

[1]) Detailed computations by, for example, Cox and Tucker (1969) show that free-free cooling dominates for $T \gtrsim 10^7$ °K in a collisionally ionized gas with normal "cosmic" abundances. In X-ray sources, photoionization by thermal photons raises the ionization level at a given temperature. This reduces the importance of cooling by line emission, and so free-free cooling may be dominant even below $\sim 10^7$ °K.

The change-over from $T_{bb}$ to $T_{mod}$ occurs at (point $B$ in Fig. 1)

$$\left(\frac{R_B}{10^6 \text{ cm}}\right) = 1.3 \times 10^3 F_{16}^{-1/9} (M/M_\odot)^{11/9} x^{8/9} y^{8/9} \quad (16)$$

and the change-over from $T_{mod}$ to $T_{ff}$ at (point $C$ in Fig.1)

$$\left(\frac{R_C}{10^6 \text{ cm}}\right) = 2.7 \times 10^2 F_{16}^{-20/9} (M/M_\odot)^{31/9} x^{16/9} y^{34/9} . \quad (17)$$

Thus it is necessary to take electron scattering into account in calculating the radiation spectrum from the disc if $R_B > R_C$, or, approximately

$$1.7 \times F_{16} > (M/M_\odot) x^{2/5} y^{5/4} . \quad (18)$$

If the factor $\mathfrak{R}$ exceeds $m_e c^2/kT$, then the outgoing spectrum is modified by the cumulative effect of the frequency shifts (each of order $\delta\nu/\nu \approx (kT/m_e c^2)^{1/2}$) in successive scatterings. This process alters the value of $T_{mod}$, and also gives rise to an extra cooling mechanism, since electrons will lose energy by scattering the photons emitted as free-free radiation. But these effects introduce only logarithmic corrections into the foregoing equations, and we shall not consider them in any further detail here.

Some words of caution are necessary at this stage. The above discussion is only valid provided that the component of the star's gravitational field perpendicular to the disc is strong enough to balance the radiation pressure gradient in that direction, i.e. provided that

$$R_6 \gtrsim 0.6 F_{16} x^{-1} . \quad (19)$$

A consequence of this is, of course, that the luminosity/mass ratio of the source cannot exceed the usual Eddington limit. If $F_{16}$ is very large, the disc would thicken until the configuration became almost spherical. The material would then fall radially inward, or be expelled, without yielding much energy in the form of radiation. Thermal gas pressure $P_g$ would itself thicken the disc until $x$ satisfied the inequality

$$(x/20) \lesssim V_c^{-1} (P_g/\varrho)^{1/2} .$$

This means that disc models are only consistent when

$$T \lesssim T_{max} = 2.4 \times 10^9 (M/M_\odot) R_6^{-1} x^2 \text{ °K} . \quad (20)$$

If $T \approx T_{ff}$, and $T_{ff}$ violates (20), then here again the disc expands into a spherical configuration. The fact that $T_{ff}$, like the right hand side of (20), also depends on $x^2$, means that the disc cannot stabilise itself against this instability merely by increasing $x$ [2]). $T_{ff}$ satisfies (20)

[2]) Note that, in disc models, $\frac{kT(R)}{m_H}$ is always $\ll GM/R$. It is because of this that we can get X-rays rather than $\gamma$-rays, even from the immediate vicinity of a collapsed object, and thereby convert the rest-mass energy of infalling gas into X-rays with high efficiency. Typical galactic sources would require accretion rates of only $\sim 10^{-10} M_\odot$ per year.

when

$$R_6 \gtrsim 5 F_{16}^{-2} (M/M_\odot)^3 y^4 . \tag{21}$$

At radii where (21) is violated, the accreted matter falls inward without necessarily releasing much further energy until (if the central object is a neutron star rather than a black hole) it impacts on the surface.

It is clear that even the idealised disc models that we have considered can display several different kinds of behaviour, depending on the particular parameters. If $x$ and $y$ are assumed independent of $R$, then the temperature increases towards the centre. If electron scattering is *un*important, then $T$ varies as $R^{-3/4}$ for $R > R_{ff}$ (Eq. (12)), each annulus radiating like a black body: for $R < R_{ff}$, $T$ varies as $R^{-2}$, and the spectrum is "flat", with free-free absorption being important only for frequencies with $h\nu \ll kT_{ff}$. If electron scattering *is* important ($\mathfrak{R} > 1$) there is an intermediate range of radii over which $T \approx T_{\text{mod}} \propto R^{-1}$. (If the electron scattering optical depth is very large, the emergent spectrum may be distorted towards a Wien law.) The dependence of $T$ on $R$ is illustrated in *Fig. 1*. The temperature $T_{ff}$ rises as $R$ decreases, and eventually the inequality (20) may be violated. This also is illustrated in the figure.

These results would be altered if $x$ and $y$ were $R$-dependent. Indeed, if this dependence were very strong even the qualitative features of Fig. 1 might change – for example $T_{ff}$ might in some circumstances be an *increasing* function of $R$. (We suggest in § 4 that this may happen in Cygnus X-1.)

In the outermost parts of the disc, where $T_{bb}$ falls below $10^4$ °K and the power generation rate $p(R)$ is low, the gas will not necessarily be completely ionized. Instead, the ionization level will adjust so that the radiation losses – primarily line emission and recombination – balance $p(R)$. As is well known for the case of the interstellar medium, the electron temperature stays close to $\sim 10^4$ °K for a wide range of values of $p(R)$, with $n_e/n$ adjusting itself somewhere between 0.01 and 1. We find $n_e/n \propto R^{-1/2} x^{1/2} y$.

The total emission spectrum $S(\nu)$ is calculated as an integral over the spectra emitted by the elemental rings that comprise the disc. Although the spectrum emitted by the material at a particular radius is thermal, the integrated spectrum from the whole disc may be of power-law form, thus mimicking a non-thermal spectrum. We shall also allow for a possible $R$-dependence of $x$ and $y$, and take $x \propto R^{\mu}$, $y \propto R^{\lambda}$. (Note, however, that is no obvious reason why $x$ and $y$ should actually obey power laws. On the other hand we have shown that a power law dependence of $T$ or $R$ would arise more naturally.) The spectra are of the form $S(\nu) \propto \nu^{-\alpha}$, where the slope $\alpha$ is easily evaluated for the different cases

*(a) Black Body Regime*

$$\alpha = -\tfrac{1}{3} .$$

*(b) Modified Black Body*

$$\alpha = \frac{2 - \dfrac{\lambda}{3} - \dfrac{2\mu}{3}}{1 - \dfrac{2\lambda}{9} - \dfrac{2\mu}{9}} - 2 .$$

*(c) Free-free*

$$\alpha = \frac{1 - (4\lambda + 2\mu)}{2 - (4\lambda + 2\mu)} . \tag{23}$$

*(d) Integrated Emission from the Partially Ionized outer Parts of the Disc*

$$\alpha = -\tfrac{2}{3} .$$

The above formulae are only valid provided that the calculated spectrum falls off, at low frequencies, *less* rapidly than the spectrum contributed by the material

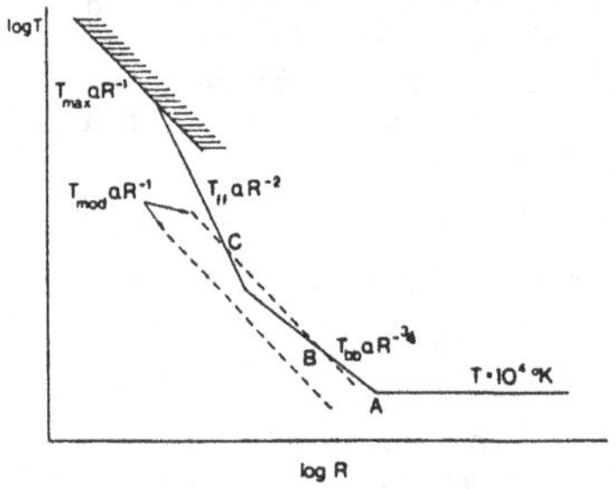


Fig. 1. Schematic representation of the dependence of $T$ on $R$ in models where $x$ and $y$ are taken as constant. In the outer parts (outside $A$) the ionization level adjusts itself to keep $T$ constant $\simeq 10^4$ °K. Inside $A$, the matter is fully ionized and the disc is at the blackbody temperature $T_{bb}$. When the disc becomes optically thin, free-free emission becomes dominant and the temperature $T_{ff}$ rises more steeply. This continues until $T = T_{max}$ where gas pressure thickens the disc and the model is no longer applicable. The temperature $T_{mod}$ due to modification of the spectra through electron scattering is shown as dotted lines. The two positions depend on the parameters $x$, $y$ and demonstrate how it may or may not be relevant

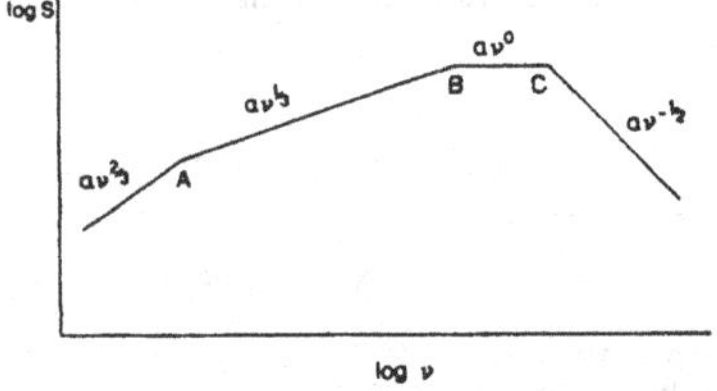


Fig. 2. A representation of the overall power spectrum from the disc in the case when $x$ and $y$ are constant, and when $T_{mod}$ is relevant (see Fig. 1). The bends on the spectrum at A, B and C correspond to the points marked similarly in Fig. 1. The spectrum would cut off exponentially at $T_{max}$

with a particular value of $R$. This has little effect on (a), (b) and (d), but in (c) must demand that the numerical value of $4\lambda + 2\mu$ does *not* lie between 1 and 2.
Figure 2 shows the type of spectrum expected in the particular case when $\lambda$ and $\mu$ are zero.

### 3. The Inner Edge of the Disc, and the Influence of a Stellar Magnetic Field

The inner edge of the disc is defined *either* by the smallest value of $R$ for which (19) and (21) hold, *or* by the radius at which the influence of the central object becomes critical. If the latter is a black hole, then (provided (19) and (21) are satisfied), the disc will extend inwards to the position of the most tightly bound stable circular orbit. Once the inward-spiralling matter reaches this point, it can be swallowed by the black hole without releasing a significant further amount of energy. The total luminosity per gram may approach the theoretical limit given by (8) – note, however, that relativistic corrections modify the detailed spectrum of the inner parts of the disc, from which most of the energy comes.
If the central object is an *unmagnetized* neutron star, then (if (19) and (21) hold) the disc terminates at $R \approx R_{ns}$, where the matter grazes the surface of the star. The integrated luminosity $L$ of the disc is given by (7). However the matter impinging on the star heats it, causing an amount of power comparable with $L$ to be radiated thermally from the stellar surface. (This power is precisely equal to $L$ if the star is not rotating, but is somewhat less if the star is spinning in the same sense as the disc.)
A neutron star – unlike an (uncharged) black hole – can possess a frozen – in magnetic field, which may be strong enough to affect the dynamics of the inflowing gas even at radii $R \gg R_{ns}$. We shall now attempt to estimate this effect. We have in mind particularly the case when the stellar field is non-axisymmetric (e.g. an oblique dipole field), since in this case the X-ray emission would be non-isotropic, giving rise to pulses with the rotation period of the star.
The matter in the disc will be highly ionized, so that once in contact with the stellar field it will only be able to move along field lines. Assuming the field to be dipolar, its magnitude is given by

$$B = 10^{12} B_{*12} \left(\frac{R}{R_{ns}}\right)^{-3} \text{G} \tag{24}$$

where $B_{*12}$ is the field at the stellar surface, at the point on the same radius vector, in units of $10^{12}$ G (the typical field strength estimated for pulsars).
The star is taken to have a rotational period $P$ seconds, and we define the corotation radius $R_\Omega$ to be the radius at which the keplerian and stellar angular velocities are equal,

$$\left(\frac{R_\Omega}{10^6\,\text{cm}}\right) = 1.5 \times 10^2\, P^{2/3} (M/M_\odot)^{1/3}. \tag{25}$$

For accretion to take place the inner edge of the disc must be within $R_\Omega$. There the angular momentum of the matter is transferred to the star via the magnetic field, and accretion along field lines results. However we must first investigate whether the star can throw off any infalling matter *before* it penetrates within $R_\Omega$.
Let us envisage a steady situation in which matter from the inner edge of the disc is being continuously ejected by the rotating magnetic field attached to the star. Consider a ring of matter, mass $m$ and circular velocity $V_c$, being accelerated by the star. By Newton's law for the ring, we may write

$$d/dt(m V_c) = \frac{B_p B_\phi}{4\pi} \times A' \tag{26}$$

where $B_\phi$ and $B_p$ are the azimuthal and poloidal parts of the magnetic field at the ring and $A'$ is the area over which contact is made. From energy considerations we see that to throw matter out we need

$$d/dt(\tfrac{1}{2} m V_c^2) \gtrsim \frac{FGM}{R}, \tag{27}$$

i.e. the rate at which kinetic energy is being supplied by the star must exceed the rate at which potential energy is being lost in the disc.
Combining the above we obtain the inequality

$$F \lesssim \frac{B_\phi B_p V_c R A'}{4\pi G M}. \tag{28}$$

The size of the area $A'$ depends on the magneto-dynamics at the interface and also on the orientation of the stellar field. However, as an order of magnitude estimate we may take

$$A' \approx 2\pi R \cdot 2b. \tag{29}$$

The poloidal field component $B_p$ is part of the stellar field, so $B_p = B_{*12}\, 10^{12} \left(\frac{R}{R_{ns}}\right)^{-3}$ G. The azimuthal component depends on the detailed magneto-dynamics at the interface: we shall leave the ratio $B_\phi/B_p$ as an unknown, hoping that it is of order unity.
Thus we find that (28) becomes

$$F_{16} \lesssim 5.5 \times 10^8 x B_{*12}^2 \left(\frac{B_\phi}{B_p}\right) (M/M_\odot)^{-1/2} \left(\frac{R_{ns}}{10^6\,\text{cm}}\right)^6 R_6^{-7/2}. \tag{30}$$

The condition that accretion be able to take place is that the reverse inequality hold at $R_\Omega$,
i.e.

$$F_{16} \gtrsim F_{\Omega 16} = 13 x B_{*12}^2 \left(\frac{B_\phi}{B_p}\right) (M/M_\odot)^{-5/3} P^{-7/3} \left(\frac{R_{ns}}{10^6\,\text{cm}}\right)^6 \tag{31}$$

An approximate upper limit to the incoming mass flux is given by the "Eddington limit", at which the pressure of outward flowing radiation on free electrons would limit accretion,

i.e. $F_{16} \lesssim F_{\mathrm{Edd}\,16} \approx 100(R_{ns}/10^6\,\mathrm{cm})$. (32)

The Eddington limit holds strictly only for a spherically symmetric situation but it is unlikely that it can be substantially exceeded in the geometry under consideration here.

By requiring that $F_\Omega \lesssim F_{\mathrm{Edd}}$ we obtain a lower limit on the rotational period $P$ if accretion is to occur:

$$P \gtrsim 0.42\, B_{*12}^{6/7} \left(\frac{B_\phi}{B_P}\right)^{3/7} \left(\frac{F_{\mathrm{Edd}\,16}}{100}\right)^{-3/7} \cdot (M/M_\odot)^{-5/7} \left(\frac{R_{ns}}{10^6\,\mathrm{cm}}\right)^{15/7} x^{\frac{3}{7}}\,\mathrm{s}. \quad (33)$$

Schwartzman (1971a) has considered the inhibition of accretion by the particle flux from pulsar-like objects. By assuming a theoretical dependence of this flux on $B$ and $\Omega$, he obtains a condition on $F$ involving these parameters. The process described here is quite distinct from that considered by Schwartzman.

For any disc model we may also define a magnetic radius $R_M$ as being the radius down to which the gas can crush the stellar field. More precisely it is the radius at which the effective pressure in the disc is equal to the external magnetic pressure (for self consistency of the accretion model we require $R_M \lesssim R_\Omega$). To obtain an estimate of the pressure $P'$ in the disc we equate the pressure gradient to the component of gravitation perpendicular to the disc,

$$P'/b \simeq \varrho \left(\frac{GM}{R^2}\right) \cdot \left(\frac{b}{R}\right). \quad (34)$$

By requiring the two pressures to be equal at $R_M$, we obtain

$$\left(\frac{R_M}{10^6\,\mathrm{cm}}\right) = 5.2 \times 10^2 B_{*12}^{4/7} F_{16}^{-2/7} (M/M_\odot)^{-1/7} \cdot x^{-2/7} y^{2/7} \left(\frac{R_{ns}}{10^6\,\mathrm{cm}}\right)^{12/7}. \quad (35)$$

The condition ($R_M \lesssim R_\Omega$) then yields a lower bound to the density of the disc, for a given value of x,

$$y^{-1} \gtrsim 76\, B_{*12}^2 F_{16}^{-1} (M/M_\odot)^{-5/3} P^{-7/3} x^{-1} \left(\frac{R_{ns}}{10^6\,\mathrm{cm}}\right)^6. \quad (36)$$

*Change of the Stellar Period*

We shall take the inner edge of the disc to be at $R_M \lesssim R_\Omega$. There the matter's angular momentum is transferred to the star, speeding the star up and thus tending to reduce $R_\Omega$ to $R_M$.

The flux of angular momentum at $R_M$ is

$$h = F(GMR_M)^{1/2} = 1.2 \times 10^{32} F_{16} (M/M_\odot)^{1/2} \left(\frac{R_M}{10^6\,\mathrm{cm}}\right)^{1/2} \mathrm{g\,cm^{-2}\,s^{-2}}. \quad (37)$$

Taking the moment of inertia of the star to be

$$I = 2/5\, M R_{ns}^2 = 8 \times 10^{44} (M/M_\odot) \left(\frac{R_{ns}}{10^6\,\mathrm{cm}}\right)^2 \mathrm{g\,cm^2}$$

we obtain

$$\dot{P}/P = -7.2 \times 10^{-7} F_{16} (M/M_\odot)^{-1/2} \cdot P \left(\frac{R_M}{10^6\,\mathrm{cm}}\right)^{1/2} \left(\frac{R_{ns}}{10^6\,\mathrm{cm}}\right)^{-2} \mathrm{year}^{-1}. \quad (38)$$

Note that the energy of the emitted radiation comes from the gravitational energy released by the infalling matter. It does *not* come from the rotational energy of the neutron star – indeed, (38) tells us that in all cases when accretion is allowed the matter falling in from the disc actually *speeds up* the stellar rotation.

*Radiation from the Star*

For $R_M/R_{ns} \gtrsim 3$ the flow near the star takes place along field lines the gas being funnelled down an almost undistorted dipole field. Thus most of the radiation will come from the stellar magnetic poles. The condition can be written.

$$B_{*12} \gtrsim 1.8 \times 10^{-4} F_{16}^{1/2} (M/M_\odot)^{1/4} \cdot \left(\frac{R_{ns}}{10^6\,\mathrm{cm}}\right)^{-5/4} x^{1/2} y^{-1/2}. \quad (39)$$

The area $A$ on the star over which accretion takes place is given approximately by

$$A \approx R_{ns}^3/R_M\,\mathrm{cm}^2.$$

We shall take

$$A = 10^{10} a^2\,\mathrm{cm}^2$$

where we see that

$$a^2 = \left(\frac{R_{ns}}{10^6\,\mathrm{cm}}\right)^2 \frac{R_{ns}}{(R_M/100)}.$$

The infalling gas has plenty of time to radiate, and the accretion will be supersonic. In fact the funnelling effect modifies Bondi's (1952) condition to be $1 < \gamma < 7/5$ where $\gamma$ is the usual ratio of specific heats.

The infall velocity is taken to be freefall, which yields the density in the column to be

$$\varrho = 6.2 \times 10^{-5} a^{-2} F_{16} (M/M_\odot)^{-1/2} R_6^{-5/2}\,\mathrm{g\,cm^{-3}}. \quad (40)$$

The kinetic energy of the gas will be dissipated when it strikes the stellar surface, the spectrum of radiation emitted being that of a hot thermal source with possibly a high energy tail. (See e.g. Zeldovich and Shakura (1969).) For a black body spectrum the temperature is given by

$$T = 1.3 \times 10^7 F_{16}^{1/4} (M/M_\odot)^{1/4} \cdot \left(\frac{R_M}{10^6\,\mathrm{cm}}\right)^{1/4} \left(\frac{R_{ns}}{10^6\,\mathrm{cm}}\right)^{-1} {}^\circ\mathrm{K}. \quad (41)$$

A large amount of both circular and linear polarization might be expected, even though the radiation is basically thermal, since it is produced in the presence of a large magnetic field. However the direction in which the radiation is emitted will be affected by the opacity of the infalling material. At these temperatures electron scattering provides the dominant opacity. The vertical optical depth from a point in the centre of the column is

$$\tau_v = 9.9\, a^{-2} F_{16} (M/M_\odot)^{-1/2} R_6^{-3/2} \tag{42}$$

whereas the *horizontal* optical depth from the same point is

$$\tau_h = 1.24\, a^{-1} F_{16} (M/M_\odot)^{-1/2} R_6^{-5/2}\,. \tag{43}$$

Thus the radiation will come out along a cone, and to an external observer would appear regularly pulsed.

## 4. Tentative Interpretations of Some Particular Sources

We may now attempt to relate the model to observations of some individual X-ray sources in the galaxy.

### *Cygnus X-1*

Cygnus X-1 is a prime candidate for being a black hole accreting matter from a disc. It belongs to a binary system in which mass transfer seems to be taking place (Webster and Murdin, 1972; Bolton, 1972), and its mass has been estimated to lie well above the limiting mass for neutron stars or white dwarfs. Also, it has an apparently non-thermal X-ray spectrum, which is what we would expect for a disc whose temperature depends on radius.

The observed spectral index $\alpha$ varies between $\sim 1.6$ and $\sim 4$ in the 1–10 keV range (Schreier *et al.*, 1971), which exceeds the value arising from free-free emission in the case $\lambda = \mu = 0$. But Eq. (23) shows that a steeper slope may occur if $\lambda$ and $\mu$ have suitable non-zero values. (For these values, however, $T_{ff}$ increases with $R$. Also, the precise value of $\alpha$ is very sensitive to $\lambda$ and $\mu$, so the variable slope is not surprising if the disc is somewhat unsteady.) The intensity fluctuates on timescales $\lesssim 1$ s. There may be short "pulse trains" (though even this is controversial (Terrell, 1972)), but there is definitely no evidence for any single preferred period. The rotation rate at the inner edge of the disc is $\sim 10^{-3} (M/M_\odot)$ s, and the observations suggest $M \simeq 4 M_\odot$. This model therefore allows variability on timescales as short as this. The infall time is $\sim 100$ times the rotation period, and this is perhaps the most likely timescale for variability. If the viscosity is primarily magnetic in origin, one could speculate that, even if $F$ were constant, "flares" might sometimes burst out of the disc, giving rise to the rapid fluctuations. Should the amplitude or timescale of the X-ray variations depend on photon energies, this would tell us something about the form of $T(R)$.

Above 10 keV, the X-ray spectrum flattens. This could be due to a change in the density gradient (i.e. in $\lambda$ or $\mu$) at some radius, or to thermalization effects. Alternatively, these hard X-rays may be due to a different mechanism, as proposed by Jackson (1972), who has attempted to explain the "anti-eclipse" (Dolan, 1971; Bolton, 1972; Webster and Murdin, 1972) observed at $\gtrsim 20$ keV [3]).

### *Centaurus X-3*

The most marked features of this source are the large amplitude pulsations, with a period of about 4.8 s. The period is steady apart from the 2.09 day modulation due to orbital motion around the inferred binary companion. The pulses certainly contain $\gtrsim 70\%$ of the X-ray energy, and may contribute as much as 99% (Schreier *et al.*, 1972).

We suggest that Centaurus X-3 is a neutron star with an oblique magnetic field and a 4.8 s rotation period, and that the X-ray pulses are due to accreted material funnelled down to the magnetic poles. The exact shape of the pulses would depend on the shape of the emission cone, the orientations of the magnetic and rotation axes, and the angle our line of sight makes with the plane of the disc. The pulse shape plotted by Schreier *et al.* (1972) gives some indications of an "interpulse", suggesting that both magnetic poles contribute to the observed pulsing.

The overall spectrum of Centaurus X-3 is thermal, as would be expected for our model (though substantial linear or circular polarization is quite possible); and it is clear from (41) that the observed temperature of $\sim 3 \times 10^7$ °K (Giacconi *et al.*, 1971) can be matched for a plausible choice of $F_{16}$ and $R_M$. Equation (38) tells us that the accretion causes the stellar rotation rate to alter. The reported decrease of 1 part in $\sim 4000$ between January and May 1971 is of the right sign and order of magnitude. The mass of the compact object in Centaurus X-3 is not yet known. Our suggestion obviously requires that it should not exceed the limiting neutron star mass. (We note, in this connection, that a "pulsar" model for Centaurus X-3 – in which the kinetic energy of a spinning neutron star provides the power – is not tenable. The rotational energy of a neutron star with a period as long as $\sim 5$ s would only be able to power the X-ray emission for $\lesssim 100$ years so a substantial *increase* in the period would already have been detected.)

It is perhaps significant that both this source and the 1.25 s period pulsating X-ray source on Hercules, designated 2U1702 +35 in the UHURU catalogue (Giacconi *et al.*, 1972) are associated with *eclipsing* binary systems. For there to be noticeable pulsations, the mag-

[3]) Jackson's model attributes the hard X-rays to inverse Compton scattering of light from the companion B star in a region $\sim 10^{11}$ cm across. It would therefore not be tenable if the hard X-rays displayed rapid variability. An alternative interpretation of the "anti-eclipse" might then be that the binary orbit is eccentric, so that the mass transfer rate, and hence $F$, varies with the 6 day binary period.

netic axis may have to make a *large* angle with the rotation axis. If this is so, and if the half-angle of the emission cone is small (say $\lesssim 45°$), this may not be a coincidence.

*Relevance of Model to Other Observations*

The sources discussed above are the two whose binary character is most firmly established. We suspect, however, that our model has more general relevance, though applications to other specific sources are at the moment somewhat conjectural.

We have seen that even our present simple model can display a wide range of behaviour for different values of the few free parameters we have included. The X-ray emission could vary on any time-scale down to $\sim 10^{-3}$ s, either irregularly (black hole) or with a steady period (spinning magnetized neutron star). X-ray spectra of either power law or exponential form are obtainable.

Violently flaring sources, such as the source in Crux observed by Lewin *et al.* (1971), may be discs with unsteady accretion rate $F$, which become unstable – because (19) or (21) is violated – for some values of $F$. It is possible that matter is alternately accreted and expelled.

Finally, we briefly indicate some reasons why searches for line emission from galactic X-ray sources may not prove fruitful. We have interpreted the X-ray pulses of Centaurus X-3 as thermal emission from "caps" around the magnetic poles of a neutron star. This radiation would have an essentially black body spectrum (but perhaps with the low energy photons attenuated by absorption), and any line features would be much less prominent than those from an optically thin plasma. In objects when the X-ray emissions comes primarily from the disc itself – as we suspect is the case in Cygnus X-1 – line emission may be comparable to free-free emission. However, the lines would be severely Doppler broadened so that $\delta v/v \simeq V_c/c$ (with $V_c$ given by (1)). Since most of the energy is liberated near the inner edge of the disc, where $V_c$ is large, we would generally expect line widths $\gtrsim 10\%$. The only exception to this would be cases when $F_{16} \gg 1$, in which case inequality (19) may be violated and the disc expanded by radiation pressure, out to values of $R$ so large that $V_c/c \ll 1$. The emission lines, especially resonance lines may be further broadened by electron scattering (Angel, 1969; Felten *et al.*, 1972).

The source whose spectrum has been searched most thoroughly for evidence of line emission is Sco X-1. So far no narrow emission lines have been seen, the upper limits falling well below estimates based on simple models. Since the binary nature of Sco X-1 is still an open question, the relevance to this source of our present discussion is unclear. However the similarity of its hard X-ray/soft $\gamma$-ray spectrum to that of Cyg X-1 is worth bearing in mind (Haymes *et al.*, 1972).

We wish to emphasize strongly that the "predicted" line strengths in Sco X-1 are based on a simple homogeneous model whose size ($\sim 10^9$ cm) and density ($n_e \sim 10^{16}$ cm$^{-3}$) are inferred by attributing the infrared cut-off in the spectrum to self-absorption by the hot X-ray-emitting plasma (see Neugebauer *et al.*, 1969). In disc models, the optical and infrared emission would come from the outer parts, and the X-rays from a more compact inner region. Indeed, it is interesting that a spectrum of the kind shown in Fig. 2 fits all the data on Sco X-1, including the infrared turnover, if the disc extends out to a radius of $\gtrsim 10^{11}$ cms. (The optical and infrared emission then comes predominantly from material at $\sim 10^4$ °K rather than $\sim 6 \times 10^7$ °K, and so that linear dimensions inferred from self-absorption ($\propto T^{-1/2}$) become $\sim 10^{11}$ cm rather than $\sim 10^9$ cm.) We would expect our simple discussion of the disc's dynamics to be applicable out to $\sim 10^{11}$ cms. Beyond that radius, the gravitational effects of the companion star would no longer be negligible.

*Note added in proof*

Because viscosity transports *energy* outwards, as well as angular momentum, the surface brightness $p(R)$ may exceed the rate of release of gravitational energy *at radius R* by a significant factor. It has been pointed out to us independently by D. Lynden-Bell and K. S. Thorne that, except at points near the inner and outer edges of the disc, this factor is 3, and we have included it in Eq. (6). The subsequent formulae are therefore somewhat inaccurate near the edges of the disc; also, our Eq. (7) is only approximate, since $L$ depends on the precise inner boundary condition.

When a neutron star accretes matter, it contracts, and this leads to an additional energy release whose magnitude is comparable with (7). This energy, however, will be radiated from the whole stellar surface and not just from the magnetic polar caps. It therefore contributes either a steady background of lower-energy X-rays, or (if the star contracts in discrete jumps) could emerge as sporadic bursts.

We have recently received a preprint by N. I. Shakura and R. A. Sunyaev which discusses accretion discs around black holes. These authors make more specific assumptions about the viscosity than the present paper, but obtain results which are fully consistent with our own.

*Acknowledgement.* J. E. Pringle acknowledges an S.R.C. Studentship.

**References**

Angel, J. R. P. 1969, *Nature* **224**, 160.
Bardeen, J. M. 1970, *Nature* **226**, 64.
Bolton, C. T. 1972, *Nature* **235**, 271.
Bondi, H. 1952, *M.N.R.A.S.* **112**, 195.
Cox, D. P., Tucker, W. H. 1969, *Ap. J.* **157**, 1157.
Dolan, J. F., 1971, *Nature* **233**, 109.
Felten, J. E., Rees, M. J. 1972, *Astr. Astrophys.* **17**, 226.

Felten, J. E., Rees, M. J., Adams, T. F. 1972, *Astr. Astrophys.* (in press).
Giacconi, R., Gursky, H., Kellogg, E., Schreier, E., Tananbaum, H. 1971, *Ap. J. Lett.* **167**, L 67.
Giacconi, R., Gursky, H., Murray, S., Schreier, E., Tananbaum, H. 1972, *Ap. J.* (in press).
Jackson, J. C. 1972, *Nature, Physical Science* **236**, 39.
Haymes, R. C., Harnden, F. R., Johnson, W. N., Prichard, H. M., Bosch, H. E. 1972, *Ap. J. Lett* **172**, L 47.
Lewin, W. H. G., McClintock, J. E., Ryckman, S. G., Smith, W. B. 1971, *Ap. J. Lett.* **166**, L 69.
Lynden-Bell, D. 1969, *Nature* **223**, 690.
Lynden-Bell, D., Rees, M. J. 1971, *M.N.R.A.S.* **152**, 461.
Neugebauer, G., Oke, J. B., Becklin, E., Garmire, G. 1969, *Ap. J.* **155**, 1.
Prendergast, K. H., Burbidge, G. R. 1968, *Ap. J. Lett.* **151**, L 83.
Schreier, E., Gursky, H., Kellogg, E., Tananbaum, H., Giacconi, R. 1971, *Ap. J. Lett.* **170**, L 21.
Schreier, E., Levinson, R., Gursky, H., Kellogg, E., Tananbaum, H., Giacconi, R. 1972, *Ap. J. Lett.* **172**, L 79.
Schwartzman, V. F. 1971a, *Sov. Astr. A. J.* **15**, 342.
Schwartzman, V. F. 1971b, *Sov. Astr. A. J.* **15**, 377.
Shklovskii, I. S. 1967, *Ap. J. Lett.* **148**, L 1.
Terrell, J. 1972, preprint.
Webster, B. L., Murdin, P. 1972, *Nature* **235**, 37.
Zeldovich, Y. B., Novikov, I. D. 1971, Relativistic Astrophysics, Stars and Relativity, Chapter 13, Univ. Press. Chicago.
Zeldovich, Y. B., Shakura, N. I. 1969, *Sov. Astr. A. J.* **13**, 175.

M. J. Rees
J. E. Pringle
Institute of Theoretical Astronomy
Madingley Road
Cambridge CB3 OEZ, U.K.

Astron. & Astrophys. 24, 337–355 (1973)

# Black Holes in Binary Systems. Observational Appearance

N. I. Shakura
Sternberg Astronomical Institute, Moscow, U.S.S.R.

R. A. Sunyaev
Institute of Applied Mathematics, Academy of Sciences, Moscow, U.S.S.R.

Received June 6, 1972

**Summary.** The outward transfer of the angular momentum of the accreting matter leads to the formation of a disk around the black hole. The structure and radiation spectrum of the disk depend, mainly on the rate of matter inflow $\dot{M}$ into the disk at its external boundary. The dependence on the efficiency of mechanisms of angular momentum transport (connected with the magnetic field and turbulence) is weaker. If $\dot{M} = 10^{-9} - 3 \cdot 10^{-8} \frac{M_\odot}{\text{year}}$ the disk around the black hole is a powerful source of X-ray radiation with $h\nu \sim 1 - 10$ keV and luminosity $L \approx 10^{37} - 10^{38}$ erg/s. If the flux of the accreting matter decreases, the effective temperature of the radiation and the luminosity will drop. On the other hand, when $\dot{M} > 10^{-9} \frac{M_\odot}{\text{year}}$ the optical luminosity of the disk exceeds the solar value. The main contribution to the optical luminosity of the black hole arises from reradiation of that part of the X-ray and ultra-violet energy which is initially produced in the central high temperature regions of the disk and which is then absorbed by the low temperature outer regions. The optical radiation spectrum of such objects must be saturated by broad recombination and resonance emission lines. Variability, connected with the character of the motion of the black hole, with gas flows in a binary system and with eclipses, is possible. Under certain conditions, the hard radiation can evaporate the gas. This can counteract the matter inflow into the disk and lead to autoregulation of the accretion.

If $\dot{M} \gg 3 \cdot 10^{-8} \frac{M_\odot}{\text{year}}$ the luminosity of the disk around the black hole is stabilized at the critical level of $L \approx 10^{38} \frac{M}{M_\odot} \frac{\text{erg}}{\text{s}}$. A small fraction of the accreting matter falls under the gravitational radius whereas the major part of it flows out with high velocity from the central regions of the disk. The outflowing matter is opaque to the disk radiation and completely transforms its spectrum. In consequence, at the supercritical regime of accretion the black hole may appear as a bright, hot, optical star with a strong outflow of matter.

**Key words:** black holes – binary systems – X-ray sources – accretion

The black hole (collapsar) does not radiate either electromagnetic or gravitational waves (Zeldovich and Novikov, 1971). Therefore, it can be found only due to its gravitational influence on the neighbouring star or on the ambient gas medium (the gas must accrete with the release of large amount of energy (Salpeter, 1964; Zeldovich, 1964)).

Many papers have suggested searching for collapsars in binary systems. It is often considered that the collapsar should appear as a "black" body which practically does not influence the total radiation of the system. In this paper, the attention of the reader is drawn to the case where the outflow of matter from the surface of the visible component and its accretion by the black hole should lead to an appreciable observational effect. In the system with an outflow of matter $\frac{dM}{dt} = \dot{M} > 10^{-12} \frac{M_\odot}{\text{year}}$, the luminosity of the disk around the black hole formed by the accreting matter can be comparable and even exceed the luminosity of the visible component. In a typical case most of the radiation is emitted in the spectral range of $h\nu \sim 100 - 10^4$ eV. However, as will be shown below, the optical and ultra-violet (responsible for the formation of a Strömgren region) luminosities are also high. Therefore, it is entirely possible that black holes are among the optical objects, soft X-ray sources and the harder X-ray sources now being intensively investigated. The radiation connected with accretion by black holes in binary systems has, in fact, distinctive features. However, they are not as astonishing as is usually assumed; the black holes may be hidden among known objects.

Truly "black" objects may be found only in remote binary systems typified by a weak stellar wind from the visible component.

## I. The General Picture

Up to 50% of stars are in binary systems (Martynov, 1971). A sufficiently massive ($M > 2M_\odot$) star, being a component of the binary system, is able to evolve up to the moment when it loses stability and to collapse[1]. In this case, it is possible that an appreciable number of binary systems will not be destroyed and the stars will remain physically bound. These statements are, of course, controversial. However if we recall that the total number of stars with $M > 2M_\odot$ which have existed in the Galaxy is of the order of $10^9$ (Zeldovich and Novikov, 1967), then it becomes clear that the number of binary systems including a black hole might be very large (up to $10^6 - 10^8$).

The outflow of matter from a star's surface – the stellar wind – is evidently one of the main properties of stars. The rate of mass loss depends upon the type of star and varies from $2 \cdot 10^{-14} \frac{M_\odot}{\text{year}}$ for the Sun up to $10^{-5} \frac{M_\odot}{\text{year}}$ for the nuclei of planetary nebulae, Wolf-Rayet stars, MI supergiants and O-stars of the main sequence (Pottasch, 1970). In binary systems, an additional strong matter outflow connected with the Roche limiting surface is possible. At a definite stage of evolution, for example after leaving the main sequence, the star begins to increase in size and after the Roche volume is filled, there is an intensive outflow of matter, mostly through the inner lagrangian point (Martynov, 1971).

What will be the consequences of the existence of a black hole in a binary system if matter flows strongly outwards from the visible star? Some fraction of the matter flowing out from the normal star must fall into the sphere of influence of the gravitational field of the black hole, accrete to it, and finally fall within its gravitational radius (Fig. 1). If the matter undergoes free radial infall (if it was initially at rest and there was no magnetic field), the cold matter accretes to the black hole without any energy release or observational effect (Zeldovich and Novikov, 1971). However, in a binary system, the matter flowing out from the normal star and falling on the black hole has considerable angular momentum relative to the latter, which prevents free fall of the matter. At some distance from the black hole centrifugal forces are comparable to gravitational ones and the matter begins to rotate in circular orbits. The matter is able to approach the gravitational radius only if there exists an effective mechanism for transporting angular momentum outward.

The magnetic field, which must exist in the matter flowing into the disk, and turbulent motions of the matter enable angular momentum to be transferred outward. The efficiency of the mechanism of angular momentum transport is characterized by parameter $\alpha = \frac{v_t}{v_s} + \frac{H^2}{4\pi \varrho v_s^2}$ where $\frac{\varrho v_s^2}{2} = \frac{3}{2} \varrho \frac{kT}{m_p} + \varepsilon_r$ is the thermal energy density of the matter, $\varepsilon_r$ is the energy density of the radiation, $v_s$ is the sound velocity and $v_t$ the turbulent velocity. In part II below we show that $\alpha \lesssim 1$. The most probable model is that of accretion with formation of a disk around the black hole (Prendergast, 1960; Gorbatsky, 1965; Burbidge and Prendergast, 1968; Lynden-Bell, 1969; Shakura, 1972; Pringle and Rees, 1972). The particles in the disk, due to friction between adjacent layers, lose their angular momentum and spiral into the black hole[2]. Gravitational energy is released during this spiraling. Part of this energy increases the kinetic energy of rotation and the other part turns into the thermal energy and is radiated from the disk surface. The total energy release and the spectrum of the outgoing radiation are determined mainly by the rate of accretion, i.e. by the rate inflow of matter into the disk[3]. The basic parameter is the

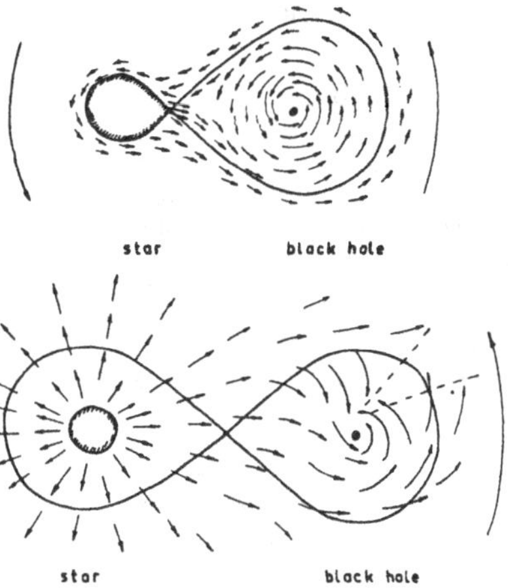


Fig. 1. Two regimes of matter capture by a collapsar: a) a normal companion fills up its Roche lobe, and the outflow goes, in the main, through the inner lagrangian point; b) the companion's size is much less than Roche lobe the outflow is connected with a stellar wind. The matter loses part of its kinetic energy in the shock wave and thereafter, gravitational capture of accreting matter becomes possible

[1]) It is possible that $M_{min}$ considerably exceeds $2M_\odot$ (Zeldovich and Novikov, 1971).

[2]) The disks formed by the matter flowing out from the second component are some times observed around one of the stars in ordinary binary systems (Kraft, 1963).

[3]) The efficiency $\alpha$ of the angular momentum transport mechanism is assumed to be constant along the disk in our calculations ($v_t$ and $H$ are varied in accordance with the change of $\varrho v_s^2$). The observational appearance of the disk (spectrum of its radiation and the effective temperature of the surface) do not strongly depend on the chosen value of $\alpha$. However, at supercritical regime, this dependence becomes dominant.

value of the flux of matter $\dot{M}_{cr}$ at which the total release of energy in the disk $L = \eta \dot{M} c^2$ is equal to the Eddington critical luminosity $L_{cr} = 10^{38} \frac{M}{M_\odot} \frac{\text{erg}}{\text{s}}$, characterized by the equality of the force of radiation pressure on the completely ionized matter and of the gravitational forces of attraction to the star ($\eta$ is the efficiency of gravitational energy release, in the case of Schwarzschild's metric $\eta \simeq 0.06$, in a Kerr black hole $\eta$ can attain 40%). For a black hole of mass $M$, the critical flux is given by $\dot{M}_{cr} = 3 \cdot 10^{-8} \frac{0.06}{\eta} \frac{M}{M_\odot} \frac{M_\odot}{\text{year}}$. This is no particular reason for considering a rate of accretion

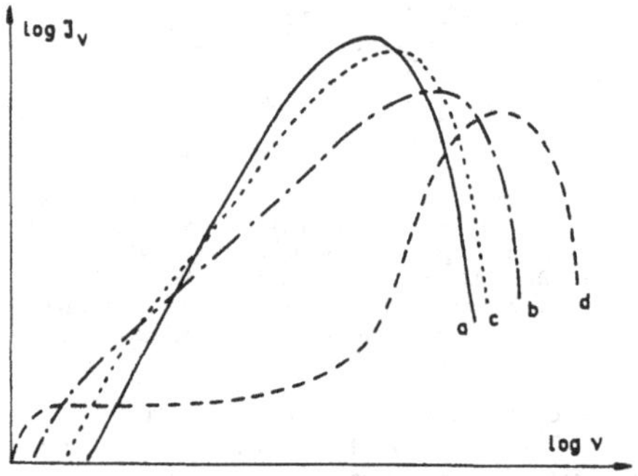


Fig. 2. Characteristic local spectra of radiation formed in the disk a) the black body spectrum $Q = bT^4$. b) the radiation spectrum of an isothermic, homogeneous medium where the main contribution to the opacity comes from scattering $Q = \text{const}\sqrt{n}\,T^{2.25}$. c) the same in an isothermal, exponential atmosphere: $Q = \text{const}\,T^{2.5}$. d) the spectrum formed as a result of comptonization $Q = \text{const}\,T^4$. The intensities are normalized so that the energy flux of radiation $Q$ is the same in all four cases. The change of effective temperature of the radiation is clearly seen

exactly equal to $\dot{M}_{cr}$. A subcritical rate of accretion to the disk is possible as well as an inflow of the matter to the disk many times exceeding the critical value.

At essentially subcritical fluxes $\dot{M} = 10^{-12} - 10^{-10} \frac{M_\odot}{\text{year}}$ the luminosity of the disk is of the order of $L = 10^{34} - 10^{36} \frac{\text{erg}}{\text{s}}$.

Maximal surface temperatures are of the order of $T_s = 3 \cdot 10^5 - 10^6\ {}^\circ$K in the inner regions of the disk where most of the energy is released. This energy is radiated mainly in the ultraviolet and soft X-ray bands, which are inaccessible to direct observations[4]. The local radiation spectrum of the disk is formed in its upper layers and depends on the distance to the black hole and the distribution of matter along $z$-coordinate. The possible forms of the local spectrum reduce to four characteristic distributions (Fig. 2). An integral spectrum (Fig. 3) is determined from the expression $J_\nu = 2\pi \int F_\nu(R) R dR$.

For the case of disk accretion, a weak dependence of the radiation intensity on frequency $\nu^{-1 \div +1/3}$ at $h\nu < kT_{max}$ is typical. As a result, the optical luminosity of the black hole may be appreciable. Estimations show (see, part II, §3) that, for black holes with $M = 10 M_\odot$, even if $\dot{M} = 10^{-9} \frac{M_\odot}{\text{year}}$, we may expect the optical luminosity to be of the order of the solar value.

In fact, the optical luminosity can be much higher. It arises from reradiation of the hard radiation of the hot central regions of the disk by the outer layers. The thickness of the disk increases with distance from the black hole (Fig. 4) This is why the outer regions of the disk effectively absorb the X-ray radiation from the

[4]) Black holes radiating in the soft X-ray and hard ultra-violet bands may significantly contribute to the galactic component of the soft X-ray background and to the thermal balance of interstellar medium. Their radiation must ionize and heat neutral interstellar hydrogen.

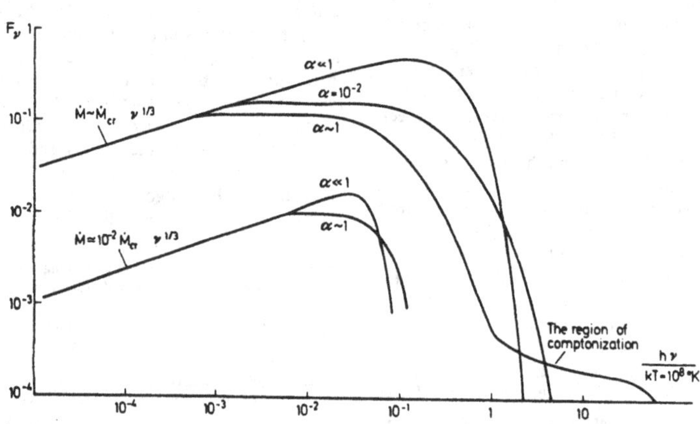


Fig. 3. The integral radiation spectrum of the disk, computed for different $\dot{M}$ and $\alpha$

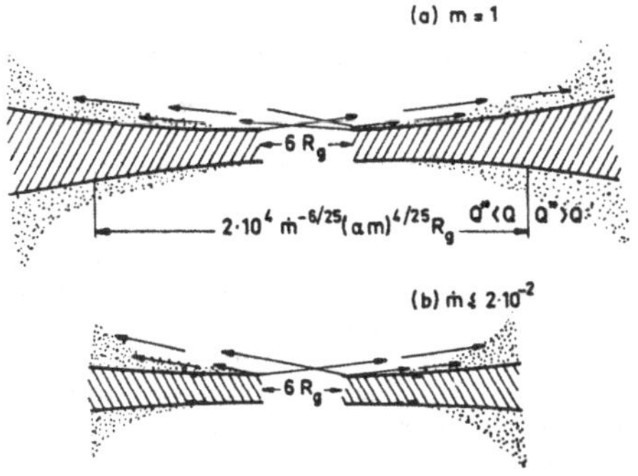


Fig. 4. The thickness of the disk as a function of the distance to the black hole: a) $\dot{M} = \dot{M}_{cr}$. b) $\dot{M} < 10^{-2}\dot{M}_{cr}$. In the central zone, $R < 3R_g$ newtonian mechanics is not applicable. Trajectories of X-ray and ultra-violet quanta which lead to evaporation and heating of the matter in the outer regions of the disk are shown by the arrows. The corona formed by the hot, evaporated matter is denoted by dots

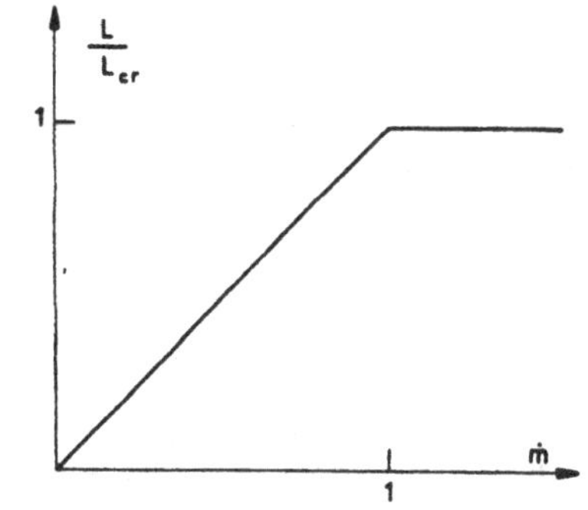


Fig. 5. Dependence of luminosity of the disk around the collapsar on the flux of matter entering its external boundary

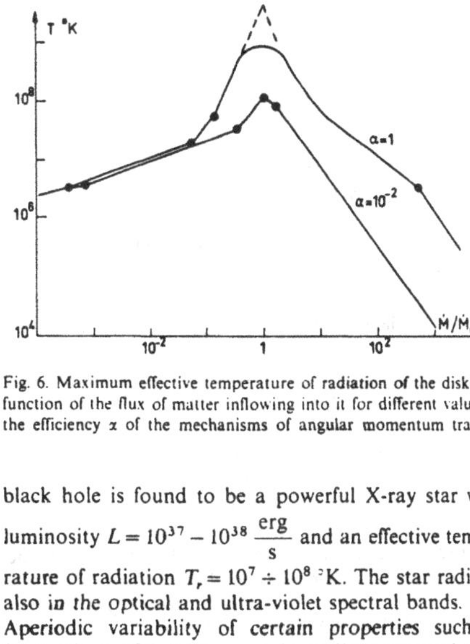


Fig. 6. Maximum effective temperature of radiation of the disk as a function of the flux of matter inflowing into it for different values of the efficiency α of the mechanisms of angular momentum transfer

central regions of the disk and reradiate the absorbed energy in the ultra-violet and optical spectral bands. Thus, from 0.1 – 10% of the total luminosity of the disk can be reradiated (see part III). The hard radiation must be reradiated both in the lines of the different elements and in the continuum.

Strong recombinational fluorescence of hydrogen must be observed with no apparent ionization source and there are possibly also lines of helium and highly ionized heavy elements. All these lines must be broad because the matter in the disk has large rotational velocities ($\gtrsim 100$ km/s). The density of matter in the disk is high and forbidden lines should be absent.

Considerable ultra-violet luminosity of the disk can lead to the formation of Strömgren region which distinguishes a black hole from normal optical stars with similar optical luminosity. In certain conditions the hard radiation of the central regions of the disk can heat the matter in the outer regions up to high temperatures and evaporate the disk, decreasing the inflow of matter into the black hole. Such an autoregulation of accretion can essentially influence the luminosity of the disk around the black hole.

In a close binary system, a significant part of the X-radiation of the black hole can hit the surface of the normal star (Shklovsky, 1967) and be reradiated by its atmosphere, which can lead to an unusual optical appearance in such a system. This effect is observed now in the HZ Her = Her X 1 system. The hemisphere of the optical component turned to X-ray source is three times brighter than opposite one and has a different spectral class (Cherepashchuk *et al.*, 1972; Lyutiy *et al.*, 1973)

When the rate of accretion increases, the luminosity grows linearly and the effective temperature of radiation rises (Figs. 5, 6). At fluxes $\dot{M} = 10^{-9} - 10^{-8}\,\frac{M_\odot}{\text{year}}$ the black hole is found to be a powerful X-ray star with luminosity $L = 10^{37} - 10^{38}\,\frac{\text{erg}}{\text{s}}$ and an effective temperature of radiation $T_r = 10^7 \div 10^8$ °K. The star radiates also in the optical and ultra-violet spectral bands.

Aperiodic variability of certain properties such as fluctuations of brightness, principally connected with the variability of infalling matter flux and its non-homogeneity, distinguish collapsars which radiate due to disk accretion of matter. In remote systems, the collapsar at the perigee of its orbit gets into the more dense part of the matter flux flowing out from the visible component. Therefore, the periodic variability of luminosity (non-sinusoidal in the general case!) should be expected. We also note the possibility of eclipses of the radiation of the central source by the disk if its plane does not coincide with the plane of rotation of the system. Such an orientation of the disk occurs, for example, when matter flows through the inner lagrangian point from the asynchronously rotating star, whose axis of rotation is inclined to the plane of rotation of the system. Taking into account the eclipse of the X-ray source by the adjacent star, the number of eclipses amounts to 3 per period of rotation. If the plane of the disk coincides with the plane of the system, then only the disk can be occulted or darkened whereas a thin disk covers only an insignificant fraction of the stellar surface. However, the most characteristic property of a black hole in a close binary system is its X-ray radiation. The detection of compact X-ray stars of mass $M > 2 \cdot M_\odot$ in binary system will be the proof of the existence of the black holes in the Galaxy.

For a neutron stars we can estimate (in order of magnitude) that the energy release of infalling matter per gramm is the same as for a black hole

$$\left(\eta \sim \frac{GM}{R_{ns}C^2} \sim 10-20\%\right)$$

However, accretion on a neutron star in a binary system has its own peculiarities.

In the case of a neutron star without a magnetic field, the disk extends to its surface. Therefore the disk radiates one half of the entire energy released at the accretion. The other half is radiated by the surface of the neutron star. Accretion to a rotating neutron star whose magnetic field does not coincide in direction with the axis of rotation can lead to the phenomenon of an X-ray pulsar (Schwartzman, 1971a; Pringle and Rees, 1972) and can explain the pulsations of Her X 1 and Cen X 3.

At a subcritical rate of flow of matter into the Roche lobe of the black hole, we may assume that most of the inflowing matter is accreted. At a supercritical value of inflow of matter into the Roche lobe of a black hole or a neutron star[5]) (there is no major difference here) there should be a qualitatively different picture. The region surrounding the black hole is apparent where an effective outflow of matter under the influence of radiation pressure takes place. The outflow begins from a radius close to that at which the forces of radiation pressure and gravitation, pressing the matter to the plane of disk, are comparable. Only the critical flux of the matter can go under the radius $R_0 = \dfrac{6GM}{C^2}$.

The integrated luminosity of such an object is limited by the value of the critical Eddington luminosity, and the band of the electromagnetic spectrum in which most of the energy is radiated depends strongly on the density of the outflowing matter. The density of the matter is, in turn, a function of $\dfrac{\dot{M}}{\dot{M}_{cr}}$ and the efficiency $\alpha$ of mechanism of the transport of angular momentum, which determines velocity of the outflowing gas. If the flux $\dot{M}$ exceeds insignificantly the critical value and $\alpha \sim 1$, the radiation of the disk is reradiated by the outflowing gas practically without changing its spectral properties, i.e. the object is a source of X-ray radiation as before. If $\dfrac{\dot{M}}{\dot{M}_{cr}}$ increases and $\alpha$ decreases the opacity of the outflowing matter grows, and the radiation is re-emitted as quanta of smaller energy. If $\dfrac{\dot{M}}{\dot{M}_{cr}} \gtrsim 3 \cdot 10^3 \left(\dfrac{\alpha M_\odot}{M}\right)^{2/3}$ the black hole turns into a bright optical star. The smaller is the parameter $\alpha$, the greater the effective radius of the radiating envelope and the smaller is the effective temperature of the radiation.

By virtue of the fact that the angular momentum of the ejected matter is conserved relative to the axis rotation of the disk a strongly anisotropic picture of matter outflow can be observed. The hot plasma is ejected with high velocity in a narrow cone about the axis of rotation. The optical depth of the outflowing gas in this cone is not great and, at a specific orientation of the binary system relative to the observer, the X-ray radiation of the black hole together with the optical should be observed.

The observational appearance of the black hole in a strongly supercritical regime of accretion can be characterized as follows: the luminosity is fixed at the Eddington critical limit $L_{cr} \approx 10^{38} \dfrac{M}{M_\odot} \dfrac{\text{erg}}{\text{s}}$; most of the energy is radiated in the ultra-violet and optical regions of the spectrum; in the upper, rarefied layers of the outflowing matter, broad emission lines are formed. In consequence, at the supercritical regime of accretion the black hole may appear as a hot, optical star. There is a strong mass outflow with velocities $v \sim \alpha \cdot 10^5 \left(\dfrac{\dot{M}_{cr}}{\dot{M}}\right)^{1/2} \dfrac{\text{km}}{\text{s}}$ and the star is surrounded by a colder disk where the accreting matter enters the collapsar. Eclipses of the black hole by the normal component are possible as well as eclipses of the star by the matter flowing out from the black hole. The latter is opaque due to Thomson's scattering to large distances from the black hole ($R_T \sim 10^{10} - 10^{12}$ cm). In the radiorange, the hot, outflowing matter becomes opaque far from the binary system. It can be a source of an appreciable thermal radiation with a smooth dependence of intensity on frequency ($J_\nu \sim \nu^{2/3}$).

In the radio range (this relates also to subcritical accretion) non-thermal radiation mechanisms connected with the existence of magnetic fields (which may achieve $H \sim 10^5 - 10^7$ Gauss) and beams of fast outflowing particles can also appear (Lynden-Bell, 1969).

Apparently, the "quiet" disk, radiating only due to thermal mechanisms, can really exist at low values of the parameter $\alpha$. If $\alpha \sim 1$, the new important effects (connected with turbulent convectivity; plasma turbulence; reconnection of magnetic field lines through neutral points, leading to solar type flares; the acceleration of particles) and non-thermal radiation can appear. The

[5]) Critical accretion to the neutron star leads to its collapse in $\sim 3 \cdot 10^{[illegible]}$ years.

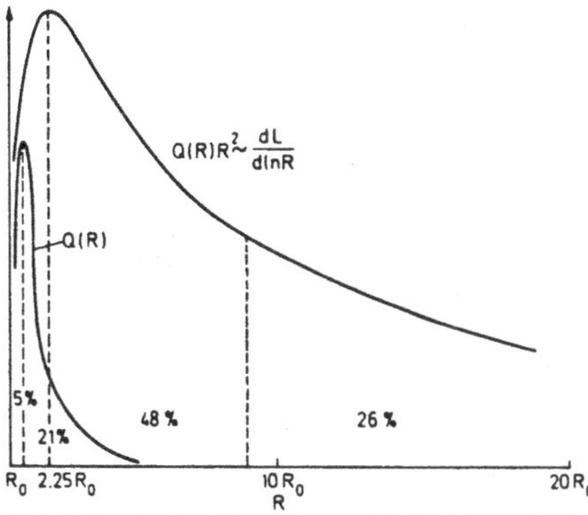

Fig. 7. The luminosity of the surface unit of the disk as a function of the radius. The function $Q(R)\,R^2$ is proportional to the luminosity of the ring with radius $R$ and $\Delta R \sim R$. The numbers illustrate the contribution of the corresponding regions to the integral luminosity of the disk

flares and hot spots on the rotating disk surface must lead to the short term fluctuations of radiation flux in some spectral bands. The variability may have a stochastic (Schwartzman, 1971b) and/or quasiperiodic nature (Sunyaev, 1972). Quasiperiod of this fluctuations must be of the order of rotational period

$$t \sim 2\frac{\pi R}{v_\varphi} \sim 6 \cdot 10^{-4}\left(\frac{R}{3R_g}\right)^{3/2}\left(\frac{M}{M_\odot}\right)\mathrm{s}$$

and depends on the distance of the hot spot from the collapsar. According to the Fig. 7, the main part of the energy released is radiated from the region with $6R_g < R < 30R_g$. At $M \sim 10\,M_\odot$ the corresponding quasiperiods $t \sim 2 \cdot 10^{-2} \div 0.2$ s are of the order of observed in Cyg X 1. Minimal period of the radiation of disk with the hot spot in the Kerr gravitational field of the rotating black hole is 8 times less, than in the Schwarzschild field of the non-rotating one with equal mass (Sunyaev, 1972).

## II. The Theory of Disk Accretion at Subcritical Regime

We consider a disk around the black hole in the case of the existence of an effective mechanism of outwards transport of angular momentum. All calculations are carried out using newtonian mechanics. It is necessary to take into account the effects of general relativity only in the region $R < 3R_g$. Near the black hole at $R < 3R_g$ stable circular orbits are not possible (Kaplan, 1949), and the motion acquires a radial character without transport of angular momentum. In the last stable orbit the binding energy is given by $0.057\,m_0c^2$ and it must escape before the particle reaches the radius $R = 3R_g$. The energy is also radiated by matter moving in a zone with $R_g < R < 3R_g$ but the contribution of this zone to the total luminosity of the disk does not appreciably alter the estimate of the energy released. In our calculations we shall assume that $0.06\,c^2$ energy per unit mass of infalling matter is released and that the total luminosity of the disk is $L_0 = 0.06\,c^2\dot{M}$. The radius $R_0 = 3R_g$ should be taken as the inner boundary of the disk, because, that in the range $R < R_0$ the matter falls without any external observable effect.

### 1. *Mechanisms of Angular Momentum Transfer*

In a differentially rotating medium, tangential stresses between adjacent layers, which are connected with existence of a magnetic field, turbulence and molecular and radiative viscosity are the mechanisms of transport of angular momentum. In the conditions of interest to us the role of molecular viscosity is negligibly small and cannot lead to disk accretion; neither can angular momentum transport by means of radiation (which itself is the consequence of accretion).

Magnetic fields exist in stars. They must also occur in the matter flowing out from the normal star and then accreting to the black hole. Depending on the initial conditions, the method of matter compression and on the degree of ordering of the magnetic field, the energy of the latter can be less than the thermal energy of matter $\varepsilon = 3\varrho\frac{kT}{m_p} + bT^4 = \varrho\frac{v_s^2}{2}$, as well as sufficiently can exceed it, reaching the value $\frac{H^2}{8\pi} \sim \varrho\frac{GM}{R}$.

In this last case, the stresses $\left(w_{r\varphi} = \left|\frac{H_r \times H_\varphi}{8\pi}\right|\right)$ and the efficiency of angular momentum transport are so high, that radial accretion will occur (Bisnovaty-Kogan and Sunyaev, 1971; Schwartzman, 1971a).

The magnetic field in the gas flowing into the disk may have a regular structure at the distances comparable to the outer radius of the disk. But as always div $H = 0$, the radial component of the magnetic field must have alternating sign. If alternation of the sign of the magnetic field takes place in the disk, then the differential rotation rapidly leads to division of the large magnetic loops into smaller ones. Hence, within the disk, the field is most likely to be chaotic and of small scale. In this case, because of plasma instabilities, reconnection of the magnetic field lines in regions of opposite polarity field, etc..., the energy of the magnetic field in the disk evidently cannot exceed the thermal energy of the matter. The latter is characterized by the velocity of sound $v_s$

$$\frac{H^2}{8\pi} < \varrho\frac{v_s^2}{2} \quad \text{and} \quad w_{r\varphi} \sim -\varrho v_s^2\left(\frac{H^2}{4\pi\varrho v_s^2}\right). \tag{1.1}$$

Now, in the absence of complete theory of turbulence on one hand, and of some observational check of existence of turbulence in the disks on the other hand, we may only assume its presence. In a differentially rotating medium with a distribution of angular momentum increasing outwards, matter is stable with respect to small shifts preserving the angular momentum. But the theory of small perturbations, being linear, can give, at best, only the condition of loss of stability of a laminar flow. The energy criterion (Vasjutinsky, 1946), which takes into account non-linearity, i.e. interaction between the turbulent pulsations at high Reynolds numbers, leads to the conclusion that in differential rotation a self-perpetuating turbulence is always possible regardless of the angular momentum distribution. Experiment (Taylor, 1937) also shows that there exists the boundary behind which fluid, rotating between two coaxial cylinders, develops turbulence (even in the case of rotation of the outer cylinder, the inner one being at rest) although conditions of stability relative to the distribution had been applied. The results obtained for an incompressible fluid are not, of course, completely applicable to the gaseous disk, but the conclusion concerning perpetuation of turbulence developed in the disk, where the Reynolds numbers are especially large, apparently remains valid. We must also bear in mind that additional sources of turbulence connected with release of gravitational energy and transfer of the radiative flux to the surface layers exist in the disk. In this case, the maximal scale of the turbulent cell $l$ is probably of the order of the disk thickness $z_0$.

In scales comparable with the radius $R$, turbulence is homogeneous and isotropic. To describe the average motions in the presence of such turbulence one may use the formulae obtained for laminar flows on replacing the molecular viscosity by the turbulent one $\eta_t = \varrho v_t l$. For tangential stresses, we have

$$w_{r\varphi} = \eta_t R \frac{d\omega}{dR} \sim -\eta_t \frac{v_\varphi}{R} \sim -\varrho v_s^2 \frac{v_t}{v_s}$$

where the disk thickness $z_0 \sim R \frac{v_s}{v_\varphi}$. Here, $v_t$ is the turbulent velocity, $v_\varphi$ is the circular keplerian velocity. Thus the efficiency of two of the most important mechanisms of angular momentum transport connected with the magnetic field (which always is present in astrophysical conditions) and turbulence (whose existence in the disk is less definite)

$$-w_{r\varphi} \sim \varrho v_s^2 \left(\frac{v_t}{v_s}\right) + \varrho v_s^2 \left(\frac{H^2}{4\pi \varrho v_s^2}\right) = \alpha \varrho v_s^2 \tag{1.2}$$

can be characterized by only one parameter, $\alpha$. In the case of a turbulent mechanism $\alpha < 1$ always. For $\alpha > 1$ turbulence must be supersonic and leads to rapid heating of the plasma and to $\alpha \lesssim 1$. For magnetic transport of angular momentum in the disk, it is likely that $\alpha < 1$. For a wide range of the initial conditions, it is possible that $\alpha \ll 1$. In addition, the parameter $\alpha$ can (and must) be a function of the disk radius[6]. Below, we point out that in the wide range

$$10^{-15} \left(\frac{\dot M}{\dot M_{cr}}\right)^2 < \alpha < 1$$

the structure of the disk is not essentially changed. This result allows us to compute the external appearance of the disk, the character of energy release, the radiation spectrum etc..., without deciding upon the mechanism of angular momentum transfer, an exact account of which is difficult.

*2. The Structure of the Disk*

To a first approximation, the matter in the disk may be assumed to rotate in circular keplerian orbits

$$v_\varphi = \sqrt{\frac{GM}{R}}, \qquad \omega = \sqrt{\frac{GM}{R^3}}. \tag{2.1}$$

The friction between the adjacent layers, connected with the existence in the disk of turbulence and chaotic, small scale magnetic fields, leads to the loss of the angular momentum of the particles. A radial component of velocity appears and the particles spiral inward to the black hole

$$\frac{u_0 d\omega R^2}{dt} = -u_0 v_r \frac{d\omega R^2}{dR} = \frac{1}{R}\frac{d}{dR} W_{r\varphi} R^2. \tag{2.2}$$

Here, $u_0 = 2\int_0^{z_0} \varrho dz$ is the surface density of matter in the disk, $W_{r\varphi}$ is the stress between adjacent layers. According to (1.2)

$$W_{r\varphi} = 2\int_0^{z_0} w_{r\varphi} dz = -\alpha u_0 v_s^2$$

In stationary conditions, $v_r < 0$ and $\dot M = 2\pi u_0 v_r R$ = const and integrating (2.2) we obtain

$$\dot M \omega R^2 = -2\pi W_{r\varphi} R^2 + C. \tag{2.3}$$

Practically the whole angular momentum is transported outward and only a small part $\sim \sqrt{\frac{3R_g}{R_1}}$ of the initial angular momentum falls together with the matter into the black hole ($R_1$ is the outer radius of the disk). The constant in (2.3) is determined by the condition, that $W_{r\varphi} \simeq 0$ on the last stable orbit ($R_0 = 3R_g$ in the Schwarzschild gravitational field of black hole or a neutron star with $R_s < R_0$, or the corresponding Kerr metric value for rotating black hole). To describe the loss of stability one must make a consequent general relativistic theory, which is worked through in Novikov and Thorne (1972) lectures.

In the case of nonrotating black hole const. in (2.3) equals to $C = \dot M \omega(R_0) R_0^2$ and

$$\dot M \omega \left[1 - \left(\frac{R_0}{R}\right)^{1/2}\right] = 2\pi \alpha u_0 v_s^2. \tag{2.4}$$

In the case of a neutron star with $R_s > R_0$ and without magnetic field the condition $W_{r\varphi} \simeq 0$ is fulfilled practically at the surface: during the slow decrease of $R$, the

[6] The dependence $\alpha(R)$ for the turbulent mechanism may be approximately evaluated from the experimental data (Taylor, 1937) on supposing the scale of the turbulence to be equal to the width of the channel between the cylinders. At the periphery of the disk, where $Z_0/R \ll 1$, the coefficient may be of the order of unity; in the vicinity of a black hole in the supercritical regime, when the accretion picture is sphereazied $Z_0/R \sim 1$ and the coefficient $\alpha \sim 10^{-3}$.

angular velocity $\omega$ first increases according to the Kepler law and then suddenly goes practically to zero in a thin layer of rotating gas supported by pressure. The maximum of $\omega$ is nearly equal to the last keplerian value $\sqrt{\frac{GM}{R_s^3}}$ at $R_s$ and const. in (2.3) equals to $\dot{M}\omega(R_s)R_s^2$ because $W_{r\varphi}(R_s) \simeq 0$.

A selfconsistent axialsymmetric picture uses injection of matter at some $R_1$. Part of matter situated at $R_1$ falls on the star but one must imagine also matter flow outward $R_1$, taking away the excess of an angular momentum.

In a direction perpendicular to the plane of the disk the normal component of the gravitational force of the star is balanced by the sum of the gradients of the gas, radiation turbulent and magnetic pressures. The equation of hydrostatic equilibrium gives the half-thickness of the disk.

$$z_0 = \frac{v_s}{v_\varphi} R . \tag{2.5}$$

In losing their angular momentum the particles also lose their gravitational energy. Part of the latter goes to increasing the kinetic energy of rotation and the other part is converted into thermal energy and can be radiated from the surface of the disk. The forces leading to the angular momentum transfer in a rotating system are also inducing the energy flow equal to $-2\pi W_{r\varphi}R^2\omega$. In keplerian motion with $\omega$ increasing inward and $W_{r\varphi} < 0$, the energy flow is directed outward. The rate of the energy dissipation in the ring between $R_2$ and $R_3$ has a term equal to divergence of this energy flow. Collecting all terms, one obtains the energy flux, radiated from surface unit of the disk in unit of the time

$$\begin{aligned} Q &= \frac{1}{2} W_{r\varphi} R \frac{d\omega}{dR} \\ &= \frac{1}{4\pi R}\frac{d}{dR}\left[\dot{M}\left(\frac{v_\varphi^2}{2} - \frac{GM}{R}\right) - 2\pi R^2 W_{r\varphi}\omega\right] \\ &= \frac{3}{8\pi}\dot{M}\frac{GM}{R^3}\left\{1 - \left(\frac{R_0}{R}\right)^{1/2}\right\}. \end{aligned} \tag{2.6}$$

At $R \gg R_0$, $R_s$ the energy flux is equal to $\frac{3}{8\pi}\frac{GM}{R^3}\dot{M}$ and the release of energy between the radii $R_2$ and $R_3$ is equal to $L = 4\pi \int QRdR = \frac{3}{2}\dot{M}GM\left(\frac{1}{R_2} - \frac{1}{R_3}\right)$. This value is increased three times as compared with the release of gravitational energy in the same region. The unexpected energy is actually released from gravity at much smaller radii, and then is transported outward mechanically by the shear stresses before being converted into heat[7]. In the contrary at $R - R_0 \ll R$ the energy flux, decreases (going to zero) near the last stable orbit (Novikov and Thorne, 1972). The maximum of $Q(R)$ takes place at $R = 1.36\ R_0$. The main contribution to integral luminosity of the disk $L(R_0)$ gives the region with $R = 2.25\ R_0$, where $Q(R)R^2$ has a maximum. At a given flux $Q$, the energy density of radiation inside the layer with the surface density $u_0$ is determined by the relation

$$\varepsilon = \frac{3}{4}\frac{Q}{c}\sigma u_0 = \frac{9}{32\pi}\dot{M}\frac{GM}{R^3}\frac{\sigma u_0}{c}\left[1 - \left(\frac{R_0}{R}\right)^{1/2}\right] \tag{2.7}$$

where $\sigma$ is the opacity of the matter. In the conditions considered, the main contribution to the opacity comes from Thomson scattering on free electrons of cross-section $6.65\ 10^{-25}\,\mathrm{cm}^2$ and free-free absorption for which $\sigma_{ff} = 0.11\ T^{-7/2} n \frac{\mathrm{cm}^2}{\mathrm{g}}$ (Zeldovich and Rayzer, 1966). Inside the disk, which is optically thick with respect to the "true" absorption, $\tau = \sigma_{ff} u_0$, or $\tau^* = \sqrt{\sigma_T \sigma_{ff}} u_0$, if $\sigma_T > \sigma_{ff}$, there exists complete thermodynamic equilibrium and the energy density of radiation is equal to $\varepsilon = bT^4$. This last expression, the dynamic Eq. (2.4) and the equation of energy balance (2.7) form a closed system of equations. Upon solving them, we find the distribution of surface density of matter $u_0(R)$ and the temperature $T(R)$ along the radius of the disk as functions of the mass flux $\dot{M}$, the mass of the black hole $M$ and the efficiency $\alpha$ of the angular momentum transport mechanism.

In the general case of rotating black hole the releasing gravitational energy is transferred mechanically from the relativistic region $\frac{1}{2}R_g < R < 3R_g$ into the nonrelativistic one $R > 3R_g$ and dissipate there. Changing the value of constant in Eq. (2.3) it is easy to take into account this contribution to $Q$ at $R > 3R_g$. According to (2.6) in the case of nonrotating black hole the energy flux from a surface unit at first increases with approaching to collapsar, reaches the maximum at $R = 1{,}36\ R_0$ and then decreases. Only 5% of the total disk luminosity is radiated at $R < 1.36R_0$. Therefore we consider this region only schematically.

The disk may be considered to be composed of a number of distinct parts:

a) the radiation pressure is dominant and $v_s^2 = \frac{\varepsilon}{3\varrho}$.

In the interaction of matter and radiation, electron scattering on free electrons plays the main rôle;

b) the pressure is determined by the gas pressure and $v_s^2 = \frac{kT}{m_p}$; electron scattering gives still the main contribution to the opacity;

c) the speed of sound is given by $v_s^2 = \frac{kT}{m_p}$ and the opacity is determined by free-free absorption and other mechanisms. Two regions c) are the extensive outermost and the very narrow closest to the black hole. In the region of the maximal energy flux the radiation

[7]) Dr. K. S. Thorne directed our attention to this increase and to the importance of the last term in square brackets in the Eq. (2.6).

pressure is dominant. Two regions b) are intermediate between a) and c).

It is convenient to introduce nondimensional parameters

$$m=\frac{M}{M_\odot}, \quad \dot{m}=\frac{\dot{M}}{\dot{M}_{cr}}=\frac{\dot{M}}{3\cdot10^{-8}\frac{M_\odot}{\mathrm{yr}}}\times\left(\frac{M_\odot}{M}\right),$$

$$r=\frac{R}{3R_g}=\frac{1}{6}\frac{Rc^2}{GM}=\frac{M_\odot}{M}\frac{R}{9\,\mathrm{km}}.$$

Let us consider the region a) $P_r \gg P_g$, $\sigma_T \gg \sigma_{ff}$. The half-thickness of the disk, corresponding to (2.5) and (2.7), is

$$z_0[\mathrm{cm}]=\frac{3}{8\pi}\frac{\sigma_T}{c}\dot{M}(1-r^{-1/2})=3.2\cdot10^6\dot{m}m\,(1-z^{-1/2}) \tag{2.8}$$

i.e. in this region at $r \gg 1$ the disk has a constant thickness along the radius, whose value depends only on the flux of accreting matter. The result (2.8) is expected in view of the fact that $z_0$ is determined by equating the force of the radiation pressure $F \sim Q \sim \dot{M}R^{-3}$ to the component of the gravitational force, normal to the plane of the disk which is also proportional to $R^{-3}$. The maximal ratio $\frac{z_0}{R}=\dot{m}$ is reached at $r=2.25$.

Substituting $v_s^2=\frac{\varepsilon}{3\varrho}$ into (2.3) and bearing in mind that $u_0=2\varrho_0 z_0$, we obtain from (2.3) and (2.7)

$$u_0\left[\frac{\mathrm{g}}{\mathrm{cm}^2}\right]=\frac{64\pi}{9\alpha}\frac{c^2}{\sigma^2}\frac{1}{\omega\dot{M}(1-r^{-1/2})} \tag{2.9}$$
$$=4.6\alpha^{-1}\dot{m}^{-1}r^{3/2}(1-r^{-1/2})^{-1},$$

$$\varepsilon\left[\frac{\mathrm{erg}}{\mathrm{cm}^3}\right]=2\frac{c}{\sigma}\omega=2.1\cdot10^{15}\alpha^{-1}m^{-1}r^{-3/2}, \tag{2.10}$$

$$\left.\begin{aligned} n[\mathrm{cm}^{-3}] &= \frac{u_0}{2m_p z_0} \\ &= 4.3\cdot10^{17}\alpha^{-1}\dot{m}^{-2}m^{-1}r^{3/2}(1-r^{-1/2})^{-2} \\ v_r\left[\frac{\mathrm{cm}}{\mathrm{s}}\right] &= \frac{\dot{M}}{2\pi u_0 R} \\ &= 7.7\cdot10^{10}\alpha\dot{m}^2r^{-5/2}(1-r^{-1/2}) \\ H[\mathrm{Gauss}] &\lesssim \sqrt{\frac{4\pi}{3}\alpha\varepsilon}=10^8m^{-1/2}r^{-3/4}. \end{aligned}\right\} \tag{2.11}$$

Assuming, that the disk is optically thick with respect to the "true" absorption, i.e. the relation $\varepsilon=bT^4$ is satisfied, we find from (2.10) the temperature of the plasma and of the radiation inside the disk

$$T=2.3\cdot10^7(\alpha m)^{-1/4}r^{-3/4}\,{}^\circ\mathrm{K}. \tag{2.12}$$

For a plasma with $\sigma_T \gg \sigma_{ff}$ a "true" optical depth with respect to absorption is determined as: $\tau^*=\sqrt{\sigma_T\sigma_{ff}}u_0$ i.e.

$$\tau^*=8.4\cdot10^{-5}\alpha^{-17/16}m^{-1/16}\dot{m}^{-2}\cdot r^{-93/32}(1-r^{-1/2})^{-2}.$$

From conditions that $\tau^*>1$, one finds that the disk is opaque if

$$r>25\,\alpha^{34/93}\dot{m}^{64/93}m^{2/93}(1-r^{-1/2})^{64/93} \tag{2.13}$$

and that the assumption of local thermodynamic equilibrium inside the disk is justified. The inequality (2.13) is only an upper limit. When $y=\frac{kT}{m_ec^2}\tau_T^2>1$ Compton scattering plays the dominant rôle in the formation of the radiation spectrum because of the Doppler shift in the frequency of the bremsstrahlung quanta (Kompaneets, 1956; Illarionov and Sunyaev, 1972). In this case an equilibrium, black body spectrum is formed when

$$\tau_T\tau_{ff}\ln^2\frac{2.35}{x_0}\gtrsim1. \tag{2.14}$$

Here

$$\left.\begin{aligned} x_0 &= \frac{h\nu_0}{kT}=3\cdot10^5\frac{n^{1/2}}{T^{9/4}}\sqrt{g(x_0)}; \\ g(x_0) &= \frac{\sqrt{3}}{\pi}\ln\frac{2.35}{x_0}. \end{aligned}\right\} \tag{2.15}$$

$\nu_0$ is a frequency near which the rates of the free-free and Compton processes are comparable. The factor $A=\frac{3}{4}\ln^2\frac{2.35}{x_0}=2.5\,g^2(x_0)=L_c^-/L_{ff}^-$ characterizes the relation of the Compton energy losses with the bremsstrahlung losses. In the physical conditions of interest to us the factor $A$ ranges from 10 to 300. Using condition (2.14), it is easy to show that, even, at $\alpha \sim 1$ and $\dot{m} \sim 1$ local thermodynamic equilibrium exists inside the disk up to $R \approx 10R_0$.

A high temperature $T \gg 10^7\,{}^\circ$K of the matter near the black hole corresponds to narrow intervals in the values of $\alpha \sim 1$ and $\dot{m} \sim 1$. The energy losses of the plasma due to radiation are limited by the low rate of production of photons due to free-free processes. Synchrotron radiation of thermal electrons in the magnetic field and plasma radiation at the Langmuire frequency are additional sources of quanta. Comptonization of the low frequency radiation leads to an increase in the energy losses as well as to a decrease in the plasma temperature (Gnedin and Sunyaev, 1972). These mechanisms of quanta production are of importance in the above mentioned conditions because, at $\alpha \sim 1$ and $r \sim 1$, the gyrofrequency $\nu_H=\frac{eH}{2\pi m_ec}=\frac{e}{2\pi m_ec}\sqrt{4\pi\alpha\varrho v_s^2}$ and Langmuire frequency $\nu_{pe}=\sqrt{\frac{e^2n}{\pi m_e}}$ exceed $\nu_0$. Estimations show that nowhere in the disk the temperature exceed $T \simeq 10^9\,{}^\circ$K.

Correlations similar to (2.8 – 2.14) can be also obtained for the remaining parts of the disk. For regions b) and c) we give only the final expressions; their derivation is similar to the one given above for the region a)

b) $P_g \gg P_r, \quad \sigma_T \gg \sigma_{ff}$

$$
\begin{aligned}
u_0 &= 1.7 \cdot 10^5 \alpha^{-4/5} \dot{m}^{3/5} m^{1/5} r^{-3/5} (1 - r^{-1/2})^{3/5} \\
T &= 3.1 \cdot 10^8 \alpha^{-1/5} \dot{m}^{2/5} m^{-1/5} r^{-9/10} (1 - r^{-1/2})^{2/5} \\
z_0 &= 1.2 \cdot 10^4 \alpha^{-1/10} \dot{m}^{1/5} m^{9/10} r^{21/20} (1 - r^{-1/2})^{1/5} \qquad (2.16) \\
n &= 4.2 \cdot 10^{24} \alpha^{-7/10} \dot{m}^{2/5} m^{-7/10} r^{-33/20} (1 - r^{-1/2})^{2/5} \\
\tau^* &= \sqrt{\sigma_{ff}\sigma_T}\, u_0 = 10^2 \alpha^{-4/5} \dot{m}^{9/10} m^{1/5} r^{3/20} (1 - r^{-1/2})^{9/10} \\
v_r &= 2 \cdot 10^6 \alpha^{4/5} \dot{m}^{2/5} m^{-1/5} r^{-2/5} (1 - r^{-1/2})^{-3/5} \\
H &\leqq 1.5 \cdot 10^9 \alpha^{1/20} \dot{m}^{2/5} m^{-9/20} r^{-51/40} (1 - r^{-1/2})^{2/5} .
\end{aligned}
$$

The boundaries between the regions a) and b) lie on the radii

$$\frac{r_{ab}}{(1 - r_{ab}^{-1/2})^{16/21}} = 150 (\alpha m)^{2/21} \dot{m}^{16/21} . \qquad (2.17)$$

From condition (2.17) we find that region a) of the disk exists only if

$$\dot{m} \gtrsim \frac{1}{170} (\alpha m)^{-1/8} \qquad (2.18)$$

c) $P_r \ll P_g, \quad \sigma_{ff} \gg \sigma_T$

$$
\begin{aligned}
u_0 &= 6.1 \cdot 10^5 \alpha^{-4/5} \dot{m}^{7/10} m^{1/5} r^{-3/4} (1 - r^{-1/2})^{7/10} \\
T &= 8.6 \cdot 10^7 \alpha^{-1/5} \dot{m}^{3/10} m^{-1/5} r^{-3/4} (1 - z^{-1/2})^{3/10} \\
z_0 &= 6.1 \cdot 10^3 \alpha^{-1/10} \dot{m}^{3/20} m^{9/10} r^{9/8} (1 - r^{-1/2})^{3/20} \qquad (2.19) \\
n &= 3 \cdot 10^{25} \alpha^{-7/10} \dot{m}^{11/20} m^{-7/10} r^{-15/8} (1 - r^{-1/2})^{11/20} \\
\tau &= \sigma_{ff} u_0 = 3.4 \cdot 10^2 \alpha^{-4/5} \dot{m}^{1/5} m^{1/5} (1 - r^{-1/2})^{1/5} \\
v_r &= 5.8 \cdot 10^5 \alpha^{4/5} \dot{m}^{3/10} m^{-1/5} r^{-1/4} (1 - r^{-1/2})^{-7/10} \\
H &\lesssim 2.1 \cdot 10^9 \alpha^{1/20} \dot{m}^{17/40} m^{-9/20} r^{-21/16} (1 - r^{-1/2})^{17/40}
\end{aligned}
$$

The boundaries between regions b) and c) lie near

$$r_{bc} = 6.3 \cdot 10^3 \dot{m}^{2/3} (1 - r_{bc}^{-1/2})^{2/3} . \qquad (2.20)$$

Our approximation is valid in the case $v_r \ll v_\varphi$. Therefore the formulae given above are valid only at $r - 1 > 10^{-6} \cdot \alpha^{8/7} \dot{m}^{3/7}$. At $r - 1 < 10^{-6}$ more complicated consideration is necessary. In addition any negligibly small energy flux from the relativistic region strongly influences the physical conditions in the vicinity of $R_0$, without any influence upon the conditions in the region $R > \frac{49}{36} R_0$, where the main part of the energy released is radiated and our consideration is applicable. Below we consider only this region. Analysing the formulae obtained above, we note the weak dependence of the thickness of the disk on the efficiency $\alpha$ of the mechanism of angular momentum transport. When $\alpha$ decreases, the surface density of the disk increases rapidly and the radial velocity of motion drops, but the disk thickness grows only as $\alpha^{-1/10}$ and is comparable with the radius only if $\alpha \sim 10^{-15} \dot{m}^2$. Disk accretion is in fact realized at sufficiently weak turbulence or at small values of the magnetic field. At extremely small $\alpha$, nuclear reactions can give some contribution to the energy release and neutrino processes can influence energy losses. We would also like to point out the weak variation of the optical depth $\tau^*$ for the "true" absorption in the region c) on changes in the accretion rate $\dot{M}$, i.e. there the disk is opaque, as a rule.

## *2a. Disk Structure along the Z-Coordinate*

In the direction perpendicular to the disk plane, hydrostatic equilibrium exists. The pressure gradient is balanced by the component of the gravitational attraction normal the disk plane (selfgravitation of the disk is negligibly small):

$$\frac{1}{\varrho} \frac{dP}{dz} = - \frac{GM}{R^3} z . \qquad (2.21)$$

The equation (2.21) together with the equation of energy balance

$$\frac{1}{\varrho} \frac{dq}{dz} = \frac{3}{4\pi} \frac{GM}{R^3} \frac{\dot{M}}{u_0} \left[ 1 - \left( \frac{R_0}{R} \right)^{1/2} \right] . \qquad (2.22)$$

and the equation of radiative transfer:

$$\frac{c}{3\sigma\varrho} \frac{d\varepsilon_r}{dz} = - q(z) \qquad (2.23)$$

form a closed system of equations. Solution of this system determines the distribution of physical quantities along the $Z$-coordinate. It is easy to integrate Eq. (2.22)

$$q = 2Q \frac{u(z)}{u_0} \qquad (2.24)$$

where

$$Q = \frac{3}{8\pi} \frac{GM}{R^3} \dot{M} \left[ 1 - \left( \frac{R_0}{R} \right)^{1/2} \right] \quad \text{and} \quad u(z) = \int_0^z \varrho(z) dz .$$

In regions a) and b) electron scattering dominates the opacity and, in (2.23), one may assume $\sigma$ to be equal to $\sigma_T = 0.4 \frac{\text{cm}^2}{\text{g}}$. Postulating thermodynamic equilibrium inside the disk, we obtain from (2.22) and (2.23):

$$T^4 = T_c^4 \left[ 1 - \frac{3Q\sigma_T u_0}{cbT_c^4} \left( \frac{u}{u_0} \right)^2 \right] .$$

Near the surface of the disk, $Q = \frac{cb}{4} T_s^4$ and $T_c^4 = T_s^4 \left[ 1 + \frac{3}{16} \sigma_T u_0 \right]$. Therefore for an opaque disk $\left( \frac{3}{16} \sigma_T u_0 \gg 1 \right)$ we obtain

$$T(u) = T_c \left[ 1 - 4 \left( \frac{u}{u_0} \right)^2 \right]^{1/4} . \qquad (2.25)$$

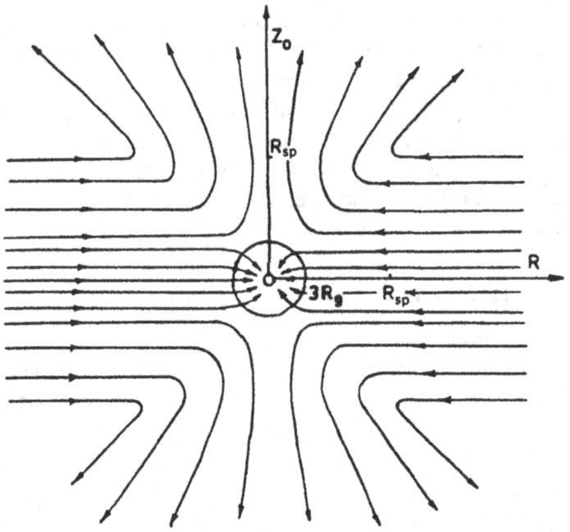


Fig. 8. Lines of matter flow at supercritical accretion (the disk section along the Z-coordinate). When $R < R_{sp}$ spherization of accretion takes place and the outflow of matter from the collapsar begins

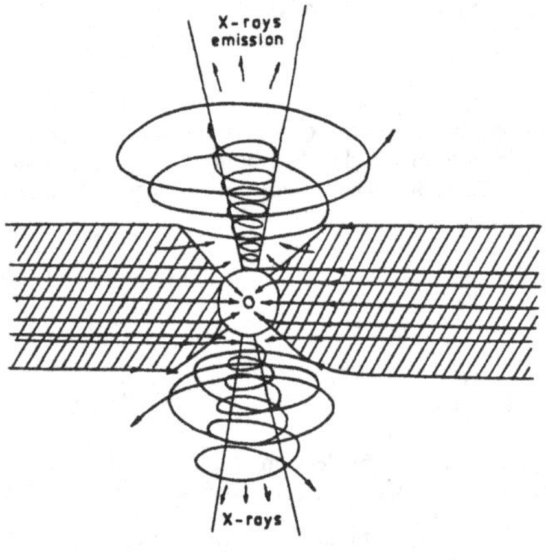


Fig. 9. The outflow of the matter from the collapsar at the supercritical regime of accretion

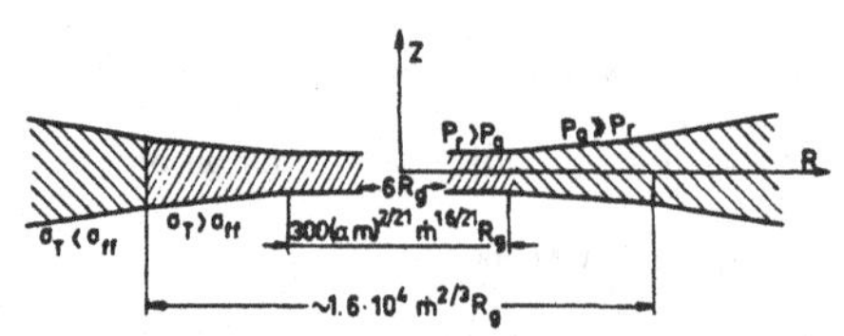


Fig. 10. The regions of disk having different physical conditions

The temperature does not vary much at low

$$u = \int_0^z \varrho\, dz \ll \frac{u_0}{2} = \int_0^\infty \varrho\, dz$$

and correspondingly at low optical depths $\tau = \sigma_T u(z)$ and low $Z$. Therefore, the disk structure may be characterized by a central temperature depending only on the coordinate $R$, and the dependence on the coordinate $Z$ may be neglected. However, the outgoing radiation spectrum, formed in the upper layer of the disk is strongly dependent on the density and temperature distribution along the coordinate $z$.

In the region a) (closest to the collapsar) radiation pressure $P_r = \frac{\varepsilon}{3}$ dominates. According to (2.21) and (2.23)

$$q(z) = \frac{c}{\sigma_T} \frac{GM}{R^3} z\,. \tag{2.26}$$

Furthermore, $q(u) = 2Q \frac{u}{u_0}$ and $\varrho = \frac{du}{dz}$, and therefore the disk must be homogeneous with a sharp (depending only on the temperature of the plasma, the turbulent and magnetic pressure) decrease of matter density at $z > z_0$.

In regions b) and c) the gas pressure $p = \varrho \frac{kT}{m_p}$ dominates.

For $\frac{u}{u_0} \ll \frac{1}{2}$, the temperature in the disk is practically constant (2.25), and the density decreases according to a gaussian curve $\varrho = \varrho_0 \exp\left[-\left(\frac{z}{z_0}\right)^2\right]$. With increasing $z$ and $u$, the temperature rapidly decreases and, according to (2.21), the density drops more rapidly.

In zone b), the outgoing radiation spectrum is formed at the depth defined by the condition $\tau^* = \int_u^\infty \sqrt{\sigma_T \sigma_{ff}} \cdot du \sim 1$. At $z > z_1$ the plasma temperature is practically constant. Therefore, according to (2.25), at $z > z_1$ we can assume the density profile

$$\varrho = \varrho(z_1) \exp\left(-\frac{z}{H_0}\right) \tag{2.28}$$

where $H_0 = \frac{R^3 kT(z_1)}{GM m_p z_1}$. The numerical solution of the system of equations (2.21 ÷ 2.23) showed that because of the rapid decrease of the temperature at $z > z_0$ for any conditions $z_1 \approx 1.2 - 1.5 z_0$. In the estimates below we shall assume $z_1 \simeq z_0$.

### 3. *Radiation Spectrum of the Disk*

a) Local Radiation Spectrum

The spectrum shape formed at the disk surface depends on its structure and temperature (which was calculated in the previous section) and, therefore, on the distance to the black hole. The local spectrum of the thermal radiation in the conditions of interest to us may be one of three typical distributions (Fig. 2): a Planck distribution (in the outer regions of the disk), a Wien distribution (in the inner regions) and a spectrum of radiation which has passed through the medium with scattering presumably playing a rôle in the opacity (intermediate region of the disk).

In the outer $r > 800\,\alpha^{4/57}\mathrm{m}^{-46/57}\dot{\mathrm{m}}^{37/57}$ regions[8]), where free-free processes (as well as free-bound processes and absorption in the lines of heavy elements broadened by the gas pressure) give the main contribution to the opacity a planckian spectrum of radiation

$$F(x) = B(x) = \frac{2\pi h}{c^2}\left(\frac{kT}{h}\right)^3 \frac{x^3}{e^x - 1}, \quad \text{where} \quad x = \frac{h\nu}{kT} \tag{3.1}$$

is formed on the disk surface (more exactly, at a depth of $\tau_{ff} \approx 1$). The corresponding flux of energy is equal to

$$Q = \int F(x)dx = \frac{c}{4} b T_s^4 \frac{\text{erg}}{\text{cm}^2\,\text{s}}$$

In the intermediate region

$$800\,\alpha^{4/57}\mathrm{m}^{-46/57}\dot{\mathrm{m}}^{37/57} > r > 25 \times \alpha^{2/9}\dot{m}^{2/3}$$

where in the opacity, Thomson scattering dominates, thermal equilibrium exists only where the "true" optical depth is large ($\tau_{ff}\tau_T > 1$). At the surface, where at sufficiently high frequencies $\varkappa_\nu \ll \sigma_T m_p$, the outgoing radiation spectrum is distorted. In the case of a homogeneous medium with a sharp boundary (Shakura, 1972; Felten and Rees, 1972)

$$F(x) = \sqrt{\frac{3\varkappa(x)n}{\sigma_T m_p}}\, B(x) \sim \text{const}\sqrt{n}\, T^{5/4} \frac{x^{3/2}e^{-x}}{(1-e^{-x})^{1/2}} \tag{3.2}$$

and $Q = 1.8 \cdot 10^{-4}\sqrt{n}\, T^{2.25} \dfrac{\text{erg}}{\text{cm}^2\,\text{s}}$.

In the case of exponential varying atmosphere $n = n(z_1)e^{-z/H_0}$ according to Zeldovich and Shakura (1969) the emerging spectrum has an appearance

$$F(x) = \left(\frac{3\varkappa(x)}{\sigma_T^2 m_p^2 H_0}\right)^{1/3} B(x) \sim \text{const}\, H_0^{-1/3}\, T^{11/6} x^2 \frac{e^{-x}}{(1-e^{-x})^{1/3}} \tag{3.3}$$

$$Q = 1.3 \cdot 10^4 H_0^{-1/3} T^{17/6} \frac{\text{erg}}{\text{cm}^2\text{s}} \sim T^{2.5}.$$

In (3.2) and (3.3) $\varkappa(x) = \dfrac{4.1 \cdot 10^{-23}(1-e^{-x})}{T^{7/2}x^3}\,\text{cm}^5$ is the coefficient of free-free absorption (Zeldovich and Rayzer, 1966). The optical depth due to Thomson scattering of the layer where $\tau^* = \sqrt{\tau_T\tau_{ff}} = 1$ is equal to $\left(\dfrac{\sigma_T m_p}{3\varkappa(x)n}\right)^{1/2}$ in the case of a homogeneous medium and to $\left(\dfrac{\sigma_T^2 m_p^2 H_0}{3\varkappa(x)}\right)^{1/3}$ in the case of exponentially varying atmosphere.

In the inner part of the disk $r < 25 \times \alpha^{2/9}\dot{m}^{2/3}$, the processes of comptonization effect strongly the shape of the emitted spectrum. The radiation spectrum formed in the layer with $y > 1$ has a Wien distribution (Illarionov and Sunyaev, 1972)

$$F(x) \sim x^3 e^{-x}, \qquad Q = \frac{cd(r)}{4} T^4, \qquad d(r) \ll b. \tag{3.4}$$

Due to the dominant role of Compton processes in a) nowhere in the disk is the radiation spectrum that of an optically thin plasma

$$F(x) = \varkappa(x)B(x) \sim e^{-x},$$

$$Q = 1.4 \cdot 10^{-27} T^{1/2} n^2 z_0 \frac{\text{erg}}{\text{cm}^2\,\text{s}}.$$

At the same time, processes connected with the existence of magnetic fields and turbulence can lead to acceleration of the particles and generation of non-thermal radiation at low frequencies (Lynden-Bell, 1969).

b) Distribution of the Surface Temperature along the Disk

The local radiation energy flux $Q$ is determined by the gravitational energy release $\dfrac{3}{8\pi}\dfrac{GM}{R^3}\dot{M}\left(1 - \left(\dfrac{R_0}{R}\right)^{1/2}\right)$ in the disk. This equation gives the surface temperature of matter as a function of radius (Fig. 12) corresponding

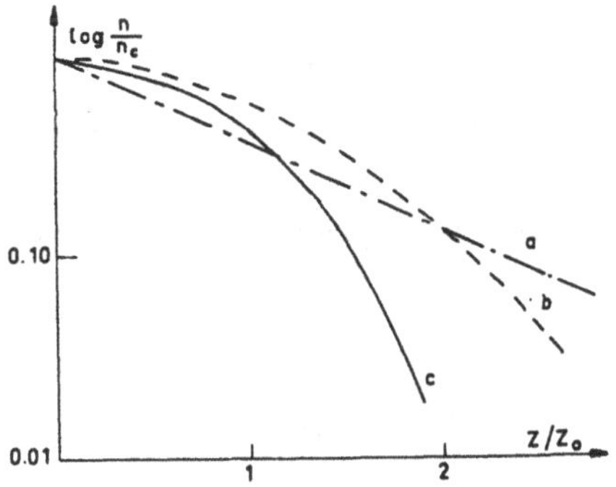


Fig. 11. The profile of the density of the matter in the disk along the Z-coordinate: a) the plane isothermal atmosphere $n \sim \exp-(Z/Z_0)$. b) gaussian atmosphere $n \sim \exp-\left[\frac{1}{2}\left(\frac{Z}{Z_0}\right)^2\right]$, c) real profile of density

[8]) This boundary is closer to the collapsar than that given above $r_{bc}$ because the surface temperature is less than the central value.

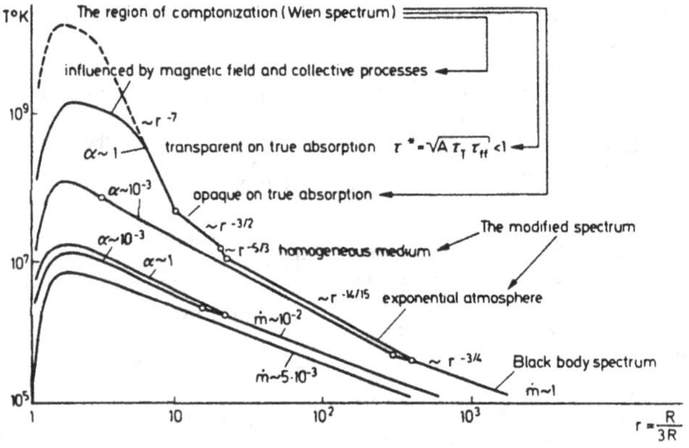


Fig. 12. Distribution of temperature along the radius of the disk if $\dot{M}$ and the parameter $\alpha$ are different

to an effective radiation temperature and mean energy of the outgoing quanta. In the outer regions of the disk

$$T_s = 3 \cdot 10^7 m^{-1/4} \dot{m}^{1/4} r^{-3/4} (1 - r^{-1/2})^{1/4}\ {}^\circ\mathrm{K}. \tag{3.5}$$

As the local radiation spectrum is planckian, then $\bar{x} = \dfrac{\int F_\nu d\nu}{\int \frac{F_\nu}{h\nu} d\nu} = 2.7$. In the intermediate region, according to (3.3), we obtain $\bar{x} = 1.66$ and

$$T_s = 10^8 \alpha^{1/75} \dot{m}^{28/75} m^{-19/75} r^{-141/150} (1 - r^{-1/2})^{28/75}\ {}^\circ\mathrm{K}. \tag{3.6}$$

The intermediate region of the disk also extends over the region with homogeneous matter density in which radiation pressure is dominant. Then, according to (3.2) $\bar{x} = 1.2$ and

$$T_s = 1.4 \cdot 10^9 \alpha^{2/9} \dot{m}^{8/9} m^{-2/9} r^{-5/3} (1 - r^{-1/2})^{8/9}\ {}^\circ\mathrm{K}. \tag{3.7}$$

In the inner region $\bar{x} = 3$. The boundary of this region, determined by condition $y = \dfrac{kT(z_1)}{m_e c^2} \tau_T^2(z_1) > 1$, was estimated assuming the atmosphere to be homogeneous. In this way we also found⁹)

$$\begin{aligned} T_s &= 1.4 \cdot 10^9 A^{-2/9} \alpha^{2/9} \dot{m}^{8/9} m^{-2/9} r^{-5/3} (1 - r^{-1/2})^{8/9} \\ &\simeq 5 \cdot 10^8 \alpha^{1/5} \dot{m}^{4/5} m^{-1/5} r^{-3/2} (1 - r^{-1/2})^{4/5}\ {}^\circ\mathrm{K}. \end{aligned} \tag{3.8}$$

In the region $\tau^*(u_0) = \sqrt{\tau_{ff} \tau_s} < 1$ we obtain

$$T_s = 10^{14} \alpha^{12/5} \dot{m}^{24/5} r^{-36/5} (1 - r^{-1/2})^{24/5}. \tag{3.9}$$

⁹) Here and below the approximation $A \approx 40(T/10^8)^{1/2}$ if $T < 4 \cdot 10^8$ °K and $A = 10^2(T/10^9)^{1/3}$, if $T > 4 \cdot 10^8$ °K is used, the dependence on density being neglected.

c) An Integral Spectrum of the Outgoing Disk Radiation

The spectral distribution of the radiation from the whole disk is obtained by integrating the local spectrum

$$I_\nu = 4\pi \int_{R_0}^{R_1} F_\nu[T_s(R)]\, R\, dR, \tag{3.10}$$

where $R_1$ is the external boundary of the disk. For a local planckian spectrum and dependence $T_s(R)$ of the form (3.5), we obtain for $\nu \ll \dfrac{kT_0}{h}$

$$I_\nu = \frac{16\pi^2 R_0^2 h}{c^2} \left(\frac{kT_0}{h}\right)^{8/3} \nu^{1/3}. \tag{3.11}$$

The spectral index of radiation $\gamma = \dfrac{\nu\, d \ln I_\nu}{d\nu}$ is equal to $\gamma = \dfrac{1}{3}$ for a spectrum of the form (3.1) and dependence of temperature on radius of the form (3.5) (Lynden-Bell, 1969; Shakura 1972). It is easy to find $\gamma$ in the case of $r \gg 1$, when the factor $(1 - r^{-1/2})$ is of no importance:

$\gamma = 0.07$ for spectrum (3.3) and temperature (3.6),

$\gamma = 0.04$ for spectrum (3.2) and temperature (3.7),

$\gamma = -1/3$ for spectrum (3.4) and temperature (3.8),

$\gamma = -1$ for spectrum (3.4) and temperature (3.9).

If $h\nu \gg kT_{\max}$ the radiation spectrum falls exponentially. The computed resulting radiation spectrum of the disk is given in Fig. 3. At these computations the factor $(1 - r^{-1/2})$ was taken into account.

In the optical region of the spectrum for a wide range of initial conditions a spectrum of the form of (3.11) is established. The optical luminosity of the disk ($3000\text{Å} < \lambda < 10000\text{Å}$) is equal to

$$L_{opt} = \int_{\nu_1}^{\nu_2} L_\nu d\nu \simeq 10^{35} m^{4/3} \dot{m}^{2/3} \frac{\text{erg}}{\text{s}} \qquad (3.12)$$

For a black hole of $M = 10 M_\odot$ and even if $\dot{M} \simeq 2 \cdot 10^{-11} \cdot \frac{M_\odot}{\text{year}}$ one may expect the optical luminosity to be of the order of the solar value with a spectrum unusual for a star. In fact the optical luminosity may be somewhat higher than given by (3.12) because a part of the hard radiation flux of the disk is reradiated by the cold periphery.

## III. Influence of the Hard Radiation of the Black Hole on the Outer Parts of the Disk and the Visible Component

### *1. Reradiation of the Hard Radiation by the External Regions of the Disk*

From the theory of disk accretion, it follows that, at $r > 150 (\alpha m)^{2/21} \dot{m}^{16/21}$ the previously plane disk begins to thicken according to the law $z_0 \sim r^{21/20}$ and, for $r > 6 \cdot 10^3 \dot{m}^{2/3}$ as $z_0 \sim r^{9/8}$. Additional thickening may occur due to the increase of the ratio $z_1/z_0$ with radius. The real form of the disk is like a saucer (Fig. 4) and so its external regions catch a certain part $L^*$ of the X-ray radiation of the inner regions $L_0$

$$L^*/L_0 \simeq \left(\frac{z_1(R_{max})}{R_{max}}\right)^2 \simeq \left(\frac{1.5\, z_0(R_{max})}{R_{max}}\right)^2 = 10^{-4} \dot{m}^{3/10} (\alpha \cdot m)^{-1/5} r_{max}^{1/4} \qquad (4.1)$$

Here, $R_{max}$ is the external boundary of the disk. The disk radiates as a plane surface, and therefore the ratio $L^*/L_0$ is proportional to the square of the angle subtended by the external regions of the disk at the central zone. In the case of a spherical source at the centre of the disk (a collapsar at a supercritical regime of accretion or a hot neutron star, for example), the ratio $L^*/L_0$ is proportional to this angle $\frac{z_1}{R_{max}}$. As the disk is optically thick to the X-ray radiation, its surface must absorb and reradiate a significant part of the incident flux of radiation $Q^* = \frac{1}{2\pi R} \frac{dL^*}{dR} = \frac{L_0}{8\pi R^2} \left(\frac{L^*}{L_0}\right)$.

At $r > 10^4 (\alpha m)^{4/25} \dot{m}^{-6/25}$ the flux, incident on the disk surface, exceeds the gravitational energy release inside the disk $Q \approx \frac{3}{8\pi} \frac{GM}{R^3} \dot{M}$.

The incident X-ray radiation is absorbed in photoionizing the heavy elements. After absorption, the elementary processes of radiation of softer quanta, partial absorption of these quanta as well as thermalization of the photoelectrons, etc... lead to important effects. Heating of the matter is accompanied by an outflow of the hot gas: the disk thickness and the absorbed fraction of the hard radiation of the central regions of the disk both increase. The effective thickness of the disk is determined by the boundary of transparency to the incident radiation. Such a swelling of the disk will take place or as long as it is evaporated completely (see discussion below), or if such density and temperature profiles across the outer layers of the disk establish, that they result in complete reradiation of the incident flux of energy. The density of matter must drop rapidly with distance from the plane of symmetry. The temperature at $z > z_0$ must also decrease as $Z$ increases providing there a radiative transfer of the gravitational energy released inside the disk. Then the absorption of the X-ray radiation leads to an increase in temperature with increasing $Z$ (these layers are transparent with respect to free-free absorption but opaque to resonance lines, photoionizing quanta, Thomson scattering, etc). The density and temperature profiles can be determined from the numerical solution of the system of equations of thermal balance, radiative transfer and hydrostatic equilibrium of the matter with the variable temperature in the gravitational field of the black hole. In turn, the profiles of density and temperature determine the fraction of the hard radiation reprocessed in the outer regions of the disk. Estimates show that this fraction $\frac{L^*}{L_0} = \left(\frac{z^*}{R}\right)^2$ can greatly exceed the value (4.1) given by the simple theory of disk accretion which does not take into account the external heating of the disk. The system of equations mentioned above can be written in the first approximation in form: $\frac{z^*}{R} = \frac{v_s}{v_\varphi} = \sqrt{\frac{kT}{m_p} \frac{R}{GM}}$ (as a consequence of the equation of hydrostatic equilibrium), $\frac{R}{z^*} \sigma_{eff} n R \sim 1$ (the condition for absorption of the hard radiation). If the temperature $T > 10^6\,°\text{K}$ and the hard radiation strongly influences the ionization balance of the plasma, one may put $\sigma_{eff} \simeq \sigma_T = 6.65 \cdot 10^{-25}\,\text{cm}^2$ in order of magnitude. $T^* = f\left(\frac{L_0}{nR^2} \frac{z^*}{R}\right) = f\left(\frac{L_0 \sigma_{eff}}{R}\right)$ is the equation of thermal balance. If $\xi = \frac{L_0}{nR^2} \frac{z^*}{R} = \frac{L_0 \sigma_{eff}}{R} > 10^4$ as is shown in the appendix, $T^* \sim T_{eff}$ and one can use the results of the computations of Tarter *et al.* (1969) to determine the temperature if $\xi < 10^4$. The factor $\frac{z^*}{R}$ in the definition of $\xi$ arises from the dependence of the radiation flux from the plane surface on direction ($\sim \cos\theta$). We always obtain $\xi > 10^{-1}$ and $T^* > 10^5\,°\text{K}$ for $10^{-3} L_{cr} < L < L_{cr}$ and $R < 10^{12}$ cm.

Therefore,

$$\frac{L^*}{L_0} = \left(\frac{z^*}{R}\right)^2 = \frac{kT}{m_p}\frac{R}{GM}$$

$$= 5\cdot 10^{-2}\left(\frac{T^*}{10^5\,{}^\circ\mathrm{K}}\right)\left(\frac{R}{10^{12}\,\mathrm{cm}}\right)\frac{M_\odot}{M}.$$

Note that $\sigma_{eff}$ increases rapidly if $\dot{M}$ and the effective temperature of the hard radiation decrease. This leads to an increase in $\xi$ and $T^*$ tends to $T_{eff}$. It would appear that at least $0.1 \div 10\%$ of the total energy released in the source must be reprocessed in the outer regions of the disk.

Fluorescence occurs, in general, in the lines of heavy elements and hydrogen. The hard quanta are able to penetrate so far into the disk that only quanta of energy less than the ionization potential of hydrogen are able to escape. Therefore a significant part (up to 10%) of the reprocessed energy leaves in the form of recombination radiation and resonance lines in the optical range of the spectrum[10]. Most of the energy reradiated is in the ultra-violet and soft X-ray bands and forms an appreciable Strömgren Zone.

*2. Autoregulation of Accretion*

Absorption of the hard radiation of the central source by the matter falling into the disk as well as by the matter on the disk periphery and its subsequent heating can lead to evaporation of this matter. It can also lead to a decrease in the accretion rate. The decrease in the accretion rate leads to a decrease in the hard radiation flux which heats the accreting matter. Such a regime of accretion will be established when the quantity of matter entering the disk is controlled by the flux of radiation emitted. The efficiency of autoregulation is determined, firstly, by the rate of mass loss from the adjacent star, by the type of outflow, by the degree to which the rotation is non-synchronous and by other parameters of the binary system. One can differentiate between two limiting cases: a) for a star filling only a small part of Roche volume matter flows out uniformly from the whole surface with a velocity greater parabolic one for this star $v_P(R_2)$ and b) outflow from a star exceeding its limiting volume takes place through the inner lagrangian point with a velocity much smaller than $v_P(R_2)$. Below we consider autoregulation for the case a) $v_{es} > v_P$ outflow is analogous to the usual "stellar wind" as in the case of single stars. Only some fraction of the outflowing matter will be captured by the black hole. In the vicinity of a collapsar, $R^* = \frac{GM_1}{v_{es}^2}$ a non-spherical shock wave appears (Salpeter, 1964) where matter loses a significant part of its kinetic energy and is gravitationally captured by the collapsar. The captured matter possesses angular momentum. Therefore, the disk is formed inside the sphere of radius $R^*$. If we know the density of the outflowing gas near radius of capture $n(R^*) = n(R_2)\left(\frac{R_2}{R_{12}}\right)^2$ one can easily express the rate of accretion in terms of the velocity of outflowing matter and parameters of the binary system

$$\dot{M}_{ac} = \dot{M}_{out}\left(\frac{M_1}{M_1+M_2}\right)^2\left(\frac{v_P(R_{12})}{v_{es}}\right)^4,$$

$$v_P = \sqrt{\frac{G(M_1+M_{22})}{R_{12}}}$$

where indices 1, 2 and 12 correspond to a black hole, a normal star and a binary system respectively. Such a picture of accretion is typical of remote binary systems. The thermal velocity of the particles $v_s$ at the radius of capture may exceed $v_{es}$, as a result of heating by X-ray radiation. In this case, the rate of inflow of matter to the disk will drop abruptly $\dot{M}'_{ac} \simeq \dot{M}_{ac}\left(\frac{v_{es}}{v_s}\right)^3$. The characteristic times of gas cooling and heating by the central X-ray source are small compared with the time of free fall of the particles at the radius of capture. Therefore, results obtained for a stationary plasma can be used in computations of the accreting gas temperature and its ionization balance. In this case the gas temperature is determined by the effective radiation temperature $T_{eff}$ and the parameter $\xi = \frac{L_0}{nR^{*2}}$. As $L_0 \simeq 0.06c^2$ $\dot{M}_{ac} = 0.06\cdot 4\pi\varrho v_{es}R^{*2}c^2$, then $\xi = 2\cdot 10^{-3}v_{es}$ if $v_{es} > v_s$ or $\zeta = 2\cdot 10^{-3}v_s$ if $v_{es} < v_s$. For $v_{es} > 100\,\frac{\mathrm{km}}{\mathrm{s}}$, we have $\zeta > 10^4$ and conditions in which Compton processes of heating and cooling play the leading role in the energy balance of the plasma (see Appendix).

In the limit of large $\xi$, when $T$ is of the order of the effective temperature $T_{eff}$ of the X-ray radiation, one can easily find the limiting rate of accretion above which it is necessary to take into account the heating: $v_{es}^2 = v_s^2 = \frac{kT_{eff}}{m_p}$ and

$$\dot{M}^* = 0.06\dot{M}_{cr}\left(\frac{v_{es}}{100\,\frac{\mathrm{km}}{\mathrm{s}}}\right)^8\left(\frac{M_1}{M_\odot}\right)^2 \qquad (5.3)$$

When $\dot{M} < \dot{M}^*$ the rate of accretion grows linearly with increasing rate of outflow from the visible star but, as soon as $\dot{M} > \dot{M}^*$, we have $\dot{M}_{ac} \sim \dot{M}_{es}^{8/11}$. For the remote systems due to the geometrical factor, the rate of accretion itself becomes small and the additional influence of autoregulation results the black holes in these systems appearing as optical and ultra-violet objects of low luminosity only.

[10] The spectral width of these lines must be of the order of

$$\frac{\Delta\nu}{\nu} \sim \frac{v_\varphi(R)}{c} \sim 3\cdot 10^{-4}\left(\frac{10^{12}\,\mathrm{cm}}{R}\right)^{1/2}\frac{M}{M_\odot}.$$

*3. The Influence on the Normal Component*

A considerable fraction of the hard radiation of the black hole may be captured by the surface of the normal component leading to unusual spectral effects (Shklovsky, 1967). Part of the energy absorbed must be transformed into thermal energy giving an additional outflow from the star surface. Another part of the absorbed energy is transformed into softer quanta in this way increasing the optical luminosity of the hemisphere of the normal star which is turned to the black hole. Due to the rotation of the whole system, the effects observed in the optical range will be unusual. At the moment when the black hole is situated between the observer and its companion, the optical spectrum has the characteristics of a gas with emission lines which are formed in the disk corona as well as in the companion's atmosphere. The absorption lines belonging to the colder side of the companion will be seen clearly expressed during total or partial eclipses of the black hole. Non-synchronous rotation of the companion and excentricity of the orbit lead asymmetry in the phases of emission and absorption spectra. The recent computations confirm this qualitative picture (Basko and Sunyaev, 1973), and show that sufficient part of X-ray energy flux absorbed by the surface of the visible component is reradiated in the optical continuum, increasing the optical luminosity of the system and leading to its variability.

We note the (somewhat improbable) possibility that, in such binary systems the rate of accretion increases to the Eddington critical value as the result of the additional outflow of matter caused by the heating of the surface of the normal star by the radiation of the black hole. The necessary condition for such a catastrophic increase is that the additional outflow should exceed the initial outflow. This becomes possible only in close binaries with

$$\frac{R_{12}}{R_2} < \left(\frac{\eta}{\gamma}\right)^{1/2} \frac{c}{v_{es}} \frac{M_1}{M_1+M_2}\left[\frac{v_P(R_{12})}{v_{es}}\right]^2$$

where $\gamma$ is the part of the radiation energy absorbed turn into the kinetic energy of the matter flowing out and $\eta$ is the efficiency of gravitational energy release.

## IV. Supercritical Regime of Disk Accretion

At the critical Eddington luminosity $L_{cr} = \frac{4\pi GM}{\sigma_T} = 10^{38}\frac{M}{M_\odot}\frac{\text{erg}}{\text{s}}$ the force of radiation pressure on the electrons $f_1 = \frac{q\sigma_T m_p}{c}$, balances the gravitational attraction of protons and nuclei $f_2 = \frac{GMmp}{R^2}\frac{R}{R}$ Therefore, in spherical accretion the flux of matter cannot exceed $\dot{M}_{cr} = \frac{L_{cr}}{\eta c^2}$ at any distance from the star if even there exist conditions favourable for the formation of a supercritical ($\dot{M} > \dot{M}_{cr}$) flux (sufficiently great density and low temperature of the matter).

The infall of matter within a narrow sector is a peculiarity of the disk accretion. During infall, the matter moves in a slowly twisting spiral in the plane perpendicular to the direction of the angular momentum of the matter flowing into the disk. The step of the spiral is determined by the efficiency of the mechanisms of angular momentum transfer. The energy release in the disk is proportional to $R^{-3}\left[1-\left(\frac{R_0}{R}\right)^{1/2}\right]$. Radiation diffuses mainly across the disk. There is no qualitative difference between the subcritical and supercritical regimes of accretion in the outer regions; the rate of energy release there is less than the critical luminosity and light pressure does not prevent infall of the matter.

A number of developments is then possible. One can imagine that periodically a supercritical flux of matter falls towards the centre of gravitation and then powerful flares of radiation reject the matter and destroy the disk completely. During accretion to the collapsar it is possible but improbable that a significant fraction of the matter flux falls into the black hole. In this case, it follows that the radiation flux is only of the order of the critical value and most of the matter and radiation are dragged into the black hole without any observable effect. It is more probable that a stationary state is established with the luminosity close to the critical value and with outflow of most of the infalling matter beyond the system (see also Schwartzman, 1971c). The value of the established luminosity may few times differ from the value of the critical luminosity, computed for the spherically symmetric case. Now let us discuss this picture in more detail.

Approaching the black hole, the energy release and the force of light pressure grow rapidly. According to (2.8), if $\dot{m} = \frac{\dot{M}}{\dot{M}_{cr}} > 1$, near the radius of spherization

$$R_{sp} \approx \frac{9}{4}\,10^6 m\dot{m}(\text{cm}); \quad r_{sp} \approx \frac{9}{4}\dot{m} \tag{7.1}$$

the thickness of the disk becomes of the order of distance to the black hole. Further infall of matter will lead to the immediate outflow of a fraction of it from the region $R < R_{sp}$.

The outflowing matter is accelerated by the difference of the forces of gravitational attraction and light pressure. In the situation under discussion, there are only two characteristic velocities: the parabolic velocity $\sqrt{2}v_\varphi(R_{sp})$ and the radial velocity of the matter in the central plane of the disk $v_r = \alpha v_\varphi(R_s)$. Apparently, autocontrol is realised leading to a velocity of outflow $v_{out} \simeq \sqrt{2}v_\varphi\sqrt{\frac{L-L_{cr}}{L_{cr}}}$ close to $v_r$. A decrease in the velocity of the outflowing matter relative to $v_r$ leads to the arrival of a considerable fraction of the inflowing matter in the region with $R < R_s$, increasing the luminosity and, consequently the velocity of the outflowing matter. If $v_{out}$ exceeds $v_r$ both the matter inflow to the region $R < R_{sp}$ and the luminosity decrease and also decreases $v_{out}$. On the other hand, if $R > R_{sp}$ the outflowing matter is subjected to the light pressure not only of the disk region which is left but to the integral radiation of the whole disk. The latter can logarithmically exceed the critical luminosity $\frac{L - L_{cr}}{L_{cr}} \simeq \ln \dot{m} \simeq \ln r_{sp}$. This must lead to the *increase* of $v_{out}$ to $v_\varphi(r \gg r_{sp}) \ll v_\varphi(r_{sp})$. Evidently, even at small $\alpha < 10^{-2}$ the velocity of the outflow may not be less than $(0.01 \div 0.1)\, v_\varphi(r_{sp})$.

The same relation $v_{out} = \alpha v_\varphi(R)$ exists for matter flowing out from any radius $R < R_{sp}$, but here the dependence on $\alpha(R)$ is dominant. The lines of matter flow are illustrated in Fig. 8, not taking into account rotation. In fact, the matter flows out in a spiral (Fig. 9). The closer to the black hole is the matter, the greater is the velocity of the outflowing matter. However, the fraction of this matter $\dot{M}(R < R_{sp}) = \frac{R}{R_{sp}} \dot{M}_0$ in the flux $\dot{M}_0$, inflowing to the external boundary of the disk decreases.

If the angular momentum is preserved relative to the disk axis, the matter is not able to enter the cylindrical region of radius less than the radius of outflow. The result is a strongly anisotropic outflow of matter at an angle $\theta$ relative to the axis of disk rotation. In a small cone near the axis, hot matter is ejected from the region close to $R \sim \frac{9}{4} R_0 \sim 7 R_g$ with high velocity and hard X-ray radiation is emitted. When $\theta$ increases, the velocity of the outflow decreases rapidly and, in most of the cone (and most of the outflowing matter), $v_{es} \sim \alpha v_\varphi(R_{sp})$. The spectrum of radiation emitted by this region depends upon the rate of matter inflow to the disk and on the parameter $\alpha$ determining the volume emission measure of the outflowing gas and the degree of reradiation of the hard radiation of the disk to the softer quanta. When $\dot{m} \gg 1$ and $\alpha(r_{sp}) \ll 1$ (the latter is rather probable because, near $r_{sp}$, the degree of compression of the matter with the frozen magnetic field is small) the outflowing matter may be opaque far from the collapsar. In this case, the collapsar may be a bright optical object. Depending on the orientation of the system, the hard or soft X-ray radiation may also be observed.

Note that, if $\dot{m} \gg 1$ and $\alpha \ll 1$, the total spherization of the accretion picture may emerge – the fast particles will be thermalized by their interaction with the main mass of the outflowing matter. The radiation pressure and an effective exchange of angular momentum between adjacent layers of the outflowing matter can also lead to spherical symmetry. In this case, no X-ray radiation can exist. We now give a rough estimate of the spectrum of the radiation (and its effective temperature) formed in a large fraction of the outflowing matter.

The density of the matter flowing out with the constant velocity $v_{out}(r_{sp})$ changes according to the law

$$n(r) = n(r_{sp}) \left( \frac{r_{sp}}{r} \right)^2 ,$$

where

$$n(r_{sp}) = 1.3 \cdot 10^{19} \alpha^{-1} \dot{m}^{-1/2} m^{-1} (\mathrm{cm}^{-3}) \tag{7.2}$$

can be found from formulae (2.11) with $r = r_{sp} = \frac{9}{4}\dot{m}$ and $(1 - r^{-1/2}) = 1/3$. The same expression can be found from the relation $\dot{M} = 2\pi m_p n(R_{sp}) \times v_{out}(R_{sp}) R_{sp}^2 z_0$. The comptonization plays an important role in this exchange of energy between the matter and radiation. At $\dot{m} < 10^3 \alpha^{34/29} A^{-16/29} m^{2/29}$, the outflowing matter is not able to reprocess the hard radiation of the disk. Therefore, the temperature of the outgoing radiation is defined by formulae (3.7 – 3.9) replacing $r$ by $r_s = \frac{9}{4}\dot{m}$ and $(1 - r^{-1/2})$ by 1/3. At larger rates of accretion, the radiation spectrum is formed in the outflowing matter. The condition $A\tau_{ff}(R_{eff})\, \tau_T(R_{eff}) \sim 1$ determines the layer where a black body spectrum is formed. The outgoing radiation spectrum is distorted because $\tau_T \gg \tau_{ff}$. Its energy density is

$$\tau_T(R_{eff}) = \int_R^\infty \sigma_T n(R)\, dR = 30 \alpha^{-1} \cdot \dot{m}^{1/2} \left( \frac{R_{sp}}{R_{eff}} \right)$$

times less than $bT^4$. Using the condition

$$L_{cr} = 10^{38} m \frac{\mathrm{erg}}{\mathrm{s}} = \frac{4\pi R_{eff}^2}{\tau_T(R_{eff})} b T_{eff}^4 \frac{c}{4}$$

one can easily determine an effective radiation temperature and the radius $R_{eff}$ of the radiating envelope

$$\begin{aligned} R_{eff} &\simeq 3 \cdot 10^2 \dot{m}^{51/22} m^{10/11} \alpha^{-17/11} A^{8/11} (\mathrm{cm}) \\ T_{eff} &\simeq 2 \cdot 10^{10} \dot{m}^{-15/11} m^{-2/11} \alpha^{10/11} A^{-6/11}\ {}^\circ\mathrm{K}\,. \end{aligned} \tag{7.3}$$

The temperature $T_{eff}$ is practically constant in a wide region with $R > R_{eff}$. At $\dot{m} < 4 \cdot 10^7 \alpha^{2/3} m^{1/9}$ the radius at which the quanta undergo their last scattering, $R(\tau_T = 1) = 9 \cdot 10^7 \dot{m}^{3/2} m \alpha^{-1}$cm exceeds greatly $R_{eff}$. If $R(\tau_T = 1) \leqq R_{eff}$, the outgoing radiation spectrum is planckian. The parameter $y$ is greater than unity and, owing to the Compton process, outgoing radiation spectrum has a Wien distribution unless $\dot{m} < 3 \cdot 10^3 \alpha^3 m^{-1/2} \cdot A^{-3/2}$. Note that the apparent radius at eclipses of the normal star by the matter flowing out from collapsar is $R(\tau_T = 1)$.

In the optical (low-frequency range at the discussed temperatures $T_{eff}$) spectral range, the radius near which the envelope becomes opaque exceeds $R_{eff}$ considerably. Putting $\tau_T \tau_{ff}(\nu, R_{opt})$ equal to unity, we find

$$R_{opt} \simeq 10^7 \alpha^{-3/4} \left( \frac{10^6\ {}^\circ\mathrm{K}}{T} \right)^{3/8} \left( \frac{10^{15}\ \mathrm{Hz}}{\nu} \right)^{1/2} \dot{m}^{9/8} m^{3/4} (\mathrm{cm})$$

and the total optical luminosity of collapsar

$$L_{\rm opt}=\int_0^{\nu_0}\frac{4\pi R_{\rm opt}^2}{\tau_T(R_{\rm opt})}\frac{2\pi k T_{\rm eff}}{c^2}\nu^2 d\nu = 3\cdot 10^{30}\left(\frac{10^6\,^\circ{\rm K}}{T}\right)^{1/8}\left(\frac{\nu_0}{10^{15}\,{\rm Hz}}\right)\left(\frac{\dot m^{3/2} m}{\alpha}\right)^{5/4}\frac{\rm erg}{\rm s} \tag{7.4}$$

$$\simeq 10^{30}\left(\frac{\nu_0}{10^{15}\,{\rm Hz}}\right)^{3/2}\left(\frac{\dot m^{3/2} m}{\alpha}\right)^{15/11}\frac{\rm erg}{\rm s}$$

is weakly dependent on $T$. The optical spectrum is unusual $F_\nu \sim \nu^{1/2}$. In fact, the other sources of opacity will strongly influence the optical spectrum of the object. The spectrum will be saturated with emission lines, which will determine the excess of optical luminosity over (7.4). The emission lines must be broad with

$$\frac{\Delta\lambda}{\lambda}\sim\frac{v_{es}(R_{sp})}{c}\sim\alpha\frac{v_\varphi(R_s)}{c}.$$

For $\alpha \sim 1$, direct exportation of radiation closed in the matter may be important in the narrow cone of angles where the matter flows out with high velocity. If $v_{\rm out} \sim v_\varphi(7R_g) \sim 10^5\,\frac{\rm km}{\rm s}$, the time for radiation exportation is comparable with the time of diffusion of radiation $t \sim \frac{7R_g}{c}\tau_T \sim \frac{200\,R_g}{c}$. Therefore the existence of a flux of fast particles ejected from the region $R \sim 7R_g$ is of independent interest and should lead to specific effects: the formation of shock waves and X-ray radiation being emitted far from the collapsar. Evidently, in case when $\dot m \gtrsim 3\cdot 10^3\left(\frac{\alpha}{m}\right)^{2/3}$ neither this flux of fact particles nor the radiation of the slowly $v_{\rm out} \sim \alpha v_\varphi(R_{sp})$ outflowing matter with $T_{\rm eff} \sim 10^5\,^\circ$K can strongly influence the rate of inflow of the matter in the outer boundary of the disk. When $\dot m > 3\cdot 10^3\left(\frac{\alpha}{m}\right)^{2/3}$ and $\dot m > 1$, the collapsar must appear as a bright optical and ultra-violet object with high mass loss (much like Wolf-Rayet stars) and possible X-ray radiation. It is surrounded by a thin disk of cold, inflowing matter in contrast to normal stars. The same situation exists when there is supercritical accretion to a neutron star in a close binary system. Only the masses and corresponding luminosities are different.

## V. Turbulence, Magnetic Fields and Temporal Characteristics of the Disk Radiation

If $\alpha \sim 1$, new, important effects connected with plasma turbulence and reconnection of magnetic field lines, through neutral points, must appear. In addition, short-term fluctuations of the radiation flux, connected with these effects, must take place.

So, for example, the flux of energy $Q' \sim \frac{\varrho v_t^3}{2}$ transferred by turbulence from the inner layer of the disk to its surface, is $\alpha^2$ times less than the radiative energy flux $Q \simeq \frac{3}{8\pi}\frac{GM\dot M}{R^3}$. Such convection must lead to granulation of the disk surface, analogous to that observed on the Sun and of scale size of the order of $z_0$. Therefore when there is disk accretion and associated turbulence, chaotic fluctuations of the observed radiation flux $\frac{\delta I}{I} \sim \frac{Z}{R}\alpha^2 \sim \dot m\alpha^2$ and of its spectrum may be expected. The typical time of these fluctuations is equal to $\Delta t_f \sim \frac{Z_0}{v_t} \sim \frac{R}{\alpha v_\varphi} \sim 10^{-4}\frac{m}{\alpha} r^{3/2}$ s. The fluctuations of the hard radiation should be the strongest, because the turbulence takes out to the disk surface high temperature clumps of plasma. In the low-frequency (optical) spectral bands, which are emitted mainly by the outer regions of the disk, the typical time of fluctuations exceeds 1 minute.

Analogous irregular activity may also be caused by chaotic magnetic fields. Note that it is difficult to receive the giant strength of the magnetic fields $H \sim 10^8$ Gauss $\left(\text{which are consequence of an assumption of equipartition } \frac{H^2}{8\pi} \sim \varrho v_s^2\right)$ in the vicinity of a collapsar as the result of a simple compression of the magnetized gas flowing out from the normal star. The maximum matter density in the disk at $\alpha \sim 1$ and $\dot m \sim 1$ are $n \sim 10^{20}$ cm$^{-3}$, i.e. only $4 \div 6$ orders of magnitude larger than the matter densities in stellar atmospheres having magnetic field strengths $H \simeq 1 - 10$ Gauss. It is possible that the magnetic field strength in the disk is many orders of magnitude less than $10^8$ Gauss. Then $\alpha$ is also small.

Equipartition could be the consequence of amplification of the magnetic field in the differentially rotating matter of the disk. Simultaneously, the magnetic field is partitioned into small loops of scale less than $z_0$. The loops with opposite directions of the magnetic field can approach under the influence of turbulent pulsations. Therefore, current sheets are formed and flares of the solar type can occur. However, the energy release may be much greater (Shakura, 1972a). The acceleration of the particles by the mechanism of Syrovatsky, which may be essential to accretion to a massive ($M > 10^6 M_\odot$) black hole (Lynden-Bell, 1969), is doubtful because the gas density is large. However, it is not excluded. The typical time of aperiodic fluctuations, connected with the flares greatly exceeds the time $\Delta t_f \sim 10^{-4}\frac{m}{\alpha^{1/2}}r^{3/2}$ s determined by the Alfven velocity and by the scale of the magnetic field $\sim z_0$. To within an order magnitude it is equal to the time $\Delta t_f \sim \frac{3R_g}{c} \sim 10^{-4}$ms deduced by Schwartzman (1971a) from different reasoning. The amplitude of the fluctuations of the integral radiation flux depends mainly on the structure of the magnetic field and drops rapidly when $\alpha$ decreases. Apparently, it cannot exceed $\frac{\delta I}{I} \sim \frac{Z}{R}\frac{H^2}{4\pi\varrho v_\varphi^2} \sim \alpha\left(\frac{Z_0}{R}\right)^3$. However, in certain spectral bands, this amplitude can be of the order of unity, because the spectra of thermal radiation and the spectra of flares are different.

At $H > 10^6$ Gauss the hot ($T \sim 10^8$ °K) plasma emits synchroton radiation with superposed resonances of cyclotron frequency $\nu_s = \frac{seH}{2\pi m_e c}$ (with $s \sim 10$) provides the black body intensity in this overlapping due to the Doppler-effect resonances and leads to the partial polarization of integral radiation of the disk in the optical and infra-red spectral bands (Gnedin and Sunyaev, 1972).

## Appendix

### The Compton Effect and the Thermal Balance of the Gas in the Vicinity of X-Ray Sources

In the vicinity $\left(\zeta = \frac{L_x}{R^2 n} > 10^4 \div 10^5 \frac{\text{erg cm}}{\text{s}}\right)$ of X-ray sources with $T_{eff} \sim 10^6 - 10^8$ °K, all elements up to neon which determine the energy balance of colder plasmas are completely ionized (Tarter *et al.*, 1969). In such conditions the stationary value of the temperature of the plasma is determined by the processes of Compton exchange of energy between radiation and free electrons and bremsstrahlung.

Equating, in turn, the rate of electron cooling due to Compton processes $P_c^- = \frac{4\sigma_T \varepsilon k T}{m_e c^2}$ and free-free radiation $P^- = 10^{-27}\sqrt{T}n$ to the rate of heating by Compton mechanism $P_c^+ = \frac{\sigma_T h}{m_e c}\int_0^\infty \varepsilon_\nu \nu d\nu$ (Levich and Sunyaev, 1971), we find for the temperature of the gas

$$T = \beta T_{eff} \quad \text{at} \quad \xi > \zeta_{cr} = \frac{3.3 \cdot 10^7}{(\beta T_{eff})^{1/2}}$$

when cooling is associated with Compton-effect, and $T = \beta T_{eff}(\zeta/\zeta_{cr})^2$ at $\zeta < \xi_{cr}$ when cooling is due to bremsstrahlung.

The coefficient $\beta$ depends on the spectrum of radiation of the source: for a black body spectrum, $\beta = 0.94$, for a Wien distribution, $\beta = 1$, and for the spectrum of the optically thin plasma, $\beta = 0.25$. If $\zeta < 10^4 \div 10^5$, the energy losses in lines of the heavy elements begin to influence the thermal balance of the plasma and calculations are very complicated (Tarter *et al.*, 1969).

The authors wish to thank A. F. Illarionov, D. Ya. Martynov, L. R. Yangurasova and Ya. B. Zeldovich for consultations and the participants of the 1972 winter school on astrophysics in Arhyz for numerous discussions.

## References


Basko, M.A., Sunyaev, R.A. 1973, preprint, *Astrophysics Space Science* (in press).

Bisnovaty-Kogan, G.S., Sunyaev, R.A. 1971, *Astron. Zh.* **48**, 881.

Cherepashchuk, A.M., Efremov, Yu.N., Kurochkin, N.E., Shakura, N.I., Sunyaev, R.A. 1972, *Inform. Bull. Var. Stars* N 720.

Felten, J.E., Rees, M.J. 1972, *Astron. & Astrophys.* **17**, 226.

Gnedin, Yu.N., Sunyaev, R.A. 1973, *Monthly Notices Roy. Astron. Soc.* (in press).

Gorbatsky, V.G. 1965, *Proceedings of Leningrad University Observatory* **22**, 16.

Illarionov, A.F., Sunyaev, R.A. 1972, *Astron. Zh.* **49**, 58.

Kaplan, S.A. 1949, *JETP* **19**, 951.

Kompaneets, A.S. 1956, *JETP* **31**, 876.

Kraft, R. 1963, *Advances in Astron. & Astrophys.* ed. Kopal, Z. **2**, 43.

Levich, E.V., Sunyaev, R.A. 1971, *Astron. Zh.* **48**, 461.

Lynden-Bell 1969, *Nature* **223**, 690.

Lyutiy, V.M., Sunyaev, R.A., Cherepashchuk, A.M. 1972, preprint 1973, *Astron. Zu.* **50**, 3.

Martynov, D.Ya. 1971, Course of General Astrophys., Nauka, Moscow.

Novikov, I.D., Thorne, K.S. 1972, Lectures in the summer school on the black holes, Les Houches, France.

Pottasch, S.R. 1970, in "*Cosmic gas dynamics*", IAU Symposium 31, Reidel, Dordrecht.

Prendergast, K.H. 1960, *Astrophys. J.* **132**, 162.

Prendergast, K.H., Burbidge, G.R. 1968, *Astrophys. J.* **151**, L 83.

Pringle, J.P., Rees, M.J. 1972, *Astronomy Aph.* **21**, 1.

Salpeter, E.E. 1964, *Astrophys. J.* **140**, 796.

Shakura, N.I. 1972a, *Astron. Zh.* **49**, 921.

Shakura, N.I. 1972b, *Astron. Zh.* **49**, 652.

Shklovsky, I.S. 1967, *Astrophys. J.* **148**, L 1, Astron. J. **44**, 930.

Schwartzman, V.F. 1971a, *Astron. Zh.* **48**, 479.

Schwartzman, V.F. 1971b, *Astron. Zh.* **48**, 438.

Schwartzman, V.F. 1971c, Thesis, Moscow University.

Sunyaev, R.A. 1972, *Astron. Zu.* **49**, 1153.

Tarter, C.B., Tucker, W.H., Salpeter, E.E. 1969, *Astrophys. J.* **156**, 943.

Taylor, G.I. 1937, *Proc. Roy. Soc. A.*, **157**, 546.

Wasiutinski, J. 1946, Studies in Hydrodynamics and Structure of Stars and Planets, Oslo.

Zeldovich, Ya.B. 1964, *Doclady Acad. Sci.*, U.S.S.R., **155**, 67.

Zeldovich, Ya.B., Novikov, I.D. 1971 The theory of the gravitation and stars evolution, Nauka, Moskow.

Zeldovich, Ya.B., Raizer, Yu.P. 1966, Physics of shock waves and high-temperature hydrodynamic phenomena, Nauka, Moscow.

Zeldovich, Ya.B., Shakura, N.I. 1969, *Astron. Zh.* **46**, 225.


N. I. Shakura
Sternberg Astronomical Institute
Moscow – U.S.S.R.

R. A. Sunyaev
Institute of Applied Mathematics
Academy of Sciences
Moscow – U.S.S.R.

THE ASTROPHYSICAL JOURNAL, 186:1015–1025, 1973 December 15


# X-RAY EMISSION FROM A NEUTRON STAR ACCRETING MATERIAL*

MARVIN L. ALME† AND JAMES R. WILSON
Lawrence Livermore Laboratory, University of California
*Received 1973 March 2; revised 1973 July 20*

## ABSTRACT

Computer calculations of the X-ray emission resulting from the accretion of material onto a neutron star have been performed. For high-mass neutron stars, the emitted spectra differ only slightly from blackbody spectra, particularly at lower luminosities. For low-mass neutron stars, the spectra differ greatly from blackbody spectra at all luminosities.
*Subject headings:* neutron stars — spectra, X-ray — X-ray sources

## I. INTRODUCTION

We have performed calculations of the X-ray emission resulting from the accretion of material onto a neutron star. Our calculations can be regarded as a more detailed study of the accretion calculation performed by Zel'dovich and Shakura (1969). Our calculational model assumes a steady-state, spherically symmetric, infall of hydrogen gas from radial infinity in a Schwarzschild metric. This gas falls onto a nonrotating neutron star. The gas is decelerated as a beam of charged particles in the atmosphere of the neutron star. In calculating this deceleration, only Coulomb collisions were considered because they dominate over nuclear collisions at the energies of interest. The neutron-star atmosphere, also assumed to be hydrogen, is heated by the infalling hydrogen gas. This atmosphere is treated as a hydrodynamical fluid characterized by temperature, density, and radial velocity. An ideal-gas equation of state is used. The atmosphere also contains a radiation field, which is described by its energy density per unit frequency. The transport of the radiation is treated in a diffusion approximation. This radiation field is not allowed to interact with the infalling hydrogen gas; i.e., there is no heating of or momentum transfer to the infalling gas by the radiation. The resulting atmosphere is sufficiently thin that gravitational acceleration of the infalling gas can be ignored in the atmosphere. The most serious limitation of our calculations is that they assume a spherically symmetric geometry. However, they can be interpreted per unit surface area and thus can be applied locally in any situation where the infall is primarily in the radial direction. An initial temperature and density distribution is assumed for the atmosphere. The atmosphere is then allowed to evolve dynamically until a steady state is achieved. The treatment of the transport of the infalling gas in the atmosphere ignores all time derivatives.

Because the atmosphere is very thin compared with the radius of the neutron star, we may ignore the self-gravitational effects of the atmosphere and set up on the surface of the neutron star a static proper reference system (coordinate lengths and times equal to proper lengths and times) (Bardeen, Press, and Teukolsky 1972). The laws of physics in this proper reference frame are those of special relativity, augmented by the gravitational acceleration

$$g = \frac{GM}{R^2(1 - 2GM/c^2R)^{1/2}} .$$

* This work was performed under the auspices of the U.S. Atomic Energy Commission.
† Fannie and John Hertz Foundation Fellow.

Further, since the internal energy density of the gas is small compared with the rest-mass energy density and the fluid velocities are small compared to $c$, we may use a Newtonian description for the hydrodynamics. However, the infalling hydrogen gas has velocities approaching $c$, so it must be treated special-relativistically. For a distant observer's interpretation of the results, see the Appendix.

## II. NOTATION AND EQUATIONS

$R$ is the value of the radial coordinate at the surface of the neutron star (Schwarzschild curvature coordinates), and $M$ is the star's gravitational mass. All other quantities occurring in the equations are the quantities as measured in the static proper reference system set up in the atmosphere. In particular, $dt$ is the differential proper time and $dr$ is the differential proper length in the radial direction. (Here $t$ and $r$ are not the Schwarzschild curvature coordinates—see Appendix.)

### a) *Equation of Motion*

The equation of motion is taken as

$$\rho \frac{dV}{dt} = -\frac{\partial P}{\partial r} - \frac{GM\rho}{R^2(1 - 2GM/Rc^2)^{1/2}} + \frac{dM/dt}{4\pi R^2}\frac{\partial}{\partial r}\left\{\frac{v}{[1 - (v/c)^2]^{1/2}}\right\} + \frac{1}{c}\int_0^\infty \rho(K_a' + K_s)F_\nu d\nu\,,$$

where $\rho$ is the density, $V$ is the radial fluid velocity, $P$ is the pressure, $dM/dt$ is the accretion rate, $v$ is the velocity of the infalling hydrogen gas at the top of the atmosphere, $\nu$ is the photon energy, $K_a'$ is the absorption opacity corrected for induced emission, $K_s$ is the Compton scattering opacity, and $F_\nu$ is the radiation flux. The gravitational constant is $G$, and the velocity of light is $c$. The $(1 - 2GM/Rc^2)^{-1/2}$ factor is included in the gravitational force to facilitate a relativistic interpretation of the results (see Appendix).

### b) *Radiation Diffusion*

The radiation field is described by the equation

$$\frac{dE_\nu}{dt} + \frac{\partial}{\partial r}(F_\nu) = c\rho K_a'(B_\nu - E_\nu) + \frac{E_\nu}{\rho}\frac{\partial \rho}{\partial t} - \frac{V}{c}\rho(K_a' + K_s)F_\nu + \frac{\nu\rho K_s}{m_e c}\frac{\partial}{\partial \nu}\left[T\left(\nu\frac{\partial E_\nu}{\partial \nu} - 3E_\nu\right) + \nu E_\nu\left(1 + \frac{\pi^4 E_\nu}{15a\nu^3}\right)\right].$$

Here $E_\nu$ is the radiation energy density per unit photon energy, $B_\nu$ is the blackbody radiation energy density per unit photon energy, and $T$ is the electron temperature. The total blackbody energy density is given by $aT^4$. The change in photon energy resulting from Compton collisions is treated in the Fokker-Planck approximation (Chang and Cooper 1970). The radiation flux is taken as

$$F_\nu = -\frac{\lambda_{\text{total}}c(\partial E_\nu/\partial r)}{3 + \lambda_{\text{total}}|\partial E_\nu/E_\nu\partial r|[1 + 3\exp(-\frac{1}{2}\lambda_{\text{total}}|\partial E_\nu/E_\nu\partial r|)]}\,,$$

where $\lambda_{\text{total}} = (\rho K_a' + \rho K_s)^{-1}$. The denominator of the diffusion coefficient has been selected so that, for optically thin systems, the radiation flux will be approximately correct. This form of the diffusion coefficient has been tested against exact solutions on slabs of one to five mean free paths in thickness, and found to have less than 10 percent error in flux and energy density.

### c) *Material Temperature*

The material temperature is determined from

$$\rho c_v \frac{\partial T}{\partial t} = c\int_0^\infty \rho K_a'(E_\nu - B_\nu)d\nu + \frac{dM/dt}{4\pi R^2}\frac{\partial}{\partial r}\left\{\frac{c^2}{[1-(v/c)^2]^{1/2}}\right\} + \frac{\partial}{\partial r}\left(K_{\text{electron}}\frac{\partial T}{\partial r}\right)$$
$$+ \frac{P}{\rho}\frac{\partial \rho}{\partial t} + \frac{\rho}{m_e c}\int_0^\infty K_s\left[T\left(\nu\frac{\partial E_\nu}{\partial \nu} - 3E_\nu\right) + \nu E_\nu\left(1 + \frac{\pi^4 E_\nu}{15 a\nu^3}\right)\right]d\nu ,$$

where $c_v$ is the material specific heat, $K_{\text{electron}}$ is the electron thermal conductivity, and $P$ is the material pressure. When the temperature gradient is superadiabatic, $K_{\text{electron}}$ is modified to include convective heat transfer in the mixing length approximation.

### d) *Opacity*

The absorption opacity is taken as

$$K_a = \frac{2.78 Z^3 \rho}{\nu^3 A^2 T^{1/2}} g^{ff} ,$$

where $g^{ff}$ is a tabular fit to the free-free Gaunt factors given by Karzas and Latter (1961), and $\nu$ and $T$ are given in keV. $K_a'$ is then

$$K_a' = K_a(1 - e^{-\nu/T}) .$$

### e) *Deceleration of Infalling Gas*

The rate of energy loss for a proton traveling in a plasma is (Shkarofsky, Johnston, and Bachynski 1966)

$$\frac{dE_p}{dt} = -[\phi(x_e) - x_e(1 + m_e/m_p)\phi'(x_e)]\frac{4\pi n_e e^4}{m_e v}\ln \Lambda_e ,$$

where we have neglected slowing down as a result of ions in the plasma. Subscript $p$ denotes the infalling protons and subscript $e$ denotes the electrons in the plasma. The number density is $n_e$, the electron charge is $e$, the proton velocity is $v$, the Coulomb logarithm is $\Lambda_e$, the error function is

$$\phi(x) = 2\pi^{-1/2}\int_0^x \exp(-z^2)dz ,$$

and $x_e = (m_e v^2/2kT_e)^{1/2}$. We take $\Lambda_e$ to be the larger of $2m_e v^3/(\gamma e^2 \omega_p)$ (Kihara and Aono 1963) and $3T^{3/2}/(e^2\omega_p m_e^{1/2})$ (Spitzer 1962). The plasma frequency is $\omega_p$, and Euler's constant is $\ln \gamma$. When $v/2c$ or $v_{\text{th}}$(electrons)$/2c$ is greater than the fine-structure constant, $\alpha$, we also use a quantum-mechanical correction for $\Lambda_e$ (Spitzer 1962), i.e., $\Lambda_e^{QM} = \Lambda_e 2\alpha c/v$. We approximate $f(x_e) = \phi(x_e) - x_e(1 + m_e/m_p)\phi'(x_e)$ as

$$f(x_e) \simeq \frac{2x_e(2x_e^2/3 - m_e/m_p)/\pi^{1/2}}{1 + 2x_e(2x_e^2/3 - m_e/m_p)/\pi^{1/2}} .$$

Then, for each gram of infalling material, we have

$$\frac{d}{dt}\frac{c^2}{[1-(v/c)^2]^{1/2}} = -f(x_e)\frac{4\pi n_e e^4}{v m_e m_p}\ln \Lambda_e \,.$$

### III. METHOD OF SOLUTION

The equation of motion, the radiation diffusion equation, and the infalling-gas deceleration equation are approximated by finite-difference equations. The atmosphere is divided into approximately 50 spatial zones. The radiation field is described by 20 frequency groups ranging from 0.5 to 150 keV. The numerical treatment used to calculate the Compton energy exchange is the scheme developed by Chang and Cooper (1970).

As a starting condition for the computer code we assume a radiation field and temperature and density distributions in the atmosphere. A rate of mass infall is specified. Then a dynamic problem is run and the atmosphere is allowed to readjust until a steady state is achieved.

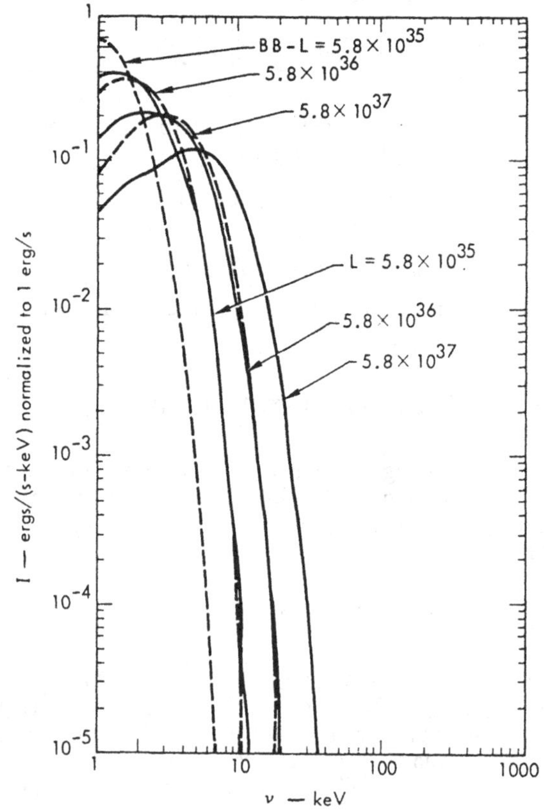


FIG. 1.—X-ray spectra for 2.4 $M_\odot$ star as seen by distant observer. Labels indicate luminosity in ergs $s^{-1}$. *Solid curves*, calculated spectra; *dashed curves*, blackbody spectra with the same luminosities.

TABLE 1

NEUTRON-STAR CONFIGURATIONS

| Mass ($M_\odot$) | Radius (km) | $(E_{\text{infall}} - Mc^2)/Mc^2$ | Critical Luminosity (ergs $s^{-1}$) |
|---|---|---|---|
| 2.4 | 10 | 0.86 | 3.0 (38) |
| 1.4 | 12 | 0.24 | 1.8 (38) |
| 0.86 | 10 | 0.16 | 1.1 (38) |
| 0.5 | 13 | 0.062 | 6.3 (37) |
| 0.3 | 20 | 0.023 | 3.8 (37) |

## IV. RESULTS

Calculations were performed for the five neutron-star configurations given in table 1. Rates of mass infall were selected to provide luminosities ranging from $10^{35}$ to $10^{38}$ ergs $s^{-1}$. The calculated X-ray spectra for these five configurations are presented in figures 1–5.

A crucial parameter in determining the departure of the radiated spectrum from a

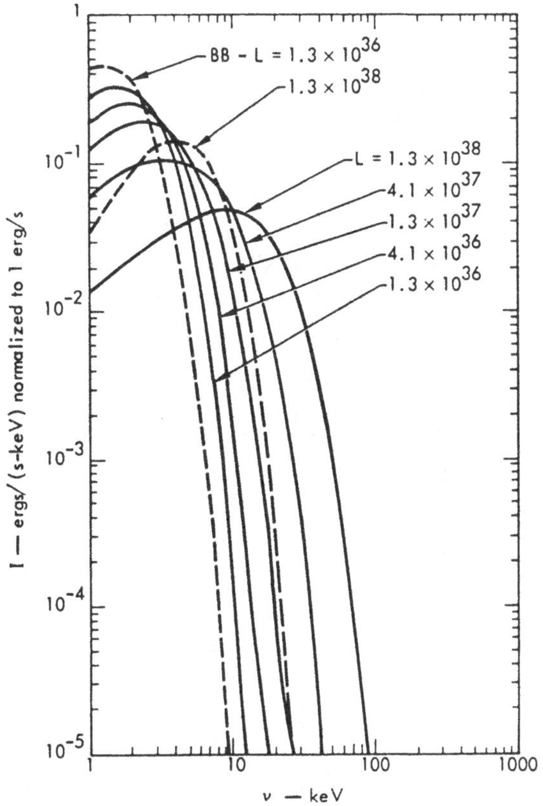


FIG. 2.—X-ray spectra for 1.4 $M_\odot$ star as seen by distant observer. Labels indicate luminosity in ergs $s^{-1}$. *Solid curves*, calculated spectra; *dashed curves*, blackbody spectra with luminosities of $1.3 \times 10^{36}$ and $1.3 \times 10^{38}$ ergs $s^{-1}$.

blackbody spectrum is the total amount of material per unit area required to decelerate the incoming proton beam. Since the stopping distance is a strong function of the infall velocity, the lower surface potentials in the smaller-mass neutron stars result in larger departures from a blackbody spectrum. Typical stopping distances are 100 g cm$^{-2}$ for the 2.4 $M_\odot$ star, 18 g cm$^{-2}$ for the 1.4 $M_\odot$ star, 8 g cm$^{-2}$ for the 0.86 $M_\odot$ star, 2–4 g cm$^{-2}$ for the 0.5 $M_\odot$ star, and 2–3 g cm$^{-2}$ for the 0.3 $M_\odot$ star. For a given star there is a weak variation in stopping distance with infall rate. Figures 6 and 7 illustrate physical conditions in the atmosphere for the 1.4 $M_\odot$ star.

Zel'dovich and Shakura (1969) have suggested that plasma oscillations may play a significant role in decelerating the infalling protons. Following this suggestion, we assume that Landau damping suppresses the two-stream instability between protons of the infalling gas and the atmospheric electrons when the electron thermal velocity is greater than the infalling gas velocity. We require the stopping power be such as to raise the local electron thermal velocity to at least equal to the local infalling-gas velocity. Incorporating this instability model, we repeated our calculation for the 1.4 $M_\odot$ neutron star with luminosity $2 \times 10^{37}$ ergs s$^{-1}$. The stopping distance was reduced from 18 g cm$^{-2}$ to 3 g cm$^{-2}$. Modifications of the spectrum and physical conditions in the atmosphere are illustrated in figures 8 through 10.

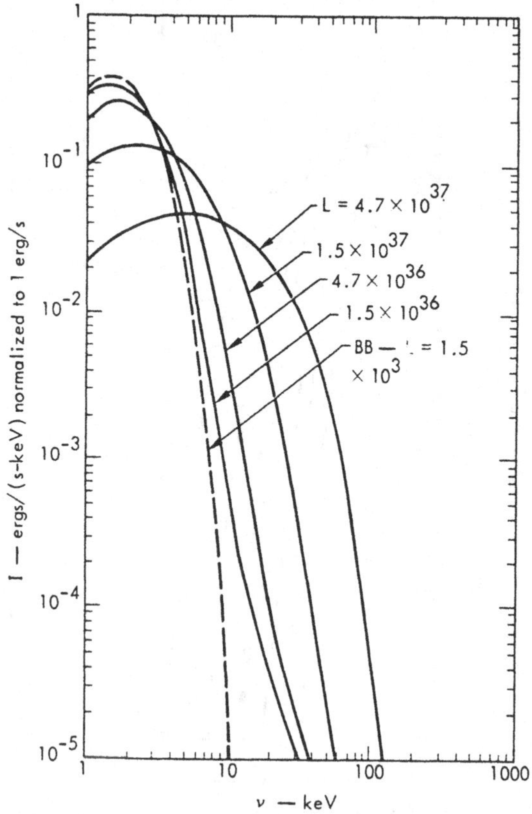


FIG. 3.—X-ray spectra for 0.86 $M_\odot$ star as seen by distant observer. Labels indicate luminosity in ergs s$^{-1}$. *Dashed curve*, a blackbody spectrum with luminosity $1.5 \times 10^{36}$ ergs s$^{-1}$.

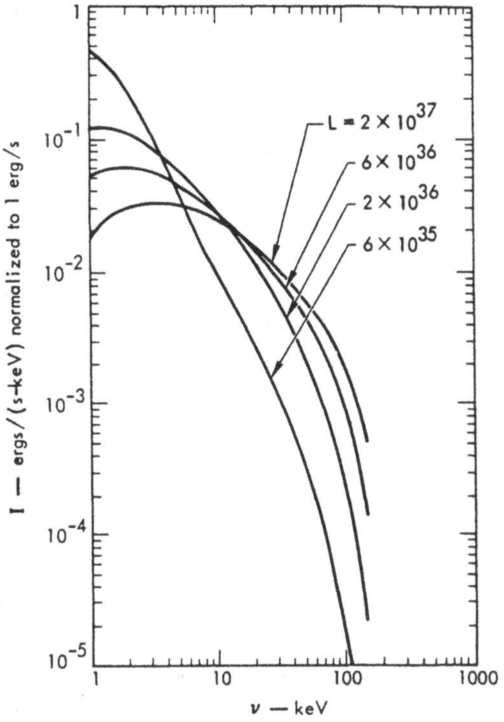


FIG. 4.—X-ray spectra for 0.5 $M_\odot$ star as seen by distant observer. Labels indicate luminosity in ergs $s^{-1}$.

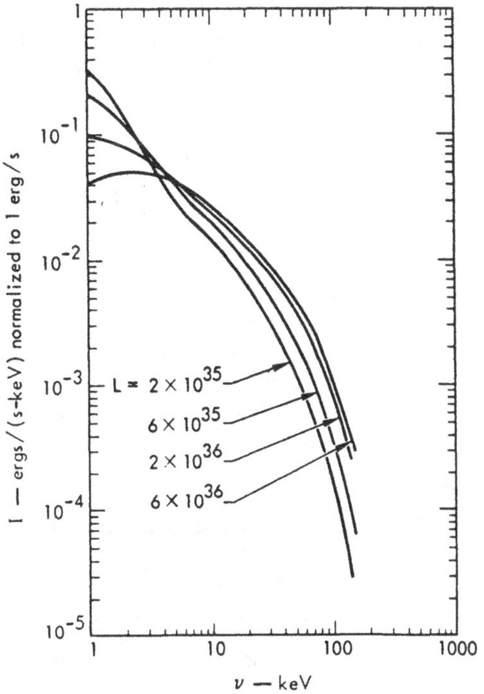


FIG. 5.—X-ray spectra for 0.3 $M_\odot$ star as seen by distant observer. Labels indicate luminosity in ergs $s^{-1}$.

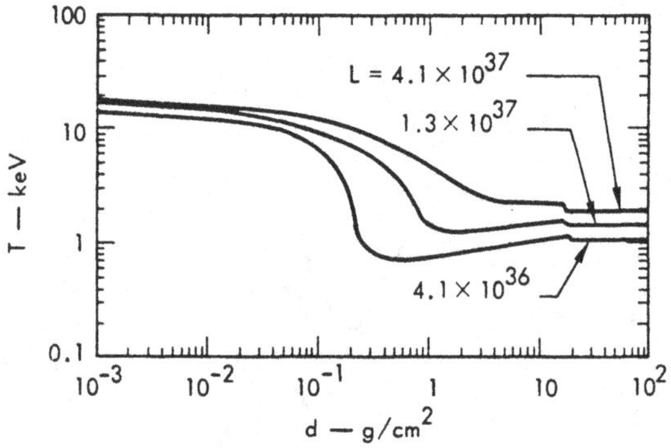


FIG. 6.—Temperature $T$ versus atmospheric depth $d$ for 1.4 $M_\odot$ star. The luminosity associated with each temperature profile is given in ergs $s^{-1}$.

We are skeptical of the existence of strong instabilities. According to Spitzer (1962) the growth rate for the most unstable wave is

$$\omega_p \frac{3^{1/2}}{2^{4/3}} \left(\frac{n_b m_e}{n_e m_b}\right)^{1/3} .$$

The Landau damping rate takes the form (Spitzer 1962)

$$-\omega_p v_{\rm th}(\text{electrons})/(v_b 3^{1/2}) .$$

We have examined the calculations that include only collisional stopping and find that

$$\frac{v_{\rm th}(\text{electrons})}{v_b} \gg \left(\frac{n_b m_e}{n_e m_b}\right)^{1/3} ,$$

except in the first few hundredths of a gram per square centimeter of the atmosphere. For example, in the 1.4 $M_\odot$ star with luminosity $2 \times 10^{37}$ ergs $s^{-1}$, the smallest value for $v_{\rm th}(\text{electrons})/v_b$ is 0.15, while 99 percent of the energy is deposited in a region such that

$$(n_b m_e / n_e m_b)^{1/3} < 0.004 .$$

Hence, the Landau damping would seem to dominate over the two-stream instability.

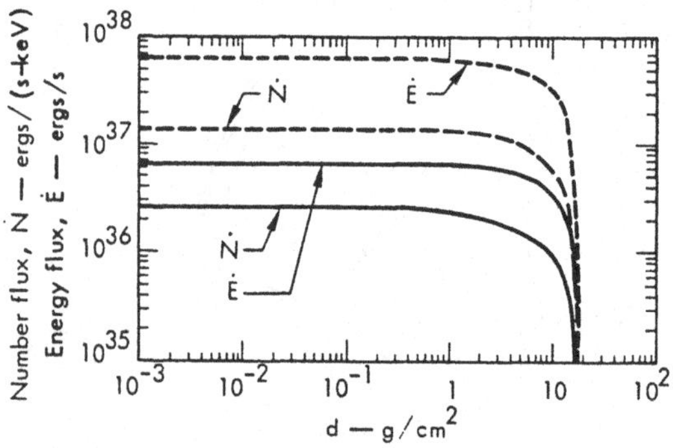


FIG. 7.—Energy flux and number flux versus atmospheric depth $d$ for 1.4 $M_\odot$ star. The solid curves are for luminosity $4.1 \times 10^{36}$ ergs $s^{-1}$ (local luminosity $6 \times 10^{36}$ ergs $s^{-1}$). The dashed curves are for luminosity $4.1 \times 10^{37}$ ergs $s^{-1}$ (local luminosity $6 \times 10^{37}$ ergs $s^{-1}$).

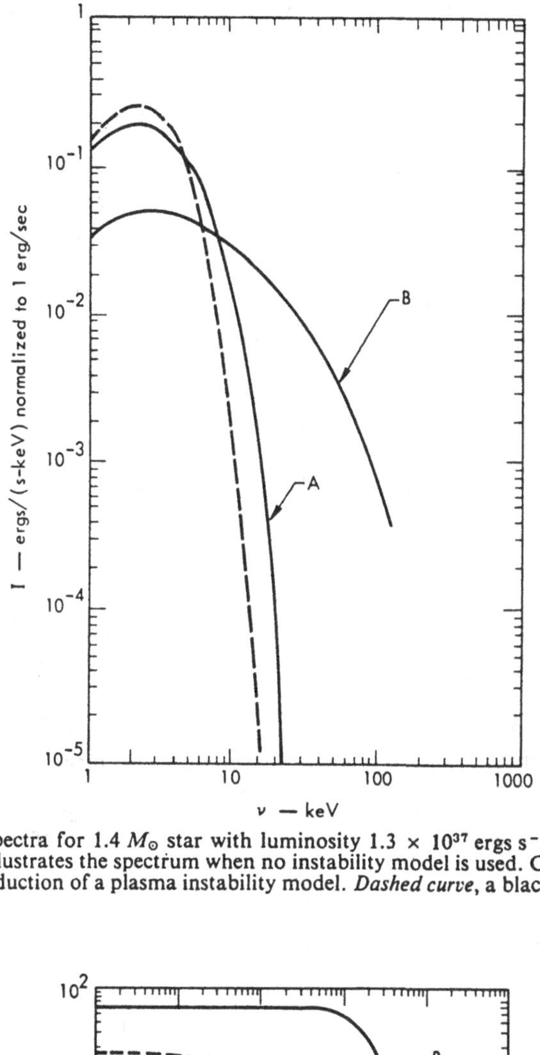


FIG. 8.—X-ray spectra for 1.4 $M_{\odot}$ star with luminosity $1.3 \times 10^{37}$ ergs $s^{-1}$ as seen by distant observer. Curve *A* illustrates the spectrum when no instability model is used. Curve *B* indicates the spectrum after introduction of a plasma instability model. *Dashed curve*, a blackbody of luminosity $1.3 \times 10^{37}$ ergs $s^{-1}$.

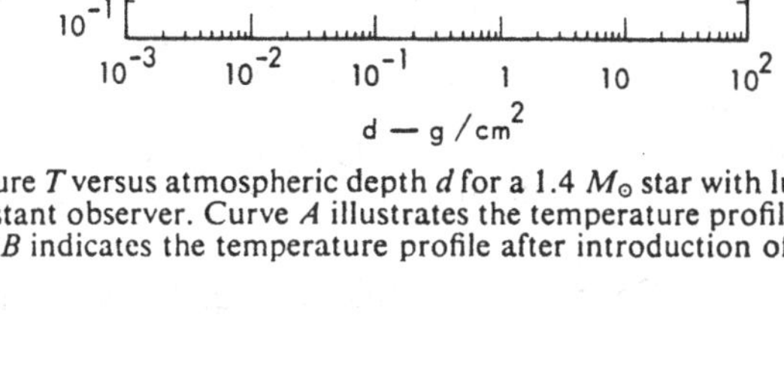

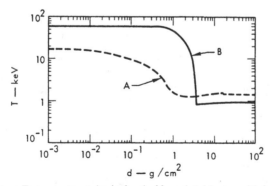

FIG. 9.—Temperature *T* versus atmospheric depth *d* for a 1.4 $M_{\odot}$ star with luminosity $1.3 \times 10^{37}$ ergs $s^{-1}$ as seen by distant observer. Curve *A* illustrates the temperature profile when no instability model is used. Curve *B* indicates the temperature profile after introduction of a plasma instability model.

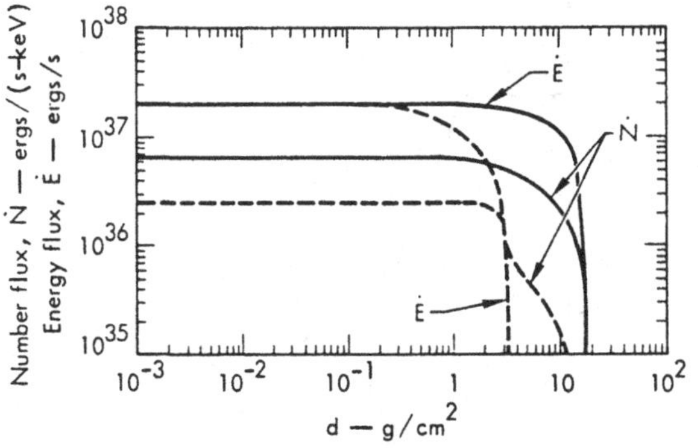


FIG. 10.—Energy flux and number flux versus atmospheric depth $d$ for a 1.4 $M_{\odot}$ star with luminosity $1.3 \times 10^{37}$ ergs s$^{-1}$ as seen by distant observer (local luminosity $2 \times 10^{37}$ ergs s$^{-1}$). The solid curves illustrate the energy and number fluxes when no instability model is used. The dashed curves indicate the energy and number fluxes after introduction of a plasma instability model.

## V. CONCLUSIONS

The temperature at the top of the atmosphere, characteristically tens of kiloelectron volts, is primarily determined by an energy balance between heating by the incoming beam and Compton cooling. The temperature at the bottom of the atmosphere, of the order of 1 keV, is set by the requirement that there be no net energy flow into the star. The atmospheric depth at which the transition from kiloelectron volts to tens of kiloelectron volts takes place is the primary parameter determining the emergent spectrum. If the bulk of the energy is deposited several Compton mean free paths into the atmosphere, the radiation must diffuse out. The photons will tend toward a Planckian distribution as a result of the scatterings. If too few photons are being emitted, the Planckian distribution will be dilute.

If the deposition takes place in only one or two Compton mean free paths, as in the 0.5 $M_{\odot}$ and 0.3 $M_{\odot}$ stars, the emitted photons stream out. As they suffer only one or two scatterings, there is only slight modification of the bremsstrahlung emission spectrum.

Our model neglects any deceleration of the infalling gas as a result of radiation pressure. This approximation obviously breaks down as the luminosity approaches the Eddington limit, $L_E = 4\pi cGM/K_s$. We also neglect any modification of the spectrum as a result of scattering in the cold infalling gas. For this assumption to have some validity, each photon should encounter less than one mean free path of material as it travels out through the infalling gas. This is a more stringent requirement than the Eddington limit; it results in $L < (v_{gas}/2c)L_E$. Because of this limitation we cannot use our model to calculate accretion onto white dwarfs.

While we have refrained from proposing specific models for some of the observed X-ray sources, we feel it is significant that this accretion model, with all its limitations, can produce a wide variety of emission spectra. Further, this can be done considering only Coulomb collisions as a stopping mechanism. Plasma instabilities need not be invoked. Particularly suggestive is the similarity in the high-energy tail observed in Cygnus X-1 and the high-energy spectra we calculate for the lower-mass neutron stars.

We wish to thank G. Cooper, J. LeBlanc, C. Lund, and D. Post for useful discussions.

## APPENDIX

Since we neglect the self-gravitation of the atmosphere, the gravitational field in the atmosphere, as well as exterior to the atmosphere, is the Schwarzschild gravitational field:

$$ds^2 = -\left(1 - \frac{2GM}{c^2\hat{r}}\right)d\hat{t}^2 + \frac{d\hat{r}^2}{(1 - 2GM/c^2\hat{r})} + \hat{r}^2(d\hat{\theta}^2 + \sin^2\hat{\theta}d\hat{\phi}^2)\,,$$

where $\hat{t}$, $\hat{r}$, $\hat{\theta}$, and $\hat{\phi}$ are the Schwarzschild curvature coordinates. Differential proper time $dt$ and differential proper radial length $dr$ in our static proper reference system are

$$dt = (1 - 2GM/c^2R)^{1/2}d\hat{t}$$

and

$$dr = (1 - 2GM/c^2R)^{-1/2}d\hat{r}\,,$$

where $R$ is the value of $\hat{r}$ at the surface of the neutron star.

In the atmosphere the relativistic hydrostatic equation is:

$$\frac{\partial P}{\partial \hat{r}} + \frac{GM}{R^2(1 - 2GM/c^2R)}\left(\rho + \frac{\rho\epsilon}{c^2} + \frac{P}{c^2}\right) = 0\,,$$

where $\rho$ is now the proper rest-mass density and $\epsilon$ is the thermal energy per unit rest mass. If we neglect the thermal energy density and pressure in comparison with the rest-mass energy density, the hydrostatic equation becomes

$$\frac{\partial P}{\partial r} + \frac{\rho GM}{R^2(1 - 2GM/c^2R)^{1/2}} = 0\,.$$

Material freely falling from radial infinity has, as measured in the static proper reference system on the surface of the star, radial velocity

$$\frac{dr}{dt} = \left(\frac{2GM}{c^2R}\right)^{1/2}$$

and total energy per unit rest mass of $c^2(1 - 2GM/c^2R)^{-1/2}$.

The accretion rate, luminosity, and photon energy used in the calculations are the quantities measured by the local observer. They are related to the same quantities measured by a distant observer as follows:

$$\frac{dM}{d\hat{t}} = (1 - 2GM/c^2R)^{1/2}\frac{dM}{dt}\,,$$

$$L_{\text{distant}} = (1 - 2GM/c^2R)L_{\text{local}}\,,$$

and

$$\nu_{\text{distant}} = (1 - 2GM/c^2R)^{1/2}\nu_{\text{local}}\,.$$

## REFERENCES


Bardeen, J. M., Press, W. H., and Teukolsky, S. A. 1972, *Ap. J.*, **178**, 347.

Chang, J. S., and Cooper, G. 1970, *J. Comp. Phys.*, **6**, 1.

Karzas, W. J., and Latter, R. 1961, *Ap. J. Suppl.*, **6**, 167.

Kihara, T., and Aono, O. 1963, *J. Phys. Soc. Japan*, **18**, 837.

Shkarofsky, I. P., Johnston, T. W., and Bachynski, M. P. 1966, *The Particle Kinetics of Plasmas* (Reading: Addison-Wesley), pp. 307–308.

Spitzer, L. 1962, *Physics of Fully Ionized Gases* (New York: Interscience).

Zel'dovich, Ya. B., and Shakura, N. I. 1969, *Soviet Astr.—AJ*, **13**, 175.

THE ASTROPHYSICAL JOURNAL, 184:271–289, 1973 August 15


# A MODEL FOR COMPACT X-RAY SOURCES: ACCRETION BY ROTATING MAGNETIC STARS

F. K. LAMB, C. J. PETHICK, AND D. PINES
Department of Physics, University of Illinois, Urbana
*Received 1973 March 5*

## ABSTRACT

The physics of accretion onto compact stars is considered, taking into account both the effects of stellar rotation and a stellar magnetic field. We show that far from the star the stellar magnetic field is screened by currents flowing in the accreting plasma, while close to the star the stellar field forces matter to corotate with the star. The location of the Alfvén surface, where the transition between the two regimes occurs, depends on the flow pattern of the accreting matter beyond the Alfvén surface and the rotation period of the compact star, as well as the mass accretion rate and the strength of the stellar magnetic field. Three types of flow pattern are considered: radial inflow toward the compact object, orbital motion about it, and streaming motion past it. For the accretion rates of interest the radius of the Alfvén surface, $r_A$, is found to be $\sim 10^7$–$10^8$ cm for a typical neutron star and $\sim 10^8$–$10^{10}$ cm for a typical magnetic degenerate dwarf. For neutron stars it is shown that inside the Alfvén surface there is comparatively little flow of matter across field lines, so that accreting matter is channeled toward the magnetic poles of the star where it forms hot spots. The resulting radiation is shown to emerge from the neighborhood of the stellar surface in a strongly anisotropic angular pattern with a spectrum which depends on the details of the accretion process; for an oblique rotator, one has a natural mechanism for the production of pulsed radiation. For X-ray luminosities of the order of $10^{37}$ ergs $s^{-1}$ it is shown that the temperature of the radiation, which is not generally expected to be blackbody in character, will be in excess of 6 keV.

Comparison of the above model with the observations suggests strongly that the X-ray stars in the pulsating binary X-ray sources Cen X-3 and Her X-1 are accreting neutron stars with their magnetic axes inclined at substantial angles to their axes of rotation. The implications of this interpretation for the observed X-ray spectra and pulse wave-forms are discussed. The change in the X-ray pulsation period due to accretion of matter is calculated; the calculated time scales for the spin-up which occurs when there is orbital inflow toward a star rotating in the same sense as the orbital motion, is in excellent agreement with those observed in Her X-1 and Cen X-3. The possibility of significant optical pulsations and the likely existence of a minimum pulsation period for accreting X-ray stars are discussed.

*Subject headings:* hydromagnetics — neutron stars — rotation, stellar — X-ray sources

## I. INTRODUCTION

The discovery of the periodic pulsating binary X-ray sources Centaurus X-3 (Schreier *et al.* 1972) and Hercules X-1 (Tananbaum *et al.* 1972) has spurred interest in the nature of compact X-ray sources. In the present paper we consider a class of models of compact X-ray sources based on the idea of a rotating magnetic star (either a neutron star or a degenerate dwarf) undergoing accretion. In the class of models described here, the energy to produce the X-ray luminosity is supplied largely by accretion, while anisotropy in the flow (which can sometimes lead to pulsed emission) is introduced by the magnetic field. In earlier studies of accretion (Bisnovatyi-Kogan and Fridman 1970; Shvartsman 1971*b*; Bisnovatyi-Kogan and Sunyaev 1972) some effects of the stellar magnetic field have been taken into account, but the combined effects of stellar rotation and the stellar magnetic field have not been considered previously. Very recently this question has been considered by Pringle and Rees (1972) and Davidson and Ostriker (1973), whose discussions came to our attention after most of the work described here had been completed. Although we shall focus our attention primarily on pulsating X-ray sources in the discussion which follows, we believe that the model of a rotating magnetic compact star undergoing accretion is also relevant for other types of compact X-ray sources.

Accretion rather than rotation is indicated as the primary energy source for Cen X-3 and Her X-1 for two reasons:

*a*) they are both members of close binary systems in which large mass flows from the secondary onto a compact primary seem likely;

*b*) if rotation with the period $P$ of the observed pulsations is assumed to be the primary energy source, then degenerate dwarfs are ruled out because, under the conditions of interest here, they cannot rotate as fast as the $1^{s}24$ period observed for Her X-1 and remain gravitationally bound, while neutron stars would have an unacceptably short slowing-down time.

The latter point can be seen from the estimated slowing-down time $T$, obtained by comparing the total rotational energy $E_{\rm rot}$ with the estimated X-ray luminosity $L(\sim 10^{36}$–$10^{37}$ ergs s$^{-1}$). The result is

$$T \lesssim E_{\rm rot}/L \sim 10/(P^2 L_{37}) \text{ years}, \tag{1}$$

where $L_{37}$ is the X-ray luminosity in units of $10^{37}$ ergs s$^{-1}$ and $P$ is the rotation period in seconds.

The luminosity $L$ provided by accretion is given by

$$L = (dM/dt)(GM/R), \tag{2}$$

where $dM/dt$ is the rate of mass accretion, $M$ is the stellar mass, and $R$ is the stellar radius. X-ray luminosities of the order of $10^{37}$ ergs s$^{-1}$ can be produced by mass-accretion rates $\sim 10^{-7}\, M_\odot$ per year onto degenerate dwarfs, or $\sim 10^{-9}\, M_\odot$ per year in the case of a neutron star.

Consider the flow of accreting matter toward the compact star. We are interested in the region near the compact star where the behavior of the accreting matter does not depend on details of the mass-transfer process. At what distances does the magnetic field of the compact star begin to influence the motion? Far from the star one expects that the magnetic field will be screened out by currents induced in the infalling material, so that the flow pattern of accreting matter is unaffected by the stellar field. On the other hand, close to the stellar surface one expects the effects of the stellar field to dominate the flow completely, enforcing corotation. We define the Alfvén surface as that surface within which the plasma is corotating: its approximate location is found by equating the energy density of the stellar magnetic field with the kinetic energy density of the accreting matter. We shall see that for typical stellar parameters, the radius $r_A$ of this surface (measured from the center of the compact star) is $\sim 10^8$ cm for a neutron star or $\sim 10^8$–$10^{10}$ cm in the case of a degenerate dwarf. For a neutron star, then, $r_A \gg R$ so that processes occurring near the Alfvén surface are clearly decoupled from those which take place near the stellar surface. This will not be the case for a degenerate dwarf in which $r_A \sim R$; the behavior of matter accreting onto such a magnetic degenerate dwarf will therefore be more complex. Accordingly, much of our later discussion will be restricted to the case of a neutron star.

In the case of a neutron star, we show that there is comparatively little flow across field lines inside the Alfvén surface, so that the accreting material will be channeled toward the magnetic poles of the star. On encountering the stellar atmosphere, it will tend to form a hot spot which has an area less than or of order $(R/r_A)R^2$. The resulting radiation emerges from the neighborhood of the stellar surface in a strongly anisotropic angular pattern with a spectrum which depends on the details of the accretion

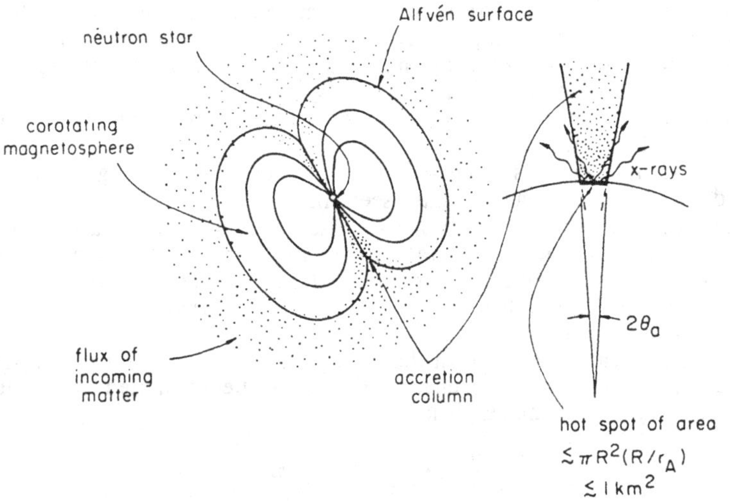


FIG. 1.—Sketch depicting some of the basic features of accretion onto a magnetic neutron star. The left side of the figure is an overall view of the magnetosphere showing the location of the Alfvén surface, the cusplike configuration of the magnetic field above the magnetic poles, the flow of accreting matter (*dots*) across the Alfvén surface, and the formation of accretion columns above the magnetic poles. To the right is a close-up view of an accretion column near the stellar surface. The accreting matter collides with the surface of the neutron star over an area $\lesssim \pi R^2(R/r_A)$, where $R$ and $r_A$ are, respectively, the radius of the star and the radius of the Alfvén surface. The result is a hot spot which leads to anisotropic X-ray emission.

process (the velocity of inflow at the stellar surface, the geometry of the flow, etc.). These general features of the accretion process are depicted in figure 1.

When rotation of the compact star is taken into account, one has, in the case of an oblique rotator, a natural mechanism for the production of pulsed radiation. As in the case of pulsars, the rotation period is to be identified with the pulsation period while the channeling of particles along field lines provides the desired anisotropy. In the case of an aligned rotator, on the other hand, no mechanism exists to produce distinct regular pulses persisting over a long time interval, although even in this case the emission might well be modulated at the rotation frequency due to inhomogeneities in the corotating magnetosphere.

The study of accretion by a rotating magnetic compact star divides logically into three distinct parts: beyond, at, and inside the Alfvén surface; and it is that order we shall follow in this paper. In § II, we consider the motion beyond the Alfvén surface; § III is devoted to a discussion of the location of the Alfvén surface, and in § IV the nature of the magnetopause (which contains the currents which screen the stellar field) is described. The motion of accreting matter inside the Alfvén surface is taken up in § V. Following this examination of the general character of the accretion process, we consider its observational implications in § VI, with particular attention to the behavior of the two pulsating sources Cen X-3 and Her X-1.

## II. BEYOND THE ALFVÉN SURFACE

In the region outside the Alfvén surface, the stellar magnetic field, being almost completely screened, exerts a negligible influence on the flow of accreting matter. In order to illustrate some of the effects to be expected, we first discuss in some detail the case of radial inflow toward the compact object and then compare this situation with that of orbital motion about the compact object and streaming motion past it. We shall pay particular attention to the effects of magnetic fields frozen into the accreting matter, since it has previously been pointed out that, in the absence of dissipation, these fields would be amplified by the flow of matter toward the compact star, and could halt the flow (Zel'dovich and Novikov 1971).

*a) Radial Inflow*

If there were small-scale magnetic fields completely frozen into freely falling matter and if there were no turbulence, then the radial component of the field, $\mathscr{B}_r$, would grow as $r^{-2}$ (Zel'dovich and Novikov 1971) while the components of the field perpendicular to $r$, $\mathscr{B}_\perp$, would grow as $r^{-1/2}$ (since an element of surface normal to $\mathscr{B}_\perp$ is compressed in the tangential direction as $r$ and extended in the radial direction as $r^{-1/2}$). The components of $\mathscr{B}$ perpendicular to $r$ cannot affect the basic features of the accretion process since they are not sufficiently amplified as the matter falls toward the compact object. However, in the absence of dissipation $\mathscr{B}_r$ could be amplified to the point where the frozen-in field would halt the inflow of matter.

This will not happen in practice since dissipative processes will cause the field to decay. Because the matter has a high electrical conductivity, ordinary ohmic losses are ineffective in reducing the field. A more efficient process is the "topological dissipation" described by Parker (1972) for which the dissipation time $\tau_d$ is $\sim \alpha l/v_A$, where $l$ is the characteristic length over which the radial field varies, $v_A$ is the Alfvén velocity $(\mathscr{B}_r^2/4\pi\rho)^{1/2}$, and $\alpha$ is a number which is of the order of one for a turbulent plasma, and somewhat larger for a quiescent plasma. The maximum strength which $\mathscr{B}_r$ can attain may be estimated by equating the dissipation time $\tau_d$ to the characteristic time for amplification of the field, $\sim r/v_{ff}$, where $v_{ff}$ is the free-fall velocity. Using the fact that $l$ is proportional to $r$, we find

$$\mathscr{B}_r^2/8\pi \lesssim \alpha^2(l_a/r_a)^2\rho v_{ff}^2/2\,; \tag{3}$$

here $l_a$ is the characteristic scale of the largest-scale embedded field at the radius $r_a$, where the average velocity of the matter first approximates radial free fall. Thus if topological dissipation occurs, the magnetic energy density is at most comparable to the kinetic energy of the bulk motion of the matter; the magnetic field is therefore too small to make the infall of matter differ appreciably from free fall. In some cases the maximum value of $\mathscr{B}_r$ can be substantially less than the value given by formula (3) (Lamb and Pethick 1973). Furthermore, the energy dissipated in the infalling matter is in turn of order $\mathscr{B}_r^2/8\pi$, so that the thermal energy per ion is less than $\sim m_i v_{ff}^2$, where $m_i$ is the ion mass. This shows that thermal pressure will not retard the infall appreciably.

So far we have concerned ourselves only with radial inflow in the absence of turbulence. If the accreting matter is also in turbulent motion, then the maximum energy density, $\mathscr{B}_t^2/8\pi$, of the small-scale fields generated by the turbulent motion is given by (Parker 1972)

$$\mathscr{B}_t^2/8\pi = \alpha[(\rho u^2/2)(\mathscr{B}_0^2/8\pi)]^{1/2}\,, \tag{4}$$

where $u$ is the velocity of the turbulent eddies, $\mathscr{B}_0$ is the magnitude of the large-scale embedded field, and where $\alpha$ is again expected to be of order unity. The turbulent velocity $u$ cannot exceed $v_{ff}$; furthermore, since $\mathscr{B}_0$ is less than $\mathscr{B}_r$, which is bounded by (3), $\mathscr{B}_0^2/8\pi$ is at most of order $\frac{1}{2}\rho v_{ff}^2$. Therefore, $\mathscr{B}_t^2/8\pi \lesssim \frac{1}{2}\rho v_{ff}^2$, so that the magnetic field generated by the turbulence does not alter the conclusion that magnetic fields present in the inflowing matter cannot halt its free fall.

In those cases where

$$\mathscr{B}^2/8\pi \ll \rho v_r^2/2 \tag{5}$$

the flow of accreting matter will be supersonic by the time it reaches the neighborhood of the Alfvén surface. In order to see this, first consider the size of the *acoustic* Mach number $\mathscr{M}_s$. When the inflow velocity of the accreting matter is given by the free-fall expression, $v_{ff} = (2GM/r)^{1/2}$, we have

$$\mathscr{M}_s(r) = v/c_s = [(2GM/r)/(\gamma kT/m_i)]^{1/2}, \tag{6}$$

where $c_s$ is the sound velocity, $T$ the temperature, and $\gamma$ the ratio of specific heats of the accreting matter. When inequality (5) is satisfied, the increase in sound velocity caused by the heating associated with magnetic field dissipation cannot keep pace with the increase in the inflow velocity and $\mathscr{M}_s \gg 1$. On the other hand, even if $T$ is as large as $10^8$ °K due to other processes (such as heating by X-rays emitted from the neighborhood of the surface [Zel'dovich and Shakura 1969]), one finds $\mathscr{M}_s(r_A) \sim 10$ for $r_A \sim 3 \times 10^8$ cm. Consider now the *Alfvén* Mach number, $\mathscr{M}_A$, which is given by

$$\mathscr{M}_A = v/v_A = [(2GM/r)/(\mathscr{B}^2/4\pi\rho)]^{1/2}, \tag{7}$$

with $v \sim v_{ff}$. Here $v_A$ is the Alfvén velocity in the embedded field $\mathscr{B}$, so that $\mathscr{M}_A \gg 1$ follows immediately when condition (5) holds.

If the radial flow of matter toward the compact object becomes supersonic (both $\mathscr{M}_s$ and $\mathscr{M}_A \gg 1$), one expects a stand-off shock to develop in the neighborhood of the Alfvén surface, which acts as a blunt obstacle in the path of the flow. The shock is likely to be collisionless because the mean free path for electron-ion collisions (Spitzer 1962),

$$\lambda = 1.3 \times 10^5 (T^2/nZ^2 \ln \Lambda) \quad \text{cm}, \tag{8}$$

where $\ln \Lambda$ is the Coulomb logarithm ($\sim$10–30 here) and $Ze$ is the ion charge, is large compared to the other microscopic lengths of interest, such as the ion Larmor radius, $r_i$. If, in addition, $\mathscr{M}_A \gg 1$, the shock will be turbulent (Sagdeev 1966). Collisionless shocks are not sufficiently well understood that one can be certain of their detailed structure, but for a spherical obstacle and very large Mach numbers the shock will be located at a radius $r_s = [1 + (\gamma' - 1)/(\gamma' + 1)]r_A$, provided that accreting matter does not accumulate between the shock front and the Alfvén surface. Here $\gamma'$ is the effective adiabatic index for matter passing through the shock. The shock will have a thickness $d \sim r_i \simeq v_{ff}/(e\mathscr{B}/m_i c)$ (Chu and Gross 1969). The density of matter immediately behind the shock is $\rho' = [(\gamma' + 1)/(\gamma' - 1)]\rho$ in terms of the preshock density $\rho$. Assuming that the directed kinetic energy of the accreting matter is transformed into random thermal energy as a consequence of passing through the shock, one has for the temperature, $T'$, of the matter immediately behind the shock,

$$kT' \sim m_i v^2 \sim GMm_i/r_s$$

or

$$T' \sim 10^9 (M/M_\odot)(10^9 \text{ cm}/r_s)^\circ \text{K}. \tag{9}$$

Except in the case of a fast oblique rotator, where the situation is less clear, the structure of the postshock ($r_s > r > r_A$) flow in the neighborhood of the magnetic poles seems certain to be different to that near the magnetic equator. One possibility is that the flow is made subsonic by the shock and then deflected away from the magnetic equator toward the magnetic poles.

The preceding considerations suggest that the accreting matter will arrive at the Alfvén surface with a temperature given approximately by formula (9) regardless of whether or not a shock forms.

### *b) Orbital Motion*

In the case of mass transfer in close binary systems, two distinct flow patterns are possible: either (i) the accreting matter has insufficient angular momentum for centrifugal forces to halt its free fall, in which case the inflow will be approximately radial; or (ii) the accreting matter approaches the compact star with sufficient angular momentum so that centrifugal forces alone would halt the matter at a radius $r_D > r_A$, in which case the matter tends to flatten and to form a disk (Prendergast and Burbidge 1968). Viscous forces in the disk will transport angular momentum, allowing a gradual inflow of matter toward the compact star. As matter approaches the compact star, the disk may thicken again due to mounting pressure, or instabilities. The flow pattern will change near the Alfvén surface, where the effects of the stellar magnetic field begin to dominate the flow.

To the extent that the motion approximates radial free fall, the previous discussion applies. However, if the inflow is significantly slowed by centrifugal forces, the effect of the stellar magnetic field will be somewhat different. This question is taken up in § III.

As with free fall, so with orbital motion any magnetic field frozen into the accreting matter will tend to be amplified, in this case by differential rotation within the disk. Considerations exactly analogous to those which led to formulae (3) and (4) again lead to the conclusion that the energy density of the embedded field is less than or of order $\frac{1}{2}\rho v^2$.

### *c) Streaming Motion*

A third possible flow pattern is streaming motion past the compact star, which may be relevant in the case of accretion fed by the stellar wind from a close binary companion. If $\mathcal{M}_* \lesssim 1$, where $\mathcal{M}_* = v_*/c_s$ is the Mach number of the compact object defined in terms of its velocity, $v_*$, relative to the accreting matter at large distances ($r \gtrsim r_a$), and the sound velocity $c_s$ at these distances, the motion of the accreting matter will be essentially radial inflow or orbital motion for $r \sim r_A$, depending on the angular momentum of the streaming matter. For somewhat larger Mach numbers, $1 \ll \mathcal{M}_* \lesssim (r_a/r_A)^{1/2}$, the Bondi-Hoyle-Lyttleton process applies with inflow occurring mainly behind the neutron star (Hoyle and Lyttleton 1939; Bondi and Hoyle 1944; Hunt 1971). As the inflowing matter approaches the Alfvén surface, one again expects the flow to become predominantly radial or orbital. (For extremely large Mach numbers, $\mathcal{M}_* \gg (r_a/r_A)^{1/2}$, a situation similar to that of the solar wind moving past the Earth's magnetosphere may occur, although such a situation appears rather unlikely.) In all cases where the streaming motion is supersonic, we expect a collisionless bow shock to form.

## III. LOCATION OF THE ALFVÉN SURFACE

The exact location of the Alfvén surface depends not only on the nature of the inflow toward the compact star but also on the rate of rotation of the star. In the present section we obtain a first estimate of the Alfvén radius by considering the case of radial free-fall toward a nonrotating star which has a dipolar magnetic field. We then examine the extent to which the position and character of the Alfvén surface wil be modified if the inflow is predominantly orbital or the star is rapidly rotating.

*a) Radial Free-Fall*

Viewed from the perspective of the infalling matter, the magnetic field of the compact star will begin to dominate the flow when the pressure of the stellar field, $B^2/8\pi$, becomes comparable to the pressure of the accreting matter, $\sim \rho v^2$ (including both ram pressure and thermal pressure). Thus $r_A$ is given approximately by

$$B(r_A)^2/8\pi = \rho(r_A)v(r_A)^2 \, . \tag{10}$$

Apart from a factor of order unity, this is equivalent to equating the energy densities of the magnetic field and the inflowing matter.

If the radius $R$ of the stellar surface is small compared with $r_A$, $B(r)$ for $r < r_A$ may be estimated by using the unscreened field of the star. If the stellar field is dipolar, then $B(r) = \mu/r^3$, where $\mu \equiv B_0 R^3$ is the magnetic moment in terms of the surface field $B_0$. [We note that neutron stars ($R \sim 10^6$ cm) with $B_0 \sim 10^{12}$ gauss and degenerate dwarfs ($R \sim 5 \times 10^8$ cm) with $B_0 = 10^4$ gauss both have magnetic moments $\mu \sim 10^{30}$ gauss cm$^3$.] The density $\rho$ of accreting matter is related to the rate of mass accretion, $dM/dt$, by the continuity equation

$$dM/dt = 4\pi\xi\rho v_r r^2 = \text{constant}\, , \tag{11}$$

where $4\pi\xi(r)$ is the solid angle subtended by the flow of accreting matter as seen from the compact star, and $v_r$ is the velocity of radial inflow. To the extent that the inflow velocity, $v_r$, of the accreting matter is equal to the free-fall velocity, it is given by

$$v_{ff}(r) = (2GM/r)^{1/2} = 1.6 \times 10^9[(M/M_\odot)/r_8]^{1/2} \text{ cm s}^{-1}\, . \tag{12}$$

The relations (2), (11), and (12) determine the energy density of accreting matter at the Alfvén radius, and using equation (10) we find

$$r_A(\text{radial}) \simeq 2.6 \times 10^8 \left[\frac{\mu_{30}{}^{4/7}(M/M_\odot)^{1/7}}{L_{37}{}^{2/7}R_6{}^{2/7}}\right] \text{ cm}\, . \tag{13}$$

Since $r_A$ is relatively insensitive to $R$ ($r_A \sim R^{-2/7}$), we see that degenerate dwarfs and neutron stars with comparable magnetic moments will have comparable Alfvén radii.

Let us now examine briefly the effect of stellar rotation on the position and structure of the Alfvén surface. For this purpose it is useful to define a centrifugal surface, $S_c$, at which the centrifugal and gravitational forces will just balance if a particle is corotating with the compact star. In order of magnitude the radius of the surface, $r_c$, is given by

$$GM/r_c^2 = r_c\Omega^2 \tag{14}$$

where $\Omega$ is the angular velocity of the compact star. In terms of the rotation period $P$, this becomes

$$r_c = 1.5 \times 10^8 P^{2/3}(M/M_\odot)^{1/3} \text{ cm}\, . \tag{15}$$

This radius $r_c$ is of course just the radius at which a particle in Keplerian circular orbit will corotate with the star.

Let us consider the effect of the stellar magnetic field on the accreting matter. At the Alfvén surface the stellar magnetic field will interact with the accreting matter, exerting a force in the tangential direction in an effort to enforce corotation. To judge from the somewhat similar situation which exists at the boundary between the solar wind and the Earth's magnetosphere, this interaction will be extremely complicated

(Parker and Ferraro 1971). In the absence of a better understanding, a crude but potentially useful first approximation may be to regard the accreting matter as unaffected for $r > r_A$ and corotating for $r < r_A$ (a similar approximation has sometimes been used in stellar-wind theory [Mestel 1968]). In any case the basic effect of the interaction seems clear: the magnetic torque will tend to spin up the accreting matter.

The accreting matter will in turn modify the stellar magnetic field near the Alfvén surface, since the inertia of the matter arriving at $r_A$ will distort the magnetic field as the field tries to enforce corotation. Studies of plasma streaming past contained magnetic fields (see Parker 1967) suggest that the boundary between plasma and field will be unsteady, at least on the small scale, that there will be a tendency for the plasma to mix with the field at the boundary, and that the plasma will tend to convect the field away. It seems reasonable to suppose that convection of the field will stop when the energy density of the field reaches that of the streaming plasma. These considerations suggest that the position of the Alfvén surface will not differ much from the value given by equation (13) unless the star is rotating so rapidly that the rotational energy density of the accreting matter at $r_A$ as seen in the frame corotating with the compact star greatly exceeds the ram pressure used to calculate the position of $r_A$ in (13), that is, unless

$$\tfrac{1}{2}\rho r_A^2\Omega^2 \gg \rho v_{ff}^2 . \tag{16}$$

An approximately equivalent condition is that $r_A$ greatly exceed $r_c$. In such a "fast rotator" case, it seems likely that $r_A$ will be determined by balancing the energy density of the stellar magnetic field, $B^2/8\pi$, not against the ram pressure, but against $\frac{1}{2}\rho r_A^2\Omega^2$. Particularly when rapid field dissipation is taken into account, it seems unlikely that corotation will be enforced much beyond this point, although distorted remnants of the stellar field may extend somewhat further. An additional complication of the fast rotator is the tendency for matter to accumulate on field lines which extend beyond $r_c$. Although this may continue for a time, eventually the density of matter will become so great that, in the absence of restraining pressure forces beyond the magnetosphere, the magnetic field will be unable to enforce corotation and the outer regions of the magnetosphere will become unstable (see Roberts and Sturrock 1973). Thus the extent to which the magnetosphere can extend beyond $r_c$ for any length of time is, at present, still an open question. On the other hand, if $r_A > r_c$ does occur, the above arguments suggest that the correct position of the Alfvén surface will be given by

$$r_A'(\text{radial}) \sim 2.6 \times 10^8 \left[\frac{\mu_{30}^{4/13} P^{4/13} (M/M_\odot)^{3/13}}{L_{37}^{2/13} R_6^{2/13}}\right] \text{cm} \qquad (r_A > r_c) . \tag{17}$$

Clearly these arguments are not conclusive, but at the present time we do not know how to make a better estimate.

### *b) Orbital Motion*

Let us now turn to the case of orbital motion of the inflowing matter, which may lead to formation of a disk. Quite generally the inner edge of the disk will be located at the Alfvén surface, where the flow of matter becomes dominated by the stellar field. As in the case of radial inflow, the stellar magnetic field will interact with the inner edge of the disk at $S_A$, exerting a torque on the matter orbiting there. In spite of the complexity of this interaction, it seems clear that the orbiting matter will tend to be spun up or spun down by the magnetic torque if the circular Keplerian velocity at $r_A$ is, respectively, less than or greater than $\Omega r_A$, where $\Omega$ is the angular velocity of the compact star. An equivalent way of stating this condition is that for spin-down of the orbiting matter $r_A < r_c$, and for spin-up $r_A > r_c$ (just as in the case of radial inflow, it is unclear whether $r_A > r_c$ can ever be realized in practice). The torque exerted by the stellar field may, by removing or adding angular momentum to the disk at $r_A$, accelerate or inhibit the accretion process.

In general, the disk matter orbiting at $S_A$ will tend to wind up the stellar magnetic field. Considerations similar to those given in the case of radial inflow suggest that $r_A$ will be determined by balancing $B^2/8\pi$ against the energy density of the orbiting matter as seen in the frame corotating with the compact star. Assuming that this is the case, the position of the Alfvén surface for orbital flow is given in terms of the position for radial inflow, equation (13), by

$$r_A(\text{orbital}) \sim (\xi v_r/v_{ff})^{2/7} r_A(\text{radial}) , \qquad r_A \ll r_c . \tag{18}$$

This expression takes into account the fact that for a given rate of mass accretion the density of matter in a disk exceeds that for the case of radial free-fall. Here $\xi r_A$ is the semithickness of the disk at $r_A$ (for a disk which thickens near $r_A$, $\xi$ will approach unity). The parameters $\xi$ and $v_r$ can be determined only if the structure of the disk is known in detail; however, even if $v_r/v_{ff}$ is as small as $10^{-2}$ and $\xi \sim 10^{-1}$, $r_A$ will be reduced from the radial inflow values only by a factor $\sim 10$. Because of the reduction of $r_A$ in the presence of a disk, very short periods ($P \lesssim 10^{-2}$ s) would be required to achieve "fast rotation" ($r_A > r_c$) in this case. Other effects, which can be neglected for longer periods, are likely to become very important for such short periods (see, e.g., § VI*d*) and hence this situation will not be discussed further here.

## IV. THE MAGNETOPAUSE

At the Alfvén surface, currents are induced in the infalling plasma by the stellar magnetic field. These currents are essentially surface currents: they completely screen the stellar field outside the Alfvén surface and enhance it inside the Alfvén surface. At the same time, the associated $\boldsymbol{j} \times \boldsymbol{B}$ forces tend to halt any flow of matter across field lines of the stellar magnetic field. Following geophysical terminology, we refer to the region containing these currents as the magnetopause (Parker and Ferraro 1971).

In order to see how a magnetopause forms, consider for a moment a model which demonstrates the basic behavior expected in the case of radial inflow near the magnetic equator of a dipole field. A neutral beam of cold plasma, initially moving with velocity $v$ in a field-free region, is incident normally on a plane surface, beyond which there is a uniform magnetic field parallel to the plane, and of strength $B$. Collisions in the plasma will be neglected, since we saw in the previous section that the mean free path for Coulomb scattering is large compared with other lengths in the problem. A charged particle entering the magnetic field performs a semicircular orbit with radius equal to its Larmor radius, before returning to the field-free region. Since the components of the electron and ion velocities parallel to the plane boundary are in opposite directions, a net electrical current will be set up in the surface region. The ions in the surface region give rise to an electrical current density parallel to the surface of the order of $n_i Zev$, where $n_i$ is the number density of ions in the incident beam. This current density extends over a thickness $\sim r_i$, and therefore the total surface current density due to the ions is $\mathcal{J}_i \simeq n_i Zevr_i \simeq n_i m_i v^2 c/B$; there is an analogous contribution from the electrons. This surface current, which is confined to a region of thickness $r_i$, produces a jump in the magnetic field of magnitude

$$4\pi(\mathcal{J}_i + \mathcal{J}_e)/c \simeq 4\pi\rho v^2/B . \tag{19}$$

If now $\rho$, $v$, and $B$ are related by the condition (10), as they are at the Alfvén surface, the jump in the magnetic field is just of order $B$, so that the assumption of screening currents induced at the Alfvén surface which cancel the stellar field at points outside the surface is self-consistent (Dessler 1968).

If the plasma has an appreciable temperature, the $\rho v^2$ factor in inequality (16) is replaced by the total kinetic energy of the plasma (including thermal motion as well as bulk motion); the $\rho v^2$ term in equation (10) is similarly modified, and the assumption of screening at the Alfvén surface is again self-consistent (Grad 1961).

The model just described is enormously oversimplified: in the case of the Earth, we know that the actual structure of the magnetopause is extremely complex (Parker and Ferraro 1971). (Some of the complications associated with plasma streaming parallel to the magnetopause have already been described in § III.) Despite the more complicated structure which the magnetopause around a compact magnetic star must surely have, we believe that the existence of a region similar in effect to the one described above is likely to be a very general feature of accretion onto such stars.

So far, we have considered only the role of the magnetopause in halting the inflow of accreting matter. How does this matter eventually cross the Alfvén surface to reach the surface of the compact star?

The stellar magnetic field appears better able to halt the inflow of matter near the magnetic equator than in the vicinity of the magnetic poles. Neutral points in magnetic polar regions are known to be subject to tearing-mode instabilities (Dungey 1958), and the Alfvén surface may well develop cusplike structures directed toward the magnetic poles, similar to those which are expected in the case of the Earth's magnetosphere (Spreiter and Summers 1967). Plasma theoretical studies indicate that such a geometry is stable (Berkowitz *et al.* 1958) and that it would lead to a rapid flow of particles across the sides of the cusp, toward the compact star (Grad 1963). These arguments are very uncertain at the present time and much more work remains to be done; nevertheless, one may reasonably expect that, in the absence of rotation, the magnetic field lines just inside the Alfvén surface above the magnetic poles are relatively more accessible to plasma flowing across the Alfvén surface toward the neutron star than are those above the magnetic equator.

Consider now the influence of rotation on the flow of matter across the Alfvén surface. As we shall see in the following section, once inside $S_A$ the matter will flow along field lines of the stellar magnetic field there. For stars rotating sufficiently slowly that the centrifugal surface $S_c$ would lie outside the Alfvén surface $S_A$, matter which has found its way onto field lines just inside $S_A$ will flow down the field lines toward the magnetic poles essentially unhindered by centrifugal forces. In the case of radial inflow toward $S_A$, matter will always be present just outside $S_A$ in the region above the magnetic poles. However, if the accreting matter forms a disk about the compact star, there may not be an appreciable density of matter above the magnetic poles unless the magnetic moment of the star lies almost in the plane of the disk; in this case the relative orientation of disk and magnetic moment may influence the rate of accretion to the stellar surface as well as characteristics of the resulting X-ray emission.

If there are stars rotating so rapidly that $S_c$ lies within $S_A$, centrifugal forces will play an important role in determining the flow of matter in the neighborhood of these stars. In the case of *oblique* fast rotators, centrifugal forces acting on matter directly above the magnetic poles near $S_A$ may be large enough to inhibit flow of matter across $S_A$ toward the stellar surface in this region; if flow across $S_A$ above the magnetic poles is sufficiently inhibited, flow across $S_A$ near the rotation axis may become important. On the other hand, for *either* aligned *or* oblique fast rotators, centrifugal forces cannot prevent matter from reaching the stellar surface once it has found its way onto field lines above the rotation axis, if it is able to fall freely along field lines. Again, matter will always be present just outside $S_A$ in the region above the rotation poles if inflow toward the compact star is radial, while the relative orientations of the rotation axis, the stellar magnetic moment, and the disk may be important in the case of disk flow.

## V. INSIDE THE ALFVÉN SURFACE

As we described in the Introduction, the interior of the magnetosphere of an accreting degenerate dwarf may well be very much more complicated than that of a neutron star. Therefore, we confine ourselves in the present section to a description of some of the basic features of the magnetosphere of an accreting, rotating neutron star.

Once inside the Alfvén surface, the accreting matter will flow along the magnetic field lines there. This follows immediately from an examination of the rate at which dissipative processes in the plasma allow flow across field lines, under the influence of the gravitational field of the compact object. Under steady-state conditions, the plasma velocity in the direction of the component of the gravitational field perpendicular to the magnetic field, $g_\perp$, is given by

$$v_\perp = (\rho c^2/\sigma B^2) g_\perp , \tag{20}$$

where $\sigma$ is the electrical conductivity in zero magnetic field. (If the collision frequency is large compared with the electron cyclotron frequency, this is the only component of the drift velocity. If the collision frequency is small, however, the drift velocity also has a large component in the direction of $\boldsymbol{g} \times \boldsymbol{B}$. Since this is in the azimuthal direction, it makes no contribution to the radial inflow.) To the extent that electron-ion collisions are the dominant source of electrical resistance, one finds

$$v_\perp^{\text{coul}} \simeq 0.2[r_8(L_{37}R_6 \ln \Lambda)^{1/2}/\mu_{30}T_8^{3/4}Z^{1/2}] \quad \text{cm s}^{-1} . \tag{21}$$

In arriving at this estimate, we assumed that $\rho$ and $v$ are related by the equation of continuity, neglecting the component of $\boldsymbol{v}$ along $\boldsymbol{B}$, i.e., by equation (11) with $v_r = v_\perp$. This clearly overestimates both $\rho$ and $v_\perp$, probably by many orders of magnitude. If the plasma is turbulent, the electrical conductivity will be reduced due to scattering of electrons by fluctuating electric fields. If the plasma is strongly turbulent (which appears unlikely, at least inside the Alfvén surface), one finds

$$v_\perp^{\text{turb}}(r) \simeq 20(M/M_\odot)(A/Z)r_8/\mu_{30} \quad \text{cm s}^{-1} , \tag{22}$$

where we have used the Bohm (1949) value for the turbulent conductivity; $A$ is the mass number of the ions. The results (21) and (22) for $v_\perp$ show that inside $S_A$, the plasma can move only a short distance across field lines, since the free-fall time at $r \sim 10^8$ cm is $\sim 0.1$ s.

If the streamlines follow the field lines exactly, the continuity equation gives the relation $\rho v/B =$ const. or, in the case of a dipole field,

$$\rho v r^3 = \text{const} . \tag{23}$$

Thus, if we know the dimensions of the accretion columns directed toward the magnetic poles and the average velocity of matter toward the stellar surface, we can estimate the density in the columns.

In order to determine the dimensions, one needs a detailed description of the Alfvén surface and, in particular, how matter crosses field lines there. However, in the absence of such a description we can estimate an upper bound on the dimensions, since it is rather unlikely that accreting matter will find its way onto field lines of the stellar field which would, even in the absence of screening currents, close inside the radius $r_A$. Since the field lines for a dipole field are given by $\sin^2 \theta/r =$ const., the foot of the last undistorted field line which would close inside $r_A$ lies at an angle $\theta_c$ determined by

$$\sin^2 \theta_c = R/r_A \,. \tag{24}$$

Using formulae (13) and (18), this relation becomes

$$\theta_c^2(\text{radial}) \simeq 4 \times 10^{-3}[R_6^{9/7}L_{37}^{2/7}/\mu_{30}^{4/7}(M/M_\odot)^{1/7}]\,, \tag{25}$$

$$\theta_c^2(\text{orbital}) \simeq \theta_c^2(\text{radial})(\xi v_r/v_{ff})^{-1/2}\,. \tag{26}$$

We write the cross-sectional area of the column of accreting plasma at the stellar surface as $\pi R^2\theta_a^2$; to the extent that the true shape of the Alfvén surface is more cusplike, we expect $\theta_a$ to be less than $\theta_c$. The area of the plasma stream at the stellar surface is therefore $\leqslant \pi R^2\theta_c^2 \simeq 10^{10}$ cm$^2$.

Consider now the average inflow velocity of matter in the accretion column. To the extent that ions which enter the magnetosphere possess a substantial magnetic moment ($\mu_i = \frac{1}{2}m_i v_\perp^2/B$, where $v_\perp$ is the component of velocity perpendicular to $\boldsymbol{B}$) and move on single-particle orbits in the dipolar field, they will tend to be reflected by the increasing magnetic field. For this to be important one needs $v_\perp^2 \sim GM/r$; we have seen earlier that this situation may be quite general, as when a shock develops near the Alfvén surface. Ions can then reach the stellar surface only by disposing of their magnetic moments. One way of doing this is by collision with electrons: the electrons are most probably very cold compared with the ions, since electrons can lose energy efficiently by radiation. [Near the Alfvén surface, electron cooling may be primarily due to bremsstrahlung; at slightly smaller radii, however, cyclotron cooling will dominate completely. If the velocity of the electrons, $v_e$, is less than the random velocity of the ions, $v_i$, then the characteristic time for ions to transfer energy to the electrons is (Spitzer 1962)

$$\tau_{i,e} \simeq m_e m_i v_i^3/[8(6\pi)^{1/2}Z^2 e^4 n_i \ln \Lambda]\,. \tag{27}$$

If one evaluates $n_i$ from the continuity equation (23), using as $v_i$ the free-fall velocity, one finds

$$\tau_{i,e} \simeq \frac{6 \times 10^{-3}}{\ln \Lambda}\left[r_8\left(\frac{A}{Z}\right)^2\left(\frac{M}{M_\odot}\right)^{20/7}(L_{37}R_6)^{-5/7}\mu_{30}^{-4/7}\right] \text{ seconds}\,. \tag{28}$$

This is to be compared with the free-fall time,

$$\tau_{ff} \simeq (r^3/GM)^{1/2} \simeq 8.6 \times 10^{-2}[r_8^{3/2}/(M/M_\odot)^{1/2}] \text{ seconds}\,. \tag{29}$$

Since the Coulomb logarithm $\ln \Lambda$ is at least 10 for the conditions existing near the Alfvén surface, the ions in this vicinity will be able to transfer their transverse kinetic energy to the electrons sufficiently rapidly that magnetic mirror effects are negligible.

We conclude, therefore, that the ions will not lose an appreciable fraction of their gravitational energy before reaching the vicinity of the stellar surface.

The ions colliding with the surface material arrive with an energy per particle of order $GMm_i/R \sim 100$ A MeV and some idea of the spectrum of the radiation emerging from the surface can be obtained by estimating the amount of matter required to stop them. An upper limit to the stopping distance can be obtained by considering only single-particle interactions. For ion energies in the range of interest, the ions are stopped primarily by distant Coulomb encounters with electrons in the surface layers. In this case the stopping time is given approximately by formula (27); an infalling ion therefore traverses a column density $\sim 8(M/M_\odot)^2(1/R_6)^2(A/Z^2)$ g cm$^{-2}$ before it loses its kinetic energy. If this estimate is correct, the layer of the neutron-star atmosphere in which the ions deposit their energy is several Thomson scatterings deep for photons coming out, since the Thomson-scattering mean free path is $\sim 2.5 \text{ g cm}^{-2}$, and the radiation is likely to emerge from the surface more or less in all directions. A lower bound on the temperature of the emitting region may be obtained by using the fact that the total luminosity cannot exceed the blackbody value. The emitting area is given by the cross-sectional area of the plasma streams at the stellar surface, $2\pi R^2\theta_a{}^2$; from this we find

$$T > 5 \times 10^7 (M/M_\odot)^{1/28} \mu_{30}{}^{1/7} L_{37}{}^{5/28} R_6{}^{-23/28} (\theta_c/\theta_a)^{1/2} \text{ °K}, \tag{30}$$

if $r_A$ is given by the value for radial inflow, equation (25). We are therefore led to associate temperatures in excess of 6 keV with radiation emitted from the surface, if the luminosity is $10^{37}$ ergs s$^{-1}$. Even if the ions are stopped as deep as 10 g cm$^{-2}$, the spectrum of the radiation which leaves the surface will be rather different from a simple thermal spectrum both because of the effects of Comptonization (Zel'dovich and Shakura 1969; Felten and Rees 1972; Illarionov and Sunyaev 1972; Alme and Wilson 1973) and because of increased emission due to the cyclotron process.

In reality, plasma collective effects are likely to play an important role in determining conditions at the surface. The direct conversion of plasma waves into radiation is probably not important since the plasma frequency at the densities of interest is less than about 1 eV. However, turbulent particle heating may well alter the temperature structure of the atmosphere. The problem of determining the spectrum of the emerging radiation when these effects are taken into account is a complex and difficult one: a detailed investigation of these questions is now in progress. The preliminary results available at the time of writing are not inconsistent with the conclusion that most of the radiation will emerge in the form of X-rays.

The angular distribution of the emitted radiation seen far from the star will depend on the strength of the stellar magnetic field. The magnetic field drastically reduces the opacity at frequencies well below the cyclotron frequency, but leaves it essentially unchanged at frequencies well above the cyclotron frequency (Canuto and Chiu 1971). The radius $r_{\rm crit}$ beyond which the cyclotron frequency is less than the photon frequency is given by

$$r_{\rm crit} = 2.3 \times 10^6 \mu_{30}{}^{1/3} \left(\frac{1 \text{ keV}}{h\nu}\right)^{1/3} \text{ cm}. \tag{31}$$

Therefore, if $\mu_{30} \leqslant \frac{1}{6}$, the magnetic field will not significantly affect the opacity for the X-ray photons of interest here ($h\nu \sim$ 2–6 keV), while if $\mu_{30}$ is substantially greater than $\frac{1}{6}$, the opacity experienced by these photons will be significantly reduced.

In the first case ($\mu_{30} \leqslant \frac{1}{6}$), the angular distribution of the emitted radiation will be influenced considerably by matter in the accretion columns above the magnetic poles. We may determine the density of matter in the columns by using the equation of continuity and the area of the columns, $2\pi(r^3/r_A)(\theta_a/\theta_c)^2$. The result is

$$\rho(r) = 7.3 \times 10^2 L_{37} R_6 (M_\odot/M)^{3/2} (\theta_c/\theta_a)^2 r_A / r^{5/2} \text{ g cm}^{-3}. \tag{32}$$

For photon energies in the keV range electron scattering is the dominant source of opacity. Using equation (32) for the density, one finds for the optical depth measured vertically from a point at radius $r$

$$\tau_v = \frac{2}{3}\left(\frac{Z}{A}\right)\frac{\rho(r)r}{2.5} = 20\left(\frac{Z}{A}\right)\left(\frac{M_\odot}{M}\right)^{3/2}\left(\frac{\theta_c}{\theta_a}\right)^2\left(\frac{r_A}{10^8 \text{ cm}}\right)^{1/2}\frac{L_{37}R_6}{r_6{}^{3/2}}. \tag{33}$$

The corresponding result for the optical depth measured horizontally from the center of the column to the edge is

$$\tau_h = \tfrac{3}{2}\tau_v\left(\frac{r}{r_A}\right)^{1/2}\left(\frac{\theta_a}{\theta_c}\right) = 3\left(\frac{Z}{A}\right)\left(\frac{M_\odot}{M}\right)^{3/2}\left(\frac{\theta_c}{\theta_a}\right)\left(\frac{r_A}{10^8\ \mathrm{cm}}\right)^{1/2}\frac{L_{37}R_6}{r_6^{1/2}}\,. \quad (34)$$

In this case, therefore, $\tau_v$ exceeds $\tau_h$ for all radii less than $\sim r_A/2$; at the stellar surface $\tau_v \gg \tau_h$. As a result, photons will be emitted preferentially perpendicular to the magnetic axis. Furthermore, since $\tau_v \gg 1$, few photons will be emitted along the magnetic axis. If the star is rotating, the observed radiation will be pulsed with a period given by the rotational period of the star.

On the other hand, if $\mu_{30} \gg \frac{1}{6}$, the opacity of matter in the accretion columns will be drastically reduced near the stellar surface, so that X-ray photons can emerge directly from the surface without interacting appreciably with matter in the columns. Again the observed radiation will be pulsed, although in this case because of the varying angle which the heated stellar surface makes with the line of sight.

If $\mu_{30} \sim \frac{1}{6}$, the radiation pattern will be intermediate between the two limiting cases discussed above. Photons will be able to make their way out of the sides of the column very easily, due to the reduction of the opacity by the magnetic field. However, photons will not be able to escape easily in directions close to the magnetic polar axis, since the vertical optical depth in the column is large compared to unity. The basic pattern of the emitted radiation will then be similar to what one expects for $\mu_{30} \gg \frac{1}{6}$ except close to the polar directions, where it will be markedly reduced as a result of occultation by the accretion column.

Since the radiation is produced and propagates in regions with a strong magnetic field, it seems quite likely that significant polarization will be present.

## VI. IMPLICATIONS FOR PULSATING SOURCES

Observations of the periodic pulsating binary X-ray sources Cen X-3 and Her X-1 are consistent with the idea that they are members of the class of sources just described, if one interprets the compact object as a neutron star with its magnetic axis inclined at a substantial angle to its axis of rotation. Consider now some of the observational implications of such an interpretation:

### *a*) *X-ray Emission*

As a result of the anisotropy of the radiation emitted by the source, rotation of the star will generally lead to the observed radiation having pulses whose period is the rotational period of the star. For most geometries one expects two pulses of different intensity for each revolution of the star. However, if the rotation axis of the star coincides with the line of sight or the magnetic axis, there will be no pulses, whereas if the rotation axis is perpendicular to the line of sight or the magnetic axis, there will be two identical pulses for each revolution of the star; in the latter cases the fundamental period of the observed pulsations will be half the rotational period of the star. The X-ray spectrum and angular distribution may be complicated by departures of the accretion column from the simple structure outlined in the previous section and by the interaction of the X-rays emitted near the stellar surface with matter elsewhere in the system. Nevertheless, the estimates of angular anisotropy and radiation temperature ($T \gtrsim 5 \times 10^7\ {}^\circ$K for $L = 10^{37}$ ergs s$^{-1}$) obtained here are consistent with the observed X-ray pulsations (Schreier *et al.* 1972; Tananbaum *et al.* 1972) and preliminary spectral data (Tananbaum 1972).

We note that if the magnetic moment of the star is of intermediate strength ($\mu_{30} \sim \frac{1}{6}$; see § V), and one of the magnetic poles passes close to the line of sight, the observed pulse wave-form will have a double-humped shape similar to that which has sometimes been observed in Her X-1 (Doxsey *et al.* 1973; Schreier and Tananbaum 1972).

### *b) Changes in Pulsation Period*

The time rate of change of the pulsation period, $\dot{P}$, is a quantity immediately accessible to observation and may provide information concerning the flow pattern of accreting matter and the magnetic and viscous torques acting just beyond the Alfvén surface. If we follow the history of an accreting particle in the case, say, of a slow rotator in which there is orbital motion in the same sense as the stellar rotation, we would find the following sequence of events (in this discussion we treat the moment of inertia of the star together with its corotating magnetosphere as a unit). When the particle approaches the Alfvén surface, where accreting matter first affects the motion of the star, it may participate in the communication of viscous torques to the star. As the particle is brought into corotation with the star, it will exert spin-up torques on the star, but these will be partially offset by the increase in the moment of inertia of the star-plus-magnetosphere caused by the addition of the particle to the magnetosphere. As the particle falls toward the stellar surface, its contribution to the moment of inertia will decrease, causing a further spin-up of the star. Finally, as the star readjusts to the addition of the particle to its surface, there will be still another decrease in the moment of inertia of the star which will cause an increase in the spin rate.

If we are only interested in $\dot{P}$ averaged over times long compared to the infall time from the Alfvén surface, then we can write a simple expression for the total effect of many such accreting particles acting in succession. Consider the star-plus-magnetosphere to be enclosed by a surface which lies just outside $S_A$. (We could in principle place this surface at any radius; the reason for choosing to place it just outside $S_A$ will become clear in the discussion below.) The net transport of angular momentum across this surface will lead to a $\dot{P}$ as described by the equation

$$\frac{\dot{P}}{P} = \frac{1}{M}\frac{dM}{dt}\left(\frac{M}{I}\frac{dI}{dM} - \frac{h_a}{h}\right) + \frac{\alpha}{I\Omega}, \qquad (35)$$

where $I$ is the moment of inertia of the star-plus-magnetosphere, $h_a$ is the component parallel to $\mathbf{\Omega}$ of the average angular momentum per unit mass of the accreting matter just outside $S_A$, $h = I\Omega/M$, and $\alpha$ denotes external magnetic and viscous torques acting just outside $S_A$. The mass-accretion rate, $dM/dt$, here represents only matter which is accreted to the stellar surface (this is consistent with our earlier definition in terms of the X-ray luminosity $L$); the effect of any matter which interacts with the magnetosphere of the star and then leaves the system without being accreted to the stellar surface is included in the term $\alpha/I\Omega$. The first term inside the parentheses in equation (35) takes into account the changing moment of inertia of the star; $dI/dM$ therefore describes the net change in $I$ after unit mass which crosses our imaginary surface just outside $S_A$ has settled onto the stellar surface and the star itself has completely readjusted. The second term inside the parentheses describes "accretion of angular momentum" by the star.

If one had purely radial inflow just outside $S_A$ ($h_a = 0$) and if there were no external torques ($\alpha = 0$), then the period would change on a time scale

$$T_{\rm radial} = \frac{M}{I}\frac{dI}{dM}T_M, \qquad (36)$$

where

$$T_M = M/(dM/dt) = 9 \times 10^8[(M/M_\odot)^2/L_{37}R_6] \text{ years} \qquad (37)$$

is the characteristic time for the mass of the star to change. For all but the very lightest neutron stars,

$$\frac{M}{I}\left(\frac{dI}{dM}\right) \simeq +1\,, \tag{38}$$

so that in this case ($h_a = 0$, $\alpha = 0$) the period would be expected to increase.

Consider now the case of orbital motion of accreting matter toward a slow rotator (the orbital motion may be in the same or opposite sense to the stellar rotation). For radii $r \gg r_A$, where the matter in the disk is essentially in a Keplerian orbit, both the accretion term and the torque term are each separately large compared to their difference; to estimate $\dot{P}$ using an imaginary surface placed at large $r$ therefore requires an extremely precise knowledge of both the angular momentum per unit mass of the inflowing matter, $h_a(r)$, and the viscous and magnetic torques, $\alpha(r)$, acting at that $r$. If one considers successively smaller radii, the angular velocity gradually decreases from the Keplerian value ($r \gg r_A$) to the angular velocity of the star ($r \lesssim r_A$); in general there will be a transition region ($r \gtrsim r_A$) of the order of $r_A$ in size in which the difference between the two terms becomes comparable to the magnitude of either of them. Thus, if we place our imaginary surface in this transition region, we may obtain an order-of-magnitude estimate for $\dot{P}$ by neglecting $\alpha(r)$ and setting $h_a = h_K(r_A)$ in (35) [here $h_K(r_A)$ is the value of $h_a(r)$ for a Keplerian circular orbit of radius $r_A$]. The resulting estimate for the time scale on which the stellar rotation changes is

$$T_{\rm orb} \sim \frac{R_g^2}{r_A^2}\frac{\Omega}{\Omega_K}\,T_{\rm radial} = 5 \times 10^4 \left(\frac{M}{M_\odot}\right)^{3/2}\left(\frac{10^8}{r_A}\right)^{1/2}\frac{(R_g/10^6)^2}{R_6 P L_{37}}\ \text{years}\,, \tag{39}$$

where $\Omega_K$ is the Keplerian angular velocity for a circular orbit at $r_A$, and $R_g[\equiv(I/M)^{1/2}]$ is the radius of gyration of the neutron star ($\sim 5$ km).

The observations show that both Her X-1 and Cen X-3 are speeding up at the present time, with a characteristic time which varies from $10^5$ to $3 \times 10^5$ years for Her X-1 (Tananbaum 1972) and from $\sim 10^3$ to $10^5$ years for Cen X-3 (Schreier *et al.* 1972). These results are in excellent agreement with the time scale (39) and indicate that the orbital motion is in the same sense as the stellar rotation. The variations in the observed values of $\dot{P}$ may reflect variations in the mass accretion rate.

### c) *Optical Pulsations*

If the pulsed X-rays are due to thermal emission at the surface of the neutron star, there will be corresponding optical emission, which might appear modulated at the rotation frequency. Unfortunately, the total optical luminosity produced in this manner is likely to be exceedingly small. Using the Rayleigh-Jeans approximation and the value of $\theta_c$ given by (25), we obtain the estimate

$$L_{\rm opt}(\text{thermal}) \sim 10^{27} T_8 R_6^{25/7} L_{37} (M_\odot/M)^{1/7}/\mu_{30}^{4/7}\ \text{ergs s}^{-1}\,. \tag{40}$$

Even if it were the only optical emission from the binary system and were unaffected by interaction with surrounding material, radiation at this level would be completely unobservable at galactic distances. More likely sources of observable optical emission are the electrons in the accretion columns, which will produce cyclotron radiation in the optical at $r_{\rm opt} \sim 10^7$ cm. Even if only 1 percent of the gravitational energy which has been gained by the accreting matter at this radius is emitted in the optical, one has

$$L_{\rm opt}(\text{cyclotron}) \sim 10^{-2}(dM/dt)(GM/r_{\rm opt})\,, \tag{41}$$

or taking as $r_{\rm opt}$ the radius at which the cyclotron frequency is $10^{15}$ Hz, $L_{\rm opt}$(cyclotron) $\sim 6 \times 10^{33} L_{37}(10^{12}/B_0)$ ergs s$^{-1}$. If $L_{37} \sim 1$ and the source is at a distance of 1 kpc, this emission is equivalent to that of a $m_v \sim 15$ mag star, and may be responsible for the modulated optical emission from Her X-1 reported by Lamb and Sorvari (1972) and Davidsen *et al.* (1972).

### *d) Minimum Pulsation Period*

An interesting question is whether there exists a maximum rotational velocity of magnetic neutron stars in close binary systems, above which mass transfer is prevented by effects related to the magnetic field. We have earlier noted the possibility that magnetic torques may prevent or at least inhibit accretion from a disk surrounding the neutron star. Here we wish to consider whether mass transfer can be prevented by pulsar-type emission of radiation and charged particles (similar arguments concerning accretion from the interstellar medium have been given by Ostriker, Rees, and Silk 1970; and from a binary companion, by Shvartsman 1971*a*).

Accretion will be impossible unless the plasma pressure $p(r_a)$ at the accretion radius $r_a$, where the gravitational field of the neutron star first begins to influence the plasma appreciably, exceeds the "pulsar" pressure, $p_{\mathrm{PSR}}(r_a)$. Therefore, the condition for accretion to begin is

$$p(r_a) = n(r_a)kT_a > p_{\mathrm{PSR}}(r_a)\,, \tag{42}$$

where $T_a$ is the temperature of the plasma at $r_a$, and $p_{\mathrm{PSR}} = (dE/dt)/(4\pi r^2 c)$ in terms of the pulsar energy-loss rate, $dE/dt$. If the poloidal field of the neutron star falls off like $(R/r)^n$, then the braking torque varies like $(\Omega R/c)^{2n-3}$ and the associated energy loss rate is $dE/dt \simeq \Omega B_0^2 R^3(\Omega R/c)^{2n-3}$ (Mestel 1971). This leads to $p_{\mathrm{PSR}}(r_a) = (B_0^2/4\pi)(R/r_a)^2(\Omega R/c)^{2n-2}$. For a pure dipole field $n = 3$, while to account for the observed time dependence of the period of the Crab pulsar one requires $n \simeq 2.7$ (Boynton *et al.* 1972). Both Cen X-3 and Her X-1 are members of binary systems, and it is interesting to ask whether condition (42) is satisfied if the accreting matter comes from the envelope of a companion. In this case $r_a$ is either of the order of the binary separation or the accretion radius defined in the usual way (Hunt 1971), if the latter is substantially smaller than the binary separation. If one takes $r_a \gtrsim 10^{11}$ cm, of the same order of magnitude as the observed orbital radii of the sources, $B_0 \lesssim 10^{12}$ gauss, and $n \gtrsim 2.5$, one finds $p_{\mathrm{PSR}} \lesssim 10^2$ dyne cm$^{-2}$ for $P \gtrsim 1$ s. Even the pressures found in supergiant envelopes are typically a few orders of magnitude greater than this (see, e.g., Van Citters and Morton 1970), and therefore accretion would not be prevented by pulsar pressure for rotational periods equal to the pulsational periods of the two sources. If the matter pressure at the accretion radius continues to satisfy condition (42) once accretion begins, one may obtain an estimate for the minimum rotation period consistent with a given luminosity due to accretion. One finds

$$P > 0.2 \times 10^{3(3-n)/(n-1)} R_6\left[\left(\frac{B_0}{10^{12}\,\mathrm{gauss}}\right)^2 \frac{R_6}{L_{37}}\left(\frac{M}{M_\odot}\right)4\pi\xi\left(\frac{10^4\,{}^\circ\mathrm{K}}{T_a}\right)^{1/2}\right]^{1/(2n-2)} \text{seconds}\,, \tag{43}$$

where $4\pi\xi r_a^2$ is the cross-sectional area of the stream of accreting matter at $r_a$. This argument suggests that if $\xi \sim \frac{1}{10}$ and $2.7 \lesssim n \lesssim 3$, the minimum period for an accreting compact X-ray source with a luminosity $\sim 10^{37}$ ergs s$^{-1}$ lies between $\sim 0.1$ s and 1 s.

In the case of radial inflow, once the condition (42) is satisfied, matter can fall freely toward the star since the ram pressure, $\rho v_{ff}^2$, of the infalling matter varies as $r^{-5/2}$, whereas the pulsar pressure varies as $r^{-2}$. Once accretion has begun, it may be possible for $P$ to decrease below the value given by inequality (43) without halting the accretion process. The question of whether or not nonradial inflow can be halted at distances much less than $r_a$ can be answered only by detailed investigation; we recall that for $P \lesssim 0.1$ s, $S_c$ probably lies within $S_A$. Clearly these questions deserve further study.

## VII. CONCLUDING REMARKS

In the preceding sections we have described a class of models for compact X-ray sources based on the idea of a compact rotating magnetic star undergoing accretion. We have seen that some of these models lead naturally to pulsed emission of X-rays. However, it must be emphasized that the basic physical mechanism is quite different from that by which X-rays are produced by some radio pulsars, such as the Crab Nebula pulsar.

The present calculations represent only a first step toward a description of accretion by compact rotating magnetic stars. They do, however, point to a number of fairly well defined problems which merit further study. Among these we mention the effect of magnetic torques on accretion from a disk, the manner in which matter crosses the Alfvén surface, the flow pattern of matter in the fast-oblique-rotator case, and the details of what goes on in the accretion column and near the stellar surface.

We have only sketched a few of the aspects of accretion onto a magnetic degenerate dwarf, where a more difficult set of problems is encountered. The fact that the Alfvén surface can lie close to the stellar surface in this case indicates that the problem of the flow of accreting matter is more complicated than in the case of a neutron star. When $r_A \sim R$, it seems quite likely that no stable well-defined pulses associated with localized hot regions are formed and that the plasma surrounding the star is turbulent; the latter point has been particularly emphasized by Ichimaru (1972).

The present work has been devoted to questions concerning matter in the vicinity of the compact star; we have not considered the role played by a binary companion and the (possibly associated) transitions between high and low states (which have been observed both in Cen X-3 and Her X-1). We have also not considered here the possibility that the collapsed object is a black hole; this has been discussed by Shvartsman (1971*b*), by Pringle and Rees (1972), and by Shakura and Sunyaev (1973).

We should like to thank Drs. Gordon Baym, Setsuo Ichimaru, Tohru Nakano, Jacob Shaham, and the participants in the seminar on compact X-ray sources at the University of Illinois for helpful discussions. We also wish to thank Dr. Harvey Tananbaum for communicating some of the results of the AS&E *Uhuru* group in advance of publication and Don Lamb for many stimulating discussions. Part of the work described here was carried out at the Aspen Center for Physics; the Aspen Workshop on Theoretical Astrophysics provided us with a valuable opportunity to discuss the problems considered here and related matters with Dr. Martin Rees and other workshop participants. F. K. Lamb is grateful to Professor D. E. Blackwell and Dr. D. W. Sciama for hospitality extended to him at the Department of Astrophysics, University of Oxford, and C. J. Pethick is grateful to NORDITA, Copenhagen, for hospitality. This work has been supported in part by the National Science Foundation, through grant GP-25855. Some of the work was carried out while F. K. Lamb held a fellowship at Magdalen College, Oxford. C. J. Pethick holds an A. P. Sloan Research Fellowship.

## REFERENCES

Alme, M., and Wilson, J. 1973, to be published.
Berkowitz, J., Grad, H., and Rubin, H. 1958, *Proceedings 2d UN International Conference on the Peaceful Uses of Atomic Energy* (Geneva: UN Publications), **31**, 177.
Bisnovatyi-Kogan, G. S., and Fridman, A. M. 1970, *Soviet Astr.—AJ*, **13**, 566.
Bisnovatyi-Kogan, G. S., and Sunyaev, R. A. 1972, *Soviet Astr.—AJ*, **15**, 697.
Bondi, H., and Hoyle, F. 1944, *M.N.R.A.S.*, **104**, 273.
Bohm, D. 1949, in *The Characteristics of Electrical Discharges in Magnetic Fields*, ed. A. Guthrie and R. K. Wakerling (New York: McGraw-Hill), Chap. 2. § 5.
Boynton, P. E., Groth, E. J., Hutchinson, D. P., Nanos, G. P., Partridge, R. B., and Wilkinson, D. T. 1972, *Ap. J.*, **175**, 217.
Canuto, V., and Chiu, H. Y. 1971, *Space Sci. Rev.*, **12**, 3.

Chu, C. K., and Gross, R. A. 1969, *Advances in Plasma Physics*, Vol. **2** (New York: Wiley), p. 139.
Davidsen, A., Henry, J. P., Middleditch, J., and Smith, H. E. 1972, *Ap. J.* (*Letters*), **177**, L97.
Davidson, K., and Ostriker, J. P. 1973 *Ap. J.*, **179**, 585.
Dessler, A. J. 1968, in *Physics of the Magnetosphere*, ed. R. F. Carovillano, J. F. McClay, and H. R. Radoski (Dordrecht: Reidel), p. 65.
Doxsey, R., Bradt, H. V., Levine, A., Murthy, G. T., Rappaport, S., and Spada, G. 1973, to be published.
Dungey, J. W. 1958, *Cosmic Electrodynamics* (Cambridge: Cambridge University Press), p. 98.
Felten, J. E., and Rees, M. J. 1972, *Astr. and Ap.*, **17**, 226.
Grad, H. 1961, *Phys. Fluids*, **4**, 1366.
———. 1963, in *Progress in Nuclear Energy*, Series XI, Vol. **2** (Oxford: Pergamon Press), p. 189.
Hoyle, F., and Lyttleton, R. A. 1939, *Proc. Cambridge Phil. Soc.*, **35**, 405.
Hunt, R. 1971, *M.N.R.A.S.*, **154**, 141.
Ichimaru, S. 1972, private communication.
Illarionov, A. F., and Sunyaev, R. A. 1972, *Astr. Zh.*, **46**, 225.
Lamb, D. Q., and Sorvari, J. M. 1972, *IAU Circ.*, No. 2422.
Lamb, F. K., and Pethick, C. J. 1973, in preparation.
Mestel, L. 1968, *M.N.R.A.S.*, **138**, 359.
———. 1971, *Nature Phys. Sci.*, **233**, 149.
Ostriker, J. P., Rees, M. J., and Silk, J. 1970, *Ap. Letters*, **6**, 179.
Parker, E. N. 1967, *J. Geophys. Res.*, **72**, 4365.
———. 1972, *Ap. J.*, **174**, 499.
Parker, E. N., and Ferraro, V. C. A. 1971, in *Handbuch der Physik*, **49**/3 (Berlin: Springer-Verlag), p. 131.
Prendergast, K. H., and Burbidge, G. R. 1968, *Ap. J.* (*Letters*), **151**, L83.
Pringle, J. E., and Rees, M. J. 1972, *Astr. and Ap.*, **21**, 1.
Roberts, D. H., and Sturrock, P. A. 1973, *Ap. J.*, **181**, 161.
Sagdeev, R. Z. 1966, *Reviews of Plasma Physics*, ed. M. A. Leontovich (New York: Consultants Bureau), **4**, 23.
Schreier, E., Levinson, R., Gursky, H., Kellogg, E., Tananbaum, H., and Giacconi, R. 1972, *Ap. J.* (*Letters*), **172**, L79.
Schreier, E., and Tananbaum, H. 1972, private communication.
Shakura, N. I., and Sunyaev, R. A. 1973, *Astr. and Ap.*, **24**, 337.
Shvartsman, V. F. 1971*a*, *Soviet Astr.—AJ*, **15**, 342.
———. 1971*b*, *ibid.*, p. 377.
Spitzer, L. 1962, *Physics of Fully Ionized Gases* (New York: Interscience).
Spreiter, J. R., and Summers, A. L. 1967, *Planet. and Space Sci.*, **15**, 787.
Tananbaum, H. 1972, private communication.
Tananbaum, H., Gursky, H., Kellogg, E. M., Levinson, R., Schreier, E., and Giacconi, R. 1972, *Ap. J.* (*Letters*), **174**, L143.
Van Citters, G. W., and Morton, D. C. 1970, *Ap. J.*, **161**, 695.
Zel'dovich, Ya. B., and Novikov, I. D. 1971, *Relativistic Astrophysics*, Vol. **1** (Chicago: University of Chicago Press).
Zel'dovich, Ya. B., and Shakura, N. I. 1969, *Soviet Astr.—AJ*, **13**, 175.

Astron. & Astrophys. 29, 179 – 184 (1973)

# Accretion onto Massive Black Holes

J. E. Pringle, M. J. Rees and A. G. Pacholczyk*
Astronomy Centre, University of Sussex

Received July 24, 1973

**Summary.** We consider the observable effects of accretion onto massive ($\gtrsim 10^4 M_\odot$) black holes, particularly in the case when angular momentum is important. Such objects could be extremely luminous, even if the surrounding density were very low.

**Key words:** black holes – accretion – intergalactic gas

## 1. Introduction

The possible importance of accretion onto massive black holes was first emphasized by Salpeter (1964). Accretion onto compact objects can be one of the most efficient means of generating radiation from matter. Lynden-Bell (1969) proposed accretion onto central black holes in galaxies as a possible model for quasars, and Schwarzman (1971) has considered the observable properties of massive accreting objects. Shapiro (1973a, b) has made detailed calculations of the radiation to be expected from lone black holes of stellar mass within our own galaxy, and finds that the expected luminosities are very low. In this article we note that, in general, *spherically symmetric* accretion onto a lone black hole is not very efficient at producing radiation; but this efficiency is however greatly enhanced if the accreted material carries even a small amount of angular momentum. In this latter case an "accretion disc" is formed, and we consider the properties of such discs in some detail. For massive black holes, these discs would be exceedingly luminous, even if the surrounding density were very low. We consider two particular examples: a $10^4 M_\odot$ black hole in the halo of our galaxy and a $10^{11} M_\odot$ black hole in a cluster of galaxies. We argue that such objects might already have been detected, and present observations in any case set significant constraints on the possible existence of massive black holes in intergalactic space.

## 2. Spherically-symmetric Accretion

We first consider spherically-symmetric accretion onto a black hole, mass $M$, embedded in a uniform medium of particle density $n_c$ cm$^{-3}$ and temperature

$$T_c = 10^4\, T_{c4} \quad {}^\circ\mathrm{K}\,.$$

* Permanent address: Steward Observatory, University of Arizona, Tucson, Arizona, USA

We suppose, following Schwarzman (1971) and Shapiro (1973b); that the infalling material contains a small magnetic field which may be amplified until the synchrotron process becomes the dominant radiation mechanism.

The accretion rate is given approximately by Bondi (1952)

$$\dot{M} \approx 2 \times 10^{10} (M/M_\odot)^2 n_c T_{c4}^{-3/2} \,\mathrm{gm\,s^{-1}}\,. \tag{1}$$

This formula applies when the infalling matter (assumed to be mainly hydrogen) is ionized out to distances $\gg R_A$. For neutral matter, the sound speed at a given temperature is lower by $2^{1/2}$, and the accretion rate consequently greater by $2^{3/2}$. If the black hole is moving through the ambient medium with a Mach number $\mu$ then the accretion rate is reduced from (1) by a factor $\sim(1+\mu^2)^{-3/2}$. Even when $\mu \gg 1$, the flow near the black hole may still be nearly spherically symmetrical (Schwarzman, 1971; Hunt, 1971) unless angular momentum is important (see §3). Within the "accretion radius" (defined as the radius $R_A$ at which the escape velocity $(2GM/R_A)^{1/2}$ equals the sound speed at infinity) the inflowing gas is approximately in free fall. The inward velocity $V_r$ at a radius $X R_S$ [where

$$R_S = 2GM/c^2 = 2 \cdot 95 \times 10^5 (M/M_\odot)\, c_M$$

is the Schwarzschild radius] is

$$V_r \approx 3 \times 10^{10} X^{-1/2} \,\mathrm{cm\,s^{-1}} \tag{2}$$

This formula is only really valid for $X \gg 1$: at radii comparable with the Schwarzschild radius, relativistic corrections should of course be included. The particle density (again for $X \gg 1$) is

$$n(X) \approx 4 \times 10^{11} n_c T_{c4}^{-3/2} X^{-3/2} \,\mathrm{cm^{-3}}\,. \tag{3}$$

Following Schwarzman (1971) we assume that the magnetic field frozen into the infalling material is occurs if recombination and line cooling (see for instance, Cox and Tucker, 1969) are included as well as free-free.

*(iii) Synchrotron Cooling*

Shearing of the material during radial infall and in the differentially rotating disc amplify an initial "seed field" in the accreted material. We suppose that in the disc

$$B^2/8\pi \approx \alpha\, nkT \tag{13}$$

where $\alpha$ is a parameter $\leqq 1$. For the cases we are considering synchrotron radiation is self-absorbed and the temperature needed to radiate the required power per unit area is

$$T_S \approx 5 \times 10^{10} \alpha^{-6/25} n_c^{6/25} T_{c4}^{-9/25} \xi_2^{6/25} X^{6/25}\ {}^{\circ}\mathrm{K}. \tag{14}$$

For the disc approximation to hold we need $c_S \lesssim V_c$,

i.e.

$$X \lesssim 80 \alpha^{6/31} n_c^{-6/31} T_{c4}^{9/31} \xi_2^{6/31}\ {}^{2)}. \tag{15}$$

Then the field strength is

$$B \approx 3 \times 10^6 \alpha^{14/25} n_c^{11/25} T_{c4}^{-33/50} \xi_2^{-14/25} X^{-39/25}\, G \tag{16}$$

and the characteristic frequency of the radiation emitted at each radius is

$$\nu_S = 3 \times 10^{14} \alpha^{2/25} n_c^{23/25} T_{c4}^{-69/50} \xi_2^{-2/25} X^{-27/25}\ \mathrm{Hz}. \tag{17}$$

The overall spectral distribution of this synchrotron radiation is $S(\nu) \propto \nu^{-2/27}$.

This region of the disc also emits some bremsstrahlung, the ratio of synchrotron luminosity to free-free luminosity at any radius being

$$L_S/L_{ff} = 10 \alpha^{14/25} \xi_2^{59/25} (M/M_\odot)^{-1} n_c^{34/25} T_{c4}^{-51/24} X^{93/80}. \tag{18}$$

*Instabilities*

For self consistency and thermal stability we require $L_S/L_{ff} \gg 1$, i.e.

$$X \gg 0.25 (M/M_\odot)^{50/93} \alpha^{-28/93} \xi_2^{-118/93} n_c^{-68/93} T_{c4}^{34/31}. \tag{19}$$

This raises the interesting possibility that the disc may emit predominently synchrotron radiation in its outer parts, but that at a radius $X_i$ the incoming gas cools rapidly from a temperature at which the electrons are relativistic to a state at which it radiates like a black body (via bremsstrahlung, etc.) at a much lower temperature.

The internal energy released during this cooling is approximately

$$L_i \sim \left(\frac{kT_i}{m_p c^2}\right) \dot{M} c^2 \sim \left(\frac{3kT_i}{m_p c^2}\right) L \tag{20}$$

[2]) This inequality applies if the electrons have a (relativistic) maxwellian distribution. The condition $c_S \lesssim V_c$ would be *more* easily satisfied if plasma processes generated a power-law spectrum (see Norman and ter Haar, 1973 and references cited therein), because the mean particle energy would then be much less than that of the particles mainly responsible for the emission.

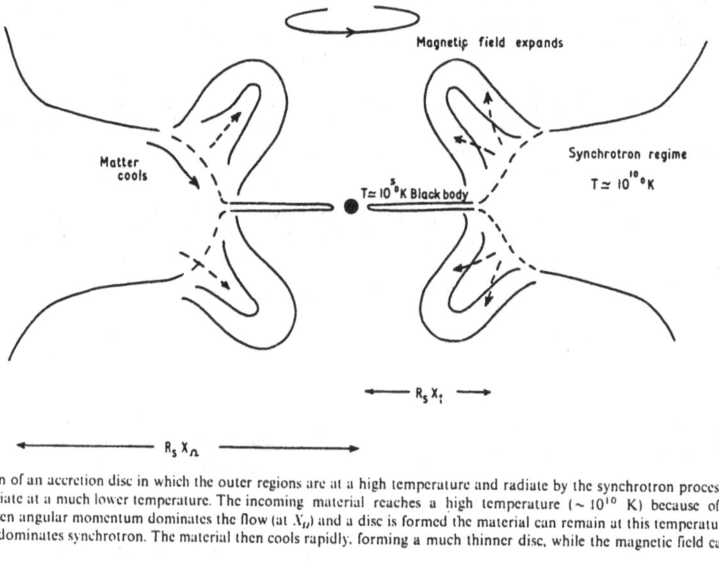


Fig. 2. Illustration of an accretion disc in which the outer regions are at a high temperature and radiate by the synchrotron process, whereas th inner regions radiate at a much lower temperature. The incoming material reaches a high temperature (~ $10^{10}$ K) because of compressio during infall. When angular momentum dominates the flow (at $X_\Omega$) and a disc is formed the material can remain at this temperature until (at $X$ free-free cooling dominates synchrotron. The material then cools rapidly, forming a much thinner disc, while the magnetic field can expand ou of the plane

quickly amplified ($B \propto X^{-2}$) until the magnetic and free fall kinetic energy densities are comparable, and that thereafter

$$B^2/8\pi \approx \alpha\, 1/2\, m_p\, n(X)\, V_r^2\,, \tag{4}$$

$\alpha$ being a parameter of order unity (i.e. $B \propto X^{-5/4}$). None of our conclusions would change drastically, however, by adopting the equally plausible assumption that $B^2/8\pi$ is proportional to the *thermal* energy density of the gas.

Adiabatic compression would render the electrons relativistic for $X \lesssim 2000$. The main cooling process competing with adiabatic heating is then synchrotron radiation. For $\alpha \approx 1$, bremsstrahlung (free-free) cooling is negligible. Radiative cooling will be unimportant relative to adiabatic heating at large values of $X$. As $X$ decreases, however, the cooling becomes relatively more important, and there may be a certain radius $X_{max} R_S$ (with $X_{max} > 1$) at which the heating and cooling rates are comparable. For $X < X_{max}$ the gas generally *cools* as it falls inward. It is then easy to see that the total energy radiated per unit mass is comparable with the *internal energy per unit mass at* $X \approx X_{max}$. Thus, if cooling is *too* efficient (and $X_{max} \gg 1$) the luminosity will be $\ll \dot{M} c^2$. The efficiency is maximal when $X_{max} \sim 1$. If cooling is unimportant even for $X = 1$ then the efficiency is again low, because the material is swallowed by the black hole before it has had time to radiate a significant fraction of the energy deriving from the "$PdV$ work" done on it during the infall.

Precise calculations of the radiation from accreting black holes must allow for the fact that much of the radiation emitted within a few Schwarzchild radii is itself swallowed by the hole. Shapiro (1973b) has carried out such calculations for stellar-mass black holes. He assumed that, for $X \lesssim 2000$, the relativistic electrons and non-relativistic protons are sufficiently well coupled that their temperature remain equal, the effective ratio of specific heats being $\gamma = 13/9$. The assumption that the electrons are compressed independently of the protons (with $\gamma = 4/3$) would alter some of his detailed results, as would allowance for possible synchrotron self-absorption. One might also question whether, in the presence of a disordered magnetic field, the relativistic electrons will really retain a maxwellian distribution, rather than developing a power law high energy tail. We have not ourselves performed detailed computations for any particular set of assumptions: the general model-independent features are, however, illustrated schematically by Fig. 1, which displays the luminosity from a black hole of given mass as a function of $n_c T_c^{-3/2}$. For optically thin synchrotron radiation from maxwellian electrons the slopes and scales are straightforwardly calculable, but for

$M \gtrsim 10^4 M_\odot$

the radiation is self-absorbed for all reasonable values of $n_c T_c^{-3/2}$ and Fig. 1 will be only schematically correct. The luminosity for a given accretion rate peaks when $X_{max} \approx 1$. For the cases of accretion of interstellar matter onto stellar-mass black holes considered by Shapiro, cooling is unimportant compared to adiabatic heating even at $X \approx 1$. A larger mass yields higher efficiency (for a given value of $n_c T_c^{-3/2}$) because the free-fall timescale is proportional to $M$.

### 3. Accretion of Material with Angular Momentum

Suppose that the black hole lies in a cloud of gas that is rotating with angular velocity $\Omega (= 2\pi/P$, where $P$ is the rotation period). The accretion rate will still be given by (1) provided that centrifugal force is unimportant at the accretion radius [1]. Within the accretion radius, gas falls inward, conserving its angular momentum as it does so. For the purpose of this exposition we assume the velocity of the black hole through the medium is such that $0 < \mu \lesssim 1$: our arguments can however, be straightforwardly adjusted to other cases.

The accreted material then comes from a cylinder of radius $\mu^{-1/2} R_A$, so a typical unit mass of accreted material has angular momentum $\sim \Omega \mu^{-1} R_A^2$ relative to the hole (the precise value depending on the direction of motion relative to the angular velocity vector). Angular momentum will start to control the flow when the azimuthal velocity is approximately keplerian. Thus occurs at a $\delta$ radius $X_\Omega R_S$, where

$$X_\Omega \approx 6 \times 10^{-6}\, T_{c4}^{-4} (M/M_\odot)^2 \left(\frac{P}{3 \times 10^8\ \mathrm{yr}}\right)^{-2} \mu^{-2}\,. \tag{5}$$

If $X_\Omega \gg 1$ we expect an "accretion disc" to form around the black hole. We now investigate the properties of such an accretion disc.

We take the inner boundary of the disc to be at the radius of the innermost stable circular orbit for a test particle ($X = 3$ for a Schwarzschild metric and at $X = 1/2$ for corotating Kerr black holes with the maximum permissible angular momentum). Outside this radius, we ignore relativistic effects and content ourselves with an approximate newtonian treatment [see Novikov and Thorne (1973) for a more rigorous discussion]. For accretion onto a "maximal" Kerr

[1]) The discussion given here is actually only applicable provided that the black hole *moves* through the gas (i.e. has a non-zero Mach number $\mu$). In this case, the accreted material comes from a cylinder, with axis along the direction of motion, whose radius is $\sim \mu^{-1/2} R_A$ if $\mu \lesssim 1$ or $\sim \mu^{-2} R_A$ if $\mu \gtrsim 1$. If $\mu \lesssim 1$, the condition that (1) still be applicable is then $\Omega^2 \ll \mu^{3/2} GM/R_A^3$. If $\Omega \neq 0$ but $\mu$ were strictly zero, there would be no true steady state: after a sufficiently long time, material would be accreted which originated arbitrarily far from the black hole, and angular momentum would *eventually* become important however small $\Omega$ was. Note also that the accreted material may have non-zero angular momentum if the massive object is moving through an *inhomogeneous* medium with *zero* vorticity.

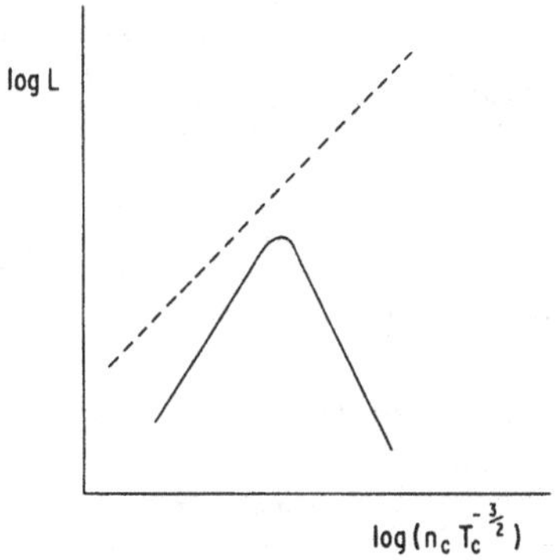


Fig. 1. A schematic plot of the luminosity $L$ released by accretion onto a black hole of given mass against $n_c T_c^{-3/2}$, where $n_c$ and $T_c$ are the particle density and temperature in the ambient medium, respectively. The full line is for spherically symmetric accretion and the dotted line for the case when angular momentum is important. The peak for spherically symmetric infall is reached when the cooling time of the infalling material and the free fall time are comparable close to the Schwarzschild radius.

metric the luminosity of the disc is

$$L \approx 0.4\,\dot{M} c^2 = 8 \times 10^{30} (M/M_{\odot})^2 n_c T_{c4}^{-3/2}\ \mathrm{erg\,s^{-1}}. \tag{5}$$

This luminosity is displayed by the dotted line in Fig. 1 as a function of $n_c T_c^{-3/2}$. It is always much greater than that obtained from spherically symmetric infall, the factor between the two luminosities is at least $10-10^2$ even when $X_{max} \approx 1$.

The disc rotates differentially, the circular velocity $V_c$ at radius $X$ being nearly Keplerian:

$$V_c = 2 \times 10^{10} X^{-1/2}\ \mathrm{cm\,s^{-1}}. \tag{6}$$

The inward mass flux is equal to the accretion rate in a steady situation, and the energy released in the disc per unit surface area is given, at least for values of $X$ where the precise boundary conditions are unimportant, by

$$p(x) = \frac{3GM\dot{M}}{4\pi R^3} = 2.5 \times 10^{19} n_c T_{c4}^{-3/2} X^{-3}\ \mathrm{erg\,s^{-1}\,cm^{-2}}. \tag{7}$$

The semi-thickness $b$ of the disc, determined by a balance between thermal pressure gradients and the component perpendicular to the disc of the central object's gravitational field, will be

$$b \approx (c_S/V_c)\, X R_S \tag{8}$$

where $c_S$ is the effective sound speed. Radiation pressure is not important for the discs considered here. The inward velocity of the material depends on the viscosity, which is the most uncertain parameter in all accretion disc models. Shakura and Sunyaev (1973) envisage the kinematic viscosity to be due to turbulence or magnetic stresses, and take its value as

$$\eta \approx \xi c_S b\,, \tag{9}$$

$\xi$ being a parameter $\ll 1$. We estimate $\xi$ to be $\sim 10^{-2}$ and define $\xi_2 = 10^2 \xi$. The inward velocity of the material is then

$$V_r \approx (\xi_2/100)\,(c_S/V_c)^2\, V_c\,. \tag{10}$$

The density of matter in the disc thus depends on the viscosity and temperature as well as on the mass flux, and is therefore uncertain and model-dependent. The properties of these discs have been described by several authors (Lynden-Bell, 1969; Pringle and Rees, 1972; Novikov and Thorne, 1973; Shakura and Sunyaev, 1973) – although mainly in somewhat different astrophysical contexts from those concerning us here – and we shall not repeat all the details.

The dominant emission mechanism, and the spectrum of the emergent radiation, depend on the particle density and magnetic field strength in the disc. There are three main possibilities.

(*i*) The disc radiates like a black body (the temperature being, of course, a function of radius).

(*ii*) Bremsstrahlung cooling is dominant, but the temperature greatly exceeds the black body temperature and reabsorption is unimportant.

(*iii*) Synchrotron cooling dominates.

*(i) Optically Thick Disc*

The temperature is

$$T_{bb}(X) = 10^6 n_c^{1/4} T_{c4}^{-3/8} X^{-3/4}\,{}^{\circ}\mathrm{K} \tag{11}$$

and the spectral distribution of the radiation, integrated over all $X$, is $S(\nu) \propto \nu^{1/3}$.

*(ii) Bremsstrahlung*

The power radiated per unit volume by this process is $\sim Cn^2 T^{1/2}$, where $C$ is a constant. If bremsstrahlung cooling dominates, the disc temperature must be such that

$$p(x) = 2bCn^2 T^{1/2}. \tag{12}$$

If the disc is supported by thermal pressure perpendicular to its plane, then we expect $b \propto T^{1/2}$. Thus, for a given mass flux and viscosity, $n \propto T^{-1/2}$. The situation is then thermally *unstable*: if $T$ increases (decreases) slightly, the disc becomes unable (more than able) to radiate the required power, so $T$ continues to increase (decrease). This instability would be aggravated if the viscosity increased with temperature – as, for example, if $\xi$ were independent of $c_S$ – because this would make $V_r$ increase with $T$, causing $n$ to depend on temperature *more* steeply than $\propto T^{-1/2}$. This thermal instability still where $T_i$ is the temperature just outside $X_i$. This energy is released by bremsstrahlung, mostly as $\gamma$-rays of 1–10 MeV. Although this cooling reduces the thermal energy of the plasma, the magnetic field energy is unaltered. Thus, unless $\alpha \ll 1$ an unstable situation arises in which flux tubes expand out of the plane, and the cooled matter drains along the field to collect in blobs in the disc, (see Fig. 2). The rate of release of magnetic energy is $\sim \alpha L_i$, though it is unclear where the magnetic flux eventually goes. It is possible that the field can be expelled in opposite directions along the rotation axis. (We shall develop and discuss elsewhere a model for double radio sources where an accretion disc provides a steady energy input into the components in this manner).

Within $X_i$, the disc is optically thick and radiates predominantly in the UV and optical.

## 4. Discussion and Specific Examples

The important distinction between the spherically symmetric situation (§2) and the case where the angular momentum of the captured matter leads to formation of an accretion disc (§3) is the following. When the matter falls *radially* inward, the efficiency with which gravitational energy is converted into radiation is high *only* when $X_{max} \approx 1$ (see Fig. 1). Moreover detailed calculations show that, even then, the conversion is only $\sim 1\%$ efficient. In contrast, when angular momentum is important ($X_\Omega \gg 1$) the efficiency is 6–40%, [the luminosity being given by (5) and the prospects for detectability correspondingly better. This was in fact noted by Salpeter (1964) who did not, however, consider in what form the radiation is emitted.

We have, of course, given merely a very crude treatment of a complex situation. We cannot predict the *spectrum* of the emergent radiation with any confidence at all. Inhomogeneties and density gradients in the gas, or turbulent motions, would complicate the dynamics. We believe that this may well have the effect of *increasing* the luminosity in situations when the mean angular momentum is zero. We regard it as unlikely, however, that these complications could substantially reduce the luminosity below (5) in cases when angular momentum *is* important. This might happen if turbulent viscosity transported the angular momentum of infalling material so efficiently that the transverse velocity never approached the Keplerian velocity, and no disc ever formed; but such a situation would plainly require *supersonic* turbulence.

We now consider two cases. (1) A black hole of $10^4 M_\odot$ in the halo of our own galaxy and (2) one of $10^{11} M_\odot$ in a cluster of galaxies. The $10^4 M_\odot$ black hole could be the remnant of an evolved globular cluster or could have been formed at the time of cluster formation. Similarly $10^{11} M_\odot$, is a typical galactic mass and could have formed at the epoch of galaxy formation.

### *1) $10^4 M_\odot$ in the Halo of our Galaxy*

We take the ambient density in the halo $n_c \sim 10^{-2}\,\mathrm{cm}^{-3}$ and suppose that the medium from which accretion takes place is rotating with the galaxy with a period $P \sim 3 \times 10^8$ y. Thus, in this case

$$X_\Omega \approx 6 \times 10^2\, T_{c4}^{-4} \left(\frac{P}{3 \times 10^8\,\mathrm{y}}\right)^{-2} \mu^{-2}. \tag{21}$$

If $\mu^{1/2} T_{c4} \lesssim 5$, an accretion disc will form. The relevant value of $P$ is, of course, the rotation period associated with a region of dimension $\sim R_A$. If the gas were turbulent, this may well be $\ll 3 \times 10^8$ y, thereby relaxing this constraint. Assuming $\mu < 1$, the luminosity would be

$$L \approx 8 \times 10^{36} \left(\frac{n_c}{10^{-2}\,\mathrm{cm}^{-2}}\right) T_{c4}^{-3/2}\,\mathrm{erg\,s^{-1}}. \tag{22}$$

The energy might all be emitted by the synchrotron process, predominantly in the far infrared, although Compton scattering of these photons would yield an optical source with $10^{-3}$ of the total luminosity. Alternatively, for a somewhat different choice of parameters, we might obtain the situation depicted in Fig. 2, with $1 \lesssim X_i \lesssim X_\Omega$ and the bulk of the luminosity emerging in the optical or ultraviolet bands. Inhomogeneities or instabilities in the disc could give rise to variability on timescales as short as $\sim 1$ s.

### *2) $10^{11} M_\odot$ in a Cluster of Galaxies*

We take the ambient density in the cluster to be

$$n_c \sim 10^{-3}\,\mathrm{cm}^{-3},$$

and the temperature in the range $10^6 - 10^8$ °K, (i.e. $T_6 = 1 - 100$). The crossing time for a cluster $\sim 3 \times 10^9$ years, and the accreting gas may well have a rotation period of this order, at least on scales of the order of the accretion radius (here 0.3 – 30 kpc). If the motion of the black hole is subsonic ($\mu < 1$), we find

$$X_\Omega \approx 6 \times 10^2\, T_6^{-4} \left(\frac{P}{3 \times 10^9\,\mathrm{y}}\right)^{-2} \mu^{-2}. \tag{23}$$

In this case the only self consistent solution is for an optically thick disc, with temperature

$$T_{bb} \approx 3 \times 10^4 \left(\frac{n_c}{10^{-3}\,\mathrm{cm}^{-3}}\right)^{1/4} T_6^{-3/8} X^{-3/4}\,{}^\circ\mathrm{K}. \tag{24}$$

If $T_{bb}$ falls below $\sim 10^4$ °K, the disc will become optically thin, with the electron temperature remaining near $\sim 10^4$ °K, and the ionization levels adjusting accordingly. The total luminosity of the source would be

$$L \approx 5 \times 10^{46} \left(\frac{n_c}{10^{-3}\,\mathrm{cm}^{-2}}\right) T_6^{-3/2}\,\mathrm{erg\,s^{-1}} \tag{25}$$

produced almost entirely in the optical and infrared. A mass $M \gg 10^{11} M_\odot$ would of course yield an even more spectacular luminosity. Note that these estimates exceed by several orders of magnitude the luminosity obtained

184 

on the assumption of spherically symmetric inflow (see Fig. 1).

We draw two main conclusions from this discussion. (i) Accreting massive black holes may already have been detected. Strittmatter *et al* (1972) have drawn attention to a class of compact objects of which the "prototype" is BL Lac. These objects are characterised by: rapid intensity variations in the radio and optical band; a continuum spectrum such that the bulk of the energy emerges in the infrared; and an absence of lines on low-dispersion optical spectra. Strittmatter *et al.* regarded these objects as akin to QSO's, but they could conceivably be either collapsed globular clusters in the galactic halo (i.e. not even extragalactic) or else could be galactic mass black holes at greater distances.

(ii) If the infalling material has even the very small amount of rotation likely to occur in the intergalactic medium the expected luminosity resulting from accretion onto galactic mass black holes is very high indeed. Existing optical and infrared observations thus already place interesting constraints on the possible occurrence of collapsed objects with masses $10^{11}-10^{14} M_{\odot}$ in clusters and/or on the properties of intergalactic gas.

*Acknowledgements.* J.E.P. acknowledges an SRC studentship. AGP is grateful for SRC support as a Senior Visiting Fellow at the University of Sussex.

## References


Bondi, H. 1952, *Monthly Notices Roy. Astron. Soc.* **112**, 195

Cox, D. P., Tucker, W. H. 1969, *Astrophys. J.* **157**, 1157

Hunt, R. 1971, *Monthly Notices Roy. Astron Soc.* **154**, 141

Lynden-Bell, D. 1969, *Nature* **223**, 690

Norman, C. A., ter Haar, D. 1973, *Astron & Astrophys.* **24**, 121

Novikov, I. D., Thorne, K. S. 1973, in "Black Holes", ed. C. de p. 243. Gordon and Breach, London

Pringle, J. E., Rees, M. J. 1972, *Astron. & Astrophys.* **21**, 1

Salpeter, E. E. 1964, *Astrophys. J.* **140**, 796

Schwarzman, V. F. 1971, *Sov. Astron. A.J.* **15**, 377

Shakura, N. I., Sunyaev, R. A. 1973, *Astron. & Astrophys.* **24**, 337

Shapiro, S. L. 1973a, *Astrophys. J.* **180**, 531

Shapiro, S. L. 1973b, *Astrophys. J.* **185**, 69

Strittmatter, P. A., Serkowski, K., Carswell, R., Stein, W. A., Merril, K. Burbidge, E. M. 1972, *Astrophys. J. Letters* **175**, L7


J. E. Pringle
A.G. Pacholczyk
Astronomy Centre
University of Sussex
Falmer, Brighton BN1 9QH, England

M. J. Rees
Institute of Astronomy
Madingley Road
Cambridge CB3 OHA, England

THE ASTROPHYSICAL JOURNAL, 198:671–682, 1975 June 15


# ACCRETION ONTO NEUTRON STARS UNDER ADIABATIC SHOCK CONDITIONS

STUART L. SHAPIRO AND EDWIN E. SALPETER
Center for Radiophysics and Space Research, Cornell University
*Received 1974 October 7*

ABSTRACT

The accretion of gas onto a neutron star is examined for the case in which a strong, adiabatic shock front forms above the stellar surface to decelerate the incident plasma stream. Steady-state, spherically symmetric flow is considered, all magnetic fields are ignored, and rapid thermalization by plasma instabilities in the shock front is assumed. The dynamical and thermal structure of the emission zone between the surface and the shock front is determined, and the emergent radiation spectrum is calculated. In many cases an appreciable fraction (about one-quarter) of the total energy is emitted in the form of $\gamma$-ray photons between 1 and 10 MeV. The possible relevance of the model calculations to observations of galactic X-ray sources and $\gamma$-ray bursts is discussed.

*Subject headings:* gamma ray bursts — neutron stars — shock waves — X-ray sources

## I. INTRODUCTION

The possibility that many galactic X-ray sources are compact objects radiating as a result of accretion has stimulated much interest in the problem of gas flow onto the surface of a neutron star. This interest has been enhanced by the suggestion that the recently detected soft $\gamma$-ray bursts (Klebesadel, Strong, and Olson 1973) may also originate from transient occurrences of accretion onto compact galactic objects (Harwit and Salpeter 1973; Lamb, Lamb, and Pines 1973*a*).

The proposal that gas accreting onto compact objects might be an important source of radiant energy was first put forth over a decade ago (Zel'dovich 1964; Salpeter 1964). However, reliable calculations of the radiation spectrum resulting from accretion, which depends on the detailed manner in which gravitational energy is converted into radiation near the stellar surface, have been performed only recently. Zel'dovich and Shakura (1969) have considered the problem of spherically symmetric accretion onto a neutron star without an intrinsic magnetic field. They have calculated the radiation spectrum emitted when a cold, incident ion beam gradually decelerates in and relinquishes its kinetic energy to a comparatively extended layer of the atmosphere of the neutron star. The atmosphere, which remains nearly static, then converts the incident energy into radiation. A more detailed treatment of the Zel'dovich-Shakura accretion model has been performed by Alme and Wilson (1974). The structure of an accretion disk around a compact object and the energy flux radiated by the disk has been examined by several authors (Pringle and Rees 1972; Shakura and Sunyaev 1973). This model is appropriate when accretion occurs in a close binary system and the infalling gas possesses significant intrinsic angular momentum with respect to the compact star. The problem of accretion onto a neutron star with a strong dipole-like magnetic field which funnels plasma along the field lines toward the magnetic polar regions has been treated by Lamb, Pethick, and Pines (1973*b*), Davidson and Ostriker (1973), and Davidson (1973). In each investigation the predicted radiation spectrum depends upon the flow geometry and the manner in which the incident gas is decelerated near the neutron star surface.

The purpose of this investigation is to examine the accretion of ionized gas onto a neutron star for the case in which a strong, standing shock wave forms near the surface of the star to decelerate the incident stream. This can arise if plasma instabilities grow rapidly, so that two interpenetrating ion beams are randomized in a few ion plasma periods. In this model, most of the kinetic energy of infalling gas is converted into thermal energy as the gas passes through the shock front. Radiation is then emitted in the narrow zone between the shock discontinuity and the surface of the neutron star. In the present analysis, we determine the dynamical and thermal structure of the emission zone and calculate the emergent radiation spectrum. A similar model has been considered by Hōshi (1973) and Aizu (1973) for accretion onto white dwarfs.

A qualitative overview of the problem treated in this report and the basic assumptions adopted are given in § II. The key equations and approximations are set forth in § III. The results of the numerical integrations are presented and analyzed for several different cases in § IV. The possible relevance of our results to observations of galactic X-ray and $\gamma$-ray sources is discussed in § V.

## II. PHYSICAL CONDITIONS AND PARAMETERS

We consider the spherically symmetric, steady-state flow of ionized gas (atomic number $A$ and charge $Z$, ion mass $m_i$ and electron mass $m_e$), onto the surface of a bare neutron star of mass $M_*$ and radius $R_*$. We assume that the mass accretion rate $\dot{M}$ is given and that magnetic fields are absent.

If matter hit the star surface under free-fall conditions and were suddenly thermalized, flow speeds would be close to the free-fall velocity $V_0$, and temperatures would be close to $T_0 = GM_*m_i/kR_* = \frac{1}{2}m_iV_0^2/k$. Numerically,

$$V_0 = 0.54c\left(\frac{M_*}{M_\odot}\right)^{1/2}\left(\frac{R_*}{10\text{ km}}\right)^{-1/2}, \qquad T_0 = 1.6 \times 10^{12}\text{ K}\left(\frac{M_*}{M_\odot}\right)\left(\frac{R_*}{10\text{ km}}\right)^{-1}A, \tag{1}$$

so that $kT_0 \sim 100$ MeV $\gg m_ec^2$.

Under steady-state conditions the total luminosity $L$ is given by $L = (GM_*/R_*)\dot{M}$. Let

$$L_E \equiv \frac{4\pi GM_*c}{\kappa_0} = 1.26 \times 10^{38}\left(\frac{M_*}{M_\odot}\right)\left(\frac{A}{Z}\right)\text{ ergs s}^{-1}, \quad \text{and} \quad \dot{M}_E \equiv \frac{4\pi cR_*}{\kappa_0} \quad \left(\kappa_0 \equiv \frac{Z}{m_i}\sigma_T\right) \tag{2}$$

be the "Eddington critical luminosity" and the corresponding "critical accretion rate" respectively, where $\sigma_T = 0.665 \times 10^{-24}$ cm$^2$ is the Thomson cross-section. We shall mainly be interested in values of $\sim 10^{-2}$ to 1 for $(L/L_E) = (\dot{M}/\dot{M}_E)$. The "effective blackbody surface temperature" (referred to the stellar surface $\sigma T_{\text{eff}}^4 = L/4\pi R_*^2$, where $\sigma$ is the Stefan-Boltzmann constant) can be written in the form

$$T_{\text{eff}} = 2.05 \times 10^7\text{ K}\left(\frac{M_*}{M_\odot}\right)^{1/4}\left(\frac{R_*}{10\text{ km}}\right)^{-1/2}\left(\frac{\dot{M}}{\dot{M}_E}\right)^{1/4}\left(\frac{A}{Z}\right)^{1/4}, \tag{3}$$

so that $kT_{\text{eff}} \sim 1$ keV $\ll m_ec^2$.

Most of the accretion energy originally resides in the incident ions with kinetic energy of order $kT_0$, but a major fraction of the total energy finally is reradiated from the vicinity of the star's surface in thermal photons with energy of order $kT_{\text{eff}}$. Since $T_0 \gg T_{\text{eff}}$, the resulting spectrum depends strongly on the mechanism for slowing down (and possibly thermalizing) the incident ions.

### *a) No Plasma Instabilities*

Consider first one extreme case where all plasma instabilities are assumed to be absent. This case (as well as intermediate cases) was first discussed by Zel'dovich and Shakura (1969) and, in more detail, by Alme and Wilson (1974) and is dominated by one further inequality: Let $t_{\text{Coul}}(\sim\nu_E^{-1}$, see eq. [17]) be the time required for the incident ions to transfer their energy to electrons and $t_{\text{Compt}}$ the time required for the thermal photons from the surface to cool a relativistic electron gas of temperature $T_0$:

$$t_{\text{Compt}} \approx \frac{3n_ekT_0}{\Lambda_c} \sim 2 \times 10^{-11}\text{ s}\left(\frac{M_*}{M_\odot}\right)^{-2}\left(\frac{R_*}{10\text{ km}}\right)^3\left(\frac{\dot{M}}{\dot{M}_E}\right)^{-1}A^{-2}Z, \tag{4}$$

where $\Lambda_c$ is the Compton cooling rate (see eq. [16]). Because $t_{\text{Compt}} \ll t_{\text{Coul}}$, the electrons never approach $T_0$, but the balance between heating by ions and Compton cooling results in an electron temperature $T_e^* \sim 10^8$ K in the "upper atmosphere" of the star where the incident ions are first slowed down. Furthermore, the ions also never approach a temperature $T_0$ as their mean streaming velocity is slowed down by collisions with electrons. In fact, one can show that the "temperature" acquired by the ions (both longitudinal and transverse) is only of order $T_e^*$. In this case no shock forms anywhere and an ambient, nearly static atmosphere is formed near the surface of the neutron star. The cold, incident ion beam gradually decelerates in this comparatively extended atmosphere, relinquishing its kinetic energy to electrons ($T_{\text{eff}} \le T_e \le T_e^*$), which convert the energy into radiation. In the lower part of the atmosphere, ion heating is balanced by free-free emission and $T_e \sim T_{\text{eff}} \sim 10^7$ K.

A key parameter in this analysis is the total amount of atmospheric material in grams cm$^{-2}$, $y_0$, required to decelerate the incoming ion beam. In the absence of plasma instabilities, $y_0 \approx \rho V_0 t_{\text{Coul}} \propto V_0^4$, where $\rho$ is the gas density and $\rho t_{\text{Coul}}$ is independent of density. When $y_0$ is large ($\gtrsim 20$), which is the case for high-mass neutron stars, the radiation spectrum is essentially a Planck curve with $T_{\text{eff}} \simeq 10^7$ K. However, when $y_0$ is small ($\lesssim 2$), which is the case for low-mass neutron stars, $T_e^*$ increases above $10^8$ K, more photons are produced in the range from 10 to 100 keV, and significant departures from a blackbody spectrum result.

Zel'dovich and Shakura also suggested that the electron-ion two-stream instability may be effective in decelerating the infalling ions and reducing $y_0$. In this case $T_e^*$ can rise to approximately $(m_e/m_i)T_0 \approx 10^9$ K. Above this temperature $v_e > V_0$, where $v_e$ is the electron thermal velocity, and the instability is stabilized by Landau damping. In the presence of plasma oscillations the radiation spectrum is characterized by a high-energy, exponential tail of soft $\gamma$-ray photons with characteristic energies near 100 keV. This radiation is produced by a thin layer of hot electrons in the upper atmosphere emitting thermal bremsstrahlung. Approximately 1 percent of the total luminosity can escape in this form. In this case there is still no shock anywhere.

### *b) Strong Plasma Instabilities*

The internal temperature of the incident stream, before being slowed down, is quite low ($\sim T_{\text{eff}}$), but the kinetic energy of an incident ion, relative to the almost stationary gas just above the stellar photosphere, is of order $kT_0$

with $T_0 \gg T_{\rm eff}$. The ion plasma frequency $\omega_{p,i}$ in the incident stream close to the star is of the form

$$\omega_{p,i} \approx 7 \times 10^{11}\ {\rm s}^{-1} \left(\frac{M}{M_\odot}\right)^{-1/4} \left(\frac{R}{10\ {\rm km}}\right)^{-1/4} \left(\frac{\dot{M}}{\dot{M}_{\rm E}}\right)^{1/2} \left(\frac{Z}{A}\right)^{1/2}, \quad (5)$$

and we have $t_{\rm Compt}\omega_{p,i} \gg 1$. If the relative velocities between two ion distributions can be randomized in a few ion plasma periods, radiative energy losses are unimportant during the thermalization process and an adiabatic shock will result.

Noerdlinger (1960) has proposed a two-stage model which leads to the growth of ion-acoustic waves which scatter the ions and randomize their motion. Estimates by Tidman (1967), based on quasi-linear calculations, suggest that randomization may indeed take place in just a few ion plasma periods and that the resulting electron temperature may be almost as high as the ion temperature. On the other hand, some one-dimensional computer simulations by McKee (1970) throw doubt on the accuracy of quasi-linear estimates for cases with high Mach number. The question thus remains open, but for the purposes of the present paper we assume that the relative motion between two ion streams is indeed randomized in a time period short compared with the electron cooling time $t_{\rm Compt}$.

Under conditions of adiabatic thermalization and steady flow, a standing shock will form at some radial distance $R_s = R_* + x_s$. We shall neglect the thickness of the shock front and any nonadiabaticity across it. We consider the ratio $\beta_s$ of electron temperature to ion temperature $T_s$ just after the shock as a parameter (obtainable, in principle, from plasma physics). For numerical work we shall choose $\beta_s = 1$ and a few slightly smaller values. For a chosen value of $\beta_s$, the value of $T_s$ will be determined numerically using the "jump" conditions (see eq. [15]), but $T_s$ will be of the order of, though slightly smaller than, $T_0$.

Numerical calculations are described below for the photon spectrum and for the "standoff distance" $x_s$ between the photosphere and the shock. However, the qualitative features can be seen easily, especially if $\beta_s \sim 1$, $A = Z = 1$, and if $10^{-2} \ll \dot{M}/\dot{M}_{\rm E} \ll 1$: Because $\dot{M} \ll \dot{M}_{\rm E}$, radiation pressure has little slowing effect on the infalling matter; because $10^{-2}\dot{M}_{\rm E} \ll \dot{M}$, we shall see that $x_s \ll R_*$. The incident velocity at the shock front is then close to $V_0$, and, because the Mach number is high, the velocity drops by almost a factor of 6 across the shock front ($\gamma \approx 13/9$ when $T \sim T_0$). Thus, most of the total accretion energy (neglecting a fraction $\sim 1/30$ in kinetic energy) appears in thermal energy (more precisely, enthalpy) immediately behind the shock front. A fraction $\beta_s(1 + \beta_s)^{-1} \sim 0.5$ of this resides in electrons at temperatures approaching $T_0 \sim 10^{12}$ K. These electrons are cooled in a layer of thickness $\sim t_{\rm Compt}V_0 \ll x_s$ by the Compton effect. In this thin layer, photospheric photons (energy $\sim kT_{\rm eff}$) are scattered at most once by the relativistic electrons and the electron's heat content appears mainly in photons of energy $\leqslant (kT_0/m_ec^2)^2 kT_{\rm eff} \sim$ (1 to 10) MeV. Half of these $\gamma$-rays move downward toward the photosphere and are thermalized via pair creation and Compton scattering, so that a fraction $\sim 0.5\beta_s(1 + \beta_s)^{-1} \sim 0.25$ of the total accretion energy finally escapes in the form of $\gamma$-rays.

The largest part of the layer between the shock front and the star's surface consists of ions at temperatures approaching $T_0$, but with electron temperature $T_e$ at some intermediate value $T_e^*$ ($T_{\rm eff} \ll T_e^* \ll T_0$) so that Compton cooling balances heat transfer from the ions. Equating the two rates, we find, approximately,

$$T_e^* \approx 2 \times 10^8\ {\rm K}(M_*/M_\odot)^{-1/5}(R_*/10\ {\rm km})^{1/5}. \quad (6)$$

The thickness of this layer $x_s$ is of order $t_{\rm Coul}V_0$, the optical depth to photon-electron scattering is moderately (but not very) large, and $kT_e^*$ is a moderate fraction of $m_ec^2$. In this layer, photospheric photons suffer a number of "mild" Compton scatterings and $\sim 0.75$ of the total accretion energy finally escapes in the form of X-rays with a broad energy distribution peaking near $kT_{\rm eff}$.

## III. METHOD AND APPROXIMATIONS

### *a) Basic Equations*

There are four basic fluid dynamical equations, expressing the conversation of mass, momentum, and energy of the electron-ion plasma. For steady-state, spherically symmetric, radial inflow, the equations are

$$4\pi\rho v r^2 = \dot{M} = {\rm const},$$

$$v\frac{dv}{dr} + \frac{1}{\rho}\frac{d}{dr}(P_e + P_i) = -\frac{GM_*}{r^2} + f_r,$$

$$v\frac{d\epsilon_i}{dr} - v\frac{(\epsilon_i + P_i)}{\rho}\frac{d\rho}{dr} = \Lambda_{\rm E},$$

$$v\frac{d\epsilon_e}{dr} - v\frac{(\epsilon_e + P_e)}{\rho}\frac{d\rho}{dr} = \Lambda_e - \Gamma_e - \Lambda_{\rm E}. \quad (7)$$

In the above equations, $\rho$ is the mass density, $P$ is the gas pressure, $v$ is the inward velocity, and $\epsilon$ is the internal energy density. The subscripts $e$ and $i$ refer to the electrons and ions, respectively. The term $f_r$ is the force on the plasma produced by radiation pressure. The quantities $\Lambda_e$ and $\Gamma_e$ represent the electron cooling and heating rates, respectively, measured in the comoving frame of the fluid. The term $\Lambda_E$ is the energy exchange rate between ions and electrons due to collisions in the plasma. In writing the above fluid equations, the gas pressure is assumed to be isotropic and given by the ideal gas law; viscosity and heat conduction are neglected. All dynamical and radiative transfer equations are correct to first order in $v/c$.

We use the following notation for the radiation quantities: $4\pi H_0(\nu)d\nu$ is the net outward radiation flux between frequency $\nu$ and $\nu + d\nu$; the subscript zero denotes a quantity measured in the comoving frame of the fluid. $H_0$ denotes the first moment of the total intensity [i.e., the integral of $H_0(\nu)$ over *all* frequencies], $K_0$ denotes the second moment, and $J_0$ the angle mean-intensity. The photon energy density $u_0$ equals $4\pi J_0/c$, and the luminosity $L_0$ equals $16\pi^2 r^2 H_0$. The radiation pressure force is given by

$$f_r = \frac{4\pi}{c}\int_0^\infty \kappa_\nu H_0(\nu)d\nu\,, \tag{8}$$

where $\kappa_\nu$ is the total opacity of the ionized gas. The stationary-frame quantity $H$ is related to $H_0$ by

$$H = H_0 - 4\pi(v/c)(J_0 + K_0)\,, \tag{9}$$

(Cassinelli and Castor 1973).

The plasma is coupled to the radiation field through the radiation pressure force and the radiative heating and cooling terms in the hydrodynamical equations. Consequently, the hydrodynamical and radiative transfer equations must be solved simultaneously. The zeroth moment of the transfer equation, integrated over frequency and combined with equations (7)–(8), gives an overall energy conservation equation,

$$L_0(r) = \dot{M}\left[\epsilon_i + \epsilon_e + \frac{(P_i + P_e)}{\rho} + \frac{v^2}{2} + \left(\frac{GM_*}{R_*} - \frac{GM_*}{r}\right) + \frac{4\pi}{\rho c}(J_0 + K_0)\right]. \tag{10}$$

Let $\bar{H}_0$ denote the integral of $H_0(\nu)$ from $\nu = 0$ to $\nu_0 = m_e c^2/h$ only (and similarly for other quantities). As discussed in § II, most of the emitted photons either have frequencies well below $\nu_0$, in which case the interaction cross section with an electron is close to $\sigma_T$, or frequencies well above $\nu_0$, in which case the cross section is much smaller than $\sigma_T$ and can be ignored. The radiation pressure force $f_r$ in equation (7) can then be approximated by

$$f_r = (4\pi/c)\kappa_0\bar{H}_0\,, \qquad \kappa_0 \equiv (Z/m_i)\sigma_T\,. \tag{11}$$

The first moment of the transfer equation integrated over frequency gives, approximately,

$$\frac{d\bar{K}}{dr} + \frac{3\bar{K} - \bar{J}}{r} = -\int_0^{\nu_0} \kappa_\nu \rho H_0(\nu)d\nu = -\kappa_0\rho\bar{H}_0\,. \tag{12}$$

For $r \gg R_*$ we assume $\bar{H}_0 = \bar{J}_0 = \bar{K}_0$ (photons moving radially outward without deflection), so that $\bar{u}_0(r) = \bar{L}_0/4\pi r^2 c$; near the stellar surface, where the optical depth due to electron scattering $\tau_{es}(r) = \int_r^\infty \sigma_T n_e dr \geq 1$, we assume $\bar{J}_0 = 3\bar{K}_0$ (isotropic radiation in this dense region). The quantity $(L_0 - \bar{L}_0)$, which together with equation (10) gives $\bar{H}_0$ for use in equations (11) and (12), can be calculated approximately as follows: Most of the photons with $\nu > \nu_0$ are emitted close to the shock front at radius $R_s$ and are undeflected unless they hit the dense medium near the stellar radius $R_*$, in which case they are absorbed (and reemitted as thermal photons). Thus $L_0 - \bar{L}_0$ is $\pi^{-1}\sin^{-1}(R_*/R_s)$ times the total radiation emitted with $\nu > \nu_0$.

The relations between the two temperatures $T_e$, $T_i$ and the quantities $\Lambda_E$, $\Lambda_e$, $\Gamma_e$, as well as the radiation field, are given in § III*b*. We define a third "photon temperature" $T_{ph} \equiv (\bar{u}_0/a)^{1/4}$, where $a$ is the radiation density constant. When integrating all the coupled equations inward, one finds that at a given radius $\rho^{-1}$ and the velocity fall rapidly toward zero and that $T_e$, $T_i$, and $T_{ph}$ all converge to the same value $T_{eff}$. The radius $r_c$ at which this convergence takes place is defined quite accurately, because the scale height $h_p$ of the atmosphere satisfies a strong inequality:

$$h_p \equiv (Z + 1)kT_{eff}/m_i g \ll R_*\,, \tag{13}$$

where $g$ is the surface gravity of the star. The numerical value $R_s$ for the radius of the standing shock front has to be chosen so that the convergence radius $r_c$ coincides with the stellar radius $R_*$. When starting the inward integration at some $r \gg R_s$ (where the velocity and density are both small), $L = L_0$ is known from the given accretion rate $\dot{M}$ but the value for $\bar{L} = \bar{L}_0$ still has to be chosen. Several iterations of the whole integration scheme are required to find the appropriate values of $R_s$ and $\bar{L}$, so that $r_c = R_*$ and $L - \bar{L}$ equals the escaping flux in photons with $\nu > \nu_0$ to sufficient accuracy.

For $r \gg R_s$ the inward flowing plasma is relatively cold, gas pressure gradients are unimportant, and the velocity can be approximated by $V_>(r)$, where

$$V_>^2(r) = \frac{2GM_*}{r}\left[1 - \frac{\bar{L}_0(r)}{L_E}\right] \tag{14}$$

and $\bar{L}_0(r)$ is determined from equations (10) and (21). The hydrodynamical gas parameters immediately behind the front are given by the Rankine-Hugoniot relations in the limit of a strong, adiabatic shock discontinuity:

$$v_s = \frac{\gamma - 1}{\gamma + 1} v_0(R_s), \qquad \rho_s = \frac{\gamma + 1}{\gamma - 1} \rho_0(R_s),$$

$$P_s \equiv \frac{\rho_s k T_s}{m_i}(1 + Z\beta_s) = \frac{2\rho_0(R_s)v_0^2(R_s)}{\gamma + 1}. \tag{15}$$

Here $\gamma$ is the ratio of specific heats in a relativistic plasma (cf. Shapiro 1973), $T_s$ is the ion temperature, and $\beta_s$ is the ratio of the electron temperature to ion temperature immediately behind the shock front.

### *b) The Radiation Spectrum and Heat Transfer*

The electron cooling term $\Lambda_e$ in equation (7) consists of the sum of three terms. The rates for electron-electron and electron-ion bremsstrahlung are taken from Maxon (1972). The third term, dominant throughout most of the emission zone between $R_*$ and $R_s$, represents inverse Compton cooling, and we approximate the rate $\Lambda_c$ by

$$\Lambda_c = 4\sigma_T c \bar{u}_0 n_e \left(\frac{kT_e}{m_e c^2}\right)\left[1 + 4\frac{kT_e}{m_e c^2}\right], \tag{16}$$

where $n_e$ is the electron number density. In the region where the electrons are relativistic, $kT_e \gg m_e c^2$, this expression is an overestimate because $\bar{u}_0$ includes all photons with frequency $\nu < \nu_0 = m_e c^2/h$ whereas photons with $\nu \gg \nu_0(m_e c^2/kT_e)$ have reduced scattering efficiency (Blumenthal and Gould 1970). Fortunately, photons in this intermediate frequency range contribute little to $\bar{u}_0$.

The exact radiative heating term $\Gamma_e$ for the electrons depends on the "color temperature" of the photons, as well as their "energy density temperature" $T_{\rm ph}$. However, these two temperatures are close to each other in the cool regions where $\Gamma_e$ is important, and we approximate $(\Lambda_e - \Gamma_e)$ in equation (7) by $\Lambda_e(T_e - T_{\rm ph})/T_e$. This expression (Zel'dovich and Shakura 1969) reduces to zero, as it should, for thermal equilibrium when all temperatures are equal. The energy exchange rate between ions and electrons $\Lambda_E$ is approximated by

$$\Lambda_E = (\tfrac{3}{2})\nu_E n_i k(T_i - T_e), \qquad \nu_E = 3.96 \times 10^{-3} n_e T_e^{-3/2} \ln \Lambda \ \mathrm{s}^{-1}, \tag{17}$$

where $n_i = n_e/Z$ is the ion number density, $\nu_E$ is the collision frequency, and $\ln \Lambda$ is the Coulomb logarithm (Spitzer 1962).

The emergent radiation spectrum is generated by very different physical processes operating in different layers of the plasma and extends over a broad energy band (100 eV–100 MeV). To compute the spectrum, we adopt a simplified scheme, using different prescriptions in various regions of the gas. The essential approximation, modeled after Zel'dovich and Shakura (1969), consists in treating the cool, nearly static, region of gas near $r \approx R_*$, where $T_e \approx T_i \approx T_{\rm ph}$, as an isothermal half-space, from which the flux can be computed analytically by solving the transfer equation in the Eddington approximation. The flux emitted by this isothermal half-space is subsequently Comptonized, first by hot electrons in the overlying emission zone and then by cold electrons in the supersonic stream above the shock front. Bremsstrahlung radiation produced in the emission zone also contributes to the outgoing flux.

The radiation flux from the isothermal half-space is

$$\begin{aligned} F_0(\nu) &= \pi B_\nu(T_{\rm ph})\left[0.5 - \frac{3h_p\kappa_0}{4\alpha_\nu^{1/3}}\frac{A_1(\alpha_\nu^{1/3}\rho_1)}{A_1'(\alpha_\nu^{1/3}\rho_1)}\right]^{-1}, \qquad & \kappa_0 \gg \kappa_\nu, \\ &= \pi B_\nu(T_{\rm ph}), & \kappa_0 \ll \kappa_\nu. \end{aligned} \tag{18}$$

Here $B_\nu$ is the Planck function, $h_p$ is the scale height (see eq. [13]), $\rho_1$ is the density at the boundary of the half-space, $\kappa_\nu$ is the opacity coefficient for free-free absorption, and $A_1$ is the Airy function of the first kind. The quantity $\alpha_\nu = 3\kappa_0\kappa_\nu h_p^2/\rho$ is independent of density. In the limit that $h_p \to \infty$, equation (18) reduces to the familiar expression for the radiation flux from an isothermal, *homogeneous* half-space in which scattering is important:

$$F_0(\nu) \approx \pi B(T_{\rm ph})\left(\frac{\kappa_\nu}{\kappa_0 + \kappa_\nu}\right)^{1/2}, \qquad h_p \to \infty \tag{19}$$

(Felten and Rees 1972).

Nonrelativistic Comptonization of the radiation flux generated in the cool half-space occurs in the nonrelativistic region of the emission zone and in the cold incident stream above the shock front. The emergent luminosity following nonrelativistic Comptonization, $L_1(\nu)$, is determined by solving the Kompaneets (1957) equation for the interaction of photons with electrons in the limit in which $m_e\langle v_e^2\rangle/2 \gg h\nu$. The result is

$$L_1(\nu) = (4\pi\tau)^{-1/2}\int_0^\infty L_0(\nu')\left(\frac{\nu}{\nu'}\right)^3 \exp\left[-\frac{(\ln(\nu/\nu') + 3\tau)^2}{4\tau}\right]\frac{d\nu'}{\nu}, \tag{20}$$

where $\tau$ measures the effective time that the photons remain in the scattering layers of the accreting plasma. The quantity $\tau$ may be computed from the relation $L_1 = L_0 + \Delta L_c = L_0 \exp(4\tau)$, where $L_0$ is obtained from equation (18) and $\Delta L_c$ is the contribution of Comptonization to the total luminosity and is determined from the total cooling rates once the temperature and density profiles in the plasma have been obtained.

In the region above the shock front where $r > R_s$ the Compton effect keeps the electron temperature relatively low, i.e., close to $T_{\rm eff}$. Nevertheless, because the flow velocity $v_>(r)$ is nonzero, the scattering of the outgoing photons by the incident electrons decelerates the incoming plasma and increases the energy content in the photons (by a fraction $\rho\kappa_0(v_>/c)dr$ over path length $dr$). The first effect leads to equation (14); the second can be expressed as

$$\bar{L}(r) = \bar{L}(r = \infty)\exp(-\dot{M}R_*/\dot{M}_E r), \tag{21}$$

where $\bar{L}$ is the stationary-frame luminosity (Cassinelli and Castor 1973) and equation (2) has been used.

In the optically thin (but important) layer just behind the shock where $kT_e \gg m_ec^2$ a photon is Compton-scattered at most once; initial frequency $\nu$ and electron energy $E_e$ lead to a higher frequency with mean value $\nu_c = (4/3)(E_e/m_ec^2)^2\nu$. We neglect the distribution of frequencies around $\nu_c$ and further replace $\nu$ by $(2.7kT_{\rm ph}/h)$, the mean value for blackbody radiation at temperature $T_{\rm ph}$. The emissivity $j(\nu)$ of the plasma due to relativistic Comptonization is then determined by the Maxwellian energy distribution of the electrons and is

$$j(\nu) = \frac{1}{192\pi}\Lambda_c\frac{\nu^{3/2}}{\bar{\nu}_c^{5/2}}\exp\left(-\frac{\nu}{\bar{\nu}_c}\right)^{1/2}, \qquad \bar{\nu}_c \equiv \frac{4}{3}\left(\frac{kT_e}{m_ec^2}\right)^2\frac{2.7kT_{\rm ph}}{h}. \tag{22}$$

As pointed out in § III*b*, this formula does not apply when the initial photon frequency $\nu \gg \nu_0(m_ec^2/kT_e)$, due to the reduced scattering cross section for these high-energy photons.

## IV. NUMERICAL RESULTS

Calculations have been performed for the four neutron star equilibrium models in Table 1. The adopted choices of mass and radius lie in the range appropriate for neutron stars. The third entry has the same surface potential as the second, but one-half the mass and radius. Neutron star models with a realistic equation of state would give parameters intermediate between the second and third entry.

The results for pure hydrogen accretion with $\beta_s = 1$ are summarized in Table 2 and Figures 1–5. In the table $x_s$ is the thickness of the emission zone, $R_s$ is the radius of the standing shock front ($R_s = R_* + x_s$), and the quantities $v_s$, $n_s$, $T_s$, and $T_{\rm ph,s}$ refer to the velocity, ion density, ion temperature (= electron temperature), and photon temperature immediately behind the shock front. The ratio $L(>100\ {\rm keV})/L$ is the fraction of the total luminosity observed above 100 keV. The ratio $L_{R_s}/L$ is the fraction of the total luminosity generated entirely in the emission zone below $r = R_s$.

The temperature profile of the emission zone is plotted in Figure 1 for the case $M_*/M_\odot = 1.1$ and $\dot{M}/\dot{M}_E = 0.1$. The quantity $x = R_s - r$ measures the depth into the emission zone from the shock front. In this zone the electron gas cools rapidly, primarily by undergoing Compton scattering with the soft X-ray photons in the outgoing radiation flux from the surface. As a result of relativistic Comptonization, the electrons lose one-half of their thermal energy in a layer less than 35 cm from the shock front. The ions heat up slightly just behind the front, due to adiabatic compression, and then cool slowly by colliding and depositing the energy with the cooler electrons.

TABLE 1

Adopted Neutron Star Models

| $M_*$ ($M_\odot$) | $R_*$ (km) | $V_0$ ($10^9$ cm s$^{-1}$) | Binding Energy (% $m_0c^2$) | $L_E$ $(A/Z)10^{38}$ ergs s$^{-1}$ | $\dot{M}_E$ $(A/Z)10^{18}$ g s$^{-1}$ |
|---|---|---|---|---|---|
| 2.4........ | 12 | 23.0 | 36.0 | 3.02 | 1.14 |
| 1.1........ | 14 | 14.4 | 12.4 | 1.38 | 1.33 |
| 0.55....... | 7 | 14.4 | 12.4 | 0.691 | 0.663 |
| 0.2........ | 16 | 5.76 | 1.86 | 0.251 | 1.52 |

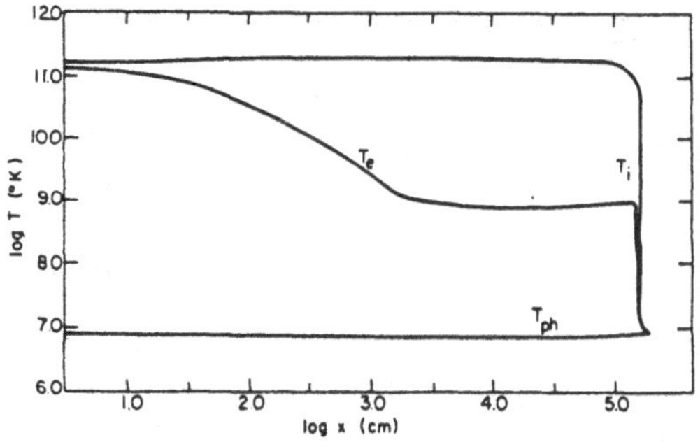


FIG. 1.—Temperature profiles in the emission zone for a 1.1 $M_\odot$ star and a hydrogen accretion rate $\dot{M}/\dot{M}_E = 0.1$ ($L = 1.38 \times 10^{37}$ ergs $s^{-1}$). The ion, electron, and photon temperatures ($T_i$, $T_e$, and $T_{ph}$, respectively) are plotted against distance from the shock front. The shock front is located at $x = 0$ and the stellar surface is located at $x = 1.64 \times 10^5$ cm.

When the electron temperature has fallen to $T_e^* \sim 10^9$ K, the heating rate of the electron gas by the ions counterbalances the cooling rate due to nonrelativistic Comptonization, and the electrons proceed inward isothermally. The extent of the emission zone is determined by the time required for the ions to transfer all of their energy to the electrons; hence

$$x_s \sim \frac{v_s}{\nu_E} \sim 0.7 \times 10^{12} \frac{T_e^{*3/2}}{n_s Z \ln \Lambda} (M_*/M_\odot)^{1/2} (R_*/10 \text{ km})^{-1/2} \text{ cm} . \tag{23}$$

At $x = x_s$ (i.e., $r = R_*$) both $T_e$ and $T_i$ drop rapidly to $T_{ph}$.

The density and velocity profiles in the emission zone are shown in Figure 2 for the case described above. The velocity decreases as the fluid approaches the stellar surface and falls rapidly to zero at the surface. The density increases as the gas proceeds inward, in accordance with the continuity equation. The abrupt decrease in velocity at $x \sim 100$ cm results from the increase in ion pressure following adiabatic compression.

The development of the outgoing radiation flux is illustrated for this case in Figure 3, where we plot the outgoing, stationary-frame luminosity at three separate points in the accreting fluid. The quantity $L_0 = 0.31 \times 10^{37}$ ergs $s^{-1}$ is the luminosity near the stellar surface and is computed from equation (22) with $T_{ph} \approx 0.80 \times 10^7$ K. The luminosity at this point has a blackbody appearance, peaking in the soft X-ray energy region near 1 keV. After undergoing Comptonization in the nonrelativistic region of the emission zone, the luminosity is given by $L_1 = 0.86 \times 10^{37}$ ergs $s^{-1}$. The effect of Comptonization is to broaden the spectral energy shape by redistributing photons: the number of hard X-ray and $\gamma$-ray photons increases at the expense of soft X-ray photons. Moreover, the peak in the spectral curve shifts upward in energy by roughly a factor of 2. The luminosity observed at infinity is $L = 1.38 \times 10^{37}$ ergs $s^{-1}$. The observed spectrum contains a high energy "shoulder" above 100 keV due to relativistic Comptonization immediately behind the shock. The spectral distribution is also broadened by further

TABLE 2

HYDROGEN ACCRETION ($\beta_s = 1$)

| $M_*$ ($M_\odot$) | $\dot{M}$ ($\dot{M}_E$) | $L$ (ergs $s^{-1}$) | $R_s/R$ | $x_s$ (km) | $v_s$ ($10^9$ cm $s^{-1}$) | $n_s$ ($10^{18}$ cm$^{-3}$) | $T_s$ ($10^{11}$ K) | $T_{ph,s}$ ($10^7$ K) | $L(>100$ keV$)/L$ | $L_{R_s}/L$ |
|---|---|---|---|---|---|---|---|---|---|---|
| 2.4...... | 0.50 | 1.51(38) | 1.01 | 7.59(−2) | 2.01 | 9.36 | 1.08 | 1.79 | 0.056 | 0.23 |
| | 0.10 | 3.02(37) | 1.21 | 2.53 | 3.44 | 0.750 | 3.22 | 0.936 | 0.22 | 0.84 |
| | 0.02 | 6.03(36) | 1.76 | 9.08 | 3.07 | 8.06(−2) | 2.57 | 0.626 | 0.20 | 0.96 |
| 1.1...... | 0.75 | 1.04(38) | 1.00 | 4.74(−3) | 0.755 | 32.0 | 0.141 | 1.95 | 0.020 | 0.071 |
| | 0.50 | 6.91(37) | 1.00 | 7.85(−2) | 1.57 | 10.3 | 0.656 | 1.38 | 0.088 | 0.35 |
| | 0.10 | 1.38(37) | 1.12 | 1.64 | 2.31 | 1.12 | 1.44 | 0.734 | 0.24 | 0.87 |
| | 0.02 | 2.77(36) | 1.57 | 8.04 | 2.08 | 0.126 | 1.17 | 0.425 | 0.22 | 0.98 |
| | 0.75(−2) | 1.04(36) | 2.16 | 16.3 | 1.80 | 2.92(−2) | 0.868 | 0.318 | 0.16 | 0.99 |
| 0.2...... | 0.50 | 1.26(37) | 1.00 | 2.62(−2) | 0.728 | 19.4 | 0.130 | 1.01 | 0.076 | 0.41 |
| | 0.10 | 2.51(36) | 1.03 | 0.524 | 1.00 | 2.65 | 0.259 | 0.517 | 0.22 | 0.87 |
| | 0.02 | 5.03(35) | 1.24 | 3.78 | 0.975 | 0.382 | 0.245 | 0.280 | 0.14 | 0.98 |
| 0.55..... | 0.10 | 6.91(36) | 1.12 | 0.829 | 2.31 | 2.23 | 1.44 | 0.871 | 0.24 | 0.87 |

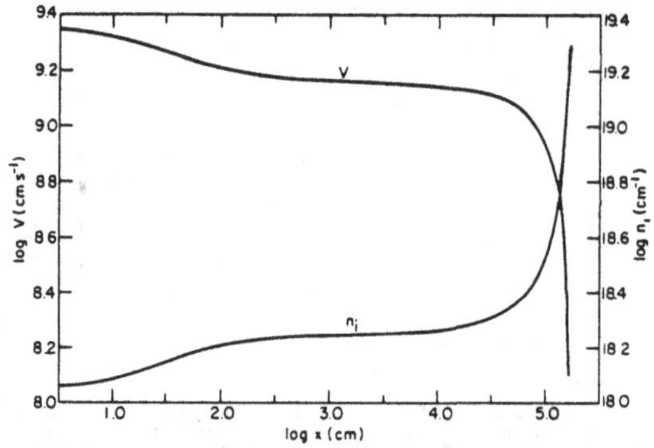


FIG. 2.—Velocity and density profiles in the emission zone for a 1.1 $M_\odot$ star and a hydrogen accretion rate $\dot{M}/\dot{M}_E = 0.1$ ($L = 1.38 \times 10^{37}$ ergs s$^{-1}$). The velocity $v$ and ion density $n_i$ are plotted against distance from the shock front. The shock front is located at $x = 0$, and the stellar surface is located at $x = 1.64 \times 10^5$ cm.

Comptonization in the cold stream above the shock front. The dashed curve labeled $L_{BB}$ represents the spectral energy distribution of a blackbody with a total luminosity equal to $L$. As discussed in § II, roughly one-quarter of the energy lies well above 100 keV.

The continuous radiation spectra for different accretion rates are plotted in Figure 4 for $M_*/M_\odot = 1.1$. For a given neutron star, the key parameter which determines the fractional energy emitted above 100 keV is the electron temperature behind the shock front, $T_{e,s} = \beta_s T_s$. This is due to the strong temperature dependence of relativistic Comptonization: $\Lambda_c \propto T_e^2$ and $\bar{\nu}_c \propto T_e^2$. For very high accretion rates approaching the Eddington limit, radiation pressure decelerates the incident stream significantly, resulting in lower velocities at the shock boundary and lower shock temperatures, according to equation (15). In this situation a significant fraction of the gravitational energy is liberated above the emission zone, as is evident from the last column in Table 1, and the $\gamma$-ray luminosity is low. For very low accretion rates, the shock forms at a large distance above the stellar surface, since $x_s \propto n_s^{-1} \propto \dot{M}^{-1}$. This also results in a low velocity at the shock surface, since $v_0(R_s) \propto R_s^{-1/2}$. Hence, the shock temperature and $\gamma$-ray generation rates are again low. The largest fractional $\gamma$-ray luminosity occurs for intermediary accretion rates near $\dot{M}/\dot{M}_E \sim 0.1$.

For the model in which $M_*/M_\odot = 1.1$ and $\dot{M}/\dot{M}_E = 0.75$ the fluid begins to decelerate *before* the fluid reaches

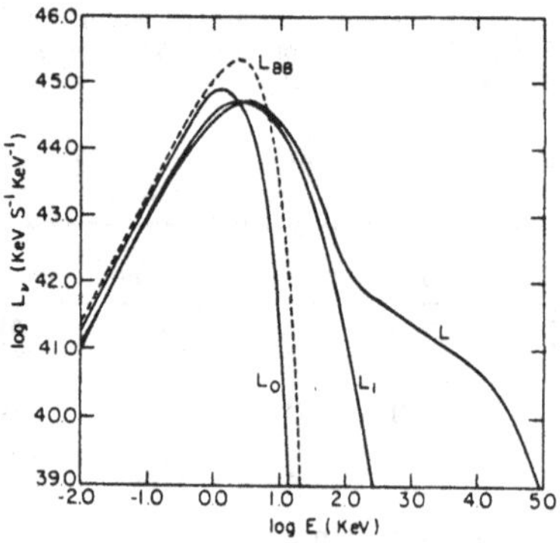


FIG. 3.—The development of the continuous emission spectrum for a 1.1 $M_\odot$ star and a hydrogen accretion rate $\dot{M}/\dot{M}_E = 0.1$. The quantity $L_0 = 0.31 \times 10^{37}$ ergs s$^{-1}$ is the stationary-frame luminosity near the stellar surface, $L_1 = 0.86 \times 10^{37}$ ergs s$^{-1}$ is the luminosity after nonrelativistic Comptonization in the emission zone, and $L = 1.38 \times 10^{37}$ ergs s$^{-1}$ is the luminosity observed at infinity. The dashed curve labeled $L_{BB}$ is a blackbody spectrum and a total luminosity equal to $L$.

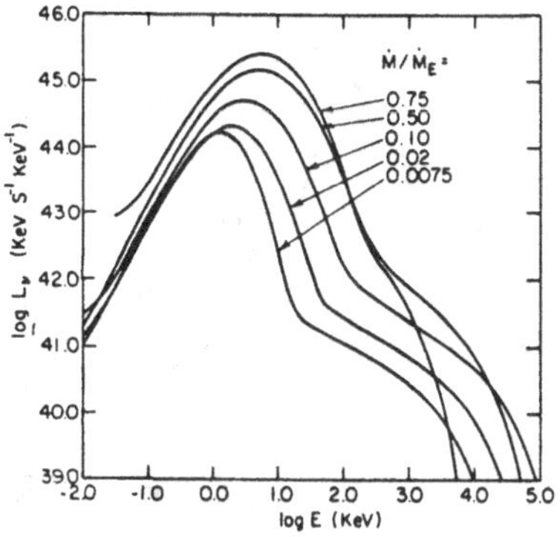


FIG. 4.—The continuous emission spectra for hydrogen accretion onto a 1.1 $M_\odot$ star

the shock front. This results from the fact that the soft X-ray luminosity as seen by the *fluid* exceeds the Eddington luminosity as the plasma approaches the neutron star, i.e., $\bar{L}_0 > L_E$, although the stationary-frame luminosity $\bar{L} < L_E$. The critical accretion rate required for radiation pressure to exactly counterbalance gravity may differ from the Eddington value $\dot{M}_E$ for *two* completely independent reasons: (1) the total luminosity may contain a significant fraction of high-energy photons with energies $h\nu > m_e c^2$, and the Klein-Nishina scattering cross section differs from $\sigma_T$; (2) the fluid-frame luminosity, $\bar{L}_0$, may differ considerably from $\bar{L}$ near the neutron star, due to abberation and Doppler effects, which become important as $v$ approaches $c$ (see eq. [9]). The first effect leads to a higher critical accretion rate (Blumenthal 1974) while the second effect leads to a lower one. In the accretion models considered here, the latter effect tends to dominate.

In Figure 5 the emitted radiation spectra for all four neutron star models are compared for the case $\dot{M}/\dot{M}_E = 0.1$. In the figure we plot the dimensionless quantity $\nu L_\nu/L$ against photon energy $E$ in keV. Here $L_\nu$ is the luminosity in keV s$^{-1}$ keV$^{-1}$ and $\nu$ is the frequency in keV. Plotted in this manner, local maxima in the curves indicate spectral regions of significant energy output. The $\gamma$-ray peak (near $\bar{\nu}_c$) exhibited by all of the curves shifts toward higher energy for models with higher surface potentials; according to equations (1), (3), and (22),

$$h\bar{\nu}_c \sim 10 \text{ MeV} \left(\frac{M_*}{M_\odot}\right)^{9/4} \left(\frac{R_*}{10 \text{ km}}\right)^{-5/2} \left(\frac{\dot{M}}{\dot{M}_E}\right)^{1/4} A^{9/4} Z^{-1/4} . \tag{24}$$

The shift in the soft X-ray peak near 1–10 keV is less pronounced since $T_{eff}$ changes more slowly with $M_*$ and $R_*$ (see eq. [3]). The computed spectra for the $M_*/M_\odot = 0.55$ and 1.1 configurations, which are constructed from different equations of state but have identical surface potentials, are similar in shape as expected. The dotted

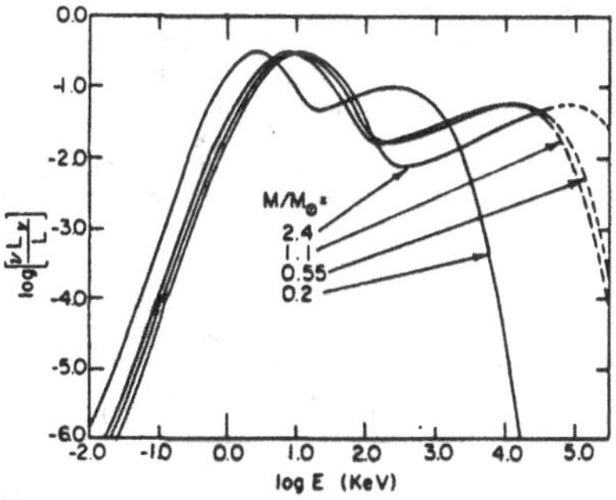


FIG. 5.—The continuous emission spectra for hydrogen accretion onto neutron stars with different masses, assuming $\dot{M}/\dot{M}_E = 0.1$. The quantity $L_\nu$ is the luminosity in keV s$^{-1}$ keV$^{-1}$, $\nu$ is the energy in keV, and $L$ is the total luminosity in keV s$^{-1}$. Local maxima in the curves indicate spectral regions of significant energy output.

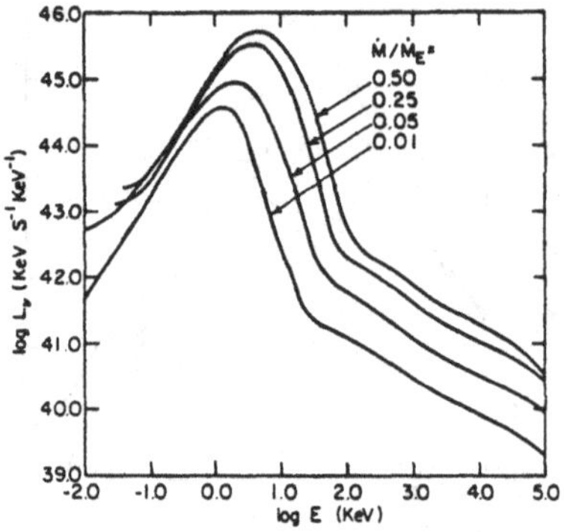


FIG. 6.—The continuous emission spectra for oxygen accretion onto a 1.1 $M_\odot$ star

portions of the spectra indicate that the computed curves above 100 MeV are inaccurate due to the neglect of some relativistic effects that have been ignored.

To study the effect of the fluid composition on the emergent radiation spectrum, we also considered the accretion of oxygen gas, which may be appropriate whenever cometary debris is captured by a neutron star. Results are summarized in Figure 6 and Table 3 for accretion onto a star with $M_*/M_\odot = 1.1$. In general, for the same accretion rate the accretion of oxygen will result in a higher fractional $\gamma$-ray luminosity than the accretion of hydrogen onto a given neutron star. This is again due in part to the higher electron temperature immediately behind the shock front, which results from the higher ion mass (see eq. [1]). In addition, a larger fraction of the infall energy behind the shock is deposited in relativistic electrons, due to the higher $n_e/n_i$ ratio. Both effects result in a higher relativistic Comptonization rate and, accordingly, a higher $\gamma$-ray luminosity. The "wiggles" which are apparent in the high-energy shoulder of the emission spectrum of oxygen are due to bremsstrahlung emission, which becomes more important at high $Z$.

Since equilibration between electron and ion temperatures immediately behind the shock may depend on the precise nature of the plasma instabilities which produce the shock transition, we considered models with different values of $\beta_s = T_{e,s}/T_s$. Results are summarized in Figure 7 and Table 4 for the accretion of hydrogen onto a neutron star with $M_*/M_\odot = 1.1$. The main effect of lowering $\beta_s$, and consequently the electron temperature behind the shock front, is to reduce the $\gamma$-ray luminosity above 100 keV and increase the soft X-ray flux, as discussed qualitatively in § II.

## V. OBSERVATIONS

In spite of the idealized nature of the problem considered and the many simplifying assumptions adopted (spherical symmetry, no magnetic fields, etc.), it is interesting to compare the results of our model calculations with observations of the galactic X-ray and transient $\gamma$-ray sources. The distinguishing feature of the accretion model analyzed in this paper is the prediction of a high-energy $\gamma$-ray tail, containing an appreciable fraction of the total luminosity, in the continuous emission spectrum of an accreting neutron star. Although extensive data exist on the flux and radiation spectrum in the soft X-ray region ($\leqslant 20$ keV) for numerous X-ray sources, the hard component of the emission spectrum cannot be obtained by a simple extrapolation of the data observed in soft X-rays. We must examine the relatively few observations performed at higher energies to evaluate our model.

TABLE 3

OXYGEN ACCRETION ($\beta_s = 1$)

| $M$ ($M_\odot$) | $\dot{M}$ ($\dot{M}_E$) | $L$ (ergs s$^{-1}$) | $R_s/R$ | $x_s$ (km) | $v_s$ ($10^9$ cm s$^{-1}$) | $n_s$ ($10^{18}$ cm$^{-3}$) | $T_s$ ($10^{11}$ K) | $T_{ph,s}$ ($10^7$ K) | $L(>100$ keV$)/L$ | $L_{R_s}/L$ |
|---|---|---|---|---|---|---|---|---|---|---|
| 1.1...... | 0.50 | 1.38(38) | 1.00 | 6.29(−3) | 1.35 | 1.49 | 2.21 | 1.60 | 0.16 | 0.38 |
| | 0.25 | 6.91(37) | 1.00 | 2.38(−2) | 1.71 | 0.588 | 3.55 | 1.30 | 0.26 | 0.62 |
| | 0.05 | 1.38(37) | 1.02 | 0.268 | 2.09 | 9.28(−2) | 5.31 | 0.699 | 0.43 | 0.94 |
| | 0.01 | 2.77(36) | 1.06 | 0.902 | 2.09 | 1.71(−2) | 5.32 | 0.525 | 0.46 | 0.98 |

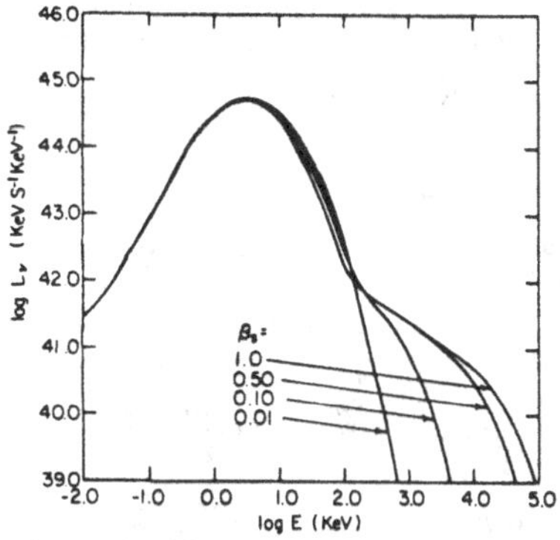


FIG. 7.—The continuous emission spectra for hydrogen accretion onto a 1.1 $M_\odot$ star. The parameter $\beta_s = T_{e,s}/T_s$ is the ratio of the electron to ion temperature immediately behind the shock front.

For the sources 2U 1223−62 and 2U 1258−61, an extrapolation of the data from the hard region between 15 and 80 keV to low energies between 1 and 10 keV gives a flux value over 100 times greater than the maximum flux reported by *Uhuru* (Fusco-Femiano and Messaro 1973). Most significant is the excess of hard X-radiation observed from Sco X-1 (Haymes *et al.* 1972; Agrawal *et al.* 1969). In the soft X-ray energy range the spectrum has a well-known exponential shape ($kT \sim 6$ keV), but a hard, nonthermal component is present in the form of a flattening of the X-ray spectrum from about 40 keV to 300 keV amounting to about 3 percent of the soft X-ray luminosity. The observed emission spectrum of Sco X-1 can be fitted out to 300 keV by both the Zel'dovich-Shakura model (with low $y_0$) and the collisionless shock model, with a suitable choice of free parameters. Observations above 1 MeV are necessary to discriminate between the two theoretical models.

The detection of a significant $\gamma$-ray flux of 0.12 photons cm$^{-2}$ s$^{-1}$ between 0.8 and 10 MeV from discrete sources in the Cygnus region (Cyg X-2, Cyg X-4), has recently been reported by Dean *et al.* (1973). If Cyg X-2 is the primary source of the observed $\gamma$-ray flux, the ratio of the $\gamma$-ray luminosity to soft X-ray luminosity, $L_\gamma/L_x \sim 10$ (Fusco-Femiano and Massaro 1973). More recent measurements by Schönfelder and Lichti (1974) indicate that the flux reported by Dean *et al.* (1973) may be too high by at least an order of magnitude. If this is the case and $L_\gamma/L_x \leqslant 1$, the soft X-ray *and* $\gamma$-ray components can both be explained by the collisionless-shock accretion model. Clearly, more data are required before one can draw definite conclusions.

Two different models involving accretion onto galactic neutron stars have been proposed to account for the recently detected soft $\gamma$-ray bursts (Klebesadel *et al.* 1973). One model involves the infall of cometary debris onto old neutron stars (Harwit and Salpeter 1973). The other involves the accretion of material produced by the flaring of a less evolved companion star in a close binary system containing a compact object (Lamb *et al.* 1973*a*). Both models can account for the characteristic time scales and energies associated with the observed bursts. Although time-dependent accretion calculations are required to explain the temporal behavior of the bursts, steady-state calculations may provide a reasonable first approximation, since the bursts typically last 1–10 s, which is significantly longer than the free-fall time at the neutron star surface ($10^{-3}$ to $10^{-4}$ s). Intensity fluctuations occurring on time scales shorter than the free-fall time are not adequately described by steady-state calculations. A comparison of the observed emission spectra of the bursts with time-independent model computations is inconclusive. The characteristic exponential spectra with $kT \approx 150$ keV observed for the bursts (Cline *et al.* 1974) is not inconsistent with the theoretical curves above 100 keV, for a suitable choice of model parameters. However, all models involving accreting neutron stars predict that the spectrum will peak in the soft X-ray region between 1 and 10 keV.

TABLE 4

HYDROGEN ACCRETION ($\beta_s \leq 1$)

| $M$ ($M_\odot$) | $\dot{M}$ ($\dot{M}_E$) | $\beta_s = T_{e,s}/T_s$ | $R_s/R$ | $x_s$ (km) | $v_s$ ($10^9$ cm s$^{-1}$) | $n_s$ ($10^{18}$ cm$^{-3}$) | $T_s$ ($10^{11}$ K) | $T_{pb,s}$ ($10^7$ K) | $L(>100$ keV$)/L$ | $L_{R,s}/L$ |
|---|---|---|---|---|---|---|---|---|---|---|
| 1.1 . . . . . . | 0.10 | 1.0 | 1.12 | 1.64 | 2.31 | 1.12 | 1.44 | 0.734 | 0.24 | 0.87 |
| | | 0.5 | 1.19 | 2.70 | 2.45 | 0.929 | 1.92 | 0.724 | 0.16 | 0.87 |
| | | 0.1 | 1.38 | 5.36 | 2.65 | 0.640 | 2.49 | 0.688 | 0.34(−1) | 0.87 |
| | | 0.01 | 1.49 | 6.85 | 2.70 | 0.543 | 2.63 | 0.661 | 0.68(−4) | 0.88 |

While the spectrum of the 1972 May 14 $\gamma$-ray event does extend to 11 keV (Wheaton *et al.* 1973), the maximum energy release appears to be in the several hundred keV region. A knowledge of the soft X-ray component of several different bursts is clearly essential in evaluating the neutron star accretion hypothesis. In addition, further theoretical work is required, including an analysis of the dynamical and radiative effects of the magnetic field of the neutron star. Crude estimates indicate that the presence of a stellar magnetic field will reduce the fraction of the total luminosity emitted above 100 keV, as cyclotron emission at lower characteristic frequencies becomes a significant cooling mechanism in the plasma.

It is a pleasure to thank S. Colgate, J. Katz, C. McKee, and M. Milgrom for stimulating discussions. This work was supported in part by NSF grant GP-36426X.

## REFERENCES

Agrawal, P. C., Biswas, S., Gokhale, G. S., Iyengar, V. S., Kunte, P. K., Manchanda, R. K., and Sreekantan, B. V. 1970, in *Non-solar X- and γ-Ray Astronomy, IAU Symposium No. 37*, ed. L. Gratton (Dordrecht: Reidel), p. 94.
Aizu, K. 1973, *Progr. Theoret. Phys.*, **49**, 1184.
Alme, M. L., and Wilson, J. R. 1973, *Ap. J.*, **186**, 1015.
Blumenthal, G. R. 1974, *Ap. J.*, **188**, 11.
Blumenthal, G. R., and Gould, R. J. 1970, *Rev. Mod. Phys.*, **42**, 237.
Cassinelli, J. P., and Castor, J. I. 1973, *Ap. J.*, **179**, 189.
Cline, T. L., Desai, U. D., Klebesadel, R. W., and Strong, I. B. 1973, *Ap. J. (Letters)*, **185**, L1.
Davidson, K. 1973, *Nature Phys. Sci.*, **246**, *5*, 1.
Davidson, K., and Ostriker, J. P. 1973, *Ap. J.*, **179**, 585.
Dean, A., Gerardi, G., DeMartinis, C., Monastero, G. F., Russo, A., and Scarsi, L. 1973, *Astr. and Ap.*, **28**, 131.
Felten, J. E., and Rees, M. J. 1972, *Astr. and Ap.*, **17**, 226.
Fusco-Femiano, R., and Massaro, E. 1973, *Ap. and Space Sci.*, **25**, 239.
Harwit, M., and Salpeter, E. E. 1973, *Ap. J. (Letters)*, **186**, L37.
Haymes, R. C., Harnden, F. R., Jr., Johnson, W. N., III, Prichard, H. M., and Bosch, H. E. 1972, *Ap. J. (Letters)*, **172**, L47.
Hōshi, R. 1973, *Progr. Theoret. Phys.*, **49**, 776.
Klebesadel, R. W., Strong, I. B., and Olson, R. A. 1973, *Ap. J. (Letters)*, **182**, L85.
Kompaneets, A. S. 1957, *Soviet Phys.—JETP*, **4**, 730.
Lamb, D. Q., Lamb, F. K., and Pines, D. 1973*a*, *Nature Phys. Sci.*, **246**, 52.
Lamb, F. K., Pethick, C. J., and Pines, D. 1973*b*, *Ap. J.*, **184**, 271.
Maxon, S. 1972, *Phys. Rev.*, **5**, 1630.
McKee, C. F. 1970, *Phys. Rev. (Letters)*, **24**, L990.
Noerdlinger, P. D. 1960, *Ap. J.*, **133**, 1034.
Pringle, J. E., and Rees, M. J. 1972, *Astr. and Ap.*, **21**, 1.
Salpeter, E. E. 1964, *Ap. J.*, **140**, 796.
Shakura, N. I., and Sunyaev, R. A. 1973, *Astr. and Ap.*, **24**, 337.
Shapiro, S. L. 1973, *Ap. J.*, **180**, 531.
Shönfelder, V., and Lichti, G. 1974, *Ap. J. (Letters)*, **192**, L1.
Spitzer, L. 1962, *Physics of Fully Ionized Gases* (New York: Interscience).
Tidman, D. A. 1967, *Phys. Fluids*, **10**, 547.
Wheaton, W. A., Ulmer, M. P., Baity, W. A., Datlowe, D. W., Elcan, M. J., Peterson, L. E., Klebesadel, R. W., Strong, I. B., Cline, T. L., and Desai, U. D. 1973, *Ap. J. (Letters)*, **185**, L57.
Zel'dovich, Ya. B. 1964, *Soviet Phys.—Doklady*, **9**, 195.
Zel'dovich, Ya. B., and Shakura, N. I. 1969, *Soviet Astr.—AJ*, **13**, 175.

EDWIN E. SALPETER and STUART L. SHAPIRO: Center for Radiophysics and Space Research, Cornell University, Ithaca, NY 14853

THE ASTROPHYSICAL JOURNAL, 187:L1–L3, 1974 January 1


# BLACK HOLES IN BINARY SYSTEMS: INSTABILITY OF DISK ACCRETION*

ALAN P. LIGHTMAN AND DOUGLAS M. EARDLEY
California Institute of Technology, Pasadena, California 91109
*Received 1973 October 15; revised 1973 November 7*

## ABSTRACT

We have tested the stability of a thin, orbiting accretion disk near a black hole. Under conditions appropriate for a binary X-ray source, with the usual (ad hoc) assumptions about viscosity, the disk is always secularly unstable on time scales of a few seconds or less. Therefore current thin-disk models for such X-ray sources are self-inconsistent. We mention possibilities for alternative models; perhaps the secular instability explains chaotic time variations in Cygnus X-1.

*Subject headings:* binaries — black holes — instabilities

Current models (Pringle and Rees 1972; Shakura and Sunyaev 1973; Novikov and Thorne 1973) for binary X-ray sources powered by accretion onto a black-hole companion envisage the gas flow near the hole as either a thin, orbiting disk or a thick, perhaps chaotic cloud. If the X-ray luminosity $L$ exceeds the Eddington limit, $L^{ED} \sim (10^{38}\ \text{ergs s}^{-1})\ (M_{BH}/M_{\odot})$, where $M_{BH} \equiv$ mass of black hole, then the cloud picture is more likely. Moreover, even at luminosities somewhat lower than the Eddington limit, say $L \gtrsim 10^{-2}\ L^{ED}$ (all figures quoted will be for typical parameters of accretion models), thermal instabilities caused by optical thinness (Pringle, Rees, and Pacholczyk 1973) may disrupt the inner region of a thin disk, transforming it into a thick cloud. We wish to point out in this *Letter* that, *with the usual (ad hoc) assumption about the viscosity*, detailed thin-disk models are *always secularly unstable* over the whole "inner region" (that region where radiation pressure dominates gas pressure, $P_R > P_G$, and the dominant opacity is electron scattering). Such an inner region exists near the hole when $L \gtrsim 10^{-4}\ L^{ED}$. Therefore these models are inconsistent. The observational consequences are great since most of the X-ray luminosity originates in the inner region.

The current thin-disk models (Pringle and Rees 1972; Shakura and Sunyaev 1973; Novikov and Thorne 1973) are *stationary* and include two key assumptions:

(*a*) Accreting matter forms a thin, orbiting, non-self-gravitating disk drifting inward on a slow time scale $t_{drift}$ (slow compared with thermal and Kepler time scales). The drift is caused by viscous stress removing angular momentum.

(*b*) Although the viscous stress $t_{\hat\phi\hat r}$ arises from intricate processes (e.g., turbulent motions on fast time scales, or magnetic fields), it may be approximated on slow time scales $\sim t_{drift}$ and longer by

$$t_{\hat\phi\hat r} = \alpha P_{tot}\,, \tag{1}$$

where $P_{tot} = P_R + P_G$ and $\alpha$ is a number believed to lie between $10^{-3}$ and 1.

To investigate stability of the above models we generalize them to allow time-dependence in the radial disk structure on the slow time scale $t_{drift}$ (a few seconds at the outer edge of the inner region; a few milliseconds at the inner edge). We shall sketch the development here. For a complete discussion of the stationary models, see Novikov and Thorne (1973) and Shakura and Sunyaev (1973). For a complete discussion of the time-dependent generalization, see Lightman (1974).

Variables describing the local, instantaneous state of the disk are surface density $\Sigma(r, t)$ (g cm$^{-2}$), total inward mass flux $\dot M(r, t)$ (g s$^{-1}$), mean half-thickness $h(r, t)$ (cm), mean pressure $P(r, t)$, mean temperature $T(r, t)$, radiative flux $F(r, t)$ (ergs cm$^{-2}$ s$^{-1}$) from top of disk (= same from bottom), and vertically-integrated viscous stress $W(r, t) \approx 2ht_{\hat\phi\hat r}$ (dyne cm$^{-1}$) (means are vertical averages). The structural equations relating these variables (ignoring relativistic corrections) are:

*Equations of radial structure:*

$$2\pi r \frac{\partial \Sigma}{\partial t} = \frac{\partial \dot M}{\partial r} \quad \text{(conservation of mass)} \tag{2a}$$

$$\frac{d(\Omega r^2)}{dr}\,\dot M = \frac{\partial}{\partial r}\,(2\pi r^2 W) \quad \text{(conservation of angular momentum);} \tag{2b}$$

here $\Omega \equiv (GM_{BH}/r^3)^{1/2}$. Equations (2) are exact.

*Equations of vertical structure (specialized to inner region):*

$$F = \tfrac{3}{4}\Omega W \quad \text{(conservation of dissipated energy),} \tag{3a}$$

$$F = \tfrac{2}{3}acT^4/(\kappa_{Compt}\,\Sigma) \quad \text{(vertical radiative diffusion, Compton opacity),} \tag{3b}$$

* Supported in part by the National Science Foundation [GP-36687X, GP-28027].

$P = \frac{1}{2}h\Omega^2\Sigma$ (vertical pressure balance against out-of-plane gravitational forces of black hole), (3c)

$P = P_R \equiv \frac{1}{3}aT^4$

(equation of state, $P_R \gg P_G$), (3d)

$W = 2\alpha hP$

(source of viscosity, eq. [1]). (3e)

Equations (3) are only approximate, because of uncertainties in averaging over vertical structure.

The stationary models are obtained by setting $\partial\Sigma/\partial t \equiv 0$ in equations (2), (3).

For time-dependent models, it is best to choose $\Sigma(r, t)$ as the sole independent variable characterizing the local, instantaneous state of the disk. Then, at each $(r, t)$, one solves equations (3) algebraically for $h$, $P$, $T$, $F$, and $W$ as functions of $(\Sigma, r)$. It is essential to determine $W(r, t)$ self-consistently in this way, rather than to fix $W$ through equation (2b) from a given $\dot{M}$, as one does in the stationary case. Equations (2) yield, as the evolution equation of $\Sigma(r, t)$,

$$\frac{\partial\Sigma}{\partial t} = \frac{\partial}{\partial r}\left[\frac{d(\Omega r^2)}{dr}\right]^{-1}\frac{\partial}{\partial r}[r^2W(\Sigma, r)]\,. \quad (4)$$

The instability arises in the inner region for the following reason: Equations (3) give

$$W(\Sigma, r) = \text{const.}/\Sigma\,. \quad (5)$$

[To justify this paradoxical result: Since $P_R \gg P_G$, $P$ is not determined directly by $\rho$ ($\rho \equiv \Sigma/2h$), but only by $T$: and in fact $T$ and $P$ turn out to be independent of $\Sigma$. Equation (3c) implies $h \propto \Sigma^{-1}$; then equation (3e) shows $W \propto \Sigma^{-1}$.] The integrated stress $W$ is here a *decreasing* function of $\Sigma$; hence the nonlinear diffusion equation for $\Sigma$, equation (4), has a *negative* effective diffusion coefficient. As a result an initially stationary disk tends to break up into rings $\Delta r \gtrsim h$, on time scales $\sim (\Delta r/r)^2\, t_{\rm drift}$; alternate rings have high-$\Sigma$/low-$W$ and low-$\Sigma$/high-$W$. The density contrast grows because matter is pushed into regions of minimum viscous stress $W$. Eventually the low-$\Sigma$ regions become optically thin and hence thermally unstable (Pringle *et al.* 1973). As $\Sigma$ grows in the high-$\Sigma$ regions, eventually a regime is reached in which the disk cannot radiate as much energy as it is generating and the vertical structure equations fail to admit a solution. Therefore the growing instability causes a complete breakdown in the thin-disk picture, assumption ($a$). These conclusions are supported by detailed analytic and numerical calculations which one of us (A.P.L.) will report elsewhere (Lightman 1974).

Definitive models must therefore await a better understanding of viscosity: we mention two quite distinct possible alternatives to current models:

1. Assumption ($a$) fails because assumption (b) is roughly correct. Around the hole forms a cloud, which is 10 to 100 times larger than the hole. If dissipation is efficient (expected, since accreting matter must still lose its angular momentum), the cloud may emit X-rays as a hot, thin plasma with Comptonization probably important (Felten and Rees 1972; Illarionov and Sunyaev 1972). Alternatively, synchrotron cooling may be important. Gross time variations, both in intensity and in spectrum, are expected on the hydrodynamical time scale of the cloud $\sim$ tens to hundreds of milliseconds and longer. If the cloud is optically thick to Compton scattering, time variations on time scales shorter than the random walk time of a photon through the cloud $\sim \tau r/c$ ($\tau$ = optical depth) may be lost (F. K. Lamb, private communication). In particular, submillisecond time variations in signal, originating very near the hole (Sunyaev 1972), might be hopelessly smeared out by scattering in the translucent cloud.

2. Assumption ($b$) is seriously wrong. With $\alpha$ a function of $\Sigma$ rather than a constant in the time-dependent case (eq. [1]), a stable, stationary, thin disk is possible if $\alpha$ falls at least as fast as $\Sigma^{-1}$ in the inner region (*less* efficient viscosity). Such an $\alpha$ leads in turn to a $\Sigma(r)$ that increases steeply toward the hole. For example, (Cunningham 1973), equation (1) might be replaced by

$$t_{\hat{r}\hat{\phi}} = \beta P_G\,, \quad \beta = \text{const.}\,, \quad (6)$$

even when $P_R \gg P_G$. (Perhaps this relation is preferable for a self-limiting magnetic viscosity, since gas is frozen to the $B$-field while radiation is not.) The stationary, thin-disk model resulting from equation (6) is stable and is much like current models except that $\Sigma$ is much greater in the inner region (typically 25 times greater at $r = 10GM_{\rm BH}/c^2$). The thickness, $2h$, is still $\lesssim 2\times 10^5$ cm. This dense disk is quite optically thick and is probably immune to thermal or magnetic disturbances on length scales $\sim h$; hence, chaotic variations in the X-ray signal are likely to be negligible.

Observations (Schreier *et al.* 1971) of Cygnus X-1 (and similar sources which have been advanced as black-hole candidates) favor alternative (1), since the observed signal is chaotic on all time scales from tens of seconds to $\sim$50 milliseconds (instrumental limit). For either alternative, we believe that the prospects of seeing characteristic ($\lesssim$ms) time variations originating very near the hole are poorer than has been generally supposed on the basis of current models (Sunyaev 1972).

The same instability arises in a disk around an unmagnetized neutron star. For a magnetized neutron star, a disk does not extend inside the magnetosphere (Pringle and Rees 1972); there is no inner region, hence there is no instability.

We are grateful to colleagues at the California Institute of Technology for discussions, especially K. S. Thorne.

## REFERENCES

Cunningham, C. 1973, unpublished Ph.D. thesis, University of Washington.
Felten, J. E., and Rees, M. J. 1972, *Astr. and Ap.*, **17**, 226.
Illarionov, A. F., and Sunyaev, R. A. 1972, *Astr. Zh.*, **49**, 58 (English transl. in *Soviet Astr.—AJ*, **16**, 45, 1972).
Lightman, A. P. 1974, paper in preparation.
Novikov, I., and Thorne, K. S. 1973, in *Black Holes, Les Houches 1973*, ed. C. DeWitt and B. S. DeWitt (New York: Gordon & Breach).
Pringle, J. E., and Rees, M. J. 1972, *Astr. and Ap.*, **21**, 1.
Pringle, J. E., Rees, M. J., and Pacholczyk, A. G. 1973, *Astr. and Ap.* (in press).
Schreier, E., Gursky, H., Kellogg, E., Tananbaum, H., and Giaconni, R. 1971, *Ap. J.*, **170**, L21.
Shakura, N. I., and Sunyaev, R. A. 1973, *Astr. and Ap.*, **24**, 337.
Sunyaev, R. A. 1972, *Astr. Zh.*, **49**, 1153 (English transl. in *Soviet Astr.—AJ*, **16**, 941, 1973).

Astron. & Astrophys. 44, 59–68 (1975)

# Radiation from Spherical Accretion onto Black Holes

P. Mészáros
Institute of Astronomy, University of Cambridge

Received May 7, 1975

**Summary.** We present calculations of the radiation properties of black holes accreting in a spherically symmetric manner, both in the case when they are isolated and accreting from the interstellar medium, and when they are in a binary system. Two new effects are included, the heating due to dissipation of turbulent motions and magnetic field line reconnection, and the radiation from cosmic rays captured with the accreted gas. The present results indicate that previous conclusions concerning the detectability of isolated black holes is too pessimistic by at least some orders of magnitude. Also, the efficiency of radiation at high accretion rates, which previously was thought to be very small, is here found to be of the order unity, due to the inclusion of the dissipative heating. For mass transfer rates comparable to that of binary X-ray source models, the spherically symmetric model gives temperatures above $10^9$ °K, and therefore can compete with disk models as a source of hard X-ray radiation.

**Key words:** spherical accretion — black holes — infrared sources — X- and $\gamma$-ray sources

## I. Introduction

Previous calculations of the accretion flow onto black holes when the angular momentum of the gas captured is not too high, so that instead of a disk the flow is spherically symmetric, have proceeded with the assumption that the presence of a frozen-in magnetic field will lead to a highly turbulent situation, where the rms value of the energy density of the random field at any radius cannot exceed the gravitational energy density (Shvartsman, 1971; Zeldovich and Novikov, 1971; Shapiro, 1973a, b). On energetic grounds, this appears a reasonable assumption, and with the value of the magnetic field given, the synchrotron losses from the gas are found to play an important role at low densities, while at higher densities bremsstrahlung also become important. The only heating considered in the above references were the $p\,dV$ work on the gas, due to the compression of a volume element as it moves inward, and photoionization from nearby stars and/or cosmic ray heating. With these assumptions, for outside densities typical of the interstellar medium, it was found that the gas heated up along an adiabat, the value of the index depending on whether electrons were coupled to protons or not, and this then led to a certain luminosity, most of which came from the layers just above the Schwarzschild radius. For outside densities $n_a \gtrsim 10^3 (T_a/10^4)^{3/2} (M/M_\odot)^{-1}\,\mathrm{cm}^{-3}$, or $\dot{M} \gtrsim 4\times 10^{12} (M/M_\odot)\,\mathrm{g\,s}^{-1}$, the cooling time became less than the free-fall time, and as the gas cooled, the $p\,dV$ work on the gas became negligible, thus reducing drastically the radiation efficiency.

In fact, the presence of a frozen-in magnetic field, implies a conversion of gravitational energy into magnetic and turbulent energy, and a certain amount of dissipation must be associated with such transformations. Both the laminar shear, and the shear induced by turbulent motions, would tend to increase the magnetic field to a point where $\langle B^2\rangle/8\pi$ would try to exceed the gravitational energy density, and this should lead to violent instabilities and reconnection of field lines, and one can estimate the joule heating produced by this. At the same time, the turbulent motions will have a random transverse component, which would tend to increase as the turbulent cells are carried inwards. These transverse components of the momentum have to be carried away to allow matter to move inwards, hence viscous dissipation must be present, perhaps in the form of hydromagnetic shocks from collisions between cells. This leads to an estimate of dissipation heat input similar to the previous (cf. Section II), $\Gamma \sim 5\times 10^{34} \cdot r^{-4} n_a (M/M_\odot)^3 \delta\,\mathrm{erg\,cm^{-3}\,s^{-1}}$, where $\delta \sim 1$ and $n_a$ is outside particle density. For the low accretion rate cases, where $t_{\mathrm{cool}} \gg t_{\mathrm{free\,fall}}$, this leads to one order of magnitude larger temperature near the Schwarzschild radius, which gives a synchrotron luminosity two orders of magnitude larger than previously. For the high accretion rate case, $t_{\mathrm{cool}} \ll t_{\mathrm{ff}}$, the temperature and luminosity near the Schwarzschild radius are many orders of magnitude larger than previously, because the dissipative heating is able to convert gravitational energy into heat at all radii, even when the gas cools fast. Thus, for all values of $n_a$ such that $t_{cool} \lesssim t_{ff}$, one achieves a radiation efficiency comparable to unity, if the magnetic and turbulent energies are in near equipartition with the gravitational energy. This is of importance for isolated black holes in galactic nuclei, and for black holes in binary systems where the accretion is spherical, rather than disk-like, which may occur if the intrinsic angular momentum of the matter captured is small. Previously it was believed that only disk systems were able efficiently to convert gravitational energy at the bottom of the potential well into heat, and thence into radiation, but the present results indicate that the spherical systems are equally able to do so. In fact the temperature reached near the Schwarzschild radius are in all cases above $10^9$ °K, so that these spherical models may be competitive as hard X-ray sources even in the higher luminosity range.

The other factor neglected in previous studies was the fact that galactic cosmic rays would be captured, together with the gas, because of the frozen-in field in the latter. One would expect that those cosmic rays whose gyroradius in the random frozen-in field is smaller than the accretion radius, from within which matter is sucked in, would be sucked in as well. These cosmic rays constitute a relativistic gas, which hardly interacts with the rest of the thermal gas, because of the very long coulomb collision times, and as the field is scooped inwards, it gets compressed along its own adiabat. Thus the total cosmic ray energy near the Schwarzschild radius is much larger than at the accretion radius, and this energy can be radiated away by different mechanisms, mainly synchrotron. The question of whether one has an *equipartition* energy density of cosmic rays is a difficult one (cf. Section VI). At any rate, the least one would expect is the value obtained assuming that the only cosmic rays captured and accelerated in the accreting flow are those from the galactic cosmic ray background, as inferred from measurements on Earth. This leads, in the case of isolated black holes accreting from the interstellar medium, to luminosities which in the X-ray band exceed by some orders of magnitude both previous calculations, and the presently computed luminosities from thermal particles.

## II. Flow Model and Energetics

We assume the angular momentum per unit mass of the matter captured to be less than $\sqrt{3}(2GM/c^2)c$, which is usually satisfied for isolated black holes, and in binary systems, for those where the mass transfer is from a stellar wind (Illarionov and Sunyaev, 1975). This ensures a spherically symmetric flow, rather than a disk. If the medium in which the black hole is immersed is turbulent, as the interstellar medium is, and as the binary transfer flow is likely to be, both observations and theory suggest that the random magnetic and turbulent energy densities will be in rough equipartition with the thermal energy, even outside the accretion radius. For the interstellar medium in the galactic plane, we know that the cosmic ray energy density also is near equipartition. The accretion radius $r_a$ is defined as the distance at which gravitational energy equals the other energy forms,

$$r_a^{-1}GM = a_s v_s^2 + a_t v_t^2 + a_B B^2(8\pi n_a \bar{m})^{-1} + a_c \varepsilon_{CR}(n_a \bar{m})^{-1}, \tag{1}$$

where $n_a$ is the thermal particle density at the accretion radius and beyond, $\bar{m}$ is average thermal particle weight, $\varepsilon_{CR}$ is cosmic ray energy density and $a_s$, $a_t$, $a_B$ and $a_c$ are constants, unity for equipartition. Previous studies have taken $a_t = a_B = a_c = 0$ in the definition of $r_a$. To characterize the flow, we define the parameters

$$\begin{aligned} \alpha &= (a_s + a_t + a_B + a_c)/4 \\ \xi &= (1 + y_a)/2 \\ \theta &= T_a/10^4 \\ m &= M/M_\odot \\ x &= r/m, \end{aligned} \tag{2}$$

where $y_a = n_{e,a}/n_a$ is relative ionization fraction and $T_a$ is temperature at $r_a$, $M$ is black hole mass, and $r$ is distance to the centre in cm. In most situations, we expect $\alpha$, $\xi$ and $\theta$ to be near unity.

The accretion radius becomes in this notation

$$x_a = 1.2 \times 10^{13}(\alpha\theta\xi)^{-1}. \tag{3}$$

With Shvartsman (1971) we assume the velocity for points $x < x_a$ to be the free-fall velocity, leading to a density distribution

$$n(x) = 4.2 \times 10^{19} n_a x^{-3/2} \tag{4}$$

and an equipartition magnetic field

$$H(x) = 2.6 \times 10^{11} n_a^{1/2} x^{-5/4} \alpha^{-5/4}(\theta\xi)^{-3/4}. \tag{5}$$

As long as the electrons remain nonrelativistic, the energy balance equation for a comoving, fully ionized volume is

$$\frac{dT}{dr} = -\frac{T}{r} + \frac{(\Lambda - \Gamma)}{3unk},$$

where $k$ = Boltzmann's constant, $u$ = free-fall velocity, $n$ = total baryon density, $\Lambda$ = radiative heat loss rate and $\Gamma$ = heat gain rate (erg cm$^{-3}$ s$^{-1}$). This is written as

$$\frac{dT}{dx} = -\frac{T}{x} + Ax^2(\Lambda - \Gamma), \tag{6}$$

where $A = 3.25 \times 10^{-18} n_a^{-1} m(\alpha\theta\xi)^{3/2}$, and the adiabatic index is $\gamma_{ad} = 5/3$ for both electrons and protons. The dissipative heating (shocks, etc.) acts on electrons and protons alike, while the cooling essentially represents energy lost by the electrons. However, we assume with Shapiro (1973) that plasma microinstabilities will couple effectively both temperatures, so that $T_e = T_p$, and a Maxwell-Boltzmann distribution would be established. Such instabilities are to be expected, for instance due to the fact that stretching of field lines with conservation of adiabatic invariants between shocks would lead to anisotropic particle distributions, causing mirror instabilities, and in the shocks, two stream instabilities are to be expected. The details of this process are unclear, however, so that $T_e \simeq T_p$ should be regarded more as a working hypothesis adopted here than as a foregone conclusion.

If this is the case, for temperatures $T > m_e c^2/k$ we shall still have $T_e \cong T_p$, and the effective $\gamma_{ad} = 13/9$, as in Shapiro's case, the energy balance equation being

$$\frac{dT}{dx} = -(2/3)\frac{T}{x} + (2/3)Ax^2(\Lambda - \Gamma). \tag{7}$$

It is worth noting that there is no universal agreement on the temperature coupling. For instance Shvartsman and Dahlbacka *et al.* (1974) assume no coupling, hence $\gamma_{ad} = 5/3$ for protons and $\gamma_{ad} = 4/3$ for electrons above $T_e = m_e c^2/k$. This leads them to higher proton temperatures near the Schwarzschild radius. Eardley *et al.* (1975) have also investigated two-temperature models for the disk case. However, the presence of dissipative heating and shocks would seem to support rather the hypothesis of effective coupling, as assumed here.

The dissipative heating can most easily be estimated, to order of magnitude correctness, by considering the turbulent energy of matter between radii $r$ and $r/2$, and requiring that this be dissipated in one free-fall time. From the equipartition argument this would give

$$\Gamma r^3 t_{ff} \simeq \bar{\varrho} v_t^2 r^3 \simeq \frac{\delta}{4\alpha}\frac{GM\bar{\varrho}}{(r/2)} r^3$$

which gives

$$\Gamma \simeq 5 \times 10^{34} n_a m^{-1} x^{-4} \delta \alpha^{-5/2} (\theta\xi)^{-3/2}, \tag{8}$$

where $\delta$ is a numerical factor of order unity, if turbulent energy is near equipartition. The same estimate is obtained if one assumes a turbulent length scale which is a fraction of $r$, and estimating the shear and turbulent viscosity for a turbulent velocity which is of the order of the free-fall velocity. Similarly, taking a magnetic field length scale a fraction of $r$, and assuming reconnection to occur on the Alfvén time-scale (which in equipartition is $\sim t_{ff}$), the same result (8) is achieved.

In the outer regions of the flow, bremsstrahlung is the principal cooling mechanism, and we used Ginzburg's (1969) expression in the non-relativistic region, and Maxon's (1972), including electron-electron and electron-proton rates, in the relativistic regime. Cyclotron cooling is important only for $x < x_p = 1.7 \times 10^8 \alpha^{-1}$, after which the cyclotron frequency $\nu_H = (eB/2\pi m_e c)$ exceeds the plasma frequency $\nu_p = (n_e e^2/\pi m_e)^{1/2}$. The loss rates for the cyclotron, and relativistic synchrotron mechanism are given by Jackson (1962) and Ginzburg (1969). For large values of $n_a$, cyclotron self-absorption becomes important, and this limits the luminosity to the blackbody value, at frequencies where $\tau_\nu > 1$. The temperatures at which this happens are usually $\gtrsim 10^8$ °K, and Trubnikov (1958), Hirshfield *et al.* (1961) and Bekefi (1966) have discussed self-absorbed cyclotron situations. Only the extraordinary mode is of importance, and for harmonics above the fifth, the absorption coefficient is nearly continuous,

$$\varkappa_\nu^c = (4\pi^2 \nu_p^2/c\nu_H)\, A \exp(-a\nu/\nu_H), \tag{9}$$

where $A$ and $a$ are slowly varying functions of $T$ ($A = 0.5$, 0.15, 0.09 and $a = 2.2$, 0.96, 0.58 at $m_e c^2/kT = 50$, 10 and 5 respectively. In the relativistic regime, Pacholczyk (1970) gives the synchrotron absorption coefficient for a relativistic Maxwell-Boltzmann distribution as

$$\varkappa_\nu^s = 2.6 \times 10^{26} n T^{-3} \nu^{-1} I(\nu/\nu_s), \tag{10}$$

where $\nu_s = 6 \times 10^{-14} HT^2$ and $I$ is a function tabulated in his book. This distribution gives a larger absorption coefficient at frequencies $\nu > \nu_s$ than is the case for power law distributions, or monoenergetic distributions. (As $E$ increases, the particle density decreases as $\exp(-E/kT)$ but the frequency $\nu_s$ goes up as $E^2$).

We deal with the self-absorbed synchrotron losses in an approximate way, by assuming that each spherical surface of radius $r$ effectively loses an amount of energy $L(r) = 4\pi r^2 \cdot \mu 2\pi(2kT/3c^2)\nu_m^3$, where $\mu$ is a numerical factor which we take to be 1/2, and $\nu_m$ is the frequency at which $\varkappa_\nu r \simeq 1$. Most of this energy is concentrated at the peak $\nu_m$, and can escape freely. We can then define an effective self-absorbed synchrotron loss

$$\Lambda_s^1 = (3/mx) \cdot (\pi 2kT/c^2)\nu_m^3 \tag{11}$$

which can be used in Eqs. (6) and (7) as if it were a volume loss. The factor $3/mx = 3/r$ is required for dimensional reasons, and one verifies that the surface losses then equ l the equivalent volume losses.

At very high densities, the Thomson optical depth can also become larger than unity, and then such complications arise as entrainment of radiation ($u/c$ terms in the radiative transfer equation) and Comptonization of the spectrum. We do not deal with these problems here, as there seems to be no easy way to handle them, but they are bound to be of great importance for any systems where the density at the accretion radius is larger than $n_a \sim 10^7$.

## III. Gas Temperatures

By thermal particles we refer to the bulk of the gas, which is assumed to obey a Maxwell-Boltzmann distribution in virtue of the plasma instabilities, even though strictly speaking it is not in thermodynamic equilibrium. The temperature characterizing them is

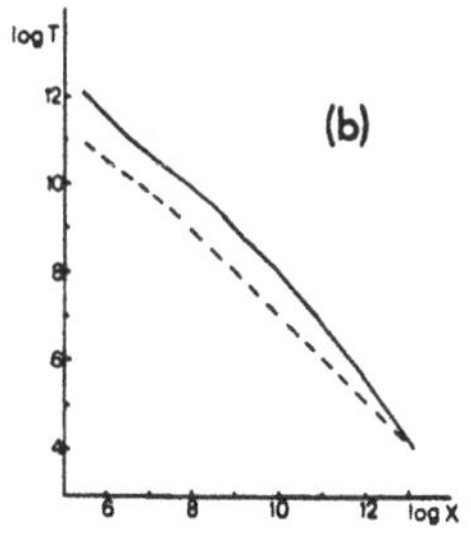


Fig. 1. The run of temperature against dimensionless radius $x(x=rm)$ for $T_a=10^4$ °K ($\theta=1$) and low densities ($t_{cool} \gg t_{ff}$). In dashed lines is shown the behaviour if only adiabatic heating were included

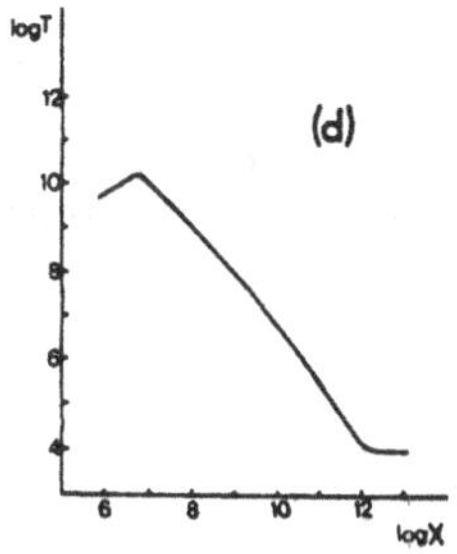


Fig. 2. The run temperatures against dimensionless radius $x$ for $T_a=10^4$ °K and a high density of $n_a=10^7$ °K ($t_{cool} \ll t_{ff}$). The gas is approximately isothermal down to $x_c=1.5\times10^{12}$, where heating starts exceeding cooling (cf. Section III)

given by Eqs. (6) and (7), as opposed to the cosmic ray particles treated in the next section, which are characterized by energies of their own, and are assumed not to affect the energy balance of the thermal particles.

For low accretion rates (low $n_a$) the cooling is negligible everywhere, compared to the adiabatic ($pdV$) and dissipative ($\Gamma$) heating, $t_{cool} \gg t_{ff}$. Writing then $\Gamma=Bx^{-4}$, the solution of (6) is

$$T=T_a(x_a/x)+(AB/x)\ln(x_a/x)$$

$$=T_a(x_a/x)[1+3.25\alpha^{-1}\log(x_a/x)] \qquad (12)$$

which differs from the adiabatic $\gamma_{ad}=5/3$ behaviour by the second term in square brackets, due to dissipative heating. The value of $x$ at which relativistic electron temperatures are reached, $x_r$, depends on the value of $T_a$, or $\theta$, and of $x_a$. For $x<x_r$, the solution of the Eq. (7) is given by

$$T=T_r(x_r/x)^{2/3}[1+(2AB/T_rx_r)\{(x_r/x)^{1/3}-1\}], \qquad (13)$$

the second term in brackets being the dissipative correction on the adiabatic $T\propto x^{-2/3}$ behaviour. For $x \ll x_r$, it reduces to

$$T\simeq T_r(2AB/T_rx_r)(x_r/x) \qquad (14)$$

which in effect is like a "$\gamma$" $=5/3$ behaviour, even though $\gamma=13/9$. This simulation of a 5/3 behaviour is caused by the dissipative heating, and is due to the fact that isotropic magnetohydrodynamic turbulence also behaves as $\gamma=5/3$. This entails a significant departure from the calculations without dissipation. At $x_s=3\times10^5$, the Schwarzschild radius, for $\theta=1$ one obtains $T_s=1.14\times10^{12}$ °K. The run of temperatures is shown in Fig. 1, where it is compared with the adiabatic case under the same assumption of $\gamma=13/9$, giving at $x_s$ a temperature of $10^{11}$ °K. The temperature obtained with dissipation at $x_s$ is even higher than the proton temperature at $x_s$ which would be obtained in a two-temperature model, with $\gamma=5/3$ for the protons ($4\times10^{11}$ K). The adiabatic temperatures mentioned here are somewhat lower than those in some previous references, because we took a smaller value for $x_a$, due to the inclusion of the other energy forms in the virial expression, but if we increased $x_a$, the non-adiabatic temperatures would also increase.

For higher accretion rates ($n_a>10^3$), such as in case (*d*) below, the cooling time is comparable or shorter than the free fall time, and a numerical solution of Eqs. (6) and (7) is required. We show this in Fig. 2, for $n_a=10^7$ cm$^{-3}$ ($\dot{M}=5\times10^{16}$ m$^2$ g s$^{-1}$), which gives, as expected, temperatures lower than when cooling is negligible, but much higher than when only adiabatic heating is present. In fact, for $n_a$ this high, the bremsstrahlung rate is higher than $\Gamma$ at the accretion radius, and remains so until $x_c=1.5\times10^{12}$, below which $\Gamma$ exceeds $\Lambda_{ff}$ ($\Gamma\propto x^{-4}$ but $\Lambda_{ff}\propto x^{-3}T^{1/2}$ in the N.R. regime). The gas is assumed to fall from $x_a=1.2\times10^{13}$ down to $x_c=1.5\times10^{12}$ at $T\cong T_a=$constant. In most situations, $T_a$ is likely to be $\sim10^4$ °K, due to ionizing radiation from the inner regions of the black hole, or the nearby environment (binary companion). Below $x_a$, $\Gamma>\Lambda$ and the temperature starts rising, but nonetheless the cooling remains non-negligible. The self absorbed synchrotron cooling does not start exceeding the bremsstrahlung losses until $x=2\times10^7$, $T=6.5\times10^9$ °K, and after this the temperature starts to saturate, turning over about $10^{10}$ °K, from which point one has, approximately, $\Gamma\sim\Lambda$ determining the temperature. In fact, for $x<5\times10^6$, another cooling mechanism comes in, namely inverse Compton cooling (since $\tau_T=n\sigma_Tr>1$ below this $x$, if $n_a=10^7$). Although it is difficult to compute the temperature in the presence of multiple Compton scatterings, it is clear that the cooling is so efficient as to be able to radiate all the energy deposited by the dissipative heating. Probably the temperature stabilizes around, or just above $T=m_ec^2/k=4\times10^9$, since both synchrotron and inverse Compton losses increase their efficiency enormously in the relativistic region ($\Lambda\propto\gamma^2$, where $\gamma=$Lorentz factor) but become relatively inefficient below $4\times10^9$.

The corresponding temperature behaviour, if dissipative heating would have been neglected, would have consisted in remaining at $T \sim T_a$ down to much smaller radii, because the adiabatic term in (6) is $\propto Tx^{-1}$ while the free-free term is $\propto Ax^2x^{-3}T^{1/2} \propto T^{1/2}x^{-1}$, until eventually the gas becomes optically thick to free-free radiation, allowing a slow rise only after this. The reson why one obtains much higher temperatures when including dissipative heating is because the term $Ax^2\Gamma \propto x^2x^{-4} \propto x^{-2}$ grows faster than the free-free term.

## IV. Radiation Spectrum from Thermal Particles

We consider here some specific examples of environments from which black holes could be accreting, typical of our galaxy and of some extragalactic situations. We take for the black hole mass 10 solar masses, $m = 10$, as being a likely value, since stellar evolutionary remnants of such mass are almost certain to be black holes, and primeval black holes, if present (cf. Mészáros, 1975) might be expected to have a mass in that neighbourhood. Most of the luminosity comes from the layers nearest to the Schwarzschild radius at $x_s = 3 \times 10^5$ ($r_s = 3 \times 10^5$ m), and a proper calculation requires taking into account gravitational escape-cone and redshift effects (Zeldovich and Novikov, 1971) as well as special relativistic effects (Shapiro, 1973). All of these effects work in the direction of reducing the luminosity that one would compute in a Newtonian calculation, and they start becoming important around $x \sim 2x_s$. To make calculations simple, we compute the luminosities in the Newtonian limit, down to a fictitious bottom layer located at $x_b = 10^6$ ($x \sim 3x_s$). The radiation from layers below that, due to the relativistic effects, probably does not introduce more than a factor 2 correction on this Newtonian luminosity, this being the same approximation made in disk calculations, e.g. Pringle and Rees, 1972, Shakura and Sunyaev, 1973.

a) $n_a = 10^{-3}\,\text{cm}^{-3}$, $\quad T_a = 10^6$ °K

These parameters might be typical of our galactic halo. The density is so low that cooling is negligible in the energy equation, and solution (12) with $\theta = 10^2$ gives $x_r = 2.6 \times 10^8$, and (13) gives the temperature at $x_b = 10^6$ as $T_b = 3.4 \times 10^{11}$. The synchrotron peak for this layer is at $3 \times 10^{11}$ Hz, and the luminosity is $L_\nu^{th} \cong (4\pi/3)r^3\Lambda_s^{th} = 1.6 \times 10^{17}\,\text{erg s}^{-1}$, exceedingly small, so that the cosmic ray luminosity computed in the next section exceeds it. The bremsstrahlung contribution extends to the MeV range, but gives a total luminosity far below the synchrotron. The synchrotron peak is optically thin, and using the fact that the low frequency behaviour of Eq. (10) is $\varkappa_\nu \propto \nu^{-5/3}$ (Pacholczyk, 1970) we find that $\tau_\nu \sim 1$ at $\nu = 10^8$ Hz. Thus the spectrum cuts off exponentially above $3 \times 10^{11}$ Hz, goes as $\nu^{1/3}$ between $10^8$ and $3 \times 10^{11}$, characteristic of the single particle spectrum, and goes as $\nu^2$ below $10^8$ Hz. Even further below, the layers above the bottom one start contributing,

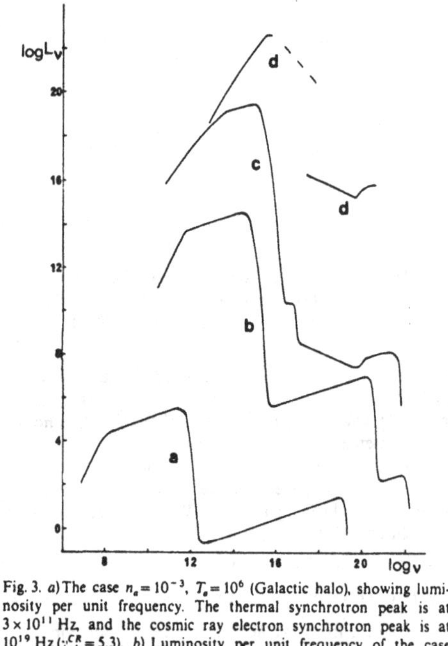


Fig. 3. *a*) The case $n_a = 10^{-3}$, $T_a = 10^6$ (Galactic halo), showing luminosity per unit frequency. The thermal synchrotron peak is at $3 \times 10^{11}$ Hz, and the cosmic ray electron synchrotron peak is at $10^{19}$ Hz ($\gamma_{ad}^{CR} = 5.3$). *b*) Luminosity per unit frequency of the case $n_a = 1$, $T_a = 10^4$ (H II region). The peak at $3.4 \times 10^{14}$ Hz is thermal synchrotron, that at $1.9 \times 10^{20}$ Hz is cosmic ray electron synchrotron ($\gamma_{ad}^{CR} = 5/3$), and that at $5 \times 10^{21}$ Hz is thermal bremsstrahlung. *c*) Luminosity per unit frequency of the case $n_a = 10^3$, $T_a = 10^4$ (dense H II region). Peak at $1.1 \times 10^{15}$ Hz is thermal synchrotron, that at $5 \times 10^{16}$ Hz is cosmic ray electron synchrotron, that at $3 \times 10^{21}$ Hz is thermal bremsstrahlung. *d*) Luminosity per unit frequency of the case $n_a = 10^7$, $T_a = 10^4$. (Galactic centre, binary systems). The peak at $5 \times 10^{15}$ Hz is thermal bremsstrahlung, that at $5 \times 10^{20}$ Hz is thermal bremsstrahlung. A plausible (but uncertain—see text) comptonized thermal synchrotron contribution is indicated in dashed lines

but with even more negligible luminosity. See however Section Va).

b) $n_a = 1\,\text{cm}^{-3}$, $\quad T_a = 10^4$ °K

These conditions could characterize normal H II regions. Here, cooling is again negligible in the energy equation, $x_r = 4.5 \times 10^8$, and at $x_b = 10^6$, $T_b = 3.4 \times 10^{11}$ again. (Notice that the accretion radius depends on $\theta$.) The synchrotron peak is at $3.4 \times 10^{14}$ Hz (the magnetic field is higher than in the previous case) and is again optically thin, becoming thick at $6.4 \times 10^{11}$ Hz, so that the spectrum is similar to (a), exponentially cut off above $3.4 \times 10^{14}$, $\nu^{1/3}$ between $3.4 \times 10^{14}$ and $6.4 \times 10^{11}$, and $\nu^2$ below that, until the upper layers start coming in, which again we can neglect. The total synchrotron luminosity is $L_s^{th} = (4\pi/3)r^3\Lambda_s = 1.6 \times 10^{29}\,\text{erg s}^{-1}$, mostly concentrated at the peak. Bremsstrahlung losses contribute

$L_{ff}^{th} = 10^{24}$ erg s$^{-1}$ with a maximum photon energy of 30 MeV. Since non-relativistic bremsstrahlung losses per unit frequency go as $\varepsilon_\nu \propto n_e^2 T^{-1/2}$ below the cutoff, and the relativistic losses as $\varepsilon_\nu \propto n_e^2(0.6 + \log\gamma)$, we have that $L_\nu^{ff} \propto x^3 x^{-3} T^{-1/2} \propto x^{1/2} \propto \nu^{-1/2}$, and $L_\nu^{ff} \propto x^3 x^{-3} \cdot [\text{const} - \log x] \propto (\text{const} + \text{Mog}\,\nu)$, as shown in Fig. 3b.

c) $n_a = 10^3$ cm$^{-3}$, $T_a = 10^4$ °K

Such conditions are encountered, for instance, in H II regions near the galactic nucleus. For these $n_a$ and $\theta$, we still have $\Gamma \gg [\Lambda_s, \Lambda_{ff}]$ in the outer regions, and $x_r = 4.5 \times 10^8$. However, by the time $T = 1.2 \times 10^{11}$ is reached, around $x = 6 \times 10^6$, one obtains $\Gamma \simeq \Lambda_s$, and $\tau_\nu \sim 1$ at the peak. Below this the gas remains approximately isothermal, since both $\Gamma$ and the optically thin synchrotron $\Lambda_s$ are proportional to $x^{-4}$ and the gas becomes progressively optically thinner at the peak, $\tau_\nu \sim 2/3$ at the peak at $x = 10^6$, located at $1.1 \times 10^{15}$ Hz. This gives a luminosity $L_s^{th} = 2 \times 10^{34}$ erg s$^{-1}$, but now the spectrum is more complicated, since the layers above $x = 6 \times 10^6$, $T = 1.2 \times 10^{11}$ are optically thick (which means that above this, the approximation of neglecting the cooling is certainly adequate because the optically thick losses are smaller than if it were thin). Thus the temperature below $1.2 \times 10^{11}$ is described by (12) and (13) and above that we take $T \sim \text{const} = 1.2 \times 10^{11}$. The spectrum is obtained numerically, determining the maximum self-absorbed frequency of each layer by means of (9) or (10), setting $x_\nu r \simeq 1$, and taking the luminosity of each layer as $L(r) = 4\pi r^2 \cdot (2\pi k T \nu_m^3/c^2)$, except for the $\tau_\nu < 1$ regions, where the same optically thin approximation as before was used. The integrated spectrum from all layers is flatter, in the optically thick region, than the $\nu^2$ behaviour of a single layer, due to the $x$ dependence of $T$ and $\nu_m$. The bremsstrahlung contribution is $L_{ff}^{th} = 4.8 \times 10^{29}$ erg s$^{-1}$, extending up to 10 MeV, and the spectrum, as in b), goes as $L_\nu^{ff} \propto \nu^{-1/2}$ (nonrelativistic) and $L_\nu^{ff} \propto (\text{const} + \log\nu)$ (relativistic).

d) $n_a = 10^7$ cm$^{-3}$, $T_a = 10^4$ °K

These may be parameters typical of a hole accreting in the denser parts of the galactic nucleus, or perhaps in the central regions of quasars or Seyferts. They would also be typical of binary stellar systems with a large mass exchange rate ($\dot{M} \simeq 5 \times 10^{16}$ m$^2$ g s$^{-1}$) due, for instance to a stellar wind. The boundary conditions are now more complicated, since, as discussed in the previous section, d), $\Lambda > \Gamma$ at $x_a$ and $\Lambda \simeq \Gamma$ only at $x_c = 1.5 \times 10^{12}$. This would cause the gas to fall inwards down to $x_c$ at $T \simeq T_a = 10^4$ °K = constant, since ionizing radiation should be plentiful to keep it at typical H II temperatures. After that the temperature rises, as computed in IIIc) and shown in Fig. 2. In this case the synchrotron radiation is optically thick at all $x$, and we compute the luminosity and spectrum numerically in the same manner as before. This integrated synchrotron spectrum has a spectral index $\sim 1.6$ below the peak, which is at $5 \times 10^{15}$ Hz, corresponding to the highest temperature achieved $T = 1.2 \times 10^{10}$ at $x = 7.5 \times 10^6$. However, the decay is not exponential above this, since there is still radiation from $7.5 \times 10^6 > x > 10^6$, where we have not computed the temperature because of difficulties in treating the relativistic Comptonization, although it probably stays above $4 \times 10^9$ °K. A very gross estimate of what this Comptonized spectrum might look like (inverse Compton scattering on synchrotron photons) is shown in dashed lines, but cannot be given much confidence. The only thing that is clear is that at $x = 10^6$, there is an energy deposition rate of $4r^3\Gamma = 2 \times 10^{38}$ erg s$^{-1}$, and this energy is easily radiated away by the synchrotron and inverse Compton mechanisms, whose rates are very high under these conditions. The bremsstrahlung spectrum from the uncomptonized layers is also straightforward, and is shown in Fig. 3d as the high energy tail extending up to $\sim 2$ MeV, again $L_\nu^{ff} \propto \nu^{-1/2}$ in the non-relativistic, and $L_\nu^{ff} \propto (\text{const} + \log\nu)$ in the relativistic region. The comptonized bremsstrahlung might fill out somewhat the trough in the spectrum around $10^{20}$ Hz, but probably does not alter much this part of the spectrum. It is worth noting that the uncomptonized bremsstrahlung, with a good degree of confidence, gives a luminosity of $L_{ff}^{th} \simeq 3 \times 10^{36}$ erg s$^{-1}$, which makes this a powerful source of $\gamma$-rays.

## V. Radiation from Cosmic Rays in the Accreting Flow

The gas around the black hole, outside $x_a$, is likely to contain cosmic rays, as well as frozen in magnetic fields. Those cosmic rays whose gyroradius $r_H = 2 \times 10^3 E_{\text{MeV}}/B_{\text{Gauss}}$ is smaller than the accretion radius, should be carried inwards with the accreted gas.
The maximum energy of cosmic rays that can be captured at $x_a$ is (demanding $r_H \leqslant r_a$)

$$E_{\text{MeV,max}} = 6 \times 10^4 m\theta^{-7/4} n_a^{1/2} . \tag{15}$$

The compression behaviour of the cosmic ray gas is given by the adiabatic exponent $\gamma_{ad}^{CR}$, so that $E \propto n_{CR}^{\gamma-1}$. For a relativistic gas, $\gamma_{ad} = 4/3$ so $E \propto n_{CR}^{1/3} \propto x^{-1/2}$ (since $n_{CR} \propto n_{th} \propto x^{-3/2}$). This would imply also that, if cosmic rays initially (at $x_a$) were in equipartition, they would fall below equipartition, as $x$ decreased. In fact, cosmic rays should be subject to the usual Fermi accelerations from random collisions with magnetic (turbulent) irregularities. This process is known to saturate (e.g. Ginzburg and Sirovatskii, 1964) when the cosmic ray energy density comes into equipartition with the energy of the turbulent cells. Hence, we should expect cosmic rays to have $\gamma_{ad} = 5/3$, just as for isotropic turbulence and for the random magnetic field. However, this may depend on the efficiency of the Fermi mechanism, and on the particular turbulent spectrum, which is uncertain, so we shall consider both $\gamma_{ad}=4/3$ and $5/3$ as delimiting the probable range of interesting cases.

The main energy loss process for cosmic ray electrons in the accretion flow is synchrotron radiation, with a peak frequency $\nu_s^e=(eB/2\pi m_e c)\gamma^2$ a loss rate per particle $(dE/dt)_e=(2c/3)(e^2/m_e c^2)^2 H_\perp^2\gamma^2=1.6\times10^{-15}H_\perp^2\gamma^2$, and a half life $t_{1/2}^e=E(dE/dt)^{-1}=5\times10^8 H_\perp^{-2}\gamma^{-1}$ s, $\gamma$ being the Lorentz factor. For cosmic ray protons, the synchrotron losses are also important, but their peak frequency is down by $(m_e/m_p)$, the loss rate is down by $(m_e/m_p)^2$, and the half life is $t_{1/2}^p=3.2\times10^{18}H_\perp^{-2}\gamma^{-1}$ s. For the protons, however, other mechanisms can also be important, for instance inelastic collisions with thermal protons leading to $\pi^0$ decay, which gives $\gamma$-rays and secondary electrons which then undergo synchrotron losses, e.g. Ginzburg and Sirovatski (1964). The cross section for this process, above 10 GeV, is about $4\times10^{-26}$ cm$^2$, and it becomes important for $n_a\gtrsim10^7$, since then the mean collision time for this process, $(n_{th}\sigma_\pi c)^{-1}$, becomes comparable to the free fall time. Also of interest are electron-positron production from proton collisions with thermal photons, and at higher photon energies, pion production from the same mechanism, e.g. Greisen (1966). The cross section for pair production is $1.8\times10^{-27}(\ln z-0.5)$ cm$^2$, where $z$ is the ratio of proton energy to threshold energy, and the fractional energy loss is $\Delta E/E\simeq10^{-3}z^{-1}$. The mean cross section for photopion production is $\sim2\times10^{-28}$ cm$^2$, and the fractional energy loss is $\Delta E/E\simeq0.13$ at threshold and rising above that.

We assume the cosmic ray proton spectrum outside the accretion radius to be of the same form as observed near the Sun at energies above 10 GeV, so that the flux is $j(E)=b\,10^5 E_{\mathrm{MeV}}^{-2.65}$ proton/cm$^2$ s sterad MeV. If we assume this spectrum to continue down to an energy $d\times10^3$ MeV, then the total cosmic ray number density is

$$n_{CRp}=4\pi c^{-1}b\,10^5\int_{d10^3}^{\infty}E^{-2.65}dE=2.4\times10^{-10}f\,\mathrm{cm}^{-3}, \tag{16}$$

where $f=d^{-1.65}$. $b$ is a factor $\sim1$ if the spectrum and density is similar to that near the Sun. We neglect the change of index to $-2$ seen to occur in the observed spectrum between 10 and 1 GeV, lumping all uncertainties in the numerical factor $f$. Essentially, we are assuming most of the cosmic rays to be concentrated at the lower end of the spectrum. The electron spectrum, by analogy with the solar neighbourhood is taken to be of the same shape as the proton flux, but down by $10^{-2}p$ in intensity, and to extend down to $g\times10$ MeV. Then the cosmic ray density is

$$n_{CRe}=5.2\times10^{-9}h\,\mathrm{cm}^{-3}, \tag{17}$$

where $h=g^{-1.65}$. $p$ is the factor which we shall take to be $\sim1$ for the galactic cosmic ray flux, with the understanding that there is considerable uncertainty in this. Both $n_{CRp}$ and $n_{CRe}$, as they are convected along with the thermal gas, behave as $n_{CR}\propto x^{-3/2}$. The electron half life, in many cases, can become shorter than the free-fall time, and for electrons that at $x_a$ had 10 MeV energy, this occurs when they have reached a Lorentz factor

$$\gamma_{1/2}=1.25\times10^{-2}n_a^{-1}x \tag{18}$$

which occurs at a value of $x_{1/2}$ depending on the cosmic ray adiabatic exponent,

$$\begin{aligned} x_{1/2}&=1.4\times10^8 n_a^{1/2} \quad \text{for} \quad \gamma_{ad}=5/3\\ x_{1/2}&=3.14\times10^6 n_a^{2/3} \quad \text{for} \quad \gamma_{ad}=4/3\,. \end{aligned} \tag{19}$$

We consider now the same four examples as in the last section.

a) $n_a=10^{-3}$ cm$^{-3}$, $T_a=10^4$ °K

From Eq. (15), the maximum energy of cosmic rays captured is 3 MeV, so probably protons are not captured at all, but electrons of the lowest energy are captured, if $g<(1/3)$. Considering, for simplicity only the most numerous electrons, those at the lower end of the spectrum, they will have a Lorentz factor $\gamma_i=5\bar{g}$, where $\bar{g}=3g\sim1$. If the cosmic ray gas has an adiabatic index $\gamma_{ad}=5/3$, then at $x=10^6$ they will have achieved a Lorentz factor $\gamma_f=\gamma_i(x_a/x_f)=6\times10^5\bar{g}$ ($x_a=1.2\times10^{11}$ here, since $\theta=10^2$) giving a synchrotron peak at $\nu=10^{19}\bar{g}^2$ Hz and a luminosity $L_s^{CRe}=3\times10^{20}\bar{g}^2h$ erg s$^{-1}$, due to the bottom layer $x=10^6$. If, on the other hand, $\gamma_{ad}=4/3$, then $\gamma_f=\gamma_i(x_a/x_f)^{1/2}=1.8\times10^3\bar{g}$ and the bottom layer has $\nu_s=5\times10^{13}\bar{g}^2$ Hz, and $L_s^{CRe}=2.6\times10^{15}\bar{g}^2h$ erg s$^{-1}$. The electron half life exceeds the free-fall time, and the spectrum behaves like the thermal synchrotron one, $L_\nu\propto\nu^{1/3}$ below the peak down to $10^8$ Hz, where the thermal electrons cause synchrotron reabsorption. Since the cosmic ray flux postulated, with $h\sim f\sim1$, is in rough equipartition with the thermal gas of $n_a=10^{-3}$, $T_a=10^6$, it is unlikely that one could raise much further the cosmic ray luminosity. But notice that, for $\gamma_{ad}=5/3$, this already gives a total luminosity 3 orders of magnitude larger than the corresponding luminosity from thermal particles, and this luminosity is concentrated at much higher frequencies.

b) $n_a=1$ cm$^{-3}$, $T_a=10^4$ °K

In this case, Eq. (15) gives $E_{\max}=6\times10^3$ MeV, so both electrons and protons will be captured, and we assume all particles to be concentrated at the lowest energy in the spectrum, $d=g=1$. The electron half-lifes become shorter than the free fall time before bottom is reached, so assuming that the turbulent thermal flow does not replenish the cosmic ray electrons, we have to compute the electron luminosities at $x_{1/2}$, but the proton luminosities at $x=10^6$. For $\gamma_{ad}=5/3$, from (18) and (19) we have $\gamma_{1/2}^e=1.75\times10^6$ at $x_{1/2}=1.4\times10^8$, and $\nu_s^e=1.4\times10^{20}g^2$ Hz, the luminosity being $L_s^e=1.5\times10^{27}g^2\cdot h$ erg s$^{-1}$. The proton synchrotron peak is at $\nu_s^p=1.7\times10^{21}d^2$, and $L_s^p=1.1\times10^{22}d^2f$ erg s$^{-1}$. For $\gamma_{ad}=4/3$, $\gamma_{1/2}^e=4\times10^4$ at $x_{1/2}=3.1\times10^6$, so $\nu_s^e=10^{19}g^2$ Hz,

$L_s^e = 3\times10^{26}g^2h$ erg s$^{-1}$, while the proton peak is $\nu_s^p = 1.2\times10^{14}d^2$, $L_s^p = 8\times10^{14}d^2 f$ erg s$^{-1}$. The spectra again go as $\nu^{1/3}$ below the peak, until reabsorption occurs. Although the total cosmic ray luminosities are smaller than the thermal luminosities, in the X-ray and $\gamma$-ray band both the $\gamma_{ad}=5/3$ and 4/3 luminosities exceed the thermal luminosities, by several orders of magnitude. We discuss this further and compare with previous calculations in the next section.

c) $n_a = 10^3$ cm$^{-3}$, $T_a = 10^4$ °K

For $\gamma_{ad}=5/3$, the electron half life equals free-fall time for a Lorentz factor $\gamma_{1/2}^e = 5.5\times10^4$ g, at $x_{1/2}=4.4\times10^9$, so that $\nu_s^e = 5\times10^{16}g^2$ Hz and $L_s^e = 10^{27}g^2h$ erg s$^{-1}$. At $x=10^6$ the protons have $\gamma_f^p = 1.2\times10^7 d$, so $\nu_s^p = 5\times10^{22}d^2$ Hz and $L_s^p = 10^{25}d^2 f$ erg s$^{-1}$. For $\gamma_{ad}=4/3$, the electrons lose their energy at $x_{1/2}=3.2\times10^8$, where $\gamma_{1/2}^e = 3.8\times10^3 g$, so that $\nu_s^e = 8.5\times10^{15}g^2$ Hz and $L_s^e = 9\times10^{25}g^2h$ erg s$^{-1}$, and for the protons at $x=10^6$, $\gamma_f^p = 3.5\times10^3 d$, $\nu_s^p = 3.8\times10^{15}d^2$ Hz, and $L_s^p = 7.7\times10^{14}d^2 f$ erg s$^{-1}$. Among the other energy loss processes, cosmic ray proton collisions with thermal protons has a collision time longer than free fall time, so the luminosity is not very high. Estimating the cosmic ray energy density at $x=10^6$ and dividing by the mean collision time, we get a $\gamma$-ray luminosity from $\pi^0$ decay of $L_\gamma \cong 10^{27} f$ erg $^{-1}$. The position of the peak is very uncertain, since the multiplicity at proton energies of $\gamma^p \sim 10^7$ is unknown, but probably photon-photon processes degrade the spectrum down to the 10–20 MeV range, e.g. Herterich (1974). Electron-positron pair formation is in principle possible, since the thermal photons are energetic enough to be above threshold in the cosmic ray proton's frame of reference, and $(n_\gamma \sigma c)^{-1} \sim t_{ff}$ at the bottom, but the fact that $\Delta E/E$ per event is $\lesssim 10^{-3}$ reduces its efficiency significantly as an energy loss mechanism. Thus, in the $\gamma_{ad}=5/3$ case, the main proton loss mechanism would by synchrotron, producing high frequency $\gamma$-rays ($10^2$ MeV). Photon-photon reactions with either the thermal photons of the source, or the galactic diffuse background are likely to degrade these photons down to 10–20 MeV before they reach the observer. Comparing with the thermal spectrum, IVc), we see that, unlike for a) and b), the thermal X- and $\gamma$-ray luminosity exceeds that from cosmic rays, except in the range above 10 MeV. The same qualitative statement holds also if the cosmic ray flux at $x_a$ is raised to $f=10^3$ (its equipartition value), unless if cosmic ray electrons are regenerated in the flow.

d) $n_a = 10^7$ cm$^{-3}$, $T_a = 10^4$ °K

For this density, the equipartition magnetic field is so high that electrons lose their energy before they have gone down significantly; for $\gamma_{ad}=5/3$, $x_{1/2}=4.4\times10^{11}$, $\gamma_{1/2}=5.5\times10^2 g$, $\nu_s^e = 2.2\times10^{12}g^2$ and $L_s^e = 2.4\times10^{25}g^2\cdot h$ erg s$^{-1}$, while for $\gamma_{ad}=4/3$, $x_{1/2}=1.5\times10^{11}$, $\gamma_{1/2}^e = 1.8\times10^2 g$, $\nu_s^e = 8\times10^{11}g^2$ and $L_s^e = 2\times10^{33}g^2$h erg s$^{-1}$. Cosmic ray protons are subject to $\pi^0$ decay from collisions with thermal protons as the main energy loss mechanism, since $(n_a\sigma_\pi c)^{-1} \sim t_{ff}$ even at the accretion radius. Since both $n_a^{-1}$ and $t_{ff}$ are proportional to $x^{3/2}$, this continues being true as the gas moves inwards, and all cosmic ray protons captured should lead to $\pi^0$ decay $\gamma$-rays, on a free-fall time scale. Computing the number of protons captured per second with the postulated flux, and converting that to 20 MeV photons (one per proton, as just above threshold the multiplicity is one) we get a $\gamma$-ray luminosity $L_\gamma \cong 3\times10^{25} b$ erg. Unlike in cases a) and b), we again find that the cosmic ray luminosities are below the thermal luminosity, except above the MeV range. This conclusion is valid if the only cosmic rays in the accreting flow are those captured at the accretion radius, with a flux similar to the solar neighbourhood's. The cosmic ray flux, in a binary mass transfer flow, or near the galactic center, may however be much higher than this, and the acceleration of cosmic rays in situ, in the accretion flow, is also a possibility, discussed briefly in the next section.

## VI. Discussion

The luminosities we have computed are in all cases higher than those of previous authors, due to both the inclusion of dissipative heating and cosmic ray radiation. The first effect gives higher temperatures near the Schwarzschild radius than obtained with adiabatic heating only. When we correct for the fact that Shvartsman took $x_a = 10^{14}\beta$, where $\beta \sim 1$, and Shapiro took $x_a = 4.8\times10^{13}$, whereas here we took $x_a = 1.2\times10^{13}$ because of the other energy forms included, we find that our own temperatures are one order of magnitude higher at the Schwarzschild radius. Our conclusions about the magnitude of the disagreement between the temperature obtained including dissipative effects, and not including it, remain valid even if we took a larger accretion radius as these authors did, since then not only the "adiabatic" temperature, but also the "dissipative" temperature would increase by similar amounts (slopes remain similar in Fig. 1, only the $x$ scale shifts.) Since the optically thin synchrotron luminosity is $\propto T^2$, our luminosity is $10^2$ higher, and the peak ($\nu_s \propto T^2$) is displaced to higher frequencies by $\sim 10^2$, into the optical range. Bremsstrahlung luminosities, being proportional to $T$, are up by one order of magnitude, as well as the maximum photon energy ($\gamma$-ray range). In the high accretion rate cases, $n_a \gg 10^3$ cm$^{-3}$, when the cooling time is much shorter than the free fall time, the difference is many orders of magnitude, because whereas with adiabatic heating only a modest temperature rise is expected (Shvartsman, 1971; Tamazawa *et al.*, 1975), giving $L \sim 10^{33}$ erg s$^{-1}$, the inclusion of dissipative heating leads to a strong temperature rise.

giving for

$n_a = 10^7\,\mathrm{cm}^{-3}$, $T_a = 10^4\,°\mathrm{K}$, $L \cong 2\times10^{38}\,\mathrm{erg\,s}^{-1}$.

These results could be subject to quantitative, though not qualitative changes, by virtue of the fact that $\delta$, in the expression for the dissipative heating $\Gamma$, is a numerical factor which is difficult to estimate properly, depending as it does on the nature of the turbulent spectrum, the field line reconnection process, and its interpretation as a viscosity. This, of course, is the same problem encountered in describing viscous disks. A value of $\delta \sim 1$ seems reasonable, and values an order of mangitude smaller might not be surprising either. However, it seems unlikely that it could go much below this, since that might require field line reconnection to proceed at a much faster rate than is allowed by the Alfvén speed, and such a process is not known at present. Some quantitative changes could also be introduced by the fact that our flow dynamics is approximated by the free-fall velocity everywhere. Previous comparisons of numerical work with this assumption have shown it to be quite good (Shapiro, 1973a), for the case of adiabatic heating only. It seems unlikely that the inclusion of dissipative heating could change this conclusion much, and basing ourselves on these previous investigations we might expect uncertainties in the velocity, density, etc., by a factor $\lesssim 2$, which would not affect significantly the spectra presented here.

The inclusion of cosmic ray effects also leads to differences. If we consider only the effect of the cosmic rays captured from the galactic diffuse background as observed near the Sun, and if the cosmic ray effective adiabatic index is 5/3 as argued (due to interaction with turbulent magnetic irregularities), then in the case a) typical of a black hole in a galactic halo of $n_a = 10^{-3}\,\mathrm{cm}^{-3}$, $T_a = 10^6\,°\mathrm{K}$, the total luminosity is five orders of magnitude larger than that which would be computed using thermal particles only and adiabatic heating. In case b), typical of a normal H II region, the X-ray and $\gamma$-ray luminosity due to cosmic rays is four orders of magnitude higher than Shapiro's or Shvartsman's if $\gamma_{ad} = 5/3$, and two orders of magnitude if $\gamma_{ad} = 4/3$. In the higher accretion rate cases c) and d), the galactic cosmic ray electrons lose all their energy, at large radii above the black hole, so in the absence of regeneration of electrons, in these two cases cosmic rays are important only in the range above $\sim 10\,\mathrm{MeV}$. The dissipative heating of thermal particles, together with a short cooling time, ensures that in any case these systems are radiating at close to unit efficiency.

The question of whether cosmic ray electrons are constantly being accelerated in the turbulent accretion flow, over and above that which is captured from the interstellar medium, is not clear. Turbulent plasmas are probably able to generate a cosmic ray spectrum (Kaplan and Tsytovich, 1972), but quantitative estimates are model dependent, and difficult to apply to our situation. The fact that in the interstellar medium, and in some (but not all) radio sources one has approximate equipartition between cosmic ray energy and other (especially magnetic) energy forms is suggestive but not conclusive. It is probably correct that the adiabatic compression law of cosmic rays corresponds to an effective index of $\gamma_{ad} = 5/3$ as argued in V. This would ensure that, if cosmic rays were in equipartition at the accretion radius, the protons at least would stay in equipartition all the way down to the Schwarzschild radius, for cases a)–c) (this latter with $f \simeq 10^3$). However in case d) even the protons would be rapidly eliminated by inelastic proton-proton collisions ($\pi^0$ decay), and would fall below equipartition unless if regenerated. If such regeneration occurred, all cosmic ray luminosities would have to be re-evaluated, and would be comparable in all cases to the thermal luminosities computed.

If one assumed that the dissipation of magnetic energy via field line reconnection, instead of heating the ambient gas, went exclusively into accelerating fast particles, one would still obtain a luminosity much larger than that found by previous investigators. Near the Schwarzschild radius, the electron half-life against synchrotron losses in an equipartition magnetic field is shorter than the free-fall time for a Lorentz factor $\gamma^e > 10^4 \theta^{3/4} n_a^{-1}$, which does not seem difficult to achieve with most acceleration mechanisms, so all the dissipated energy would be converted into radiation. If protons were accelerated as well, for $\gamma^p > 5\times10^{13}\cdot \theta^{3/2} n_a^{-1}$ their half-life against synchrotron losses would also be less than the free-fall time, and at least for large values of $n_a$, the relativistic proton energy made available by the dissipative process is also radiated away. Other mechnisms, such as pair-production, etc., may be of importance as well, but in the absence of a definite theory of acceleration, neither the cosmic ray energy spectrum, nor the radiation spectrum are easy to estimate. Since, however, it appears that at least the synchrotron mechanism is sufficient to radiate away the cosmic ray energy, the total luminosity would be $L \sim 4(3r_s)^3 \cdot \Gamma(3r_s)\,\mathrm{erg\,s}^{-1}$, similarly to the case when dissipation goes into heating the ambient gas.

Calculations on a spherical model with field line reconnection ressembling ours are mentioned by Illarionov and Sunyaev (1975), without however giving details. As no preprint seems yet to be available, a comparison with our results is difficult. Their reported $\gamma$-ray luminosities of $L_\gamma \sim 10^{34}$–$10^{37}\,\mathrm{erg\,s}^{-1}$ in binary systems, much larger than ours, may be achieved for instance if one assumes proton temperatures uncoupled from that of the electrons, as in Dahlbacka *et al.* (1974). In this case thermal protons have $\gamma_{ad} = 5/3$, and could heat up to temperatures $T_p \gtrsim 100\,\mathrm{MeV}$, which then gives through $\pi^0$ decay a $\gamma$-ray luminosity well in excess of that computed in Section V, d for captured cosmic rays (with our assumption $T_e \simeq T_p$, $\alpha \simeq 1$, thermal protons stay well below 100 MeV, especially for $n_a \gtrsim 10^3\theta^{3/2}$ m$^{-1}$ cm$^{-3}$).

The possibility of observing isolated black holes is dependent on their distance, and on the density and temperature of the gas out of which they are accreting. For a black hole in a gas of $n_a = 10^{-3}$ cm$^{-3}$, $T = 10^6$ °K, which might be in the galactic halo, we expect a luminosity of $L = 3 \times 10^{20}$ erg s$^{-1}$ at a frequency of $10^{19}$ Hz, which extends down to radio frequencies with a spectrum $L_\nu \propto \nu^{1/3}$, so that at $10^{14}$ Hz we have $L_\nu \sim 0.6$ erg s$^{-1}$ Hz$^{-1}$. This is so low that probably it is impossible to detect these objects in the halo with present techniques, even if one considers the integrated luminosity of $\sim 10^{12}$ black holes distributed throughout the halo. For a black hole in the galactic plane, or near the galactic center, the situation is more favourable. For $n_a = 1$ cm$^{-3}$, $T_a = 10^4$ °K, the luminosity has one component with $L^{th} \simeq 2 \times 10^{29}$ erg s$^{-1}$ at a frequency $\nu_s \sim 3 \times 10^{14}$, and another component (cosmic ray electrons with $\gamma_{ad} = 5/3$) of $L^{CR} \simeq 2 \times 10^{27}$ erg s$^{-1}$ at $\nu_s^{CR} \simeq 10^{20}$ Hz, both behaving as $L_\nu \propto \nu^{1/3}$ at low frequencies. At a distance of 1 kpc, this would give $10^{-28}$ erg cm$^{-2}$ s$^{-1}$ Hz$^{-1}$ in the $K$ or $L$ band (2.2 μ, 3.4 μ), and $4 \times 10^{-19}$ erg cm$^{-2}$ s$^{-1}$ in the 2–10 keV band. These infrared bands are accesible from Earth, and should be observable, but the X-ray flux is probably too low for individual sources to be picked out, although the integrated flux from many sources may be measurable if there are enough of them. The characteristic spectrum $L_\nu \propto \nu^{1/3}$ may be useful in distinguishing the black holes from other stellar or diffuse components. A black hole submerged in a medium with $n_a = 10^3$ cm$^{-3}$, $T_a = 10^4$ °K, has a component of $L^{th} \simeq 2 \times 10^{34}$ erg s$^{-1}$ at $\nu_s \simeq 1.1 \times 10^{15}$ Hz, with spectrum of slope $L_\nu \propto \nu^{1.3}$ at lower frequencies, flattening out gradually and dropping exponentially above the peak. It has also a $\gamma$-ray component of $5 \times 10^{29}$ erg s$^{-1}$ extending to 10 MeV with a logarithmically rising spectrum below the peak. In the 2–10 keV band, the luminosity is $10^{26}$ erg s$^{-1}$ with a spectrum $L_\nu \propto \nu^{-1/2}$. The X- and $\gamma$-ray luminosities are again low, but the infrared fluxes should be measurable without difficulty.

Finally, black holes immersed in a medium with $n_a = 10^7$ cm$^{-3}$, $T_a = 10^4$ °K, may have been observed already, in the form of binary X-ray sources and/or $\gamma$-ray sources, since the luminosities are in the right range. The interesting feature that emerges from this calculation, is the optical and infrared spectrum that should be associated with these X-ray sources, if they are represented by the model of spherically symmetrically accreting black holes. For these parameters, the IR and optical spectrum goes as $L_\nu \propto \nu^{1.6}$ up to $\nu \sim 5 \times 10^{15}$ Hz, giving $L \sim 8 \times 10^{37}$ erg s$^{-1}$. The X-ray flux is $2 \times 10^{38}$ erg s$^{-1}$, but the spectrum is difficult to compute because of the presence of relativistic comptonization. The $\gamma$-ray flux, however, is straightforward, and gives a luminosity of $3 \times 10^{36}$ erg s$^{-1}$, extending to $\sim 2$ MeV with a spectrum $L_\nu \propto \nu^{-1/2}$ below 0.5 MeV, and $L_\nu \propto (\text{const} + \log \nu)$ above that. If $\delta$ in Eq. (8) were less than unity, one might reduce the maximum temperature achieved to $T \lesssim 4 \times 10^9$, and the $\gamma$-ray spectrum would than only have the $\nu^{-1/2}$ portion, followed by an exponential drop-off. Whether the comptonized portion of the spectrum extends to $\gamma$-ray energies is not clear, at present. A very naive picture of the comptonization process would lead one to expect a neative power law of index about $-1$ in the X-ray range, followed by an exponential drop-off. Cygnus X-1, and several other sources as well, seem to have spectra steeper than $\nu^{-1/2}$ at energies below 100 keV, so one may conclude that to reproduce the observations of these binary X-ray sources, a Thomson optical depth greater than unity is required in the spherical accretion models. In a paper to appear in Nature, a spherical model with comptonization for Cygnus X-1 is discussed further.

*Acknowledgements.* The author is grateful to Prof. M. J. Rees for discussions and advice on this problem, and has benefited from helpful comments from A. C. Fabian, D. Lynden-Bell, J. E. Pringle, E. T. Scharlemann, E. E. Salpeter, E. Spiegel and J. A. J. Whelan.

## References

Bekefi, G. 1966, Radiation processes in plasmas, Wiley, New York
Dahlbacka, G. H., Chapline, G. F., Weaver, T. A. 1974, *Nature* **250**, 36
Eardley, D. M., Lightman, A. P., Shapiro, S. L. 1975 (preprint)
Ginzburg, V. L., Syrovatskii, S. E. 1964, The Origin of Cosmic Rays, MacMillan, New York
Ginzburg, V. L., Syrovatskii, S. I. 1964, *J.E.T.P.* **18**, 245
Ginzburg, V. L. 1969, Elementary processes for cosmic ray astrophysics, Gordon and Breach, New York
Greisen, K. 1966, *Phys. Rev. Letters* **16**, 748
Herterich, K. 1974, *Nature* **250**, 311
Hirschfield, J. L., Baldwin, D. E., Brown, S. C. 1961, *Phys. Fluids* **4**, 198
Illarionov, A. F., Sunyaev, R. A. 1975, *Astron. & Astrophys.* **39**, 185
Jackson, J. D. 1962, Classical Electrodynamics, Wiley, New York
Kaplan, S. A., Tsytovich, V. N. 1973, Plasma Astrophysics, Pergamon, Oxford
Maxon, S. 1972, *Phys. Rev.* **5**, 1630
Mészáros, P. 1975, *Astron. & Astrophys.* **38**, 5
Pacholczyk, A. G. 1970, Radio Astrophysics, Freeman, San Francisco
Pringle, J. E., Rees, M. J. 1972, *Astron. & Astrophys.* **21**, 1
Shvartsman, V. F. 1971, *Soviet Astron. J.* **15**, 37
Shakura, N. J., Sunyaev, R. A., 1973, *Astron. & Astrophys.* **24**, 337
Shapiro, S. L. 1973a, *Astrophys. J.* **180**, 531
Shapiro, S. L. 1973b, *Astrophys. J.* **185**, 69
Tamazawa, S., Toyama, K., Kaneko, N., Ono, Y. 1975, *Astrophys. Space Sci.* **32**, 403
Trubnikov, B. A. 1958, *Soviet Phys. Doklady* **3**, 136
Zeldovich, Ya. B., Novikov, I. D. 1971, Relativistic Astrophysics, Vol. I, U. of Chicago Press, Chicago

P. Mészáros
Max-Planck-Institut für Physik und Astrophysik
D-8000 München 40
Föhringer Ring 6
Federal Republic of Germany

*Mon. Not. R. astr. Soc.* (1976) **175**, 43–60.

# X-RAY EMISSION FROM ACCRETION ON TO WHITE DWARFS

*A. C. Fabian, J. E. Pringle and M. J. Rees*

Institute of Astronomy, Madingley Road, Cambridge

(Received 1975 October 24; in original form September 25)

SUMMARY

We have extended calculations by Hoshi and by Aizu to produce a self-consistent model for X-radiation from accreting, possibly magnetized, white dwarfs. To generate keV X-rays the flow must be radial on to the stellar surface. We expect X-ray luminosities to be in the range $10^{32}$–$10^{36}$ erg s$^{-1}$, and the spectra to be quasi-bremsstrahlung with $kT \sim 30$–$100$ keV and with a substantial low energy cut-off. The optical (bolometric) luminosity should be comparable to that emitted in the X-rays. If the white dwarf is magnetized a comparable, or greater, amount can be radiated as cyclotron emission in the infrared or optical.

## 1. INTRODUCTION

The emission of X-radiation from accretion on to compact objects has been intensively studied over the past few years. X-ray temperatures ($T \gtrsim 10^7$ K) are reached if the kinetic energy due to infall of gas on to the surface of an object of solar mass, and radius $R \lesssim 3 \times 10^{10}$ cm, is converted via a shock into internal energy. White dwarfs, neutron stars and the possibly more massive black holes are therefore all candidates. The weaker gravitational field at the surface of a white dwarf, compared to a more compact object, means that a much higher mass flux ($\times 10^3$) is required to produce the same luminosity. The much larger surface area makes the blackbody temperatures of even luminous sources, $L_X \sim 10^{38}$ erg s$^{-1}$, less than $10^6$ K, and so in order to radiate X-rays the infalling matter must pass through a shock and emit *optically thin* bremsstrahlung. Consequently, most theoretical work on the detected compact X-ray sources has concentrated on neutron star or black hole accretion.

We investigate X-ray emission from accretion on to white dwarfs without regard to specific observed sources, extending earlier work by Hoshi (1973) and Aizu (1973). Shapiro & Salpeter (1975) have recently discussed the analogous problem for neutron stars, though without including a magnetic field. Electron scattering and absorption in the infalling material are considered, as well as cyclotron emission in the cases where strong ($B \gtrsim 10^6$ G) magnetic fields are present. We find that the X-ray luminosities are unlikely to exceed $\sim 10^{36}$ erg s$^{-1}$ and that the bremsstrahlung temperatures are comparatively high ($kT \sim 50$ keV). Accreting white dwarfs may produce X-ray sources that have not yet been detected.

In Section 2 we review and discuss the general properties of accretion shocks above the surface of white dwarfs when magnetic fields are negligible. The magnetic fields may, however, be strong enough to channel the inflow towards the magnetic polar caps, and in this situation *cyclotron cooling* is competitive with bremsstrahlung.

The influence of cyclotron emission on the location of the shock and on the expected spectrum is considered in Section 3; and in Section 4 we discuss how the magnetic field affects the accretion flow and calculate the ranges of parameters which permit 'self-consistent' X-ray emission from accreting white dwarfs. The details of Section 4 are summarized in Figs 2 and 3. Applications to various observed systems —e.g. DQ Her, Mira B and transient X-ray sources—are made in Section 5.

## 2. SHOCK STRUCTURE WITH BREMSSTRAHLUNG COOLING

Consider material accreting on to a white dwarf of mass $M$ and radius $R$, at a rate $F$. Then the luminosity due to accretion alone is

$$L = \frac{FGM}{R} = 1{\cdot}3 \times 10^{37}\, F_{20} \left(\frac{M}{M_\odot}\right) R_9 \text{ erg s}^{-1} \qquad (1)$$

where $R_9$ is the stellar radius in units of $10^9$ cm and $F_{20}$ is the accretion rate in units of $10^{20}$ g s$^{-1}$ = $1{\cdot}5 \times 10^{-6}\, M_\odot$ yr$^{-1}$. From the mass–radius relation for white dwarfs given by Webbink (1975) and the assumption that the white dwarf is composed of pure helium, the luminosity per gram can be derived as a function of white dwarf mass (Fig. 1). For generality we define $f$ as the fraction of the stellar surface area on to which the material accretes. In the spherically symmetric situation $f = 1$, but if the white dwarf has a strong enough magnetic field to affect the accretion flow then in general $f < 1$. The velocity of matter falling freely on to the

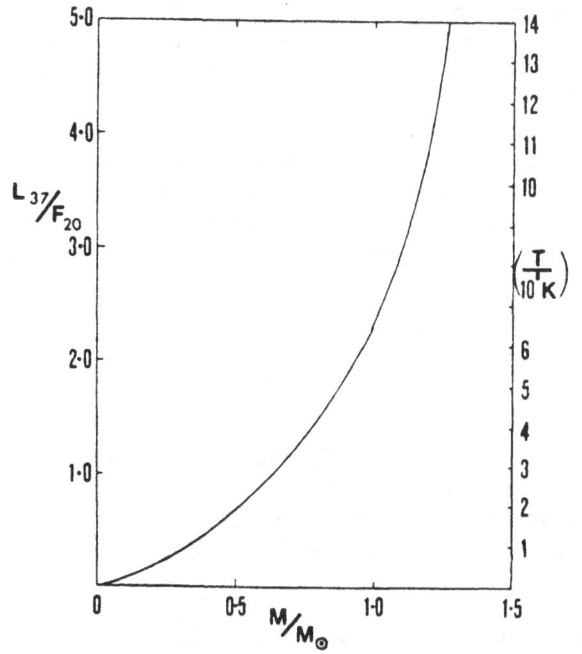


FIG. 1. *The ratio of luminosity to mass flow rate, $L_{37}/F_{20}$ (left-hand scale) and shock temperature, $T_s$, in units of $10^8$ K (right-hand scale) plotted as a function of white dwarf mass. The mass–radius relation for (cold) white dwarfs of Webbink (1975) has been assumed.*

star is

$$V = \left(\frac{2GM}{R}\right)^{1/2} = 5\times10^{8}\left(\frac{M}{M_{\odot}}\right)^{1/2} R_9^{-1/2}\ \mathrm{cm\ s^{-1}} \tag{2}$$

and, if the mass flux is uniform and radial over the accreting area, the density is

$$\rho_1 = \frac{F}{4\pi R^2 . f . V} = 1{\cdot}5\times10^{-8} F_{20} f^{-1}\left(\frac{M}{M_{\odot}}\right)^{-1/2} R_9^{-3/2}\ \mathrm{g\ cm^{-3}}. \tag{3}$$

At a distance $D$ above the surface of the star, a standing shock forms (see, e.g. Hoshi 1973; Aizu 1973). Before encountering the shock, the gas will in general be so cool that its inflow velocity is highly supersonic. The shock is therefore ' strong ', causing the density to increase by 4, and the inward velocity to decrease by the same factor. This is so provided that the optical depth through the shock thickness is less than unity. Otherwise there is also a solution in which the accretion energy is thermalized and re-radiated by the white dwarf atmosphere. If we take the shock thickness to correspond to $\sim 10^{-2}\ \mathrm{g\ cm^{-2}}$ then this has an optical depth of unity if the atmospheric temperature $\sim T_{cr} = 2{\cdot}5\times10^4$ K (Allen 1973). Thus for a standing shock and possible X-ray emission we require $T_{cr} < T_b$, where $T_b$ is defined by equation (7), i.e.

$$F_{20} > 2{\cdot}1\times10^{-5} f (M/M_{\odot})^{-1} R_9^{3}.$$

The value of $D$, and the density and velocity profile in the subsonic region, are determined by the condition that the gas should have time to cool as it ' settles ' on to the stellar surface. $D$ varies inversely with $\rho$, and for most cases of interest $Ff^{-1}$ will be high enough to ensure that $D \ll R$. The density of the shocked region is then $\rho_2 = 4\rho_1$, where $\rho_1$ is given by (3) for $R = R_*$. Note, however, that even if $D \gtrsim R$ most of the energy is still released at radii between $R$ and $2R$, and the temperature and density structure in this region is hardly changed even if $D \gg R$ (see Appendix).

When the dominant radiation process for the shocked material is bremsstrahlung it is possible to find an approximate analytic profile for the flow below the shock (Aizu 1973). For our purposes, however, it is adequate to assume that the material in the region between the shock and the stellar surface has uniform density and temperature. We take the temperature to be given by the shock temperature (see Fig. 1):

$$T_s = \frac{3}{8}\frac{GMm_p\mu}{kR}$$

$$= 3{\cdot}7\times10^{8}\left(\frac{M}{M_{\odot}}\right) R_9^{-1}\ \mathrm{K} \tag{4}$$

where $m_p$ is the proton mass, $k$ is Boltzmann's constant, and we have taken the mean molecular weight $\mu = 0{\cdot}615$ corresponding to a hydrogen mass fraction of $X = 0{\cdot}7$ in the accreting material. When bremsstrahlung is the dominant post-shock cooling mechanism, the distance $D_{ff}$ is given by

$$L = f . 4\pi R^2 . D_{ff}\epsilon_{ff} \tag{5}$$

where $\epsilon_{ff}$ is the bremsstrahlung cooling rate. Hence

$$D_{ff} = 3{\cdot}3\times10^{7} F_{20}^{-1} f (M/M_{\odot})^{3/2} R_9^{1/2}\ \mathrm{cm}. \tag{6}$$

This agrees well with the more detailed calculation of Aizu (1973).

The bremsstrahlung cooling time is *longer* than the time scale for equalizing electron and proton temperatures via coulomb encounters. Thus we are here justified in assuming that the electron temperature does indeed attain a value $T_s$, and that the shock thickness is $\ll D_{ff}$, even if there is no enhanced coupling between protons and electrons due to plasma oscillations. This is not necessarily still true, however, if cyclotron or Compton cooling exceed bremsstrahlung, and we discuss this further in the next section.

Thermal conduction of energy from the shocked region into the star can be neglected. The fraction of radiated luminosity lost to thermal conduction is approximately

$$\frac{L_c}{L_x} = f.4\pi R^2.K.TD^{-1}L_x^{-1}$$

$$= 2{\cdot}8\times 10^{-2}\, F_{20}{}^{7/2}f^{-7/2}\left(\frac{M}{M_\odot}\right)^{5/2} R_9{}^{-5/2},$$

where $K = 1{\cdot}0\times 10^{-6}\, T^{5/2}$ is the thermal conductivity (Allen 1973). This is much less than unity for the cases of interest. There remains the possibility, however, that there is another *self-consistent* solution with $D$ much smaller than above in which the dominant energy loss from the shocked region is due to thermal conduction down into the star. Presumably this energy would emerge as thermalized radiation from the stellar surface. If the radiation released were radiated as a black body, the temperature of the radiating region would be

$$T_b = \frac{L}{f.4\pi R^2.\sigma} = 3{\cdot}5\times 10^5\, F_{20}{}^{1/4}f^{-1/4}\left(\frac{M}{M_\odot}\right)^{1/4} R_9{}^{-3/4}\,\mathrm{K} \qquad (7)$$

where $\sigma$ is Stefan's constant. Thus, for the cases we shall be interested in ($F_{20}\lesssim 1$), $T_b \ll T_s$ and the emitting region is optically thin to bremsstrahlung. However, as a number of authors (de Gregoria 1974; Hayakawa 1974) have pointed out, the spectrum of the emerging radiation can be seriously affected by its outward passage through the accreting material. The optical depth vertically through the emitting region due to electron scattering is

$$\tau_{es}(D) = 0{\cdot}2(1+X)\,\rho_2 D_{ff}$$

$$= 0{\cdot}7(M/M_\odot)\, R_9{}^{-1} \qquad (8)$$

independent of the accretion rate. We may therefore neglect the effect of electron scattering in the radiating region. The value to infinity along a radius is

$$\tau_{es}(\infty) = \int_R^\infty 0{\cdot}2(1+X)\,\rho_1(r)\,dr$$

$$= 10F_{20}f^{-1}(M/M_\odot)^{-1/2}\, R_9{}^{-1/2}. \qquad (9)$$

As the calculations of de Gregoria (1974) show, when $\tau_{es}(\infty) > 1$, inelastic Compton scattering degrades all photons with energies

$$E \gtrsim m_e c^2/\tau_{es}{}^2(\infty).$$

For $f < 1$, the photons can escape sideways through the accretion column. The radius of the column $d \approx \sqrt{2f}R$ and then the optical depth sideways through the column is

$$\tau_{es}(d) = 7{\cdot}4F_{20}f^{-1/2}(M/M_\odot)^{-1/2}\, R_9{}^{-1/2}. \qquad (10)$$

When $D \ll d$, the photons must all escape through the cool unshocked material and the inelastic scattering criterion can be applied using $\tau_{es}(d)$. (In general $D \ll d$ for $\tau_{es}(d) \gtrsim 1$.) Usually, however, the X-ray flux is insufficient to maintain the heavy elements in the incoming material ionized (Hayakawa 1974) and photo-electric absorption cannot be ignored. We discuss this process quantitatively in Section 5.4.

## 3. INFLUENCE OF MAGNETIC FIELD

We now consider in more detail the case ($f < 1$) when the accretion flow is funnelled down the magnetic field on to the magnetic poles of a white dwarf. We take the strength of the surface field at the poles to be $B = 10^6 B_6$ gauss and initially treat $f$ and $B$ as independent parameters. We estimate later (Section 4) how $f$ and $B$ can be related. As noted by Ichimaru & Nakano (1973), cyclotron losses may now dominate the emission process. Inoue & Hoshi (1975) considered the case $f < 1$, but did not take account of the cyclotron process. The cyclotron emission rate is

$$\epsilon_c = n_e \frac{B^2}{8\pi} 2\sigma_T c \left(\frac{v}{c}\right)^2 \tag{11}$$

where $\sigma_T$ is the Thomson cross-section and $v = 6 \times 10^5 T_e^{1/2}$ cm s$^{-1}$ is the rms electron velocity. We may then define a dimensionless ratio

$$\gamma = \frac{\epsilon_c}{\epsilon_{ff}} = 1 \cdot 9 \times 10^2 F_{20}^{-1} f \left(\frac{M}{M_\odot}\right) R_9 B_6^2. \tag{12}$$

If the emitting region is optically thin to cyclotron radiation, cyclotron losses may be ignored only if $\gamma < 1$, that is if

$$F_{20} > 1 \cdot 9 \times 10^2 f(M/M_\odot) R_9 B_6^2. \tag{13}$$

The optical depth to cyclotron emission is not easy to assess, since the details of the radiative transfer are difficult to calculate. For our present purposes, however, we may obtain a rough estimate in the following manner. Define a cyclotron emission 'temperature' $T_c$ by

$$T_c = h\nu_c/k = 1 \cdot 3 \times 10^2 B_6 \text{ K}$$

where $\nu_c = 2 \cdot 8 \times 10^{12} B_6$ Hz is the cyclotron frequency. Then the requirement that the intensity of radiation at frequency $\nu_c$ be less than that given by a Planck spectrum with temperature $T_c$ yields an approximate optical depth†

$$\tau_c = T_b^4 / T_e T_c^3. \tag{14}$$

When $D \gtrsim d$, we should take account of the fact that the effective surface area of the emitting region is increased by a factor $\phi = (1 + 2D/d)$, and modify (7) accordingly by including an extra factor $\phi^{-1/4}$. The opacity to cyclotron radiation then exceeds unity (that is $\tau_c > 1$) if

$$F_{20} > 4 \cdot 8 \times 10^{-8} f R_9^2 B_6^3 \phi. \tag{15}$$

† This estimate for $\tau_c$ depends on the shape and breadth of the cyclotron emission spectrum. The spread of magnetic field strength in the emitting region, and thermal doppler broadening, widen the cyclotron line such that $\Delta\nu_c/\nu_c \gtrsim 0 \cdot 3$. Thus the approximation of the cyclotron emission spectrum by a Planck curve with temperature $T_c$ is probably not unreasonable.

When $\tau_c > 1$, the effective ratio of cyclotron to free–free luminosity is

$$\gamma_1 = \gamma\tau_c^{-1}$$
$$= 9{\cdot}3\times 10^{-6} f^2 F_{20}^{-2}(M/M_\odot)\,R_9^3\phi.$$

For X-ray emission to dominate, we require $\gamma_1 < 1$, that is

$$F_{20} > 3\times 10^{-3} f(M/M_\odot)^{1/2}\,R_9^{3/2}B_6^{5/2}\phi^{1/2}. \tag{16}$$

We must now check our assumption that $T_e = T_s$. It is, for example, conceivable that cyclotron emission is so efficient that the ion and electron temperatures can never reach equality after the shock. By comparing the relevant time scales we find that $T_e = T_s$ provided that

$$F_{20} > 8\times 10^{-4} f(M/M_\odot)\,R_9B_6^{5/2}\phi^{1/2}. \tag{17}$$

When $\gamma_1 = 1$, $T_e = T_s$ provided that

$$(M/M_\odot)\,R_9 > 6{\cdot}3\times 10^{-2}, \tag{18}$$

which is satisfied for white dwarfs.

When $T_e < T_s$ the efficiency of cyclotron emission ($\propto T_e$) relative to bremsstrahlung emission ($\propto T_e^{1/2}$) is reduced, although, for the parameters we consider, the cyclotron is always optically thick. The electron temperature $T_e$ is calculated by balancing the (optically thick) cyclotron emission loss with the rate at which the ions heat the electrons by collisional heating. We find

$$T_e = 2{\cdot}2\times 10^{10}F_{20}^{4/7}f^{-4/7}(M/M_\odot)^{3/7}\,R_9^{-4/7}B_6^{-10/7}\phi^{-2/7}. \tag{19}$$

In this situation the ratio of cyclotron to bremsstrahlung emission is given by

$$\gamma_2 = \gamma_1\left(\frac{T_e}{T_s}\right)^{1/2}$$
$$= 4{\cdot}4\times 10^{-3}\,F_{20}^{-8/7}f^{8/7}\left(\frac{M}{M_\odot}\right)^{1/7}R_9^{15/7}B_6^{20/7}\phi^{6/7}. \tag{20}$$

Note that when $T_e = T_s$ (and $\phi = 1$), $\gamma_1 = \gamma_2 = 15(M/M_\odot)^{-1}\,R_9$.

Electron energy loss due to scattering of the cyclotron photons can be ignored for the values of the parameters we consider here. About half of the bremsstrahlung radiation is emitted downwards and is absorbed in, and re-radiated thermally from, the stellar surface. Thus we expect the X-ray luminosity to be roughly $L_x = \frac{1}{2}\Gamma^{-1}L$ where

$$\Gamma = \begin{cases} 1 & \gamma_1 < 1 \\ \gamma_1 & \gamma_1 > 1 \text{ and } T_e = T_s \\ \gamma_2 & T_e < T_s \text{ (in this case the X-rays have a softer spectrum).} \end{cases}$$

Similarly there is an additional contribution to the luminosity from the stellar magnetic poles at a temperature $T_* = (2\Gamma)^{-1/4}\,T_b$, provided $T_* > T_c$.

We now estimate the parameter $\phi = 1 + 2D/d$. When bremsstrahlung is the dominant emission mechanism behind the shock, the height of the shock above the stellar surface $D = D_{ff}$ (defined in equation (6)). However, when cyclotron emission dominates, we have $D = D_{ff}\Gamma^{-1}$. It should be noted that for $\Gamma > 1$, $\Gamma$ and hence $D$ is a function of $\phi$, and the equation $\phi = 1 + 2D(\phi)/d$ must be solved properly to obtain $\phi$.

For self-consistency we must now check that, even when bremsstrahlung dominates over cyclotron emission, it also dominates over Compton scattering losses with ambient photons. The time scale for energy loss by an electron of energy $E$ to the Compton process is approximately $t_c = E/\langle\dot{E}\rangle$ where

$$\langle\dot{E}\rangle \simeq -\frac{4}{3}\sigma_T \frac{2c\rho_r}{m_e c^2} E. \tag{21}$$

Here $\rho_r$ is the radiation energy density. This is to be compared with $t_b$, the time scale for energy loss to bremsstrahlung, and we find

$$\frac{t_c}{t_b} \simeq \frac{7 \cdot 5 \times 10^{-5}\, n_e}{\rho_r T^{1/2}}. \tag{22}$$

We may neglect Comptonization if $t_c \ll t_b$.

The dominant source of ambient photons is the hot surface of the star where about one-half of the emitted X-ray photons are absorbed and thermalized. In this case

$$\rho_r \simeq \frac{L}{2f.4\pi R^2.c}$$

and we find $t_c/t_b \simeq 2 \cdot 4(M/M_\odot)^{-2}$, where we have assumed that the luminosity $L$ is *solely* due to accretion. This contrasts with the neutron star case (*cf.* Shapiro & Salpeter 1975) where $L/F$ and $T$ are higher than for white dwarf accretion. Thus we are justified in neglecting Comptonization for our purposes but it should not be ignored in more detailed calculations. If, however, the dwarf is intrinsically highly luminous and, in particular, if the accreted nuclear fuel is burnt in a continuous fashion, Comptonization is the *dominant* mechanism for electron energy loss (see also, Katz & Salpeter 1974) and the resulting hard X-ray flux is considerably reduced. We assume throughout, therefore, that nuclear burning of accreted matter is not taking place continuously (for a fuller discussion of the plausibility of this assumption see Bath *et al.* 1974).

## 4. ESTIMATE OF ALFVÉN RADIUS AND SELF-CONSISTENT MODELS

We now show how $f$ and $B$ may be related. In common with previous work on accretion by magnetized stars (for example, Lamb, Pethick & Pines 1973) we assume that the stellar field is dipolar and that it controls the accretion flow within the Alfvén radius $R_M$. If inside $R_M$ material flows only along those field lines that are open at $R_M$, it is easy to show that $d/R \approx (R/R_M)^{1/2}$ and thus $f \approx R/2R_M$. The precise way in which infalling matter distorts the stellar field and diffuses across field lines is exceedingly complex, although Arons & Lea (1976) find that the above relationships are valid in at least an order of magnitude sense. If the accretion is spherically symmetric outside $R_M$ we can obtain an estimate for $R_M$ by balancing the magnetic pressure and the ram pressure of infall at that radius.

This yields

$$\frac{R_M}{R} = 2 \cdot 3\, F_{20}^{-2/7} \left(\frac{M}{M_\odot}\right)^{-1/7} R_9^{5/7} B_6^{4/7}$$

or

$$f_{sph} = 0 \cdot 2\, F_{20}^{2/7} \left(\frac{M}{M_\odot}\right)^{1/7} R_9^{-5/7} B_6^{-4/7}.$$

If, however, the flow outside $R_M$ is in the form of an accretion disc, this is probably an overestimate of $R_M$. Using the estimate of Bath, Evans & Pringle (1974) for this case we find

$$f_{disc} = 0.2\, F_{20}^{7/27}(M/M_\odot)^{7/27}\, R_9^{-7/9} B_6^{-16/27}. \tag{23}$$

In addition, if the magnetized white dwarf rotates with a period $P$ (min), the 'corotation radius' $R_\Omega$ at which the centrifugal force of matter attached to field lines balances the gravitational pull of the white dwarf is given by

$$\frac{R_\Omega}{R} = 2.3\, P^{2/3} \left(\frac{M}{M_\odot}\right)^{1/3} R_9^{-1}. \tag{24}$$

Roughly speaking, accretion may only take place if $R_M < R_\Omega$ (see Section 5.3). We define

$$f_\Omega = \frac{R}{2R_\Omega} = 0.2\, P^{-2/3} \left(\frac{M}{M_\odot}\right)^{-1/3} R_9. \tag{25}$$

Thus the condition that a white dwarf accretes material and is a pulsing source can be written as $f_\Omega < f < 1$. This is possible if

$$P \gtrsim 0.4(M/M_\odot)^{-1/2}\, R_9^{3/2} \qquad (f = f_{disc}) \tag{26}$$

or

$$P \gtrsim 0.1(M/M_\odot)^{-1/2}\, R_9^{3/2} \qquad (f = f_{sph}). \tag{27}$$

We can now use the above estimates of $f$ to construct some models of accreting magnetized white dwarfs, and the radiation they emit, which are at least self-consistent in the context of our simplifying assumptions. The results are summarized in Figs 2 and 3.

(a) *Spherically symmetric flow outside $R_M$*

The conditions derived in this section are illustrated in Fig. 2 for a 1 $M_\odot$ white dwarf. The condition $\gamma_1 = 1$, that cyclotron and bremsstrahlung emission balance, is

$$F_{20} = 3.5 \times 10^{-5}(M/M_\odot)^{9/10} R_9^{11/10} B_6^{27/10}, \tag{28}$$

provided that $\phi = 1$. The regions in which $T_e = T_s$ and $T_e < T_s$ are separated by the line $\gamma_1 = \gamma_2$ which (for $\phi = 1$) is

$$F_{20} = 5 \times 10^{-6}(M/M_\odot)^{6/5}\, R_9^{2/7} B_6^{27/10}. \tag{29}$$

On both these lines $\phi = 2$ when $D_{ff} = 0.5$ d, that is when

$$F_{20} = 1.1 \times 10^{-2}(M/M_\odot)^{11/6}\, R_9^{-17/12} B_6^{-1/3}. \tag{30}$$

When $\phi \gg 1$, $\gamma_1$ = constant implies that $F \propto B^{25/16}$, and $\gamma_1 = \gamma_2$ implies that $F \propto B^{1123/488}$.

In the region $T_e < T_s$, the lines $\gamma_2$ = constant are given (for $\phi = 1$) by

$$F_{20} = 1.5 \times 10^{-4}(M/M_\odot)^{3/8}\, R_9^{13/8} B_6^{27/10} \gamma_2^{-49/40}. \tag{31}$$

On these lines $\phi = 2$ when $D_{ff} = 0.5\gamma_2$ d which implies $F \propto B^{-61}$. For $\phi \gg 1$ (and $T_e < T_s$) $\gamma_2$ = constant implies $F_2 \propto B^{144/121}$. Note that, in the region $\phi \gg 1$, lines of constant cyclotron to bremsstrahlung ratio can switch between $\gamma_1$ = constant and $\gamma_2$ = constant by crossing the line $\gamma_1 = \gamma_2$.

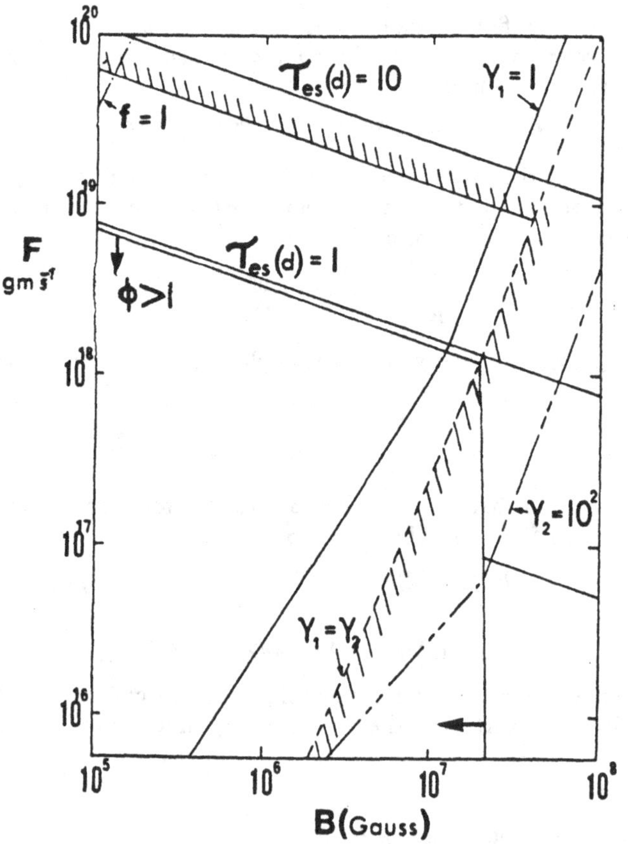


FIG. 2. *The mass-flow-rate F, plotted as a function of surface magnetic field, for various criteria (see Section (4)). The accretion flow beyond the Alfvén radius is assumed spherically symmetrical, and the white dwarf is taken to have mass $M_\odot$ and to be non-rotating. The magnetic field is dynamically important at the surface (a necessary condition for pulsed emission) to the right of the $f = 1$ line (equation (33)). X-rays above 10 keV are degraded above the $\tau_{es}(d) = 7$ line (equation (32)). In the region marked $\phi > 1$, the shock in the accretion column occurs at a height $\gtrsim f^{1/2}R$ above the surface, so the radiation escapes predominantly through the sides of the shocked region. To the right of the line $\gamma_1 = 1$ (equations (16) and (28)) cyclotron cooling dominates bremsstrahlung, and the efficiency of hard X-ray emission is $\sim \frac{1}{3}\gamma_1^{-1}$. To the right of the $\gamma_1 = \gamma_2$ line (equation (29)) cyclotron cooling is so efficient that it prevents the electrons from ever attaining the shock temperature $T_s$ and thus suppresses hard X-ray emission (unless collective effects enhance the coupling between electrons and ions). On the line $\gamma_2 = 100$, the ratio of cyclotron and bremsstrahlung cooling, calculated for the 'self-consistent' electron temperature $T_e < T_s$, is 100:1. These considerations lead us to expect hard X-ray emission only when F and B lie within the hatched region on the diagram. We emphasize, however, that the equations derived in this paper undoubtedly oversimplify the real situation, so the lines in this figure (and in Fig. 3) are of an approximate character.*

The lines $\tau_{es}(d) = \tau =$ constant are given by

$$F_{20} = 4 \times 10^{-2} (M/M_\odot)^{2/3} R_9^{1/6} B_6^{-1/3} \tau^{7/6}. \tag{32}$$

These are only relevant if $\phi \approx 1$: if $\phi \gg 1$ the radiation escapes predominantly from the sides of the shocked region.

For accretion to be appreciably channelled by the field we require $f_{\rm sph} < 1$, that is

$$F_{20} < 2 \times 10^{2}(M/M_{\odot})^{-2} R_9^{5/2} B_6^{2}. \tag{33}$$

If the white dwarf rotates, the rough condition that accretion takes place, $f_{\rm sph} > f_{\Omega}$, is

$$F_{20} > 1{\cdot}1(M/M_{\odot})^{-5/3} R_9^{6} B_6^{2} P^{-7/3}. \tag{34}$$

At the edge of this region, $f_{\rm sph} = f_{\Omega} = 0{\cdot}2(M/M_{\odot})^{-1/3} R_9 P^{-2/3}$.

(b) *Accretion disc flow outside $R_M$*

The conditions derived in this section are illustrated in Fig. 3 for a white dwarf of 1 $M_{\odot}$. The condition $\gamma_1 = 1$ ($\phi = 1$) is

$$F_{20} = 2 \times 10^{-5}(M/M_{\odot}) R_9 B^{18/7}. \tag{35}$$

The line $\gamma_1 = \gamma_2$ ($\phi = 1$) is

$$F_{20} = 3 \times 10^{-5}(M/M_{\odot})^{5/3} R_9^{1/3} B^{18/7}. \tag{36}$$

On both these lines $\phi = 2$ when $D_{ff} = 0{\cdot}5$ d, that is when

$$F_{20} = 2 \times 10^{-2}(M/M_{\odot})^{13/7} R_9^{-1} B_6^{-16/49}. \tag{37}$$

When $\phi \gg 1$, $\gamma_1 =$ constant implies $F \propto B^{200/133}$ and $\gamma_1 = \gamma_2$ implies $F \propto B^{9784/4485}$.

In the region $T_e < T_s$ the lines $\gamma_2 =$ constant ($\phi = 1$) are given by

$$F_{20} = 8 \times 10^{-4}(M/M_{\odot})^{1/2} R_9^{3/2} B_6^{18/7} \gamma_2^{-7/6}. \tag{38}$$

On these lines $\phi = 2$ implies $F \propto B^{-976/7}$. For $\phi \gg 1$ (and $T_e < T_s$) $\gamma_2 =$ constant implies $F \propto B^{2304/2037}$.

The lines $\tau_{es}(\mathrm{d}) = \tau =$ constant are given by

$$F_{20} = 7 \times 10^{-2}(M/M_{\odot})^{5/7} R_9^{1/7} B_6^{-16/49} \tau^{8/7} \tag{39}$$

and are, again, only relevant for $\phi \approx 1$.

The condition $f_{\rm disc} < 1$ is

$$F_{20} < 10(M/M_{\odot})^{-1} R_9^{3} B_6^{16/7} \tag{40}$$

and the condition $f_{\rm disc} > f_{\Omega}$ is

$$F_{20} > 0{\cdot}15(M/M_{\odot})^{-7/3} R_9^{7} B_6^{16/7} P^{-8/3}.$$

## 5. THE MASS TRANSFER PROCESS

In the previous sections we have considered the processes occurring at the surface of the white dwarf, where most of the primary emission takes place. The gas flowing in the binary system can, however, have severe effects on the observable properties of the accretion energy, for example by reflection (including absorption and re-emission) and by absorption, particularly at non-X-ray wavelengths. In this section we consider the process of mass transfer and gas flow and estimate some of its observable consequences.

The matter accreted is generally considered to be transferred from the normal (non-compact) companion either by Roche lobe overflow or by a stellar wind. In

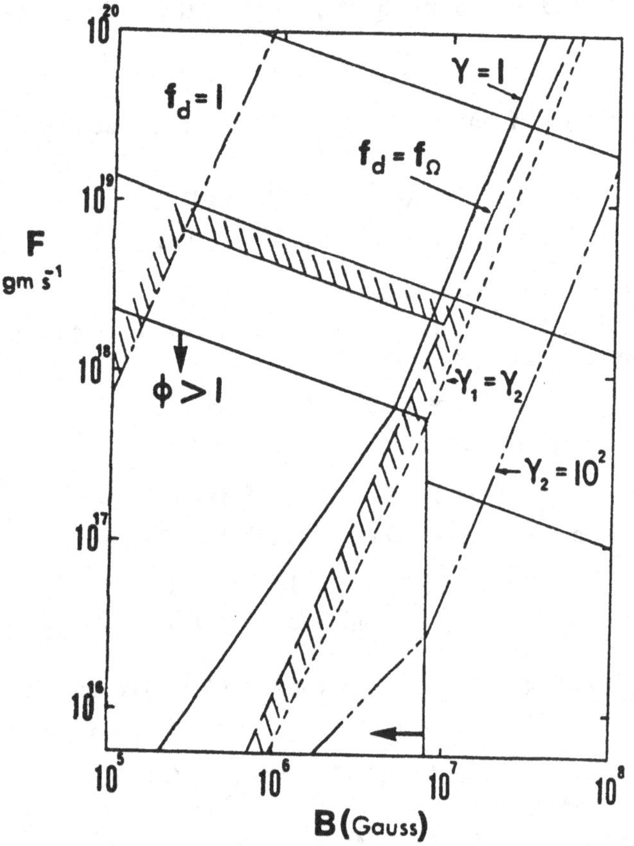


FIG. 3. *The same criteria as in Fig. 2 are here plotted for a system such as DQ Her where the accreted material is expected to form a disc outside the Alfvén radius (case (b) in Section 4). Note that a somewhat different value of B is now needed to ensure $f < 1$ (cf. equations (33) and (40)). This is a necessary condition for hard X-ray emission, because otherwise the disc extends right down to the stellar surface and no shock can form. For a white dwarf of mass $M_\odot$ spinning with a period $2\pi/\Omega = 142$ s (the appropriate value for DQ Her), the maximum value of B compatible with accretion is given by the line $f_d = f_\Omega$. For DQ Her (where the corresponding value of f is 0·062) it is this constraint, rather than the condition $\gamma_1 = \gamma_2$, which determines the maximum magnetic field compatible with hard X-ray emission. Hard X-rays (necessarily pulsed) are therefore expected only within the hatched area of the B–F plane.*

reality, of course, these are two extremes of a spectrum of possibilities and the transfer process may well be caused by a mixture of both. For simplicity, we consider each extreme possibility separately.

### 5.1 *Roche lobe overflow*

We use the term 'Roche lobe' loosely in as much as we make no distinction as to whether or not the mass-losing component corotates with the binary period, and do not necessarily demand that the orbit should be circular. In other words, we just demand in this case that some parts of the outer layers of the mass-transferring star are 'sufficiently' attracted to the accreting star for a 'large enough' fraction of the orbital period. There are, however, strong reasons for supposing that if mass transfer takes place the orbit becomes circular very rapidly (Heggie & Pringle, in preparation). In this case the initial velocity of the mass transfer flow in the direction of the accreting star is approximately the speed of sound in the outer layers of the mass-losing component, which is in general much less than the orbital or rotational velocities. For this reason we expect accretion to take place by means of an accretion disc. If the disc extends down to the surface of the accreting star, the accretion energy is radiated in two approximately equal components. First, the radiation emitted in the disc itself, with a maximum temperature of about $2\times 10^5 F_{20}^{1/4}(M/M_\odot)^{1/4} R_9^{-1/4}$ K for an optically thick disc, and second the radiation emitted from the boundary layer formed where the disc grazes the stellar surface, which is likely to be emitted at a higher temperature (Lynden-Bell & Pringle 1974), possibly in the soft ($\sim 0{\cdot}1$ keV) X-ray range. Hard X-rays are not likely to be emitted from such a disc even for luminosities approaching the Eddington limit.

A dwarf nova is a binary system in which Roche lobe transfer on to a white dwarf is taking place at a rate of $\sim 10^{16}$–$10^{18}$ g s$^{-1}$ and in which the optical radiation from the system is usually dominated by the accretion energy. No hard ($\gtrsim 2$ keV) X-rays have been observed from these systems, although there are reports of soft X-ray emission detected during outburst when the accretion rate is greatest (Rappaport *et al.* 1974; Heise *et al.* 1975).

If, however, the accreting white dwarf has a surface magnetic field strong enough to disrupt the disc out to say, radii $\gtrsim 3$ times the stellar radius, then accretion can take place along the field lines in a more or less radial fashion. A shock can form above the stellar surface and the accretion energy can be emitted as hard X-rays.

### 5.2 *Application to DQ Herculis*

The old nova (1934) DQ Herculis is known to be a binary system which contains an accreting magnetized white dwarf (Bath, Evans & Pringle 1974; Katz 1975; Lamb 1974; Kemp, Swedland & Wolstencroft 1974; Swedland, Kemp & Wolstencroft 1974) which is rotating with a period of 142 s or 2·37 min. The mass of the white dwarf is about 1 $M_\odot$ (Robinson 1976; Webbink 1975) and the rotation period is decreasing on a time scale of $P/\dot{P} \sim 10^7$ yr (Herbst, Hesser & Ostriker 1974). If this speed-up of the rotation rate is caused by accretion of angular momentum, then the accretion rate is $\sim 3\times 10^{16}$ g s$^{-1}$. This is consistent with the observed upper limits to the rate of change of orbital period (Pringle 1975). For these parameters we see from Fig. 3 that for accretion to take place we require $B \lesssim 10^6$ G and that the dominant emission mechanism is then likely to be bremsstrahlung (although cyclotron emission could dominate by a factor 10 or so). The total luminosity due to accretion is $\sim 10^{34}$ erg s$^{-1}$. The assumption of $7^m$ for the distance modulus (Kraft 1963) yields a distance $\mathscr{D} \sim 250$ pc. If bremsstrahlung emission dominates, we expect an X-ray flux at Earth of $\sim 7\times 10^{-10}$ erg cm$^{-2}$ with a flat (quasi-bremsstrahlung) spectrum and $kT \sim 60$ keV. This flux could, however, be considerably reduced by absorption and scattering in the material flowing in the white dwarf's Roche lobe: there is evidence that much of the light in the system is due to reflection effects. There remains the additional possibility that much, but not all, of the luminosity can be radiated as cyclotron radiation at $\sim 100$ $\mu$m. This would be modulated with a period of 142 s.

5.3 *Stellar wind*

A variety of stars are observed to emit stellar winds at rates of $10^{-5}$–$10^{-7}\,M_\odot\,\mathrm{yr}^{-1}$ and a compact object orbiting such a star can accrete a fraction of it. The accretion energy so produced is not necessarily the dominant contribution to the luminosity of the system (mainly because the star must be very luminous to produce such a strong wind) but can be readily observable if it is emitted at wavelengths different from the bulk of the luminosity.

A number of authors (Davidson & Ostriker 1973; Illarionov & Sunyaev 1975; Shapiro & Lightman 1976) have considered the accretion process in this case. We briefly repeat their arguments here. Consider a binary system, consisting of stars masses $M_1$ and $M_2$, and with semi-major axis $a$. Assume for simplicity that the orbit is circular. Let $M_2$, with radius $R_2$, be a compact accreting object and $M_1$, with radius $R_1$, emit a stellar wind with velocity

$$V_w(R) = (\lambda(R)\, 2GM_1/R_1)^{1/2}$$

where $R$ is the distance from the centre of $M_1$ and $\lambda$ is a parameter, probably of order unity for $R \sim a$. If we assume that the wind velocity is larger than both the speed of sound in the wind and the orbital velocities of the two stars (a good approximation provided that the mass ratio is not too extreme and the system is not too close), we may define an accretion radius $R_A$ around $M_2$ approximately by

$$R_A = \frac{\alpha G M_2}{V_w^2} = \frac{\alpha M_2 R_1}{2\lambda(a)\, M_1} \tag{41}$$

where $\alpha$ is another parameter of order unity.

In general $R_A \ll a$ and thus the fraction of the wind that is accreted by the white dwarf is

$$\frac{F}{F_w} = \frac{\pi R_A^2}{4\pi a^2} = \left(\frac{\alpha}{\lambda(a)}\right)^2 \frac{M_2^2}{16 M_1^2}\left(\frac{R_1}{a}\right)^2. \tag{42}$$

The angular momentum per unit mass $h$ of the accreted material is given roughly by (Illarionov & Sunyaev 1975)

$$h = \tfrac{1}{4}\Omega R_A^2 \tag{43}$$

where $\Omega$ is the orbital angular velocity. Therefore the angular momentum of the accreted material begins to dominate the flow—that is, an accretion disc starts to form—at the circularization radius $R_c$, where $h = (GM_2R_c)^{1/2}$. Thus

$$\frac{R_c}{a} = \left(\frac{\alpha}{\lambda(a)}\right)^4 \frac{M_2^3(M_1+M_2)}{256 M_1^4}\left(\frac{R_1}{a}\right)^4. \tag{44}$$

If the accreting object does not have a strong magnetic field, then the occurrence or not of radial infall on to the surface depends on the relative sizes of $R_c$ and $R_2$.

We see at once that the specific angular momentum of matter accreted from a wind is much less (by a factor $\sim (R_A/a)^2$ than that of matter transferred by Roche lobe overflow and that therefore a disc is less likely to form when a stellar wind provides the main mass transfer. We stress that the value of $R_c$ estimated above is very uncertain. Shapiro & Lightman (1976) have made a more detailed estimate, but nevertheless the main uncertainty lies in the function $\lambda(R)$. Theoretical estimates suggest (Lucy & Solomon 1970) that in a steady wind driven by radiation pressure $\lambda(R) = 1-(R_1/R)$, although this could be seriously wrong if the accretion luminosity gave rise to significant ionization in the acceleration zone of the wind (Hatchett, Buff & McCray 1975). (We consider the effect of radiation on the accretion radius in the next section.) On the other hand, observational data (Kuan & Kuhi 1975) suggest that, for the star P Cygni, $\lambda = (R_1/R)$, in which case $F/F_w$ and $R_c/a$ are ***independent of the binary separation.*** If the mass-loss rate in the wind is variable, further complications occur since the velocity and density structure of the wind can then vary independently.

We conclude therefore that radial and disc accretion can arise from stellar wind transfer and that it is not a trivial matter to decide which in fact occurs. Moreover, since $R_c$ is so sensitive to the wind parameters the flow can switch to and fro between radial (hard X-rays emitted) and disc (soft or no X-rays) on the same time scale as the variation of the wind flow (*cf.* Shapiro & Lightman 1975). This may be of some importance in the understanding of the transient X-ray sources.

One further point is worth making at this stage. We mentioned in Section 4 that if $R_M > R_\Omega$ the accretion flow is obstructed from latching on to the magnetic field lines by a centrifugal barrier. If, however, $R_c < R_\Omega < R_M$ some accretion can nevertheless take place. In particular, the matter that falls on to the magnetosphere within a distance $R_\Omega$ of the rotation axis (that is $\sim R_\Omega^2/4R_M^2$ of the accretion flux) can be accreted. The rest is expelled by centrifugal forces. It is even possible that the expelled material could form a centrally-driven accretion disc (*cf.* Lynden-Bell & Pringle 1974) and so extract a considerable amount of angular momentum from the accreting object. In this case the rotation rate of the star could be decreasing, even though accretion is taking place (*cf.* Fabian 1975).

### 5.4 *Interactions between the outgoing radiation and infalling matter*

The accretion radius, $R_A$, is modified if the radiative heating of the ambient gas is sufficient to produce thermal velocities comparable with the wind velocity, $V_w$, near $R_A$. It may be that no static self-consistent solution exists and that the accretion flow becomes unstable (Buff & McCray 1974). The gas temperature can be estimated from the parameter $\xi = L/nr^2$, where $L$ is the accretion luminosity, $n$ the number density of the gas and $r$ the distance from the accreting object (Tarter, Tucker & Salpeter 1969; Buff & McCray 1974). At the accretion radius,

$$\xi(R_A) \approx 2.2 \left(\frac{V_w}{30 \text{ km s}^{-1}}\right)\left(\frac{M}{M_\odot}\right) R_9^{-1}, \tag{45}$$

while the thermal velocity in the wind is given by

$$\frac{V_s}{30 \text{ km s}^{-1}} \simeq 1.6 \left(\frac{T}{10^5 \text{ K}}\right)^{1/2}.$$

Fig. 1 of Buff & McCray (1974) shows that

$$\frac{T}{10^5 \text{ K}} \simeq \left(\frac{\xi}{30}\right)^2 \qquad (10 < \xi < 10^4) \tag{46}$$

for a thermal spectrum ($kT \sim 10$ keV) with no low energy cut-off ($\epsilon_{min} = 0$), so

$$\frac{V_s}{V_w} \simeq 0.12 \left(\frac{M}{M_\odot}\right) R_9^{-1} \tag{47}$$

and stable accretion can occur. Note that, for radial infall within $R_A$, the flow speed $\propto r^{-1/2}$ (free-fall) and the sound speed $V_s \propto \xi \propto r^{-1/2}$ and the flow is self-consistent for all $r < R_A$. If $\xi > 10^4$ or if $\epsilon_{min} > 0$, (46) is an overestimate for $T$ and stable accretion occurs *a fortiori*. When the wind speed is low, we see from (45) that $\xi < 10$. We find approximately that, when $V_w \lesssim 15$ km s$^{-1}$ and when $\epsilon_{min}$ is small, $V_s \gtrsim V_w$ and the above can break down, with the accretion flow becoming non-static.

If the accretion flow is radial, for $r < R_A$,

$$\xi(r) \simeq 4 \times 10^2 \left(\frac{r}{10^9\ \text{cm}}\right)^{-1/2} \left(\frac{M}{M_\odot}\right)^{3/2} R_9^{-1}. \tag{48}$$

If appreciable funnelling takes place this is an overestimate of the effective $\xi$ at some radii, since (a) $n$ increases more steeply with decreasing radius and (b) the full accretion luminosity does not reach all points of the flow. Hatchett, Buff & McCray (1976) have shown that the elements C, N and O are completely ionized if $\xi \gtrsim 200$, and that to ionize the heavier elements (S and Fe) requires $\xi \gtrsim 10^3$. Thus the elements producing absorption in the keV range retain their K shell electrons during infall.

Note that the dynamical effects of the outgoing radiation on the gas are negligible, even in the spherically symmetric case, provided the effective cross-section $\sigma$ per electron is less than $(L_{edd}/L)\ \sigma_T$, $L_{edd}$ being the 'Eddington limit' $(4\pi GMm_pc)/\sigma_T$. The precise value of $\sigma$ depends on the state of ionization of the gas, particularly the H and He (*cf.* Hatchett *et al.* 1976). We have shown that hard X-rays can escape only when $L \lesssim 10^{-2} L_{edd}$, and under these conditions we expect that the neglect of these dynamical effects is indeed self-consistent.

If $D \lesssim d$ the emergent X-rays must pass through a column density $N$ of absorbing matter,

$$N \simeq 2 \times 10^{25} F_{20} f^{-1/2} (M/M_\odot)^{-1/2} R_9^{1/2}\ \text{cm}^{-2} \tag{49}$$

where the $f$-dependence allows for the escape of X-rays from the sides of the accretion column. Of course, if $\phi \gtrsim 2$, X-rays escape from the sides of the emitting region and do not have to pass through unshocked, un-ionized material. If we crudely approximate the absorption cross-section $\sigma_p(E)$ of Brown & Gould (1969) by $\sigma_p(E) = 4 \times 10^{-22} E^{-3}$ cm$^2$, and define a cut-off energy $E_a$ by $\sigma_p(E_a) N = 1$, we find

$$E_a \simeq 19 F_{20}^{1/3} f^{-1/6} (M/M_\odot)^{-1/6} R_9^{-1/6}\ \text{keV}. \tag{50}$$

This holds provided that $E_a \gtrsim 3$ keV, where absorption by C, N and O is negligible. If equation (50) gives formally $E_a < 3$ keV, the low-energy cut-off depends more critically on the ionization state of the lighter elements and the secondary radiation emitted by the infalling gas should be considered. In general, however, the dimensionless parameter $\gamma$ defined by Tarter & Salpeter (1969) is less than unity and we expect (50) to provide a reasonable estimate of $E_a$, although fluorescence lines and other features may become evident.

### 5.5 *Application to Mira and similar systems*

Mira B is thought to be a white dwarf which is accreting material from Mira itself (Warner 1972). The accretion radius is of comparable size to Mira, allowing a comparatively large fraction (~1 per cent) of the mass in the wind to be accreted, despite the relatively large separation (~$10^{15}$ cm). Following the discussion in Section 5.3, a disc forms at an approximate radius of

$$R_c \simeq 4 \times 10^{10} \left(\frac{P_b}{261\ \text{yr}}\right)^{-2} \left(\frac{V_w}{10\ \text{km s}^{-1}}\right)^{-4}\ \text{cm}. \tag{51}$$

Here $P_b$ is the binary orbital period. The steep dependence of $R_c$ on $V_w$ means that spherically symmetric accretion may occur at least some of the time. As mentioned in Section 5.3, density gradients in the highly variable wind from Mira can act to substantially reduce the rate of accretion of angular momentum per unit mass. Ionization instabilities can occur as outlined in Section 5.4. We therefore envisage the possibility that Mira B is a sporadic X-ray source, and it may be responsible for the transient sources Cet X-1 and Cet X-2 (Fabian, Pringle & Webbink 1975a; P. J. N. Davison, private communication). There are at present no indications of a magnetic field on the dwarf, but of course if there is a strong enough one to disrupt the disk, radial infall can occur all the time. SY Fornacis is a system similar to Mira (Feast 1975) and we note that the Mira-type variable RS Centauri has been proposed as a candidate for A1118-61 (Fabian *et al.* 1975b).

The accretion rate on to Mira B also depends strongly on the wind velocity, $V_w$, and consequently it is difficult to make any precise estimate of the X-ray luminosity, $L_X$. If we take a value, $L_X \sim 10^{33}$ erg s$^{-1}$, similar to the optical luminosity estimated by Warner (1972), we derive an X-ray flux at the Earth of $\sim 3 \times 10^{-9}$ erg cm$^{-2}$ s$^{-1}$ for a distance of $\sim 50$ pc. A relatively hard bremsstrahlung spectrum coupled with strong absorption in the infalling material and wind could render such a source undetectable by, for example, the *Uhuru* satellite. The value of $L_X$ used above could at times be a serious underestimate. Some of the optical light might originate in an accretion disc, with the contribution from the white dwarf itself emerging in the ultraviolet. Moreover, Mira B is only clearly seen when Mira is near minimum.

## 6. SUMMARY

We have extended the calculations of Hoshi (1973) and Aizu (1973) to produce a self-consistent model for X-radiation from accreting, possibly magnetized, white dwarfs. An accretion rate of $\gtrsim 10^{15}$ g s$^{-1}$ and a quasi-radial accretion flow is required to produce a stand-off shock which heats the material to X-ray temperatures ($\sim 6 \times 10^8$ K for a 1 $M_\odot$ white dwarf (see Fig. 1)). The gas radiates X-rays by optically thin bremsstrahlung. Radial flow to the stellar surface is likely when the accretion takes place from a stellar wind and/or when the dwarf has a strong enough magnetic field to channel the flow on to the magnetic polar caps. In the latter case cyclotron emission can dominate (see Figs 2 and 3). If the X-ray luminosity exceeds $\sim 10^{36}$ erg s$^{-1}$, electron scattering in the infalling material is sufficient to degrade the keV X-rays and in any case the low-energy absorption cut-off is likely to be substantial, particularly for the more luminous sources.

Thus the characteristics of the sources we describe here are as follows. The X-ray luminosities lie roughly in the range $10^{32}$–$10^{36}$ erg s$^{-1}$ (and cannot therefore account for the bright variable sources seen by *Uhuru*). The X-ray spectrum is quasi-bremsstrahlung with $kT \sim 30$–$100$ keV and with a substantial low-energy cut-off (note that little or none of the X-ray emission need fall in the range detectable by *Uhuru* and similar instruments). The optical (or ultraviolet) luminosity should be at least comparable to that emitted as X-rays. It is possible that a comparable, or greater, luminosity is radiated by the cyclotron mechanism

## ACKNOWLEDGMENTS

Part of this work was done while JEP and MJR were participating in the astrophysics workshop at the Aspen Center for Physics. Discussions with other participants, particularly R. A. McCray, are gratefully acknowledged. JEP also wishes to thank the Harvard–Smithsonian Center for Astrophysics for hospitality during 1975 June.

## REFERENCES

Aizu, K., 1973. *Prog. theor. Phys.*, **49**, 1184.
Allen, C. W., 1973. *Astrophysical quantities*, 3rd edition, Athlone Press, London.
Arons, J. & Lea, S. M., 1976. *Astrophys. J.*, in press.
Bath, G. T., Evans, W. D., Papaloizou, J. & Pringle, J. E., 1974. *Mon. Not. R. astr. Soc.*, **169**, 447.
Bath, G. T., Evans, W. D. & Pringle, J. E., 1974. *Mon. Not. R. astr. Soc.*, **166**, 113.
Brown, R. L. & Gould, R. J., 1969. *Phys. Rev. D.*, **1**, 2852.
Buff, J. M. & McCray, R., 1974. *Astrophys. J.*, **189**, 147.
Davidson, K. & Ostriker, J. P., 1973. *Astrophys. J.*, **179**, 585.
Fabian, A. C., 1975. *Mon. Not. R. astr. Soc.*, **173**, 161.
Fabian, A. C., Pringle, J. E. & Webbink, R. F., 1975a. *Astrophys. Space Sci.*, in press.
Fabian, A. C., Pringle, J. E. & Webbink, R. F., 1975b. *Nature*, **255**, 208.
Feast, M. W., 1975. *Observatory*, **95**, 19.
Gregoria, A. J. de, 1974. *Astrophys. J. Lett.*, **189**, 555.
Hatchett, S., Buff, J. & McCray, R., 1976. *Astrophys. J.*, in press.
Hayakawa, S., 1974. *Prog. theor. Phys.*, **50**, 459.
Heise, J., Brinkman, A. C., Schrijver, J., Mewe, R., Groneschild, E. & Den Boggende, A., 1975. *Astrophys. Space Sci.*, in press.
Herbst, W., Hesser, J. E. & Ostriker, J. P., 1974. *Astrophys. J.*, **193**, 679.
Hoshi, R., 1973. *Prog. theor. Phys.*, **49**, 776.
Ichimaru, S. & Nakano, T., 1973. *Prog. theor. Phys.*, **50**, 1867.
Illarionov, A. F. & Sunyaev, R. A., 1975. *Astr. Astrophys.*, **39**, 185.
Inoue, H. & Hoshi, R., 1975. *Prog. theor. Phys.*, **54**, 415.
Katz, J. I., 1975. *Astrophys. J.*, **200**, 298.
Katz, J. I. & Salpeter, E. E., 1974. *Astrophys. J.*, **193**, 429.
Kraft, R. P., 1963. *Adv. Astr. Astrophys.*, **2**, 43.
Kemp, J. C., Swedland, J. B. & Wolstencroft, R. D., 1975. *Astrophys. J. Lett.*, **193**, L15.
Kuan, P. & Kuhi, L. V., 1975. *Astrophys. J.*, **199**, 148.
Lamb, D. Q., 1974. *Astrophys. J.*, **192**, 129.
Lamb, F. K., Pethick, C. J. & Pines, D., 1973. *Astrophys. J.*, **184**, 271.
Lucy, L. B. & Solomon, P. M., 1970. *Astrophys. J.*, **159**, 879.
Lynden-Bell, D. & Pringle, J. E., 1974. *Mon. Not. R. astr. Soc.*, **168**, 603.
Pringle, J. E., 1975. *Mon. Not. R. astr. Soc.*, **170**, 633.
Rappaport, S., Cash, W., Doxsey, R., McClintock, J. & Moore, G., 1974. *Astrophys. J. Lett.*, **187**, L5.
Robinson, E. L., 1976. *Astrophys. J.*, in press.
Shapiro, S. L. & Lightman, A. P., 1976. *Astrophys. J.*, in press.
Shapiro, S. L. & Salpeter, E. E., 1975. *Astrophys. J.*, **198**, 671.
Swedland, J. B., Kemp, J. C. & Wolstencroft, R. D., 1975. *Astrophys. J. Lett.*, **193**, L11.
Tarter, C. B. & Salpeter, E. E., 1969. *Astrophys. J.*, **156**, 953.
Tarter, C. B., Tucker, W. H. & Salpeter, E. E., 1969. *Astrophys. J.*, **156**, 943.
Warner, B., 1972. *Mon. Not. R. astr. Soc.*, **159**, 95.
Webbink, R. F., 1975. *PhD thesis*, University of Cambridge.

## APPENDIX

When $F_{20} \lesssim 3 \times 10^{-2} f(M/M_{\odot})^{3/2} R_9^{-1/2}$, the height of the shock above the stellar surface $D \gtrsim R$. In this case the temperature behind the shock $\propto (R+D)^{-1}$ is lower than that used in previous calculations and at first sight it seems that the above analysis is not applicable. Most of the energy, however, is still liberated close to the star, within the region $R \leqslant r \lesssim 2R$ where $r$ is the radial coordinate. We obtain below an approximate solution for the shocked gas which, as we shall see, is a good approximation in the region $R \leqslant r \lesssim 2R$ when $D \gg R$.

Since the flow velocities behind the shock are subsonic we may assume the approximate validity of the hydrostatic equation

$$\frac{dp}{dr} = -\frac{GM\rho}{r^2}. \tag{A1}$$

For simplicity we assume spherical symmetry. Well below the shock (where most of the energy is released) we may assume roughly that the energy liberated is radiated *locally* by bremsstrahlung at a rate $\epsilon\rho^2 T^{1/2}$, where $\epsilon$ is a constant (apart from the temperature dependence of the Gaunt factor). Thus we may write

$$\frac{FGM}{r^2} = 4\pi r^2 . \epsilon\rho^2 T^{1/2}. \tag{A2}$$

Solving these, together with the gas equation

$$p = \frac{\mathscr{R}\rho T}{\mu} \tag{A3}$$

we obtain

$$p^{4/3} - p_s{}^{4/3} = \frac{4}{11}(GM)^{5/3}\left(\frac{\mathscr{R}}{\mu}\right)^{1/3}\left(\frac{F}{4\pi\epsilon}\right)^{2/3}\left[\frac{1}{r^{11/3}} - \frac{1}{(R+D)^{11/3}}\right]$$

where $p_s$ is the pressure just after the shock at $r = R+D$. For $R \lesssim r \ll R+D$, we find

$$T \approx \frac{4}{11}\left(\frac{\mu}{\mathscr{R}}\right)\frac{GM}{r} \tag{A4}$$

and $p \propto r^{-11/4}$, $\rho \propto r^{-7/4}$ where the constants of proportionality *are independent of the position of the shock.*

This is not a surprising result. The amount of energy to be liberated at $r \sim R$ and the temperature at $r \sim R$ have been fixed. These, together with the assumption that bremsstrahlung is the dominant cooling mechanism, imply a density for the region. Thus the temperature and density of the radiating region for $D \gg R$ do not differ appreciably from the solutions with $D \lesssim R$ and the above analysis is approximately valid for this case too.

*Mon. Not. R. astr. Soc.* (1982) **199**, 267–280

# The evolution of viscous discs – II. Viscous variations

G. T. Bath *Department of Astrophysics, South Parks Road, Oxford OX1 3RQ*
J. E. Pringle *Institute of Astronomy, Madingley Road, Cambridge CB3 0HA*

Received 1981 September 16; in original form 1981 July 30

**Summary.** The necessary conditions for mass flow through an accretion disc to be modulated by viscous variations are examined. We show that outburst behaviour is possible if the disc undergoes a global limit-cycle in which the viscosity is a two-valued function of disc structure. We attempt to model the eruptions of dwarf nova discs as a limit-cycle. The resulting light curves exhibit discontinuous changes in luminosity. If the viscosity is imposed, by some external agency, as a function of time then improved light curves are obtained. Nonetheless, it is not possible to obtain the whole range of dwarf nova light curves, and those which are fitted require a restricted range of conditions to be satisfied.

## 1 Introduction

Unstable eruptions and spectral variations occur in a wide variety of interacting binary stars and in many cases are most easily explained by fluctuations in the accretion rate on to the more compact binary component. In general the cause of accretion variations is not settled, though in the case of the well-studied dwarf novae two specific models have been proposed:

(1) Variable mass transfer due to dynamical instabilities driven by surface ionization zones in the cool companion (Bath 1975; Papaloizou & Bath 1975; Wood 1977) (or some other mechanism which varies the mass transfer rate appropriately).

(2) Disc instabilities which, through some mechanism still to be isolated, modulate the mass accretion flow through the accretion disc (Osaki 1974; Paczynski 1977; Madej & Paczynski 1977).

In a previous paper (Bath & Pringle 1981; hereafter Paper 1) we presented numerical computations of the time-dependent behaviour of accretion discs in dwarf novae, with the particular aim of modelling the outburst behaviour. In the first class of model, discussed in Paper 1 and shown there to produce reasonable agreement with the observations, the rate at which mass is put into the disc (that is, transferred from the companion star) is varied in a prescribed manner. The second method of varying the accretion rate is to vary the viscosity in the disc in a prescribed manner. This class of model is discussed here.

In Paper 1, we used the Shakura & Sunyaev (1973) viscosity parametrization (the $\alpha$-prescription) and modelled the outbursts by varying the mass input rate, $\dot{M}$, and keeping $\alpha$ constant. Our aim was to use comparison with observations to obtain restrictions on $\alpha$ and on the way in which $\dot{M}$ could be varied. The main restriction on the value of $\alpha$, which must hold for all accretion models for dwarf nova outbursts, was that the time-scale on which the brightness is observed to vary during outburst implies that during outburst $\alpha$ must be close to unity, or equivalently, the disc viscosity must be close to its maximum physically plausible value.

In this paper we experiment with the alternative point of view and attempt to model the outbursts by keeping $\dot{M}$ constant and varying $\alpha$. Since we have no physical understanding of the viscosity in accretion discs, we have *a priori* no physical constraints on the manner in which $\alpha$ may be varied. Rather we attempt to address the more general problem of asking what constraints the observations can put on the possible variations of $\alpha$.

## 2 Methodology and simple models

The method and assumptions used here are as in Paper 1 and the reader is referred to that paper for more details. We make the thin disc approximation appropriate to discs in cataclysmic variables. The basic equation describing disc evolution is

$$\frac{\partial \Sigma}{\partial t} = \frac{3}{R}\frac{\partial}{\partial R}\left[R^{1/2}\frac{\partial}{\partial R}(\nu \Sigma R^{1/2})\right], \tag{1}$$

where $\Sigma(R, t)$ is the disc surface density at radius $R$ and time $t$, and $\nu(R, t)$ is the kinematic viscosity. If we consider the flow to take place between two radii $R_1$, $R_N$, we must introduce the relevant boundary conditions. At the inner radius we assume that any matter arriving is accepted by the central star, and therefore set $\Sigma(R_1, t) = 0$. At the outer boundary we assume no material is lost, but that an appropriate torque is applied there to assure this (cf. Papaloizou & Pringle 1977). Thus we take $\partial\Sigma/\partial R = 0$ at $R = R_N$. In fact the exact nature of the boundary conditions is irrelevant to our general conclusions.

### 2.1 PROPERTIES OF THE EVOLUTION EQUATION

Before proceeding with the numerical solution of equation (1) it is instructive to consider its general properties from an analytic point of view. For these purposes the convenient way of allowing material to be injected into the disc is to introduce a source term $S(R, t)$ in equation (1). We write

$$\frac{\partial \Sigma}{\partial t} = \frac{3}{R}\frac{\partial}{\partial R}\left[R^{1/2}\frac{\partial}{\partial R}(\mu R^{1/2})\right] + S(R, t), \tag{2}$$

where $\mu(\equiv \nu\Sigma)$ is $z$-integrated viscosity and $S(R, t)$ represents the mass input rate per unit area at radius $R$, the mass being injected with specific angular momentum $(GM_1R)^{1/2}$ where $M_1$ is the mass of the accreting object at the disc centre. For definiteness we take

$$S(R, t) = \frac{\dot{M}(t)}{2\pi R_K}\,\delta(R - R_K) \tag{3}$$

which represents an input rate of $\dot{M}$ at radius $R = R_K$, where $R_1 < R_K < R_N$. Since equation (2) is a diffusion equation then, if $\partial S/\partial t = 0$ and provided that $\mu$ is a suitably behaved function of $\Sigma$, the evolution of equation (2) is *always* towards the steady-state given by $\partial\Sigma/\partial t = 0$.

When $\partial\Sigma/\partial t = 0$, equation (2) defines $\mu$ not $\Sigma$ as a function of $R$. We let the steady-state solution be given by $\mu = \mu_{SS}(R)$ and note that $\mu_{SS}$ is a *unique* function of radius. In the case we consider, we find

$$\mu_{SS} = \frac{\dot{M}}{3\pi} \times \begin{cases} 1 & R_1 < R < R_K \\ (R_K/R)^{1/2} & R_K < R < R_N \end{cases}, \tag{4}$$

If, at each radius, the surface density $\Sigma$ is a single-valued function of $\mu$, there is a corresponding unique equilibrium solution $\Sigma = \Sigma_{SS}(R)$. The equilibrium is stable provided that $\partial\mu/\partial\Sigma > 0$ (Lightman & Eardley 1974).

A number of authors have invoked the instability that occurs when $\partial\mu/\partial\Sigma < 0$ (the (Lightman-Eardley instability) as a source of bursting behaviour in accretion discs. We stress here simply that the instability is not physically realizable, since in any physical situation the unstable equilibrium state cannot be attained. Rather the existence of an instability of the equilibrium state implies that the assumptions made about the dependence of viscosity on local parameters are not consistent with the assumed existence of a steady disc. What happens if the viscosity law is such that $\partial\mu/\partial\Sigma < 0$ has yet to be established. This point is discussed further by Pringle (1981).

Thus if $\mu(\Sigma, R)$ is a monotonically increasing function of $\Sigma$, then for every time-independent input function $S(\equiv\dot{M})$ there is a corresponding equilibrium steady-state towards which the disc evolves on a viscous time-scale. In this case no cyclic behaviour is possible, and the time-scale for approach to the steady state is determined by the value of the viscosity. Conversely, if some variation in the viscosity produces the outbursts and if the viscosity at each point in the disc is a function only of local disc parameters (without the intervention of some outside agency) then, at fixed $R$, $\mu$ cannot be a continuous monotonic function of $\Sigma$.

In Fig. 1 we consider three representative examples of the function $\mu(\Sigma)$. In Fig. 1(a) $\mu$ is a monotonically increasing function of $\Sigma$, so that there is a stable disc configuration for each value of $\mu$. In Fig. 1(b) there is a range of $\mu$ in which there are two possible stable solutions for $\Sigma$. Which of these is attained depends on the past history of the disc (that is on the previous behaviour of $\dot{M}(t)$). Such a function $\mu(\Sigma)$ cannot give rise to bursting behaviour itself, but with carefully chosen parameters it can be used to amplify the effects of small variations in $\dot{M}$. In Fig. 1(c) there is a range of values of $\mu$ with no corresponding stable value of $\Sigma$. If $\mu$ is in this range then such a function $\mu(\Sigma)$ gives rise to limit-cycle behaviour with the diffusion equation continually hunting for, but never finding, a stable disc configuration. This is the simplest example in which, despite the viscosity being defined locally in the disc, a global limit-cycle involving the whole disc can be achieved. With reference to the nomenclature of Fig. 1(c), the cycle proceeds as follows. In the low-viscosity state the disc grows towards the desired equilibrium and $\Sigma$ increases. When $\Sigma$ reaches $\Sigma_1$, $\mu$ jumps to higher branch of the curve and the viscosity is now so high that $\Sigma$ decreases. When $\Sigma$ reaches $\Sigma_2$, a jump of $\mu$ to the lower branch occurs and the cycle repeats. In this way a cycle can be

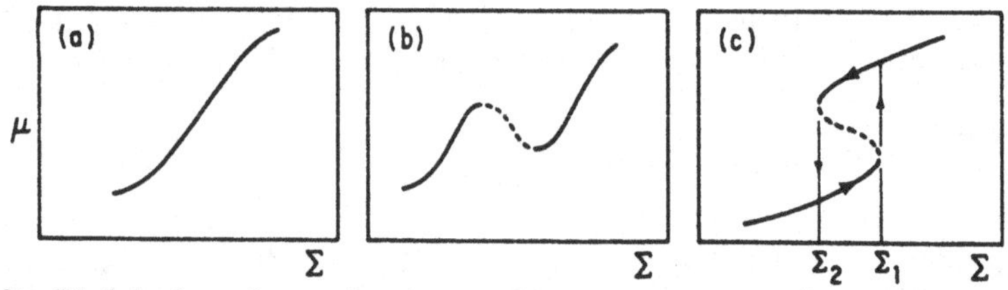


**Figure 1. Possible behaviour of $\mu$ as a function of $\Sigma$ and the origin of limit-cycle behaviour. See text for details.**

generated in which the disc evolves towards a high disc-mass state (high $\Sigma$) on a long time-scale (low $\mu$), corresponding to quiescence, and then evolves rapidly (high $\mu$) towards a low-mass state (low $\Sigma$), corresponding to outburst.

### 2.2 NUMERICAL MODELS

We now apply the above ideas and investigate the properties of discs undergoing such cyclic instabilities, using the numerical procedure to calculate disc evolution which was described in Paper 1. This procedure differs from the discussion in Section 2.1 in that numerically we are able to put mass into the disc in a more realistic way, which mimics mixing of the mass transfer stream with disc material. We assume dynamical mixing, and sharing of angular momentum of the input stream with material in the outer disc region. If $\Delta M^i$ is the mass transferred at the $i$th time step, the mass is shared between the $J$th and $J-1$th zone as

$$\overline{\Delta M_J^i} = \frac{D_J - X_{J-1}}{X_J - X_{J-1}} \left(\Delta M^i + \sum_J^N \Delta M_n^i\right)$$

$$\overline{\Delta M_{J-1}^i} = \frac{X_J - D_J}{X_J - X_{J-1}} \left(\Delta M^i + \sum_J^N \Delta M_n^i\right) + \Delta M_{J-1}^i, \tag{5}$$

where

$$D_J = \frac{X_K \Delta M^i + \sum_J^N X_n \Delta M_n^i}{\Delta M^i + \sum_J^N \Delta M_n^i} > X_{J-1} \tag{6}$$

and $\Delta M_n^i$ is the mass in a zone at radius $R_n$, and $X_n = 2R_n^{1/2}$. (We write this here explicitly because equations (2.15) of Paper 1 are misprinted.)

We define the kinematic viscosity as $\nu = (2/3)\alpha H^2 \Omega$ (Paper 1) and examine disc evolution following *ad hoc* viscous changes which we take to depend on the global state of the disc. This is equivalent to assuming that $\mu(\Sigma, R)$ is of the form shown in Fig. 1(c) at all radii in the disc. Viscous changes are imposed as changes in $\alpha$, with $\alpha$ taken to be constant over the whole disc. The simplest behaviour is a two-valued $\alpha$ with a low value, $\alpha_1$, in the quiescent phase and a high value, $\alpha_2$, in the outburst phase. The condition that the viscosity changes simultaneously at all disc radii when the disc reaches a critical mass requires that $\Sigma_1$ and $\Sigma_2$ are assumed to vary suitably over the disc. In the quiescent phase the disc evolves towards a steady-state corresponding to a high disc-mass on a long time-scale, and in the outburst phase the disc evolves towards a steady-state corresponding to a low disc-mass on a short time-scale.

This behaviour is demonstrated by the model in Fig. 2. In the case shown the disc is gaining matter at a constant rate $\dot{M} = 10^{16}\,\mathrm{g\,s^{-1}}$. The two values of $\alpha$ are $\alpha_1 = 0.1$ and $\alpha_2 = 1.0$, as indicated at the top of the figure. We let $\alpha_1$ switch to $\alpha_2$ when the disc mass reaches $1.9 \times 10^{22}$ g, and $\alpha_2$ switch to $\alpha_1$ when the disc mass drops to $5.0 \times 10^{21}$ g.

Initially, with $\alpha = 0.1$, the disc mass, $M_d$, increases, and matter is transferred into a slowly spreading torus. The luminosity emitted by one side, $L_D$, of the torus rises toward the steady-state value of $L_{SS} = GM_1\dot{M}/4R_1$ (Note that $L_D$ is the disc luminosity emitted through one disc face). When the disc mass reaches the value $1.9 \times 10^{22}$ g, $\alpha$ switches to 1.0. Providing, as is the case here, that the disc mass before instability occurs is greater than the steady-

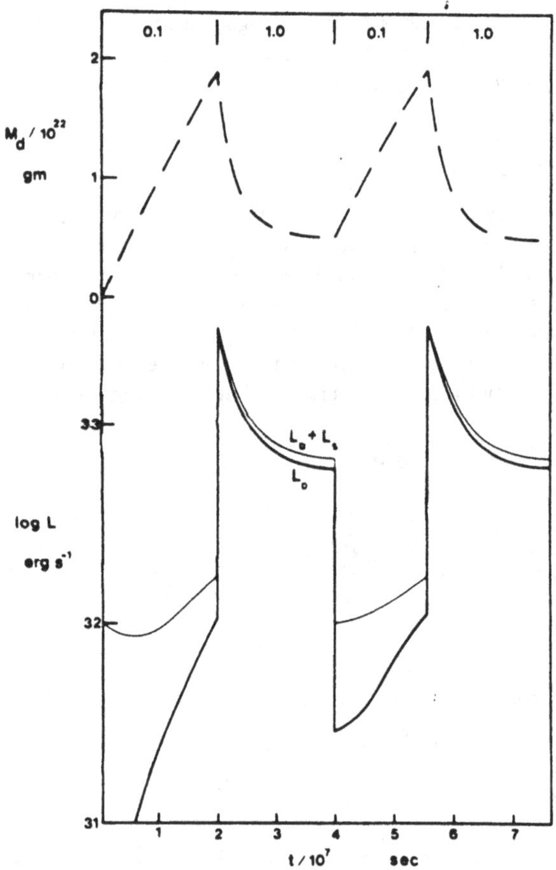


**Figure 2.** Variation of disc mass (broken curve), and disc and disc + spot luminosity (continuous curves) for a disc in which $\alpha$ switches between 0.1 and 1.0. The disc cycles about the steady-state luminosity ($\log L_{SS} = 32.7$) with two diffusion regimes above and below $L_{SS}$, and a discontinuous jump when $\alpha$ switches.

state disc mass with $\alpha = \alpha_2$, then the torus spreads into a disc, and matter drains onto the accreting central star. The luminosity rises rapidly at the switching point due to the sudden increase in viscous dissipation, and then decays towards the steady-state value. A cycle is produced if $\alpha$ reverts back to $\alpha_1$ when the disc mass drops to a lower critical value, in this case $5 \times 10^{21}$ g.

In both the states $\alpha = \alpha_1$ and $\alpha = \alpha_2$, the disc evolves towards a steady-state. In the low $\alpha$ state it evolves by building up toward a steady disc on a long time-scale and the luminosity grows slowly toward the steady-state value. In the outburst state the mass of the disc is larger than the corresponding equilibrium mass. Rapid diffusion of material out of the disc produces a light curve which decays rapidly toward the steady-state luminosity.

The decline is a two-stage process. The first stage is a decay due to diffusion of matter out of the disc, the second stage is the discontinuous jump as $\alpha$ switches and the dissipation rate falls. The overall behaviour is a cycle about the steady-state luminosity, with two stages in both the rise and decay of the light curves – a diffusion-controlled growth or decay, followed (in the case of a discontinuous switch) by a discontinuous jump in luminosity.

## 3 Limit cycles as a model of dwarf nova eruptions

### 3.1 A SIMPLE MODEL

We now examine the way in which cyclic oscillations of the disc about the equilibrium steady-state could act to generate dwarf nova eruptions. Our initial aim has been the limited one of obtaining a successful fit to the properties of the optical light curve. For the time being we ignore such problems as the simultaneous fitting of the optical, ultraviolet and X-ray light curves, the variation of the continuum spectral distribution and the behaviour in the two-colour diagram. As will become apparent, we find difficulties in fitting the optical light curve alone.

In Figs 3–7 we show results for the simplest limit-cycle model in which $\alpha$ is a two-valued function of the disc state, as described in the previous section. In this case $\alpha_1 = 0.01, \alpha_2 = 1.0$, $\dot{M} = 5 \times 10^{16}\,\mathrm{g\,s^{-1}}$, $M_1 = 1\,M_\odot$, $R_1 = 5 \times 10^8\,\mathrm{cm}$, $R_K = 2 \times 10^{10}\,\mathrm{cm}$, and $R_N = 8 \times 10^{10}\,\mathrm{cm}$. Switching from $\alpha_1$ to $\alpha_2$ occurs when the disc mass is $3.5 \times 10^{23}\,\mathrm{g}$ and from $\alpha_2$ to $\alpha_1$ when the disc mass is $2.5 \times 10^{23}\,\mathrm{g}$.

At time $t = 0$ the disc contains no mass. The initial build-up of mass in the disc, and the successive oscillations between the upper and lower values are shown in Fig. 3. The associated cyclic switches of $\alpha$ are shown in Fig. 4.

The luminosity and magnitude variations of the disc which the fluctuation in viscosity generates are shown in Figs 5 and 6. In the initial phase, as a torus grows and slowly spreads, the luminosity of the disc is negligible compared to the spot luminosity ($L_S/L_D > 10$). The torus orbits at a radius $R_K$, since there is no material to inhibit penetration to a Keplerian radius with the same specific angular momentum as the mass transfer stream. The first, preliminary outburst is therefore short, since the radius $R_K$ is close to the white dwarf surface ($R_K \sim R_N/4$) and, with $\alpha_2 = 1.0$, matter is rapidly transported on to the white dwarf surface.

This preliminary outburst is followed by a quiescent phase in which the spot luminosity again dominates. A torus now forms somewhat further out than $R_K$, due to the sharing of angular momentum between the incoming stream and disc material pushed out to $R_N$ in

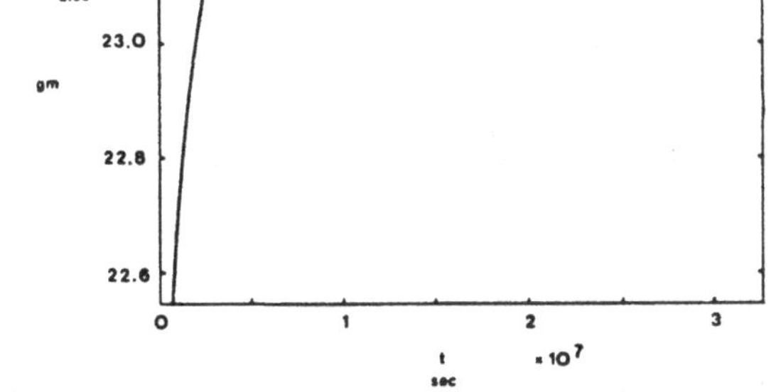

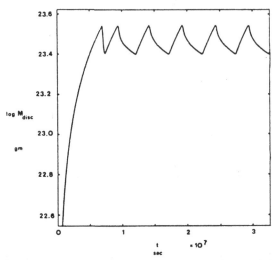

**Figure 3.** Variation of disc mass with time in a two-valued $\alpha$ model. $\alpha$ switches at values of $M_d = 3.5 \times 10^{23}\,\mathrm{g}$ ($0.01 \to 1.0$) and $M_d = 2.5 \times 10^{23}\,\mathrm{g}$ ($1.0 \to 0.01$).

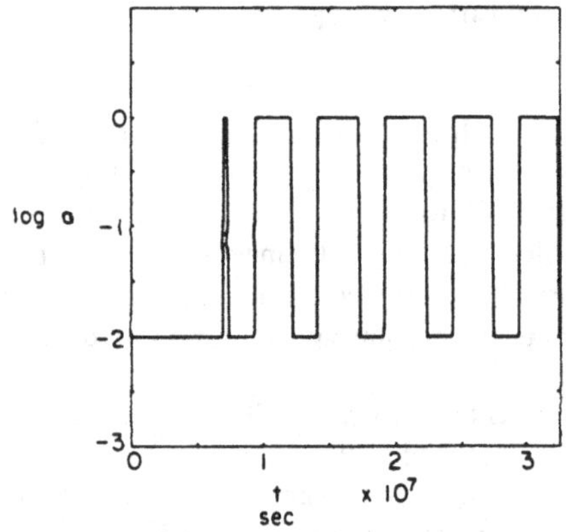


**Figure 4.** Variation of α with time in a two-valued α model.

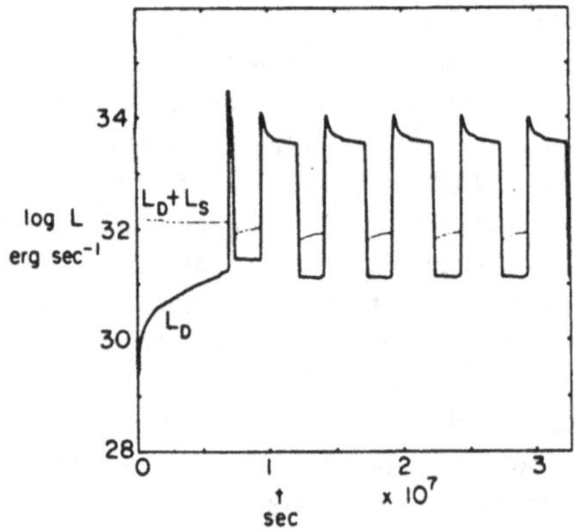


**Figure 5.** Luminosity changes of the disc and disc + spot for a two valued α model. In this model with $\alpha_1 = 0.01$, the spot dominates in quiescence. Since the torus is shri. king, the spot luminosity increases between outbursts.

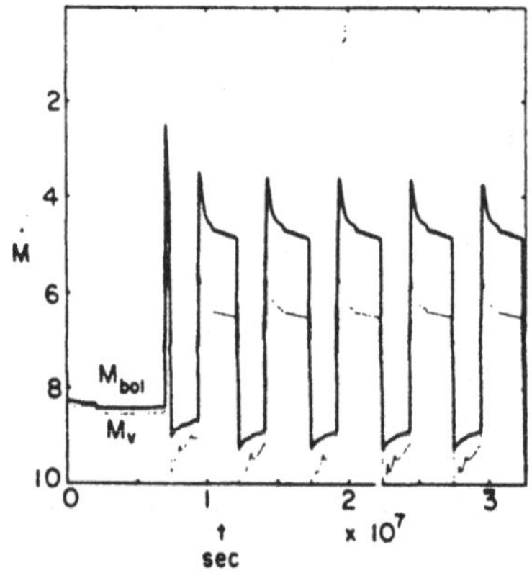


**Figure 6.** Resulting bolometric and visual light curve for a two-valued viscosity model. Two stages exist on the decline – diffusion toward the steady-state luminosity, followed by a discontinuous jump as α switches to the quiescent state value.

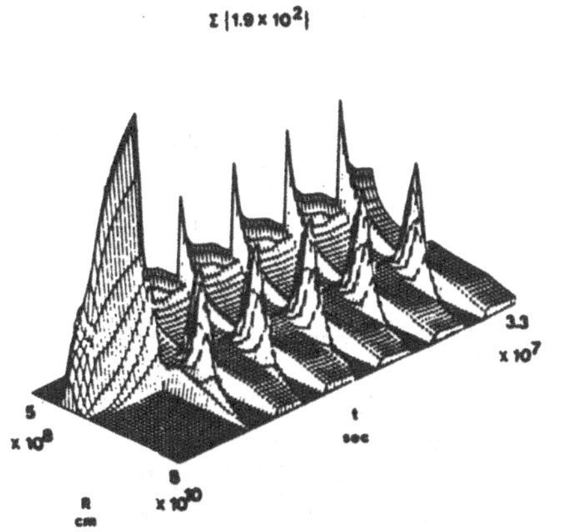


**Figure 7.** Evolution of surface density over the disc (in g $\mathrm{cm}^{-2}$). Once a limit cycle has been set up, two evolution phases are evident. First, a storage phase ($\alpha_1 = 0.01$) in which an outer torus grows in density but shrinks inward, as material with lower specific angular momentum is transferred from the companion. Second, spreading of the ring ($\alpha_2 = 1.0$) – out to the tidal radius and down to the surface of the white dwarf, where the densities rise rapidly and then decay toward a steady state. Note that the inner disc regions are never 'empty'. The maximum value of Σ is given above. The vertical scale is linear and the time coordinate starts at $t = 0$.

the previous outburst. As more matter impinges on the disc the torus shrinks and the spot luminosity consequently increases. With $\alpha_1 = 0.01$ the disc hardly evolves in the inter-outburst state and the spot luminosity dominates. Evolution of the disc through successive cycles is shown in the surface density plot of Fig. 7, which illustrates the basic cycle between a disc and disc + torus. Note that the inner disc region is *not* depleted of matter in quiescence. The lowest densities are at outburst, since only then is the viscosity large. With lower viscosity at quiescence the *whole disc* builds up to a higher density, with an outer torus growing fastest.

The final light curve of this model is shown in Fig. 6. A two-stage division in the decline is evident, as discussed in the previous section. Note that the outburst length, and the outburst repetition period depend sensitively on the upper and lower critical masses, and the values of $\alpha_1$ and $\alpha_2$.

### 3.2 FURTHER CONSIDERATIONS

The observed eruption light curves of dwarf novae bear little resemblance to the models described above. Normal dwarf nova outbursts show a smooth, roughly linear decline from outburst maximum, and the maximum itself is either flat-topped or triangular.

In an attempt to fit the observed light curve we next experiment with the functional form of $\alpha(t)$. Since no instability has been proposed which has a specific $\alpha$ variation, we consider ourselves free to choose the functional form of $\alpha(t)$ and ask what constraints can be placed on such a form by the observations. In doing this we are abandoning the considerations of Section 2 where we set about obtaining a limit cycle with viscosity being simply a function of local disc variables. We now invoke the intervention of some outside agency to give $\alpha$ as a function of time.

We first attempt to remove the discontinuous luminosity jump, generated by the discontinuous change in $\alpha$. This is easily achieved – a smooth variation in $\alpha$ generates a smooth variation of the light curve.

Figs 8–13 illustrate such a model. Fig. 8 shows the assumed functional form for $\alpha(t)$. We

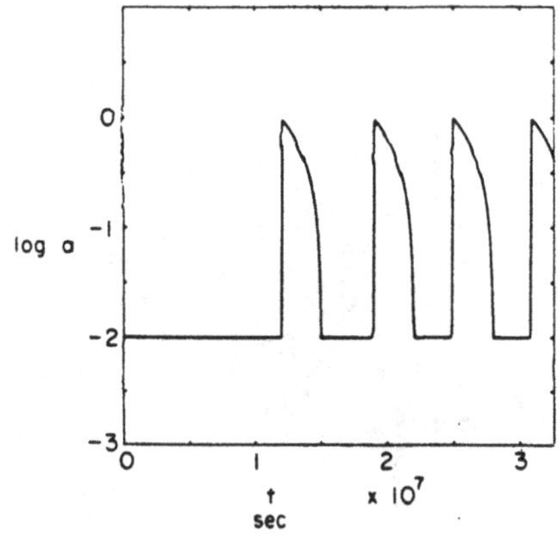


**Figure 8.** Changes of α with time for the basic model. α increases to 1.0 when $M_d = 6 \times 10^{23}$ g, and then falls off linearly to 0.01 in $3 \times 10^6$ s.

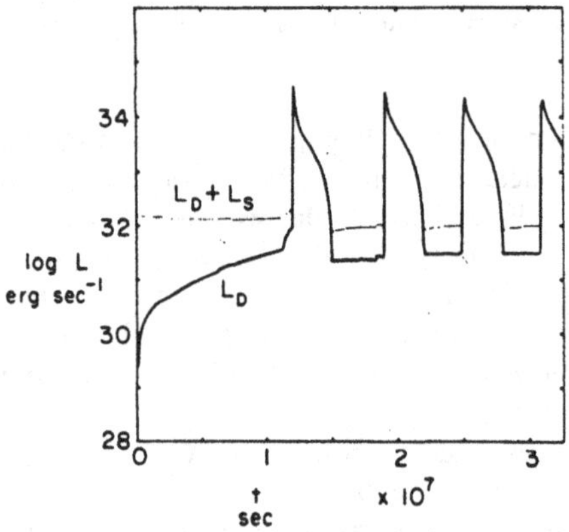


**Figure 9.** Luminosity curve of the basic model. The initial decay is due to diffusion of matter out of the torus. The later decay of the light curve is due to the rate of dissipation falling as α falls.

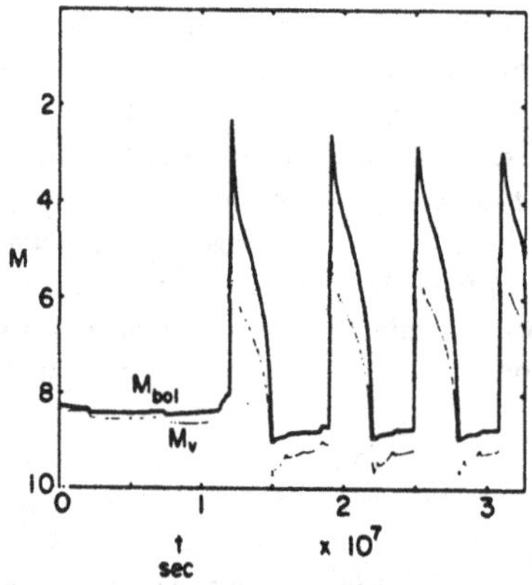


**Figure 10.** Bolometric and visual light curves of the basic model.

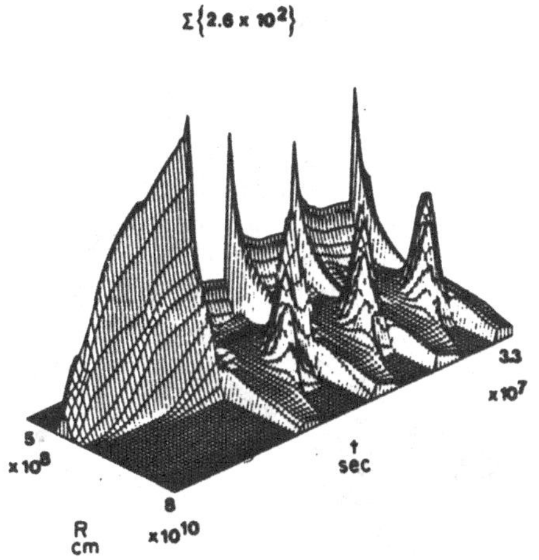


**Figure 11.** Evolution of surface density over the disc (in g cm$^{-2}$) showing the two stages of torus-storage and diffusion down onto the white dwarf surface. The maximum value of Σ is shown above. The vertical scale is linear and the time coordinate starts at $t$ = 0.0.

assumed that α increases instantly from 0.01 to 1.0 and then falls back linearly to 0.01 on a fall-off time of $3 \times 10^6$ s when the disc mass reaches an upper limit of $6 \times 10^{23}$ g. The luminosity variations are shown in Fig. 9. This model gives a smooth decline from maximum light back to the quiescent state, apparent in the visual light curve of Fig. 10. These light curves look more similar to those of dwarf novae.

The changes in disc structure are shown in Figs 11–13. The change in surface density (Fig. 11) is fundamentally the same as in the previous model. The smooth change in α causes the disc to freeze-up progressively after outburst. The inner disc region shows a clearly visible density *increase* at quiescence.

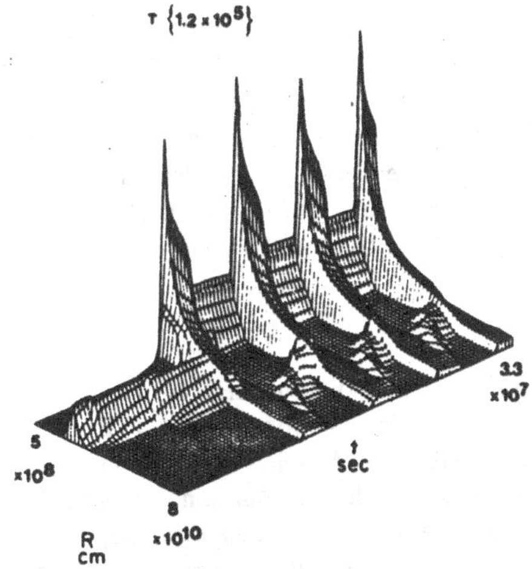


**Figure 12.** Evolution of central disc temperature (in K). In the storage phase only the hot spot and innermost disc regions are at $T > 10^4$. In the outburst phase the disc has a roughly steady-state distribution. The maximum value of $T$ is shown above. The vertical scale is linear and the time coordinate starts at $t$ = 0.0.

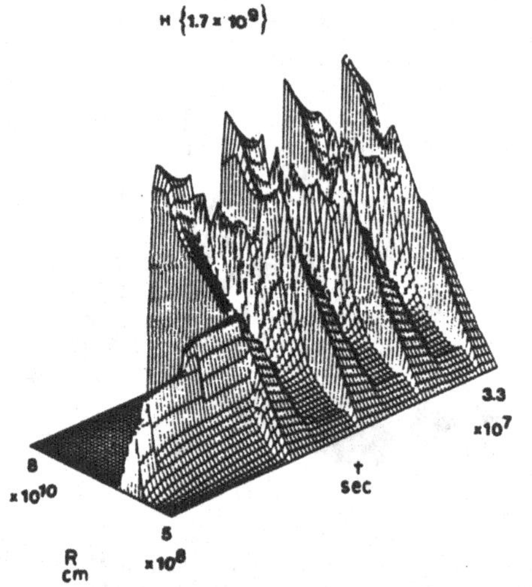


**Figure 13.** Evolution of disc semi-thickness (in cm). The maximum value of $H$ is given above and the time coordinate starts at $t = 0.0$.

The associated changes in central ($z = 0$) temperature are shown in Fig. 12. Increased dissipation at outburst produces a temperature increase over the whole disc region. At quiescence only the extreme innermost region is hot, together with the bright-spot collision region. Note that the temperature of the inner region would be cooler if the quiescent value of $\alpha$ were less than 0.01. In general we find $T < 10^4$ K and most of the inner region optically thin, if $\alpha_1$ is low enough to produce a torus of stored material in typical dwarf nova eruption periods. Changes in disc semi-thickness are shown in Fig. 13.

The effect of increasing the quiescent value of $\alpha$ is shown in the next model (Figs 14–16) which has $\alpha_1 = 0.1$, and the same critical maximum mass. The variations in $\alpha$ are shown in Fig. 14 and the resulting light curve in Fig. 16. The effect of increasing $\alpha_1$ to 0.1 is to allow the disc time in which to relax towards the equilibrium luminosity at quiescence (Fig. 15). Thus the quiescent state is not at constant luminosity, and the disc makes a significant contribution to the luminosity above the bright spot at quiescence.

This model is beginning to 'leak' at quiescence and could not provide an adequate fit to the observed light curves. The earlier model, and indeed any similar model with $\alpha_1 < 0.01$, does not experience this problem, but note that in that case the inner disc temperatures have $T < 10^4$ K, are optically thin and likely to be the strongest emission-line region.

## 4 Discussion

We have examined the possibility that the outbursts of dwarf novae are produced by variations of viscosity alone. Initial models (Sections 2 and 3.1) were restricted to those in which the viscosity at each point in the disc is solely a function of disc properties. We showed quite generally that, since the disc evolution equation is parabolic in character, it will evolve towards an equilibrium configuration. Thus, if $\dot{M}$ does not vary and there exists a stable steady-state configuration, the disc will find it and stay there. To circumvent this we suggested a functional form for the viscosity in which there is no stable steady configuration, and showed that this gave rise to limit-cycle behaviour in which the disc is continuously hunting for a stable state which is unachievable. Any solution of this form, however, must

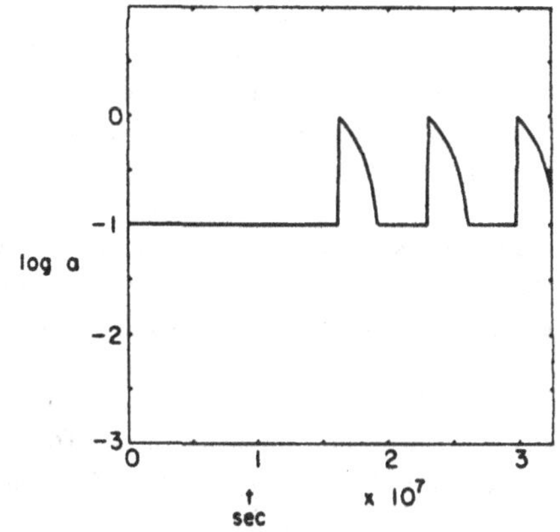


**Figure 14.** α variation of a model with the quiescent value of $\alpha_1$ increased to 0.1.

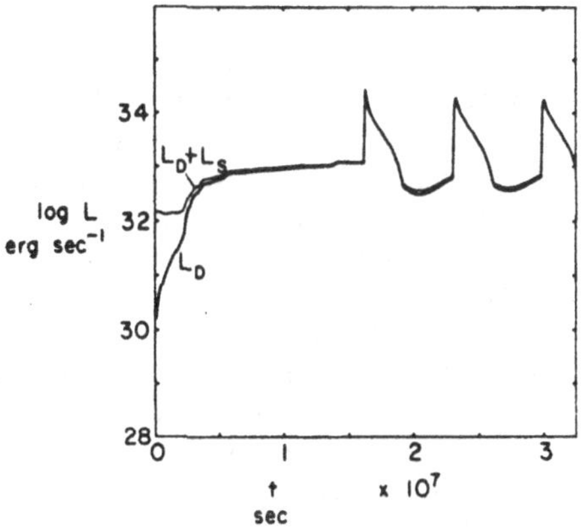


**Figure 15.** Luminosity changes in the model with $\alpha_1 = 0.1$. The torus is spreading significantly into a disc structure and now 'leaks' in the quiescent phase.

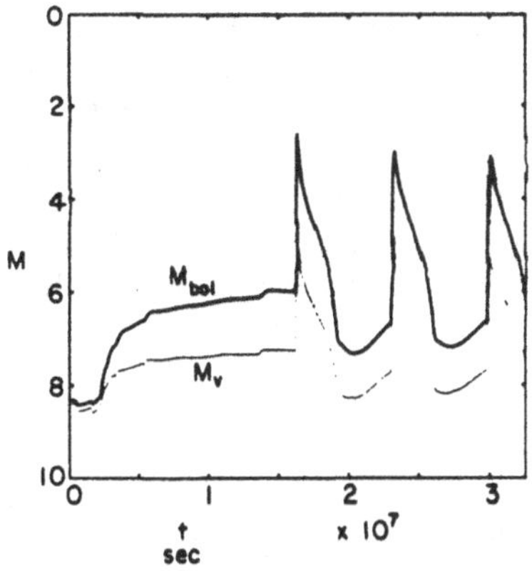


**Figure 16.** Light curve of the 'leaky' $\alpha = 0.1$ model.

involve discontinuous jumps in viscosity at stages of the limit cycle. Numerical models of such behaviour (Sections 2.2 and 3.1) show corresponding discontinuous jumps in the outburst light curve, not only on the rise but also on the decline (Figs 2 and 6).

In Section 3.2 the investigation was broadened to consider models in which the viscosity is determined by some outside agency as a given function of time. We therefore prescribe $\alpha(t)$ and ask what restrictions the observations place on the functional form of $\alpha$. We noted in the introduction that, in all models, in order to account for the variation time-scales in outburst, $\alpha$ must be of order unity at outburst. Two models have been considered, one (Figs 8–13) with the quiescent value of $\alpha$ being $\alpha_1 = 0.01$, and one (Figs 14–16) with $\alpha_1 = 0.1$.

The basic property of such discs is that, since $\dot{M}$ is fixed, the disc always tries to evolve towards an equilibrium steady-state of given luminosity, $L_{SS} = (GM_1 \dot{M}/4R_1)$. Then, during outburst, we must have $L > L_{SS}$ and hence $\alpha_2 > \alpha_1$. This implies immediately that the outburst must be sharply peaked (Figs 10 and 16). The flat- or smooth-topped outbursts often seen in dwarf novae cannot be produced by this mechanism, because a constant light curve implies a balance between the mass input rate ($\dot{M}$) and the mass through-put rate (dependent on $\alpha$). This is in contrast to the models discussed in Paper 1 where $\dot{M}$ increases during outburst.

Similar considerations apply to quiescence, when $L < L_{SS}$. Here the observations indicate a steady brightness, but a variable-viscosity model predicts a slow evolution, with $L$ increasing towards $L_{SS}$. The obvious way to circumvent this problem is to choose $\alpha_1$ sufficiently small that (a) little disc evolution occurs during quiescence and (b) the bright spot luminosity dominates by several orders of magnitude the disc luminosity. For the model parameters illustrated here, a value of $\alpha_1 = 0.1$ (Fig. 16) is too large, but $\alpha_1 \lesssim 0.01$ (Figs 9, 10) is sufficient for this purpose.

We have also found the choice of $\alpha(t)$ is important, in particular the way in which it varies from its maximum outburst value $\alpha_2$ to its quiescent value $\alpha_1$. The observed outburst light curves show a steady decline from maximum to minimum light. If $\alpha$ remains close to $\alpha_2$ for a time comparable to the viscous evolution at $\alpha_2$, say $t_\nu$ $(\alpha_2)$, then the luminosity $L$ is able to approach $L_{SS}$ by diffusion and the light curve flattens (see Fig. 2) toward $L_{SS}$. In order to obtain a steady decline from the region $L > L_{SS}$ to the region $L < L_{SS}$ it is necessary that $|\alpha/\dot{\alpha}| \sim t_\nu(\alpha)$. Once $L$ has dropped below $L_{SS}$, disc evolution essentially stops and $L$ depends almost solely on $\alpha$. At this stage $|\alpha/\dot{\alpha}| < t_\nu(\alpha)$. The shape of the light curve at the late stages of the decline is completely governed by the given function $\alpha(t)$. In general, in order for the light curve to be approximately linear over the whole decline, we require the rate of change of $\alpha$ as $\alpha$ approaches $\alpha_1$, $|\alpha/\dot{\alpha}| = t_\nu(\alpha_2)$.

If the changes in viscosity do not occur simultaneously at all disc radii, but $\alpha$ is a general function of space and time, $\alpha(R, t)$, then a variety of complex behaviour is obtained. At this stage we ignore this more general class of model in order to illustrate the fundamental features of the evolution equation. We note that, in general, variations of $\alpha$ with radius lead to a disc response-time which is determined by the lowest value of $\alpha(R)$. Low $\alpha$ at some radius $R_1 < R_S < R_N$ produces a local delay in the infall rate which acts to block inflow further in. This reduces the time-scale for global changes inside that radius to the local time-scale at $R_S$.

## 5 Conclusions

We have shown that global viscous variations of the limit-cycle type in which the viscosity is a function of local disc variables give rise to discontinuous jumps in viscosity, and hence luminosity, which do not match observed dwarf nova outburst light curves. Attempts to remove these discontinuities by allowing the viscosity to vary with time in an arbitrary manner are partly successful. For simple dynamical reasons, variations of viscosity always lead to outbursts which are sharply peaked; flat- or smooth-topped outbursts can be achieved only by allowing the mass input rate to vary. To achieve the smooth decline of the observed outburst light curve requires that the viscosity decline smoothly during outburst on a time-scale comparable to that of the luminosity decline. Too slow a fall of $\alpha$ produces a flat shoulder part-way down, followed by subsequent decline on a long time-scale, and too fast a fall produces a steep luminosity drop. Finally, the viscosity in quiescence must be sufficiently low that no disc evolution takes place between outburst. In this case the disc is so faint that the bright spot dominates as a luminosity source (generally by more than an order of magnitude).

We conclude that increased mass transfer must be occurring in at least some of the outbursts of dwarf novae, and that if viscous variations are causing some of the outbursts then the viscosity must be varying globally over the disc in a way which is not a simple function of local disc variables, but yet is strongly constrained.

## Acknowledgments

GTB thanks the National Radio Astronomy Observatory for hospitality while this paper was written.

## References

Bath, G. T., 1975. *Mon. Not. R. astr. Soc.*, **171**, 311.
Bath, G. T. & Pringle, J. E., 1981. *Mon. Not. R. astr. Soc.*, **194**, 967.
Lightman, A. P. & Eardley, D. M., 1974. *Astrophys. J.*, **187**, L1.
Madej, J. & Paczynski, B., 1977. *Nonstationary Evolution of Close Binaries, IInd Symp. Physics and Evolution of Stars*, ed. Zytkow, A., Polish Scientific Publishers, Warsaw.
Osaki, Y., 1974. *Publs astr. Soc. Japan*, **26**, 429.
Paczynski, B., 1977. *Nonstationary Evolution of Close Binaries, IInd Symp. Physics and Evolution of the Stars*, ed. Zytkow, A., Polish Scientific Publishers, Warsaw.
Papaloizou, J. C. B. & Bath, G. T., 1975. *Mon. Not. R. astr. Soc.*, **172**, 339.
Papaloizou, J. C. B. & Pringle, J. E., 1977. *Mon. Not. R. astr. Soc.*, **181**, 441.
Pringle, J. E., 1981. *Ann. Rev. Astr. Astrophys.*, **19**, 137.
Shakura, N. J. & Sunyaev, R. A., 1973. *Astr. Astrophys.*, **24**, 337.
Wood, P. R., 1977. *Astrophys. J.*, **217**, 530.

# Chapter 4 SOME FURTHER DEVELOPMENTS

THE ASTROPHYSICAL JOURNAL, 206:295–300, 1976 May 15


# THE EFFECT OF RADIATION PRESSURE ON ACCRETION DISKS AROUND BLACK HOLES

LAURA MARASCHI, CESARE REINA, AND ALDO TREVES
Istituto di Fisica dell'Università, and Laboratorio di Fisica Cosmica e Tecnologie, Relative del C.N.R., Milano
*Received 1975 September 29; revised 1975 November 7*

## ABSTRACT

Stationary disk accretion onto a black hole is studied for high accretion rates $\dot{M} \gtrsim \dot{M}_c = 2r_0L_E/GM$ ($L_E$ is the Eddington luminosity) for which the dynamic effect of radiation pressure is important. The rotation of the disk is not assumed to be Keplerian but is considered as an unknown in the Newtonian dynamic equation. The problem is reduced to a set of two differential equations which are solved numerically. It is found that stationary solutions without mass-outflow exist for $\dot{M} > \dot{M}_c$. The radiated luminosity, however, is always of the order of the Eddington luminosity. For increasing accretion rates, the kinetic energy swallowed by the hole and the size of the radiating region increase.

*Subject headings:* black holes — stars: accretion — X-rays: sources

## I. INTRODUCTION

Accretion of a black hole by matter endowed with angular momentum has been treated in connection with X-ray sources by Pringle and Rees (1972), and Shakura and Sunyaev (SS) (1973). Novikov and Thorne (1973) have considered general-relativistic corrections, Lightman (1974*b*) has studied the time-dependent problem. For further references see Lamb (1974).

It was noted by several authors (Salpeter 1972; Dilworth, Maraschi, and Reina 1973; Margon and Ostriker 1973) that the luminosity function of X-ray sources has a cutoff above $10^{38}$–$10^{39}$ ergs s$^{-1}$, which is of the same order of the Eddington luminosity $L_E = 1.2 \times 10^{38}$ ergs s$^{-1}$ of a 1 $M_\odot$ star. This indicates that, in the brightest X-ray sources, radiation pressure plays an important role.

When radiation pressure is important with respect to gravitation, the approximation adopted so far, that the rotation of the gas is Keplerian, needs to be reconsidered. In this paper we take the angular velocity as an unknown of the problem and show that in the inner region of the disk the Keplerian approximation becomes inadequate.

The flow of gas through the disk is treated with a number of simplifying assumptions. Thermal pressure is neglected compared with radiation pressure. Radiative transfer is described as a diffusive process, dominated by Thomson scattering. The consistency of these approximations is discussed *a posteriori*. Angular momentum transfer is treated as in SS, assuming that the shear is proportional to the total energy density. Newtonian dynamics is used throughout. Although this is not adequate near the inner boundary, relativistic corrections should not alter the qualitative features of the results (Novikov and Thorne 1973).

The resulting set of equations can be reduced to two coupled differential equations for the angular and radial velocities of the matter, which are solved numerically.

The main result of the paper is that a steady solution exists for every value of the accretion rate $\dot{M}$. However, for high values of $\dot{M}$ the luminosity radiated by the disk tends to an upper bound of the order of $L_E$, being no longer proportional to the accretion rate. The reason is that, under these conditions, the kinetic energy of the gas at the inner boundary is greater than in the Keplerian case. This energy is then swallowed by the hole.

## II. EQUATIONS

The relevant equations for our problem are: (1) mass conservation; (2) Euler equation; (3) hydrostatic equilibrium in the vertical direction; (4) conservation of angular momentum; (5) definition of tangential stress; (6) energy dissipation; (7) energy conservation; (8) photon diffusion.

With respect to the set considered by SS, equations (2), (7), and (8) have been added. The analytic expression of the equations is:

$$\frac{1}{r}\frac{\partial}{\partial r}(\rho r v_r) + \frac{\partial}{\partial z}(\rho v_r) = 0\,, \tag{1}$$

$$v_r\frac{\partial v_r}{\partial r} = -\frac{GM}{r^2} - \frac{1}{\rho}\frac{\partial}{\partial r}\frac{\epsilon}{3} + r\omega^2\,, \tag{2}$$

$$0 = -\frac{GM}{r^3}z - \frac{1}{\rho}\frac{\partial}{\partial z}\left(\frac{\epsilon}{3}\right)\,, \tag{3}$$

$$\rho v \frac{\partial}{\partial r}(r^2\omega) = -\frac{1}{r}\frac{\partial}{\partial r}(r^2 w_{r\varphi}), \tag{4}$$

$$w_{r\varphi} = -\alpha\epsilon, \tag{5}$$

$$q = w_{r\varphi} r \frac{\partial\omega}{\partial r}, \tag{6}$$

$$\nabla \cdot F = q, \tag{7}$$

$$F = -\frac{cm_p}{\sigma_T\rho}\nabla\left(\frac{\epsilon}{3}\right), \tag{8}$$

where $\rho$ is the density, $v_r$ is the inward velocity, $\omega$ is the angular velocity, $\epsilon$ is the photon energy density, $w$ is the shear stress, $q$ is the power dissipated per unit volume, $F$ is the radiative flux, $c$ is the velocity of light, $m_p$ is the proton mass, $\sigma_T$ is the Thomson cross section, and $\alpha$ is a constant describing the viscosity.

The previous equations can be simplified by integrating over the vertical coordinate (see, e.g., Lightman 1974*a*). Introducing the new variables:

$$\Sigma = \int_0^\infty \rho dz, \qquad \epsilon = \int_0^\infty \epsilon dz,$$

$$W_{r\varphi} = \int_0^\infty w_{r\varphi} dz, \qquad Q = \int_0^\infty q dz,$$

and the scale height

$$h = \left(\frac{\epsilon}{3\Sigma}\frac{r^3}{GM}\right)^{1/2},$$

assuming that $v_r$ is independent of $z$ and $v_z \ll v_r = v$, the equations become:

$$4\pi r\Sigma v = \dot{M}, \tag{1a}$$

$$v\frac{dv}{dr} = -\frac{GM}{r^2} - \frac{1}{\Sigma}\frac{d}{dr}\left(\frac{\epsilon}{3}\right) + r\omega^2, \tag{2a}$$

$$W_{r\varphi} = -\frac{\dot{M}}{4\pi}\left[\omega - \omega_0\left(\frac{r_0}{r}\right)^2\right], \tag{4a}$$

$$W_{r\varphi} = -\alpha\epsilon, \tag{5a}$$

$$Q = W_{r\varphi} r \frac{d\omega}{dr}, \tag{6a}$$

$$\frac{1}{r}\frac{\partial}{\partial r}\left[r\left(-v\frac{dv}{dr} - \frac{GM}{r^2} + r\omega^2\right)\right] - \frac{GM}{r^3} = \frac{\sigma_T}{cm_p}\frac{Q}{h}, \tag{7a}$$

where $\dot{M}$ is the accretion rate and $r_0$ is the radius at which the shear stress $w_{r\varphi}$ is null. This boundary condition allows one to obtain (4a) from integration of (4). In the case treated by SS, $r_0$ corresponds to the radius of the last stable orbit, inside which the flow becomes essentially radial. Equation (7a) has been obtained from equations (2), (7), and (8), assuming $q = Q/h$.

The system (1)–(7) can be reduced to two differential equations for $\omega$ and $v$ which read

$$\frac{\partial\omega}{\partial r} = -\frac{3\alpha}{rv}\left(v\frac{dv}{dr} + \frac{GM}{r^2} - \omega^2 r\right) - 2\omega_0\frac{r_0^2}{r^3}, \tag{1}$$

$$\frac{\partial}{\partial r}\left(r^2\omega^2 - rv\frac{dv}{dr}\right) = \frac{\sigma_T}{cm_p}\frac{\dot{M}}{4\pi}(3\alpha GM)^{1/2}\left(\frac{\omega r^2 - \omega_0 r_0^2}{r^2 v^2}\right)^{1/2}\frac{d\omega}{dr}. \tag{2}$$

Introducing nondimensional variables

$$x = r/r_0, \qquad A = v(GM/r_0)^{-1/2}, \qquad B = \omega(GM/r_0^3)^{-1/2},$$

the equations become

$$\frac{dB}{dx} = -\frac{3\alpha}{Ax}\left(A\frac{dA}{dx} + \frac{1}{x^2} - xB^2\right) - \frac{2B}{x^3} \tag{1}$$

$$\frac{d^2A}{dx^2} = \frac{1}{xA}\left[\frac{d}{dx}(x^2B^2) - A\frac{dA}{dx} - x\left(\frac{dA}{dx}\right)^2 - 2(3\alpha)^{1/2}\frac{\dot{M}}{\dot{M}_c}\left(\frac{B - B_0x^{-2}}{A}\right)\frac{dB}{dx}\right], \tag{2}$$

where $B_0 = B(1)$, $\dot{M}_c = 2r_0L_E/GM$, and $L_E = 4\pi GMm_pc/\sigma_T$ is the Eddington luminosity.

When the effects of radiation pressure are negligible, i.e., $\dot{M} \ll \dot{M}_c = 10^{-8}\ M_\odot\ \mathrm{yr}^{-1}$, one has $v \ll \omega r$ and on the last stable orbit $\omega_0 = \omega_{0k} = (GM/r_0^3)^{1/2}$. In this case the energy released by the infalling gas is $GM\dot{M}/2r_0$.

## III. SOLUTION OF THE EQUATIONS

We have solved the system numerically starting from an asymptotic solution for large values of $x$. The dominant terms in the asymptotic solution are $B \propto x^{-1.5}$, $A \propto Hx^{-2.5}$, where $H = 27\alpha(\dot{M}/\dot{M}_c)^2$.

In order to expand the right-hand term of equation (2) to the relevant order $(x^{-9/2})$, the asymptotic solution must be determined for $B$ up to terms $O(x^{-5.5})$ and for $A$ up to terms $O(x^{-6.5})$.

The resulting expansions are

$$B \propto x^{-1.5} - \frac{H}{4\alpha}x^{-3.5} + \frac{7}{12}\frac{HB_0}{\alpha}x^{-4} - \frac{1}{3}\frac{HB_0^2}{\alpha}x^{-4.5} - \left(\frac{9}{32}\frac{H^2}{\alpha^2} + \frac{5}{4}H^2\right)x^{-5.5} + o(x^{-5.5}),$$

$$A \propto H\left[x^{-2.5} - B_0x^{-3} + \frac{19}{12}\frac{H}{\alpha}x^{-4.5} - \frac{227}{36}\frac{H}{\alpha}B_0x^{-5} + \frac{145}{18}\frac{H}{\alpha}B_0^2x^{-5.5} - \frac{10}{3}\frac{H}{\alpha}B_0^3x^{-6} + \left(-\frac{125}{12}H^2 + \frac{1667}{288}\frac{H^2}{\alpha^2}\right)x^{-6.5} + o(x^{-6.5})\right].$$

The standard finite-difference method has been used for constructing numerical solutions. For $\dot{M} > \dot{M}_c$, solutions are obtained which are insensitive to variations of the initial point and step of the integration. $B_0$, which appears

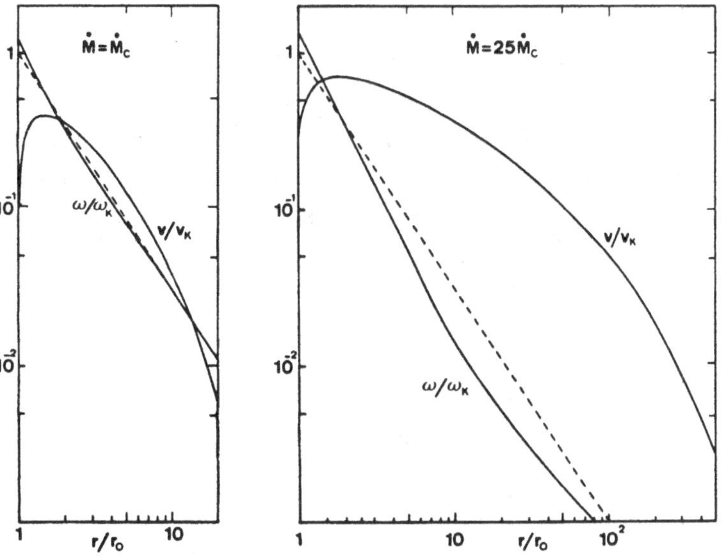


FIG. 1a  FIG. 1b

FIGS. 1a, 1b.—Angular velocity $\omega$, and radial velocity $v$, of the infalling gas, versus distance for different values of the accretion rate $\dot{M}$. $\omega_k = (GM/r_0^3)^{1/2}$, $v_k = (GM/r_0)^{1/2}$, and $\dot{M}_c = 2r_0L_E/GM$. The dotted line corresponds to the Keplerian angular velocity.

as an eigenvalue, is determined by an iterative process. For $\dot{M} < \dot{M}_c$, numerical instabilities at large values of $x$ prevent the construction of solutions by this crude method. The same happens for values of $\alpha < 1$.

In the following we discuss results for $\alpha = 1$ and $1 < \dot{M}/\dot{M}_c < 25$, the upper limit being imposed by computing time considerations. We think that this range of values is sufficiently representative to allow interesting conclusions.

## IV. RESULTS AND DISCUSSION

In Figure 1 the angular and radial velocities are shown for $\dot{M}/\dot{M}_c = 1$ and 25. For large values of $r$ the numerical solutions tend to the Keplerian values. Approaching the black hole, the angular velocity is found to be significantly smaller than in the Keplerian approximation. For increasing accretion rates the deviations become larger and start from larger values of $r$. In the very vicinity of the hole the tendency is opposite. The angular velocity grows very rapidly, approaching the law $\omega \propto r^{-2}$, and for $r \to r_0$ it becomes larger than the Keplerian value $\omega_{0k} = \sqrt{(GM/r_0^3)}$. The eigenvalues for $\dot{M}/\dot{M}_c = 2, 5, 25$ are $\omega_0 = 1.2\omega_{0k}, 1.3\omega_{0k}, 1.35\omega_{0k}$, respectively.

The physical reason underlying this behavior is that, since no energy is dissipated beyond the boundary $r_0$, the pressure gradient is directed inward for $r \sim r_0$, while for larger radii it is directed outward.

The total energy radiated by the infalling gas is

$$L = \dot{M}GM/r_0 - \tfrac{1}{2}(\dot{M}\omega_0^2 r_0^2 + \dot{M}v^2),$$

where the term in brackets represents the kinetic energy at the last stable orbit, which is swallowed by the hole and therefore does not contribute to the luminosity. In the Keplerian approximation $\omega_0 = \omega_{0k}$, $v \approx 0$, and therefore the radiated energy is $\tfrac{1}{2}(\dot{M}GM/r_0)$. In the case considered here, since $\omega_0 > \omega_{0k}$ and $v \neq 0$, the luminosity is smaller and does not depend linearly on the accretion rate. In Figure 2, $L$ is given versus the accretion rate. The saturation effect is clearly visible. It seems that the curve is upper-bounded, but from the limited range of accretion rates for which the solution has been computed it is not possible to determine the value of the limiting luminosity.

In Figure 3 the thickness $h$ and the mean density $\rho = \Sigma/h$ of the disk are given. For large values of $x$ a larger thickness corresponds to a larger accretion rate in agreement with the asymptotic expressions (see SS). Near the hole the thickness decreases rapidly and, for fixed values of $r$, the disk is thicker for smaller values of the accretion rate. Note that $h/r$ is always less than one; i.e., no region of spherization is found, as was suggested by SS. The energy radiated by the disk within a radius $r$,

$$L(r) = \int_{r_0}^{r} 4\pi r Q dr,$$

is given in Figure 4. It is apparent that the region where most of the luminosity is produced increases with increasing accretion rate.

On the basis of the numerical results we can discuss the consistency of our approximations. The disk is thick to Thomson scattering; and free-free opacity, calculated at the minimum blackbody temperature, is indeed negligible with respect to Thomson opacity, therefore justifying the approximations used in the treatment of the radiation transfer.

We now consider the assumption that radiation pressure dominates with respect to kinetic pressure. It turns out that this is the case if $T \leqslant 10^{10}$ K. In the case $\dot{M} = 25\,\dot{M}_c$ this is always true, as in the optically thick regions, $(\tau_{es}\tau_{ff})^{1/2} \geqslant 1$, the blackbody temperature is less than $10^7$ K and in the intermediate optically thin region, $10 \leqslant r/r_0 \leqslant 100$, the free-free temperature is less than $10^9$ K. In the case $\dot{M} = \dot{M}_c$ the disk is almost everywhere

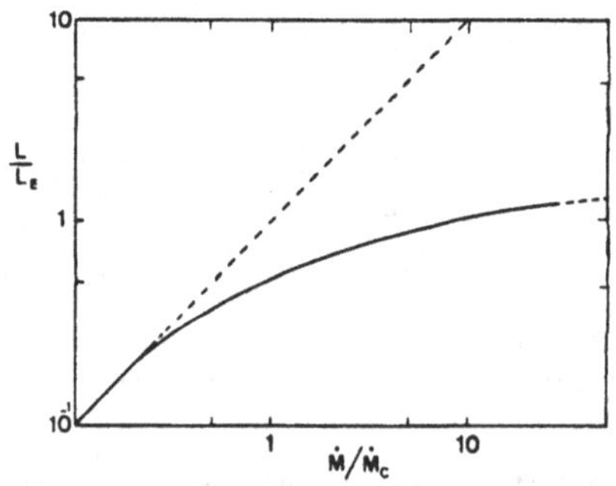


FIG. 2.—Total luminosity radiated by the disk versus accretion rate, $L_E = 4\pi GMm_p c/\sigma_T$, $\dot{M}_c = 2r_0L_E/GM$

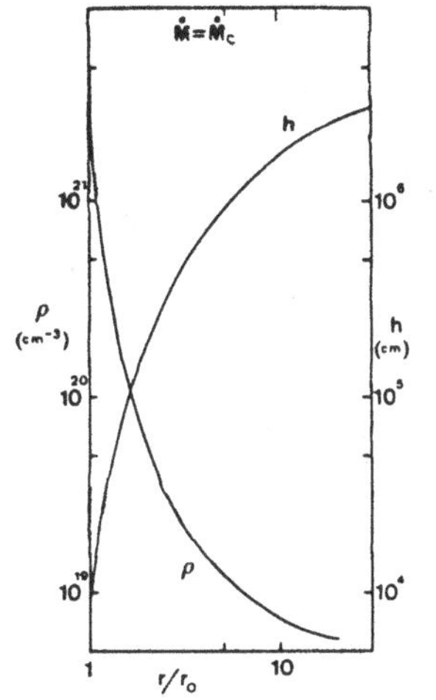


FIG. 3a

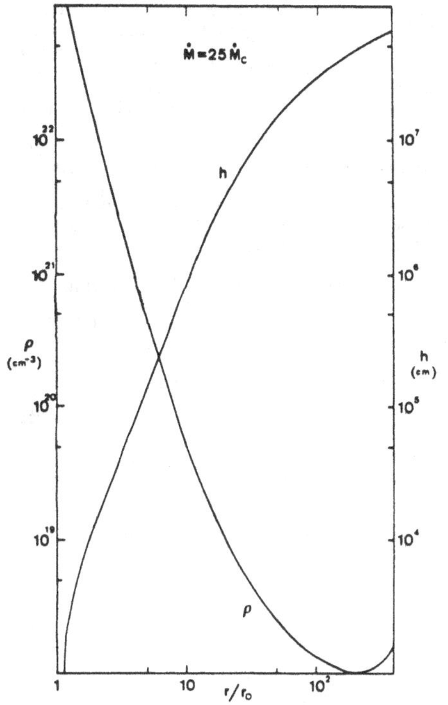


FIG. 3b

FIGS. 3a, 3b.—Half-thickness $h$ and mean density $\rho$ of the disk versus distance for different values of the accretion rate $\dot{M}$

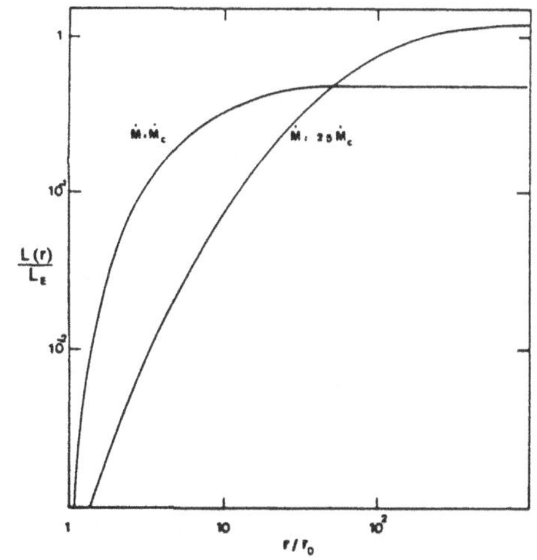


FIG. 4.—Luminosity radiated by the disk within distance $r$ for different values of the accretion rate $M$

optically thin in the region of interest; and if one calculates the free-free temperature, one gets values between $10^{10}$ and $10^{12}$ K for $1.5 \leq r/r_0 \leq 10$. However, since Thomson scattering opacity is appreciable ($\tau_{es} > 1$), at such high temperatures energy losses by Compton scattering will be important, resulting in a substantial reduction of the temperature (by a factor $10^2$–$10^4$) as discussed by SS. Therefore also in the case $\dot{M} = \dot{M}_c$ the solution is consistent with the assumption that the disk region under consideration is radiation pressure dominated.

A final point is the discussion of the actual value of the inner radius of the disk $r_0$, which in the numerical estimates has been assumed to be $r_0 = 6GM/c^2$. This value, which corresponds to the last stable orbit of a test particle in a Schwarzschild field, is generally taken as the inner boundary of accretion disks around black holes (SS). In the present case the effect of radiation pressure, calculated in the Newtonian approximation, is that of increasing the angular momentum of the stable orbits near the inner boundary, and therefore the value of $r_0$ could conceivably be altered. However, a boundary, where the shear stress $w_{r\phi} \simeq 0$, should exist at $r \geqslant 2GM/c^2$, and the main results of our treatment depend only on the existence of such a boundary. In particular the dependence of the radiated luminosity $L$ on the accretion rate $\dot{M}$ should not change essentially, even if $r_0$ is a function of $\dot{M}$. In fact, given a black boundary, radiation pressure would add to gravitation, allowing more kinetic energy to be swallowed by the hole than in the case where radiation pressure is neglected. This argument would remain valid even if relativistic corrections were taken into account.

We are grateful to Professor M. Rees and Dr. A. Lightman for valuable comments.

## REFERENCES

Dilworth, C., Maraschi, L., and Reina, C. 1973, *Astr. and Ap.*, **28**, 71.
Lamb, F. K. 1974, Rapporteur Paper Int. Conference on X-Rays in Space, Calgary.
Lightman, A. P. 1974*a*, *Ap. J.*, **194**, 419.
———. 1974*b*, *ibid.*, p. 429.
Margon, B., and Ostriker, J. P. 1973, *Ap. J.*, **186**, 9.
Novikov, I., and Thorne, K. S. 1973, in *Black Holes*, ed. C. and B. deWitt (London: Gordon & Breach).
Pringle, J. E., and Rees, M. J. 1972, *Astr. and Ap.*, **21**, 1.
Salpeter, E. E. 1973, in *IAU Symposium No. 55, X- and Gamma-Ray Astronomy*, ed. H. Bradt and R. Giacconi (Dordrecht: Reidel).
Shakura, N. I., and Sunyaev, R. A. 1973, *Astr. and Ap.*, **24**, 337 (SS).

Laura Maraschi, Cesare Reina, and Aldo Treves: Istituto di Fisica, Via Celoria 16, 20133 Milano, Italy

THE ASTROPHYSICAL JOURNAL 242: 772–788, 1980 December 1


# THICK ACCRETION DISKS WITH SUPER-EDDINGTON LUMINOSITIES[1]

MAREK A. ABRAMOWICZ[2]
Center for Relativity, Department of Physics, The University of Texas at Austin
MASSIMO CALVANI
Istituto di Astronomia, Università degli Studi di Padova, Italy
AND
LUCIANO NOBILI
Istituto di Fisica Galileo Galilei, Università degli Studi di Padova, Italy
*Received 1980 February 25; accepted 1980 June 3*

## ABSTRACT

We describe a Newtonian version of the theory of thick accretion disks orbiting black holes. In view of the present inadequate knowledge of microscopic viscosity processes, this theory adopts a macroscopic approach. All the uncertainties are absorbed in currently arbitrary functions $\lambda(\xi)$ and $f(\xi)$ which describe the (non-Keplerian) angular momentum along the disk, and the outgoing radiation flux in units of the critical flux. Assuming surface distributions of angular momentum and radiation flux, one can construct a model of a thick, non-Keplerian accretion disk with no knowledge of the viscosity mechanism. Thick accretion disks can have luminosities 100 times above the Eddington limit. We study the self-consistency constraints for the theory of thick accretion disks with super-Eddington luminosities.

*Subject headings:* black holes — galaxies: nuclei — quasars

## I. INTRODUCTION

In the *standard* theory of accretion disks (Pringle and Rees 1972; Shakura and Sunyaev 1973; Novikov and Thorne 1973; Lynden-Bell and Pringle 1974), one assumes *a priori* that the angular momentum distribution is Keplerian and that the dissipative processes can be described by the so-called $\alpha$-viscosity. In the standard theory, accretion disks are optically thick but geometrically very thin in the vertical direction.

Recently Paczyński and Wiita (1980, hereafter PW), and Jaroszyński, Abramowicz, and Paczyński (1980, hereafter JAP) have proposed a new type of thick accretion disk model. In their approach one assumes *a priori two* structure functions: the surface distributions of the angular momentum and of the flux of radiation. Having assumed these two functions, one can compute not only the total luminosity $L$ and the accretion rate $\dot{M}$, but also other disk characteristics (such as shape) with *no* reference to the disk interior at all.

For thin accretion disks a rough estimate of the maximum possible luminosity is the *Eddington* one, i.e., the luminosity for which the pull of gravity on accreting matter is precisely counterbalanced by the outward force of photons that scatter on the matter's electrons. Many authors have found that if a disk's luminosity exceeds the Eddington value, then some material would be blown off by the pressure of the supercritical radiation flux (Bisnovatyi-Kogan and Blinnikov 1977; Liang and Price 1977). This has sometimes led to the opinion that no accretion disk will be able to have a luminosity much in excess of the Eddington one. This opinion, however, is based on our experience with the standard theory of geometrically thin disks, and it is exactly the assumption that the thickness is small which prevents the disks from having highly supercritical luminosities. PW and JAP have constructed thick accretion disk models which are in mechanical equilibrium and have luminosities greater than the critical one by more than an order of magnitude.

The aim of this paper is to study all the physical limitations to the applicability of the theory of thick accretion disks. We shall prove that the absolute upper limit for the luminosity of non-self-gravitating disks orbiting supermassive black holes decreases with the hole mass. For a $10^8\ M_\odot$ black hole this limit is about 100 Eddington luminosities, i.e., $10^{48}$ ergs s$^{-1}$.

The Newtonian theory of gravitation will be used, and this deserves a few comments. First, there are no conceptually difficult problems connected with a general-relativistic version of the theory. Second, adopting the realistic point of view of Blandford and Thorne (1979), sophisticated relativistic calculations will probably not play a crucial role in theoretical models of accretion disks for some time to come, because of the extreme uncertainties inherent in the microphysics. There

[1] Work supported by Italian Council of Research and U.S. National Science Foundation, grant AST 7923166.
[2] On leave from N. Copernicus Astronomical Center, Warszawa, Poland.

are some purely general relativistic effects whose influence is important, but most of them can be easily modeled in Newtonian theory by some modification of the potential. While one is interested in non-self-gravitating disks, such modifications are not damaging to the self-consistency of the models.

## II. THE EDDINGTON LUMINOSITY

In mechanical equilibrium the flux of radiation, $F$, emitted locally from the surface of a stationary and electromagnetically neutral body reaches its maximal (critical) value when the effective gravitational force is balanced by the radiation-pressure-gradient force:

$$F_{\rm crit} = -\frac{c}{\kappa} g_{\rm eff} \,. \tag{2.1}$$

Here $\kappa$ is the opacity per unit mass and $g_{\rm eff}$ is the effective gravity (acceleration) on the surface. Therefore, the maximal luminosity is

$$L_{\rm max} = -\frac{c}{\kappa}\int_\Sigma g_{\rm eff}\cdot d\Sigma \,, \tag{2.2}$$

where $\Sigma$ is the surface of the body and $d\Sigma$ is the oriented surface element. If the body does not rotate, $g_{\rm eff} = \nabla\Phi$ with $\Phi$ being the gravitational potential. Let us assume that the body consists of a solid "nucleus" surrounded by a gaseous "atmosphere" such that $\Sigma$ is an equipotential surface, $\Phi$ = const. As there are no restrictions on the shape of the nucleus, the surface $\Sigma$ can be highly nonspherical (in both geometry and topology). However, regardless of this, Gauss's theorem guarantees that

$$\int_\Sigma \nabla\Phi\cdot d\Sigma = -4\pi G M_\Sigma, \tag{2.3}$$

where $M_\Sigma$ is the mass inside $\Sigma$. Thus, the maximal luminosity of *any* nonrotating, stationary, electromagnetically neutral object (no matter how complicated a shape it has) cannot exceed the Eddington luminosity,

$$L_{\rm Edd} = \frac{4\pi G M c}{\kappa}, \tag{2.4}$$

with $M \geq M_\Sigma$ being the *total* mass of the object.

*Only rotating objects can have* $L > L_{\rm Edd}$, and the reason for this is quite simple: in the case of a rotating object we have $\nabla\cdot g_{\rm eff} = -4\pi G\epsilon - 2\sigma^2 + 2\omega^2$, where $\epsilon$ is the mass density, $\sigma$ is the shear, and $\omega$ is the vorticity. The surface integral $\int_\Sigma g_{\rm eff}\cdot d\Sigma$ can be transformed to a volume integral $-\int_v (4\pi G\epsilon + 2\sigma^2 - 2\omega^2)dV$. The matter contribution alone gives precisely the Eddington limit. The rotation contribution in the case $\omega = 0$ (i.e., almost constant angular momentum) may be very large. Thus, objects with big shear, small vorticity, and small density (in the sense $2\pi G\epsilon/\sigma^2 \ll 1$) may have super-Eddington luminosities. We shall see that in the case of thick accretion disks one has indeed $\omega^2/\sigma^2 \ll 1$, $2\pi G\epsilon/\sigma^2 \ll 1$ and $L/L_{\rm Edd} \gg 1$.

If Thomson scattering provides the main source for opacity and the relevant material is fully ionized, one has $\kappa = \sigma_{\rm T}/m_p$, where $m_p$ is the mass of the proton and $\sigma_{\rm T}$ is the Thomson cross section. In this case,

$$L_{\rm Edd} = \frac{4\pi G c m_p M}{\sigma_{\rm T}} = 1.257\times 10^{38}\frac{M}{M_\odot}\ {\rm ergs\ s^{-1}}\,. \tag{2.5}$$

As Gunn pointed out some time ago, no object can radiate more energy than $E = Mc^2$ in a time shorter than $t = r_g/c$, where $r_g \equiv 2GM/c^2$ is the gravitational radius of the object. According to these fundamental limitations no object can have its luminosity, $L = E/t$, greater than

$$L_{\rm Gunn} \equiv c^5/G \equiv 3.629\times 10^{59}\ {\rm ergs\ s^{-1}}\,. \tag{2.6}$$

Note that

$$L_{\rm Edd} = L_{\rm Gunn}\left(\frac{\pi r_g^2}{N\sigma_{\rm T}}\right), \tag{2.7}$$

with the "efficiency factor" equal to the "total gravitational cross-section" of the star, i.e., $\pi r_g^2$, divided by the "total scattering cross-section," i.e., $\sigma_{\rm T}N$, where $N = M/m_p$ is the number of electrons in the star. From (2.5) and (2.6) it follows that there is an absolute upper limit to the mass of objects which can radiate at the Eddington rate:

$$M_* = \frac{c^4\sigma_{\rm T}}{4G^2 m_p} = \left(\frac{\sigma_{\rm T}}{r_{gp}^2}\right)m_p = 1.078\times 10^{79} m_p = 1.804\times 10^{55}[{\rm g}] = 0.907\times 10^{22} M_\odot\,, \tag{2.8}$$

where $r_{gp} \equiv 2Gm_p/c^2 = 2.4838 \times 10^{-52}$ cm is the "gravitational radius of the proton."[3] Note that $M_*/m_p \approx 10^{80}$ is one of Dirac's *big numbers*. According to the big number hypothesis $M_* \approx 10^{80} m_p$ is close to the "total mass of the universe" (Bondi 1960), so the limit (2.8) is practically not restrictive.

## III. THE MODEL

We employ cylindrical coordinates $r, z, \phi$, with $z = 0$ being the equatorial symmetry plane and $r = 0$ being the axis of rotation. Our models are axisymmetric and stationary, so that any physical or geometrical quantity $X$ can be expressed as $X = X(r, z)$. The equation for the surface of the disk follows from the fact that on the surface pressure vanishes, $p = p(r, z) = 0$. This equation, when solved with respect to $z$, gives the half-thickness of the disk, $z = h(r)$. The *surface distribution* of $X$ is therefore a function of *one* variable only: $X = X[r, h(r)] = X(r)$. Following PW and JAP, we shall assume that one knows *a priori* the surface distributions of the angular momentum, $l = l(r)$, and of the flux of outgoing radiation in terms of the critical flux: $f = f(r) \equiv |F(r)|/|F_{crit}(r)|$, where $F(r)$ is the surface flux in ergs cm$^{-2}$ s$^{-1}$. According to stability criteria one has $dl/dr \geq 0$; and because the angular momentum has to be transported outward to make accretion possible, one has $d(l/r^2)/dr \equiv d\Omega/dr < 0$, where $\Omega$ is the angular velocity. The function $f(r)$ should obey the condition $0 \leq |f| < 1$.

We assume that the disk does not contribute to the gravitational field of the system, i.e., that the gravitational potential $\Phi$ is due to the central body alone. The necessary condition for this is $m/M \ll 1$, where $m$ is the mass of the disk and $M$ the total mass of the system. Although our approach is valid for any potential, $\Phi = \Phi(r, z)$, we assume for simplicity

$$\Phi = -\frac{GM}{(r^2 + z^2)^{1/2}} . \tag{3.1}$$

The angular momentum of free test particles on circular orbits in the equatorial plane,

$$l(r) = l_K(r) \equiv [r^3(\partial\Phi/\partial r)_{z=0}]^{1/2} = (GMr)^{1/2} , \tag{3.2}$$

defines the Keplerian distribution, $l_K(r)$.

In order to exclude the possibility of self-gravity instabilities, we assume that everywhere in the disk the density $\epsilon$ is well below the "Roche limit," $M/r^3$ (see Paczyński 1978).

Except in a small region near the inner edge, mechanical equilibrium is assumed everywhere in the disk; i.e., it is assumed that the velocity of the accretion flow, $v$, is highly subsonic: $v \ll v_s \equiv (\partial p/\partial \epsilon)^{1/2}$. (Close to $r_{in}$ the accretion flow is transonic; see JAP.)

We assume that there are no stresses operating through the "lower" and "upper" surfaces of the disk. (There *is* nonzero torque, $t_{out}$, at the outer edge; otherwise accretion would not be possible.)

The last assumption is that the flux of internal energy (or enthalpy) through the inner edge of the disk is small in comparison with the flux of the outgoing radiation. This means that we assume that the transport of energy in the "vertical" direction is sufficiently efficient.

The main properties of the model are pictured in Figure 1.

Introducing six dimensionless and small quantities, $\mu$, one can list our assumptions in the following way:

$$\left\{\begin{matrix}\text{self-gravity} \\ \text{not important}\end{matrix}\right\} \Rightarrow \frac{m}{M} \equiv \mu_M \ll 1 , \tag{3.3}$$

$$\left\{\begin{matrix}\text{no self-gravity} \\ \text{instabilities}\end{matrix}\right\} \Rightarrow \frac{\epsilon}{(M/r^3)} \equiv \mu_\epsilon \ll 1 , \tag{3.4}$$

$$\left\{\begin{matrix}\text{mechanical} \\ \text{equilibrium}\end{matrix}\right\} \Rightarrow \frac{v}{v_s} \equiv \mu_v \ll 1 , \tag{3.5}$$

$$\left\{\begin{matrix}\text{no stresses operating} \\ \text{through disk's surface}\end{matrix}\right\} \Rightarrow \frac{t_{surf}}{t_{out}} \equiv \mu_t \ll 1 , \tag{3.6}$$

$$\left\{\begin{matrix}\text{flux of internal energy} \\ \text{small through } r_{in}\end{matrix}\right\} \Rightarrow \frac{L_{in}}{L} \equiv \mu_L \ll 1 , \tag{3.7}$$

$$\left\{\begin{matrix}\text{general relativistic} \\ \text{effects not important}\end{matrix}\right\} \Rightarrow \frac{r_g}{r_{in}} \equiv \mu_{GR} \ll 1 . \tag{3.8}$$

Here $r_g \equiv 2MG/c^2$ is the gravitational radius of the central object, and $v = (v_r^2 + v_z^2)^{1/2}$.

[3] One can define $L_{Edd}$ and $M_*$ not only for Thomson scattering but for *any* opacity mechanism. Denoting by $\kappa$ the opacity per unit mass one can write $L_{Edd} = L_{Gunn}(\pi r_g^2/M\kappa)$ and $M_* = \kappa c^4/4G^2$.

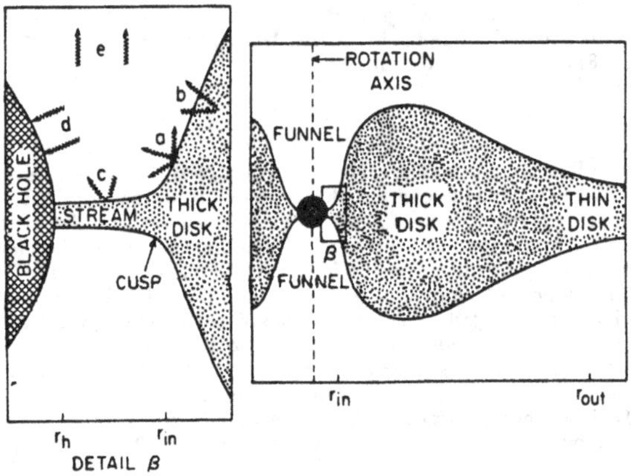


FIG. 1 (adopted from JAP).—The main properties of our model are pictured in this meridional section (not to scale). Outside $r_{out}$ the disk is *thin* in the sense that its height $h(r)$ is small in comparison with $r$: $h/r \ll 1$. This means that for $r > r_{out}$ all kinematical characteristics of the matter motion (e.g., angular velocity, angular momentum, etc.) are equal to their Keplerian values. Between $r_{out}$ and $r_{in}$ the disk is *thick*. Its shape follows from the assumed, non-Keplerian, angular momentum distribution on the surface. This distribution can be calculated if a viscosity mechanism is known. Since we left such a mechanism unspecified, the distribution of angular momentum can be almost arbitrarily chosen. It is assumed that on the surface of the disk the effective gravity is balanced by the radiation pressure. The flow in the stream can be approximated by free-fall. Close to the central object the walls of the disk are very steep and form *funnels* along the rotation axis. Radiation emitted somewhere ($a$) can be absorbed and reemitted ($b$, $c$), or swallowed by the central object (black hole) ($d$). The net result of these processes is strong collimation of the radiation flux (Lynden-Bell 1978; Abramowicz and Piran 1980) in the direction of the axis ($e$).

For any assumed pair of functions, $l(r)$ and $f(r)$, which satisfy the conditions $dl/dr \geq 0, d\Omega/dr \leq 0, 0 \leq f \leq 1$ one can construct a model and check *a posteriori* whether the inequalities (3.3)–(3.8) are fulfilled.

Let us introduce dimensionless coordinates

$$\xi \equiv r/r_{in}, \quad \eta \equiv z/r_{in}, \quad \text{with} \quad q = 1/\xi_{out} = r_{in}/r_{out}, \tag{3.9}$$

and define the following dimensionless functions:

$$\lambda(\xi) \equiv l(r)/l_K(r_{in}) \equiv l(r)/(GMr_{in})^{1/2}, \quad \chi(\xi) \equiv h(r)/r_{in}, \tag{3.10a}$$

$$S(\xi) \equiv \frac{r_{in}}{GM}\left[\int_{r_{in}}^{r} l^2(r)r^{-3}dr - \frac{GM}{r_{in}}\right] \equiv \int_1^{\xi} \lambda^2(\xi)\xi^{-3}d\xi - 1, \tag{3.10b}$$

$$Q(q) \equiv 3(1-q)/2(1-q^{3/2}). \tag{3.10c}$$

In terms of $S(\xi)$ the conditions $dl/dr \geq 0$, $d\Omega/dr \leq 0$ are:

$$S'' + 3\xi^{-1}S' \geq 0, \quad -S'' + \xi^{-1}S' \geq 0, \tag{3.11}$$

where a prime stands for $d/d\xi$. In the limit of very big disks $q \to 0$.

### *a) Mechanical Equilibrium*

The equation for the surface of the disk follows from the fact that on the surface the pressure vanishes: $p(r, z) = 0$. This equation, when solved with respect to $z$, gives the half-thickness of the disk: $z = h(r)$. Because $\partial p/\partial r$ and $\partial p/\partial z$ are given by the Euler equation, one can compute $dh/dr$:

$$-\frac{dh}{dr} \equiv \left[\frac{(\partial p/\partial r)}{(\partial p/\partial z)}\right]_{z=h} = \left[\frac{(\partial\Phi/\partial r) - l^2(r)/r^3}{(\partial\Phi/\partial z)}\right]_{z=h}. \tag{3.12}$$

Note that a barytropic equation of state is *not* assumed. For any given $\Phi = \Phi(r, z)$ and $l = l(r)$ the right-hand side of this equation is an explicitly known function of $r$ and $h$. Therefore, (3.12) can be integrated. If $\Phi(r, z)$ is given by (3.1), the result is:

$$h(r) = \left\{\left[\frac{1}{GM}\int_{r_{in}}^{r} l^2(r)r^{-3}dr - (r_{in}^2 + h_{in}^2)^{-1/2}\right]^{-2} - r^2\right\}^{1/2}, \tag{3.13}$$

where $h_{in} = h(r_{in})$. In general $h_{in}/r_{in}$ is small but nonzero. If $h_{out}/r_{out}$ is also small, then:

$$\left(\frac{h}{r}\right)^2_{out} = \frac{r_{out}}{r_{in}}\left\{\frac{2r_{in}}{GM}\int_{r_{in}}^{r_{out}} r^{-3}[l^2(r) - l_K^2(r)]dr + \left(\frac{h}{r}\right)^2_{in}\right\} + O^4\left(\frac{h}{r}\right)_{in,out} \quad (3.14)$$

Assuming $(h/r)_{in}$ is sufficiently small, the *necessary and sufficient condition* for a big disk ($r_{out} \gg r_{in}$) to be thin in its outer edge is

$$\int_{r_{in}}^{r_{out}} r^{-3}[l^2(r) - l_K^2(r)]dr = 0\,, \qquad \text{or} \qquad S(1/q) + q = 0\,. \quad (3.15)$$

In practice, the error induced by putting $(h/r)_{in} = 0$ (instead of keeping it small but nonzero) is negligibly small in the computation of *global* disk properties such as total luminosity, total mass, etc. Only the *local* analysis of the details of accretion (close to the inner edge) is sensitive to the precise value of $(h/r)_{in}$. Therefore, when computing global properties we shall put $h_{in} = 0$. This simplifies formula (3.13) for the half-thickness of the disk:

$$\chi(\xi) = [1/S^2(\xi) - \xi^2]^{1/2}\,, \quad (3.16)$$

as well as many other formulae. Equations (3.15) and (3.16) contain all the information about the mechanical equilibrium which is needed to construct a model. Topological properties of the equipressure surfaces, $p = p(r, z)$, are discussed in Appendix A.

### *b) Luminosity*

According to our assumption, the flux of radiation on the disk's surface is expressed by

$$F(r) = -\frac{c}{\kappa}f(r)g_{eff}$$

with the function $f(r)$ given *a priori*. The effective gravity, $g_{eff}$, is orthogonal to the surface of the disk and can be calculated when the shape of the disk is known. Therefore $F(r)$ can also be calculated:

$$F(\xi) = F_0 S^2(\xi)\left[1 + \frac{1}{S^4(\xi)}\left(\frac{dS}{d\xi}\right)^2 + \frac{2\xi}{S(\xi)}\left(\frac{dS}{d\xi}\right)\right]^{1/2} f(\xi)\,, \quad (3.18)$$

where $F_0 \equiv cGM/\kappa r_{in}^2$. (We give only the normal component of $F$.) Integrating this over the whole surface of the disk, one gets (after a considerable amount of algebra) the total luminosity, $L$:

$$\Lambda(q) \equiv \frac{L}{L_{Edd}} = \langle f\rangle\frac{m}{M} - \int_1^{1/q}\chi(\xi)\frac{d}{d\xi}\left[\frac{\lambda^2(\xi)}{\xi^2}\right]d\xi\,, \quad (3.19)$$

where $\langle f\rangle$ denotes a mean value of $f(\xi)$ on the surface. The maximum possible value of $\Lambda$ is associated with $f(\xi) \equiv 1$. For non-self-gravitating disks ($m/M = 0$) one has:

$$\Lambda_{max} = -\int_1^{1/q}[S^{-2}(\xi) - \xi^2]^{1/2}\frac{d}{d\xi}\left(\xi\frac{dS}{d\xi}\right)d\xi \quad (3.20)$$

### *c) Accretion Rate*

The efficiency of accretion may be defined as

$$\mathscr{E} \equiv L/\dot{M}\,, \quad (3.21)$$

where $L$ is the total luminosity and $\dot{M}$ is the accretion rate. The Newtonian expression for $\mathscr{E}$ in the case of a thin accretion disk, $\mathscr{E}_\infty(r_{in}) = GM/2r_{in}$, was obtained long ago and is well known. PW in the Newtonian theory (and JAP in general relativity) have computed $\mathscr{E}$ employing the global conservation laws for energy and angular momentum:

$$\frac{\mathscr{E}}{\mathscr{E}_\infty} = 2(1 - q) - \lambda_{in}^2\left[q^2\left(\frac{\lambda_{out}}{\lambda_{in}}\right)^2 - 2q^2\left(\frac{\lambda_{out}}{\lambda_{in}}\right) + 1\right]. \quad (3.22)$$

This formula is not symmetric with respect to $r_{in}$ and $r_{out}$ because the torque in the inner edge vanishes (we shall prove this later), whereas it is nonzero in the other edge. In the case of constant angular momentum disks, $\mathscr{E} = 0$. The efficiency is maximal (for fixed $q$) when $\lambda_{in} = 1$, $\lambda_{out} = 1/(q)^{1/2}$, i.e., for disks which are Keplerian on both edges. In this case

$$\mathscr{E}^*(q) = 2(1 - q^{3/2})(Q - 1)\,. \quad (3.23)$$

Define the *critical accretion rate* by

$$\dot{M}_{\rm crit} = \frac{L_{\rm Edd}}{\mathscr{E}_\infty} = \frac{8\pi c r_{\rm in}}{\kappa} . \tag{3.24}$$

If Thomson scattering is the main source of opacity, then

$$\dot{M}_{\rm crit} \equiv \frac{16\pi G m_p M}{c\sigma_{\rm T}} \mu_{\rm GR} = 5.594 \times 10^{17} \mu_{\rm GR} \frac{M}{M_\odot} \,{\rm g\,s^{-1}}$$

$$= 0.888 \times 10^{-8} \mu_{\rm GR} \frac{M}{M_\odot} M_\odot\,{\rm yr}^{-1} . \tag{3.25}$$

Note that because $\mathscr{E} < \mathscr{E}_\infty$, super-Eddington luminosities are possible only for a disk with $\dot{M} > \dot{M}_{\rm crit}$ (typically $\dot{M} \gg \dot{M}_{\rm crit}$).

## IV. JAROSZYŃSKI'S DISTRIBUTION OF ANGULAR MOMENTUM

Let us ask for which type of surface angular momentum distribution, $l(r)$, the total luminosity $L$ is maximal, assuming that the disk extends between the inner radius, $r_{\rm in}$, and the outer radius, $r_{\rm out}$, that it is Keplerian in both its edges, and that the mechanical equilibrium condition and the other conditions discussed earlier are fulfilled. The variational problem connected with this question is the following (cf. eqs. [3.11], [3.15], [3.20]):

Find a function $S = S(\xi)$ which extremizes the functional

$$\Lambda_{\max} = -\int_1^{1/q} F(S'', S', S) d\xi \equiv -\int_1^{1/q} (S^{-2} - \xi^2)^{1/2} (\xi S')' d\xi , \tag{4.1a}$$

with fixed $q$ and with additional constraints:

$$\{\text{condition (3.15)}\} \Rightarrow S(1/q) + q = 0 , \tag{4.1b}$$

$$\{l' \geq 0\} \Rightarrow S'' + 3S'/\xi > 0 , \tag{4.1c}$$

$$\{\Omega' \leq 0\} \Rightarrow -S'' + S'/\xi > 0 . \tag{4.1d}$$

$$\{\text{Keplerian on edges}\} \Rightarrow S'(1) = 1 , \quad S'(1/q) = q^2 . \tag{4.1e}$$

There is no general theory for variational problems of such a kind. Jaroszyński (1979, unpublished) suggested that because only non-Keplerian disks have $L > L_{\rm E}$, the extreme non-Keplerian one will have the maximal luminosity. The integral

$$\int_{r_{\rm in}}^{r_{\rm out}} r^{-3} |l^2 - l_k{}^2| dr$$

measures how non-Keplerian a disk is. It is extreme for the special angular momentum distribution in which either the angular velocity or the angular momentum is constant. (See Fig. 2.) We shall call this distribution the *Jaroszyński*

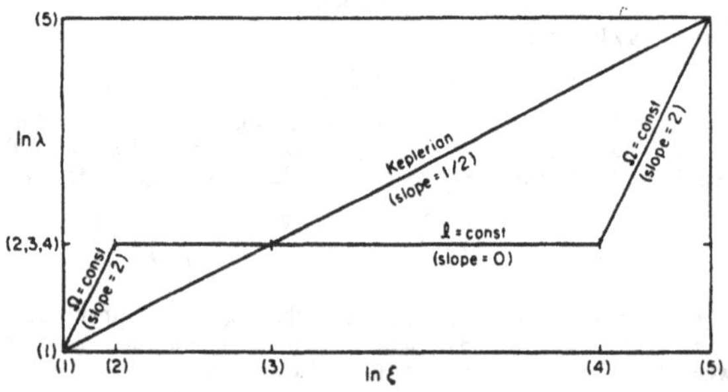


FIG. 2.—The Jaroszyński distribution of angular momentum.

TABLE 1

LUMINOSITY OF JAROSZYŃSKI'S DISKS

| | | | | | | | | | | | | | |
|---|---|---|---|---|---|---|---|---|---|---|---|---|---|
| $\log q^{-1}$ ................ | 0.104 | 0.013 | 0.33 | 0.50 | 0.85 | 0.93 | 1.11 | 1.28 | 2.17 | 4.34 | 6.51 | 8.69 | 10.86 |
| $\Lambda = L/L_{Edd}$ .......... | 0.005 | 0.045 | 0.276 | 0.631 | 1.681 | 1.11 | 2.65 | 3.36 | 7.26 | 17.16 | 27.15 | 37.15 | 47.15 |

*distribution.* For the Jaroszyński distribution one gets from (4.1b) and (4.1e):

$$\xi_1 = 1\,, \quad \xi_2 = Q^{1/2}\,, \quad \xi_3 = Q^2\,, \quad \xi_4 = Q^{1/2}/q^{3/4}\,, \quad \xi_5 = q^{-1}\,, \tag{4.2a}$$

$$\lambda_1 = 1\,, \quad \lambda_2 = Q\,, \quad \lambda_3 = Q\,, \quad \lambda_4 = Q\,, \quad \lambda_5 = q^{-1/2}\,. \tag{4.2b}$$

The function $S(\xi)$ is given by

$$\begin{aligned} S(\xi) &= -\tfrac{1}{2}(3 - \xi^2) && \text{for } 1 \le \xi \le \xi_2\,, \\ &= -\tfrac{1}{2}(3 - 2Q + Q^2/\xi^2) && \text{for } \xi_2 \le \xi \le \xi_4\,, \\ &= -\frac{q}{2}(3 - q^2\xi^2) && \text{for } \xi_4 \le \xi \le q^{-1}\,. \end{aligned} \tag{4.3}$$

After a considerable amount of nontrivial algebra, from (4.3) and (4.1a) one can compute analytically an asymptotic (big disks, $q \to 0$) value for the luminosity of Jaroszyński's disk:

$$(\Lambda_{max})_{JAR} \approx -2 \ln q - 2.44\,, \qquad q \lesssim 10^{-2}\,. \tag{4.4}$$

For any finite value of $q^{-1}$, $\Lambda_{max}$ can be easily computed numerically (see Table 1). Examples of the shapes of two Jaroszyński disks are shown in Figure 3, and the critical flux emitted from their surfaces is shown in Figure 4.

Note that for a given $q$ all Jaroszyński disks are self-similar (there is no natural length-scale in Newtonian theory). Relativistic Jaroszyński disks are not self-similar.

We leave the proof that Jaroszyński's distribution extremizes the luminosity to Appendix B.

## V. SELF-CONSISTENCY CONSTRAINTS

Because in Jaroszyński's disks either $l$ = const. or $\Omega$ = const., from the Poincaré theorem (see Tassoul 1978), it follows that the pressure and the density are connected by a barytropic relation, $p = p(\epsilon)$. In this case one can introduce the total potential $W$:

$$\int_0^p \frac{dp}{\epsilon(p)} \equiv W(r, z) = -\frac{GM}{(r^2 + z^2)^{1/2}} - \int_{r_{in}}^r l^2(r)r^{-3}dr + \frac{GM}{r_{in}}\,. \tag{5.1}$$

The usually assumed model for barytropy in astrophysical applications is the polytropic equation of state:

$$p = K\epsilon^{1+1/n}\,. \tag{5.2}$$

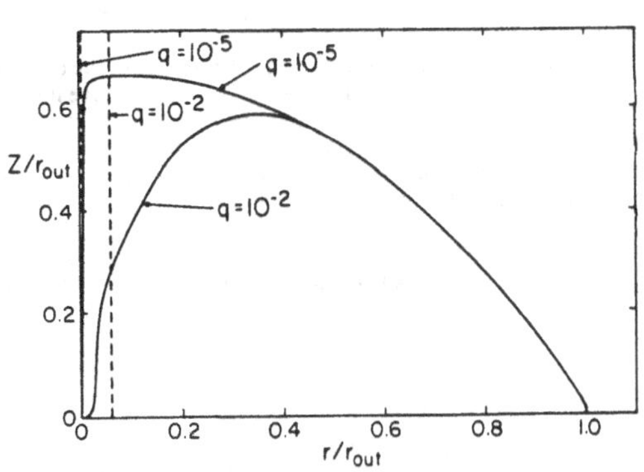


FIG. 3.—The shape of two Jaroszyński models. Note how steep the funnels are in the case of big disks.

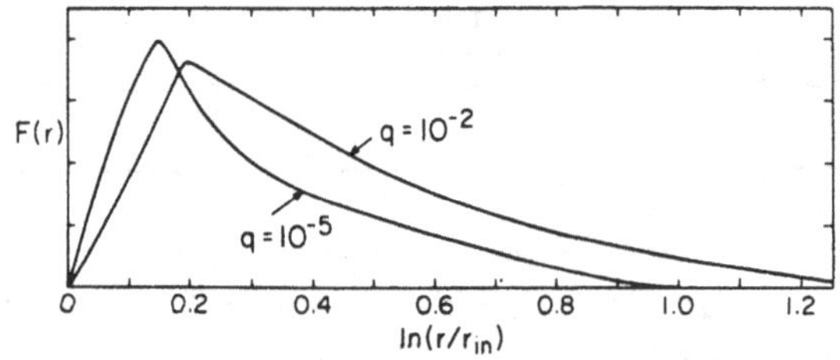


FIG. 4.—The critical flux (in arbitrary units) emitted from the surfaces of the disks shown in Fig. 3. The distance from the axis is plotted in the logarithmic scale here. The whole region shown in this figure is contained in the range indicated by dashed lines to the left of Fig. 3.

Let us define $W_s \equiv W(r_{in}, h_{in})$, $W_{in} \equiv W(r_{in}, 0)$, and $W_c = W(r_c, 0)$, where $r_c$ denotes the location of the maximal pressure in the equatorial plane. The thickness of the inner edge will be measured by the normalized potential difference

$$\beta \equiv \frac{W_s - W_{in}}{W_s - W_c} \ll 1 . \tag{5.3}$$

Let us evaluate the accretion rate near the inner edge. It is defined as the flux of matter, $v\epsilon$ [here $v = (v_r^2 + v_z^2)^{1/2}$], integrated over the surface of the inner edge, i.e., $4\pi r_{in}^2 (h/r)_{in}$. The density cannot be greater than (typically) $M/r_{in}^3$ times a dimensionless, model-dependent factor which is less than unity, because otherwise either the mass of the disk would be greater than the mass of the central object, or some instabilities due to self-gravity would be present, or both. Similarly, the velocity of the accretion flow, $v$, cannot be greater than $(GM/r_{in})^{1/2}$ times a model-dependent, less-than-unity, dimensionless factor. Therefore, $\dot{M}/\dot{M}_{crit} \leq (M_*/M)\mu_{GR}^{5/2}(h/r)_{in}$ times a model-dependent factor.

The mechanism of the accretion flow through the inner edge (discussed in detail later) is very similar to that which powers accretion in the case of Roche-lobe overflow in a close binary. Such a mechanism has been studied in detail by many authors after the significant work of Paczyński and Sienkiewicz (1972), and it is well known that $\dot{M} \sim (h/r)_{in} \sim \beta^{1+n}$. Finally, the thickness in the inner edge $(h/r)_{in} \sim \beta^{1+n}$ must be small according to our assumption that the internal energy flux through the edge is small with respect to the flux of outgoing radiation. Therefore,

$$\Lambda \leq \dot{M}/\dot{M}_{crit} \leq (M_*/M)_{GR}^{5/2}\beta^{1+n}\mu A_\mu(q) , \qquad \beta \leq B(q)\mu_L . \tag{5.4}$$

In the first inequality $M_*$ is given by (2.8); $\mu$ is one of the quantities $\mu_M, \mu_\epsilon, \mu_v, \mu_l$; and $A_\mu(q)$ is a dimensionless function, different for each $\mu$. In the second inequality $B(q)$ is a dimensionless function.

### *a) Paczyński's Mechanism for Accretion through $r_{in}$*

Accretion into the central object through the vicinity of the inner edge is driven by a slight overflowing of a critical equipotential surface, $W = W_{in}$, by the surface of the disk (Fig. 5), i.e., by a small violation of mechanical equilibrium. In this case no viscosity is needed to support the accretion. Such a mechanism was suggested by Paczyński (1978, unpublished). There are three conditions for Paczyński's mechanism to work:

1. The surface $W = W_{in}$ should intersect itself in the inner edge, the central object should be *inside* the region bounded by $W = W_\infty$, and $|W(r_{in}, 0)| < |W_s| < |W_\infty|$.
2. There should be no infinite potential barrier between $r_{in}$ and the surface of the central object, i.e., $|W(r, 0)| < \infty$.
3. From the inner edge to the surface of the central object the matter should move freely, i.e., with $l =$ const. Between the inner edge and the surface of the central object there should be *no* stable, circular orbits for free particles.

These conditions cannot be satisfied in Newtonian mechanics with the potential $\Phi = -GM/(r^2 + z^2)^{1/2}$: Paczyński's mechanism is a purely general-relativistic one. The physical reason for this is that in Newtonian mechanics the Keplerian distribution of angular momentum is monotonic, while in general relativity it has a minimum at the marginally stable circular orbit, $r = r_{ms}$.

Paczyński (1978, unpublished) notices that all the relevant properties of these physical quantities which determine the accretion process near the inner edge can be reproduced in Newtonian mechanics by assuming that the gravitational potential has the form

$$\Phi(r, z) = -\frac{GM}{(r^2 + z^2)^{1/2} - r_g} \equiv \Phi_p . \tag{5.5}$$

Such a potential implies, through the Poisson equation, unphysical density,

$$\epsilon(r, 0) = \frac{M}{r^3}\left(\frac{r_g}{r}\right)\left(1 - \frac{r_g}{r}\right)^3 .$$

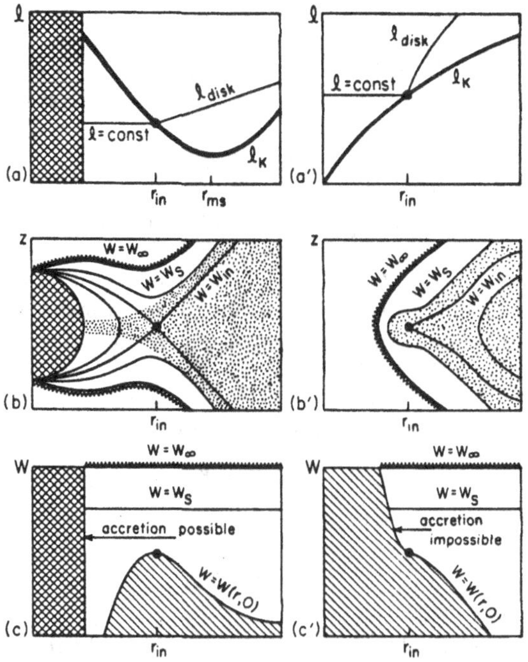


FIG. 5.—Paczyński's mechanism for accretion through the inner edge of an accretion disk. The left column refers to the relativistic case; the right column, to the Newtonian one.

However, as far as self-gravity is not taken into account, this is not a serious problem. In the potential (5.5) the Keplerian distribution of angular momentum has a minimum on the innermost stable circular orbit located at $r = r_{ms} = 3r_g$. The binding energy of a test particle for this orbit is $e(r_{ms}) = 0.0625 m_0 c^2$. The corresponding values in the Schwarzschild geometry are $r_{ms} = 3r_g$ and $e(r_{ms}) = 0.057 m_0 c^2$. The binding energy is zero on the innermost bound circular orbit $r = r_{mb}$. For both potential (5.5) and the Schwarzschild geometry one has $r_{mb} = 2r_g$. Therefore, Paczyński's mechanism for accretion works in Newtonian mechanics with the potential (5.5) *exactly* in the same way as it does in general relativity. This may simplify many calculations in which the required accuracy is 10% or less.[4] We shall use $\Phi_p$ in the computation of the accretion rate.

From the Bernoulli equation, $K(1 + n)\epsilon^{1/n} + \frac{1}{2}v^2 + W = W_s$, it follows that the flux of mass, $\epsilon v$, is maximal when the drift velocity, $v$, is equal to the velocity of sound, $v_s$. In this case:

$$(\epsilon v)_{max} = \left(\frac{W_s - W}{n + \frac{1}{2}}\right)^{n+1/2} \left[\frac{n}{K(n+1)}\right]^n . \tag{5.6}$$

We have put the Bernoulli constant equal to $W_s$ for *all* the flow lines. The physical justification for doing this is that far from the inner edge mechanical equilibrium holds and at the surface of the disk both $p$ and $v$ are equal to zero.

Note that one can expect $v \approx v_s$ close to the inner edge and therefore also $\epsilon v \approx (\epsilon v)_{max}$. Thus, the actual accretion rate will be slightly less than the $M_{max}$ connected with the matter flux (5.6):

$$\dot{M} \leq \dot{M}_{max} = 4\pi r_{in} \int_0^{h_{in}} (\epsilon v)_{max} dz = 4\pi r_{in} \int_{W_{in}}^{W_s} \left(\frac{W_s - W}{n + \frac{1}{2}}\right)^{n+1/2} \left[\frac{n}{K(1+n)}\right]^n \left(\frac{dz}{dW}\right) dW . \tag{5.7}$$

[4] Note that for *static* spacetimes the Lagrangian describing the geodesic motion is given by $L = \frac{1}{2}(g_{ik}/g_{tt})\dot{x}^i\dot{x}^k - (\frac{1}{2}g_{tt})$. Comparing this with the Lagrangian in Newtonian mechanics, $L = v^2/2 - \Phi$, one can claim that $\Phi = \frac{1}{2}g_{tt} = \Phi_p$ will indeed be the best generalization of the potential. See also Landau and Lifshitz (1962) and Damour (1978).

Because $dW/dz = 0$, for $r = r_{in}$ and $z = 0$, one has

$$W - W_{in} = \frac{1}{2}\left(\frac{\partial^2 W}{\partial z^2}\right)_{in} z^2 + \cdots, \qquad \frac{dz}{dW} = \frac{1}{(2)^{1/2}\Omega_K(r_{in})(W - W_{in})^{1/2}}, \tag{5.8}$$

*independently* of the angular momentum distribution in the vicinity of $r_{in}$. Putting (5.8) into (5.7), one gets, after a little algebra,

$$\dot{M} \leq \dot{M}_{max} = H(n)K^{-n}(W_s - W_{in})^{n+1}, \tag{5.9}$$

where the numerical factor $H(n)$ is defined by

$$H(n) \equiv 2\pi(2\pi)^{1/2}\frac{(1 + 1/n)^{-n}}{(n + 1/2)^{n+1/2}}\frac{\Gamma(n + 3/n)}{\Gamma(n + 2)} \tag{5.10}$$

and $\Gamma$ denotes the Euler gamma function. In the most important cases $n = 0, \frac{3}{2}, 3$, we have:

$$H(0) = 19.7264, \qquad H(\tfrac{3}{2}) = 1.1018, \qquad H(3) = 0.04016. \tag{5.11}$$

Kozłowski, Jaroszyński, and Abramowicz (1978) have obtained a similar expression for $\dot{M}$ in general relativity.

Repeating the same calculations for the internal energy (enthalpy) flux

$$(\text{specific enthalpy})v = (n + 1)K\epsilon^{1/n}v, \tag{5.12}$$

one easily gets the amount of internal energy outflowing through the inner edge:

$$L_{in} = \frac{n + \frac{3}{2}}{n + 2}(W_s - W_{in})\dot{M}. \tag{5.13}$$

Therefore, using (3.7), (3.21), (5.9), (5.10), and (5.13) we can write:

$$\beta \leq \frac{n + 2}{n + \frac{3}{2}}\frac{\mathcal{E}}{(W_s - W_c)}\mu_L. \tag{5.14}$$

For Jaroszyński's distribution:

$$\mathcal{E} = \tfrac{3}{4}(1 - q)(Q - 1)Q^{-1}c^2\mu_{GR}, \qquad W_s - W_c = \tfrac{1}{2}(Q - 1)^2(Q + \tfrac{1}{2})Q^{-2}c^2\mu_{GR}, \tag{5.15}$$

and one can write in this case

$$\beta \leq \frac{3n + 2}{2n + \frac{3}{2}}\frac{(1 - q)Q}{(Q - 1)(Q + \frac{1}{2})}\mu_L. \tag{5.16}$$

For $n = 3$ and $\mu_L < 0.1$ this formula gives $\beta < 0.25$ in the most important case of big disks.

Because $v \approx v_s$ near the inner edge and $v > v_s$ for $r < r_{in}$, the matter inside the circle $r = r_{in}$ cannot affect the matter in the disk. Therefore, there is no torque acting on the inner edge of the disk.

### b) *Central Density*, $\epsilon_c$

In order to prevent self-gravity instabilities one should assume

$$\epsilon_c < \mu_c\frac{3M}{4\pi r_c^3}, \tag{5.17}$$

where $\mu_c \approx 0.1$ (Paczyński 1978). From the Euler equation it follows that:

$$\epsilon_c = \left[\frac{W_s - W_c}{K(1 + n)}\right]^n. \tag{5.18}$$

Using this formula, one can eliminate $K$ from (5.9), and then employing (3.21) and (5.3) one finds that the *minimum possible* central density associated with the luminosity $L$ is:

$$(\epsilon_c)_{min} = L\left[H(n)\left(\frac{r_{in}^5}{GM}\right)^{1/2}\beta^{1+n}(W_s - W_c)\mathcal{E}\right]^{-1}. \tag{5.19}$$

Because $\epsilon_c \geq (\epsilon_c)_{min}$, in order to avoid self-gravity instabilities, the luminosity obeys:

$$L \leq \frac{3M}{4\pi r_c^3}H(n)\left(\frac{r_{in}^5}{GM}\right)^{1/2}\tfrac{1}{2}(W_s - W_c)\mathcal{E}\mu_c. \tag{5.20}$$

In the case of Jaroszyński's disks one gets $\Lambda \le \Lambda_\epsilon^*$, where (cf. [4.2] and [5.15]):

$$\begin{aligned}\Lambda_\epsilon^* &= 4.65 \times 10^{20}(M_\odot/M)\mu_{GR}{}^{5/2}\mu_\epsilon\beta^4 G(q) \qquad (n = 3)\\ &= 5.4 \times 10^{20}(M_\odot/M)\mu_{GR}{}^{5/2}\mu_\epsilon\beta^{3.5} G(q) \qquad (n = \tfrac{5}{2})\\ &= 6.4 \times 10^{20}(M_\odot/M)\mu_{GR}{}^{5/2}\mu_\epsilon\beta^3 G(q) \qquad (n = 2)\,, \end{aligned} \tag{5.21}$$

with the function $G(q)$ defined by:

$$G(q) \equiv (Q-1)^3(Q+\tfrac{1}{3})(1-q)Q^{-9}\,. \tag{5.22}$$

*c) The Mass of the Disk*

Define a dimensionless function

$$E(q, n) \equiv 4\pi \int_1^{1/q}\int_0^{x(\xi)} [(\xi^2+\eta^2)^{-1/2} + S(\xi)]^n \xi d\eta d\xi\,. \tag{5.23}$$

After much algebra one can transform the expression for the mass of the disk into a very simple form:

$$m = \left(\frac{GM}{r_{in}}\right)^n (W_s - W_c)^n r_{in}{}^3 E(q, n)\epsilon_c\,. \tag{5.24}$$

In the special case $n = 3$ one has

$$E(q, 3) = \int_1^{1/q} A(\xi)\xi d\xi\,, \tag{5.25}$$

$$A(\xi) \equiv S^2\{(1-y^2)^{1/2}(y^{-2}+1) + 3y^{-1}\,\text{Arctan}\,(y^{-2}-1)^{1/2} + 3\ln\tfrac{1}{2}(1+|y^{-1}|)\}\,; \qquad y \equiv \xi S\,. \tag{5.26}$$

Using (5.19) and (3.3), one concludes that if $m \le M$ for a disk, then its luminosity should be less than the limit, $\Lambda_M^*$, given by:

$$\begin{aligned}\Lambda_M^* &= 1.3 \times 10^{17}(M_\odot/M)\mu_{GR}{}^{5/2}\mu_M\beta^4 E^{-1}(q, 3) \qquad (n = 3)\\ &= 3.1 \times 10^{17}(M_\odot/M)\mu_{GR}{}_{5/2}\mu_M\beta^{3.5} E^{-1}(q, 2.5) \qquad (n = \tfrac{5}{2})\\ &= 7.8 \times 10^{17}(M_\odot/M)\mu_{GR}{}^{5/2}\mu_M\beta^3 E^{-1}(q, 2) \qquad (n = 2)\,. \end{aligned} \tag{5.27}$$

The general asymptotic ($q \ll 1$) formula for $E(q, n)$ is

$$E(q, n) = \frac{a}{(3-n)^2}\left[\frac{1-\exp(-3b+3n)}{2^{(3-n)}} - \exp(3b - bn) + 1\right], \qquad a = 0.24\,, \qquad b = 3.8\,, \qquad n \neq 3\,, \tag{5.28}$$

$$E(q, 3) = ab\ln q^{-1} - 2ab\,, \qquad n = 3\,. \tag{5.29}$$

The function $E(q, n)$ is shown in Figure 6.

*d) Mechanical Equilibrium*

In the standard theory of thin accretion disks, the kinematic viscosity $\nu$ is described in terms of a dimensionless parameter $\alpha$:

$$\nu = \alpha\frac{p}{\epsilon}\left(-\frac{d\Omega}{dr}r\right)^{-1}\,. \tag{5.30}$$

The standard theory assumes (quite arbitrarily) that $\alpha$ is constant. For small $\alpha$, the $\phi$-component of the Navier-Stokes equation yields:

$$r\,\epsilon\,\nu\frac{d}{dr}(\Omega r^2) = \frac{\partial}{\partial r}(r^2 p\alpha) \tag{5.31}$$

for a barytropic accretion disk (note that mechanical equilibrium is *not* assumed now!). This equation can be integrated, and the result is:

$$\frac{\dot{M}}{4\pi}(l - l_{in}) - r^2\int_0^{h(r)} p(r, z)\alpha(r, z)dz = 0\,. \tag{5.32}$$

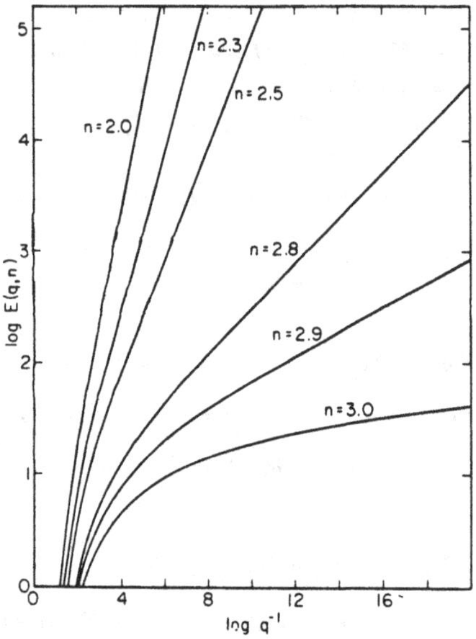


FIG. 6.—The function $E(q, n)$.

Defining for any physical quantity $X$ its "mean value" on a cylinder $r =$ const. by the formula

$$\bar{X}(r) = \frac{1}{h(r)} \int_0^{h(r)} X(r, z) dz \,, \tag{5.33}$$

one can write (5.32) in the form

$$\mu_v \approx \bar{v}/\bar{v}_s \approx \bar{\alpha} r \bar{v}_s/(l - l_{in}) \,. \tag{5.34}$$

In mechanical equilibrium ($\mu_v \ll 1$) the angular momentum distribution cannot be too flat. Assuming it is steep enough, we have $l - l_{in} \approx r v_\phi$; and because $\bar{v}_s/v_\phi \sim h/r$ we also have $\mu_v \approx \bar{\alpha} h/r$. For a thick disk $h \approx r$ and therefore $\bar{\alpha} \ll 1$. (The same conclusion was reached by JAP.) Because the total luminosity is proportional to $\bar{\alpha}$ via the volume integral

$$L = \int_V p\alpha \left( -r \frac{d\Omega}{dr} \right) dV \,, \tag{5.35}$$

one can worry that for $\bar{\alpha} \ll 1$, it is not possible to get super-Eddington luminosities. However, from (5.35) and $\bar{\alpha} \ll 1$ one gets (in the limit $q \to 0$ and for $n = 3$):

$$\Lambda \leq \Lambda_r^* = 1.23 \times 10^{22} \mu_v \mu_M \mu_{GR}{}^{5/2} \frac{M_\odot}{M} \,, \tag{5.36}$$

and a still weaker condition connected with $\mu_\epsilon$. Thus, for Jaroszyński's disks, the condition for mechanical equilibrium is much weaker than the previous ones and gives no additional restriction on the total luminosity (we are grateful to Dr. B. Paczyński for pointing this out to us). For some special (unphysical?) types of disks this condition may be restrictive (Wiita 1980, preprint).

## VI. DISCUSSION

The most important constraint for the self-consistency of the models is given by (5.27). The corresponding limits for the total luminosity of the Jaroszyński disks are shown in Figure 7.

The absolute upper limit for the luminosity of Jaroszyński's disks is of the order of $10^{48}$ ergs s$^{-1}$, i.e., *about a million times less* than the same limit estimated by JAP ($2 \times 10^{54}$ ergs s$^{-1}$). However, there is no contradiction between JAP's and our results. JAP did not use *all* of our self-consistency constraints. Their limit corresponds only to our condition

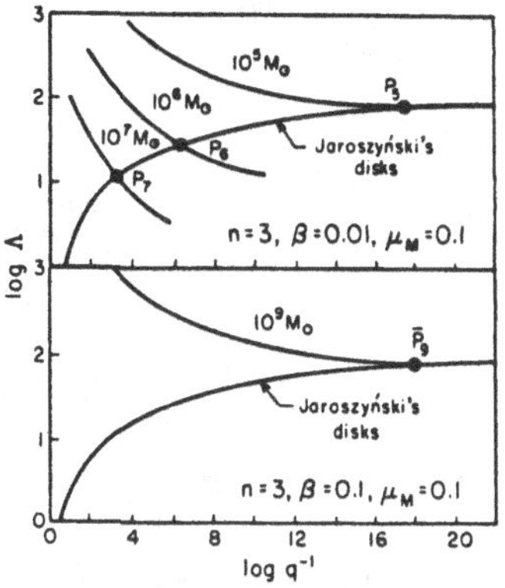


FIG. 7.—Examples of how the constraint (5.27) works. The points $P_7$, $P_6$, $P_5$ indicate the maximum possible values of Λ for a $10^7 M_\odot$, $10^6 M_\odot$, $10^5 M_\odot$ central object in the case $n = 3$, $\beta = 0.01$, $\mu_M = 0.1$, $\mu_{GR} = 1$. The demand that $\beta \leq 0.01$ is very strong (and probably much too strong). In a more realistic situation $\beta \leq 0.1$.

(5.21) which prevents the central density from being too big. This condition, as we have shown, is *much weaker* than our condition (5.27) which demands $m < M$. Note that for big disks, $q \approx 0$, it is possible to have a finite and small central density and almost infinite volume, i.e., almost infinite mass. This follows formally from the asymptotic values of $G(q)$ and $E(q, n)$:

$$G(0) = 5.2 \times 10^{-3}, \qquad E(0, n) = \infty . \tag{6.1}$$

Therefore (5.27) must be much stronger than (5.21). Putting $n = 3$, $\mu_e = 1$, $\beta = 0.25$, $q = 0$, and (6.1) in our formula (5.21), i.e., the same numbers as those which have been used by JAP, we obtain the limit $1.2 \times 10^{55}$ ergs $\text{s}^{-1}$, in excellent agreement with JAP's order-of-magnitude estimate.

In the vicinity of the inner edge the velocity of the accretion flow, $v$, is *transonic* (see JAP). Liang and Thompson (1980) noticed that because the transonic solution for $v$ is unique, this implies that the accretion rate $\dot{M}$ must be related to the mass of the central black hole $M$, exactly as in the case of spherical accretion (Bondi 1952; Carter *et al.* 1976). Such a relation will also depend on such parameters as density, temperature, and angular momentum at the sonic point, $r_s$. As Liang and Thompson pointed out, such uniqueness has extremely important observational consequences. Note that a more detailed study of Paczyński's mechanism could substantially narrow the range of the additional parameters in the Liang-Thompson relation $\dot{M} = \dot{M}(M)$, or $\dot{M} = \dot{M}(M, a)$ for rotating black holes, because (1) the sonic point will be, for realistic, self-consistent models, *always* very close to the inner cusp: $r_s \approx r_{in}$ (Abramowicz and Żurek 1980, work in progress); (2) the location of the cusp is a unique function of $l_{in}$, $M$, and $a$ ($a$ is the specific angular momentum of the hole), $r_{in} = r_{in}(l_{in}, M, a)$; (3) near the inner cusp the angular momentum is almost constant, independent of the viscosity mechanism, $l \approx l_{in}$ = const. (see Kozlowski, Jaroszyński, and Abramowicz 1978).

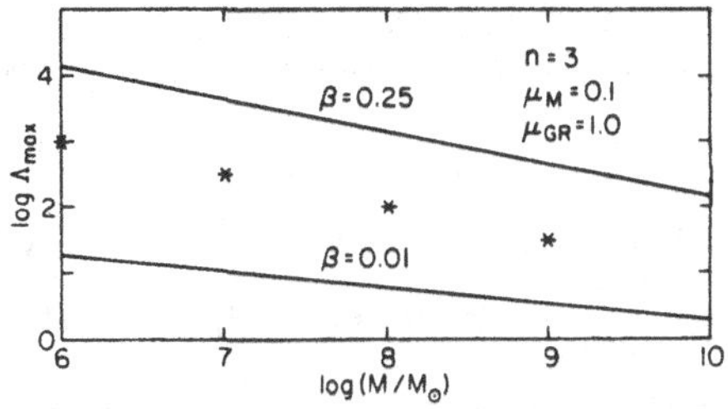


FIG. 8.—The maximum possible luminosity of non-self-gravitating thick accretion disks orbiting a central black hole depends on the mass of the hole. The most realistic values for $n$, $\mu_M$, $\mu_{GR}$, and $\beta$ are indicated by asterisks.

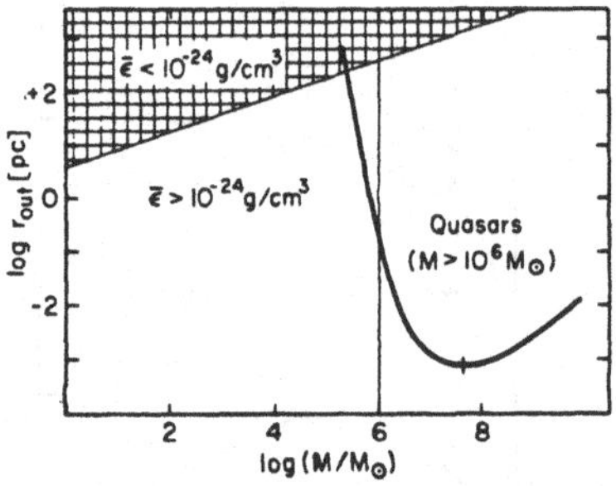


FIG. 9.—The outer radius for the Jaroszyński disks with maximal luminosity. (For $n = 3$, $\mu_M = 0.1$, $\beta = 0.01$). The shaded region in the upper left corner is forbidden (mean density too low).

We are grateful to Dr. B. Carter for discussing the importance of the transonic solution with us and for helpful advice.

In the discussion of the self-consistency limits of our approach we have used all the constraints except that given by (3.6), i.e., except the condition that the stresses operating through the surface of the disk are small, $\mu_t \ll 1$. Preliminary results of Sikora (1980) show that the effects of the surface stresses cannot lower luminosity by a substantial factor.

We close the discussion on self-consistency with a note about the location of the outer edge of the disk. From Figures 7 and 8 it follows that $\xi_{out}$ is a *decreasing* function of $M$ (for models with maximal luminosity). But $r_{out} = \xi_{out} \cdot r_{in}$ and, assuming that $r_{in} \approx r_g$, $r_{in}$ is an increasing function of $M$. It is therefore not strange that $r_{out} = r_{out}(M)$ is not monotonic but has a minimum. The function $r_{out}(M)$ is plotted in Figure 9. Consider a line of constant "mean density of the disk," $\bar{\epsilon} \approx m/r_{out}^3 = \mu_M M/r_{out}^3$, in this figure. Such a line has a constant slope of 3/1. Of course, the mean density of the disk cannot be too small—it must be greater than $10^{-24}$ g cm$^{-3}$, say. This gives an additional constraint for the self-consistency for small-mass black holes ($M < 10^5\ M_\odot$) and does not affect quasar models ($M > 10^6\ M_\odot$).

Note that from Figure 9 it follows that the Jaroszyński disk around a "typical" supermassive black hole, $M = 10^8\ M_\odot$, has an outer radius of the order of $10^{-3}$ pc $\approx$ 100 AU. This number is in nice agreement with what is generally expected from observations of quasars and active galactic nuclei.

This work has benefited from discussions with many colleagues. In particular, it is a pleasure to thank Brandon Carter, Wojtek Dziembowski, Michał Jaroszyński, Bohdan Paczynski, Tsvi Piran, Martin Rees, Dennis Sciama, Nigel Sharp, Marek Sikora, and Wojciech Zurek for particularly helpful suggestions and remarks. We would like also to thank Joyce Patton and Nigel Sharp for their help in editing the manuscript. This work was begun while M.A.A. was visiting Groupe d'Astrophysique Relativiste, Observatoire de Paris—Meudon (France) and continued during a visit of M.A.A. at Istituto di Fisica Galileo Galilei in Padova (Italy) and M.C. at N. Copernicus Center in Warszawa (Poland). The final version of the paper was written at the Center for Relativity, The University of Texas at Austin.

## APPENDIX A

## MECHANICAL EQUILIBRIUM

Let us assume, for simplicity of presentation, either that $\Omega = \Omega(r)$ in the whole volume or that $p = p(\epsilon)$. In both cases the centrifugal force has a potential $\psi$, and the total potential, $\Phi + \psi$, equals (up to a constant) the specific enthalpy, $W(p) = \int \epsilon(p) dp$,

$$W(p) = -GM/(r^2 + z^2)^{1/2} - \int l^2(r) r^{-3} dr \,. \tag{A1}$$

Thus, the equation for the half-thickness of any equipressure surface, $p(r, z) = \text{const.}$, has the form (cf. eq. [3.13])

$$h(r) = \left\{ \left[ \frac{1}{GM} \int_{r_{in}}^{r} l^2(r) r^{-3} dr - (r_0^2 + h_0^2)^{-1/2} \right]^{-2} - r^2 \right\}^{1/2} , \tag{A2}$$

where $h_0 = h(r_0)$ and $r_0$ is an arbitrary point. In the special case of the surface of the disk, $p(r, z) = 0$, $r_0 = r_{in}$ can be chosen to be the inner edge of the disk.

It is obvious that all the *neutral points* (i.e., points at which the pressure gradient vanishes) can be located only on the equatorial plane, $z = 0$. At a neutral point the gravitational force should precisely balance the centrifugal one. Off the equatorial plane, however, this cannot happen as the gravitational force has a nonzero $z$-component while the centrifugal force has not. [For a similar reason $h = h(r)$ can be locally parallel to the axis of rotation only if $h = 0$.] On the equatorial plane the $r$-component of the Euler equation takes the form:

$$\left(\frac{\partial W}{\partial r}\right)_{z=0} = -\frac{1}{\epsilon}\left(\frac{\partial p}{\partial r}\right)_{z=0} = \frac{1}{r^3}[l_K{}^2(r) - l^2(r)] . \tag{A3}$$

Therefore, the locations of all the neutral points are given by the solution of the equation $l_K(r) = l(r)$. Denote by $r = r_0$ the circle in which an equipressure surface crosses the equatorial plane. From (A2) it follows that

$$\frac{dh}{dr} = \frac{r}{h}\frac{l^2 - l_K{}^2}{l_K{}^2} + \frac{l^2}{l_K{}^2}\frac{h}{r} + O^2\left(\frac{h}{r}\right), \qquad r = r_0 . \tag{A4}$$

Note that it is only in the case in which there is a neutral point at $r = r_0$ that it may be possible to have *finite* $dh/dr$. If $l \neq l_K$ at $r_0$, the quantity $(r/h)(l^2 - l_K{}^2)$ will always blow up; if $l = l_K$ at $r_0$, we have $dh/dr = 0/0$, and calculations should be made more precise. Write $r = r_0 + \Delta r$, with $|\Delta r|/r_0 \ll 1$. Using (A2) and the following expansions:

$$l^2(r_0 \pm \Delta r) = l_K{}^2(r_0 \pm \Delta r) \pm [2\theta(r_0) - 1]GM\Delta r + O^2\left(\frac{\Delta r}{r_0}\right), \tag{A5}$$

$$r_0 \pm \Delta r l^2(r) r^{-3} dr = \pm\frac{\Delta r}{2r_0}\left\{2\frac{\Delta r}{r_0}[2\theta(r_0) - 3] + O^2\left(\frac{\Delta r}{r_0}\right)\right\}, \tag{A6}$$

$$\theta(r) \equiv \frac{d\ln l}{d\ln r}, \qquad 0 \le \theta \le 2, \qquad \theta_K(r) = \tfrac{1}{2}, \tag{A7}$$

one can express the angular opening, $\gamma$, between the "upper" and "lower" parts of an equipressure surface which crosses the equatorial plane at a neutral point $r = r_0$, by $\gamma = \text{Arctan}\,(dh/dr)$, where:

$$\left(\frac{dh}{dr}\right)_{r_0} = \pm[2\theta(r_0) - 1]^{1/2} . \tag{A8}$$

There are exactly four types of *isolated* neutral points which correspond to the four possible types of crossing between $l = l(r)$ and $l = l_K(r)$ curves. They are shown in Figure 10.

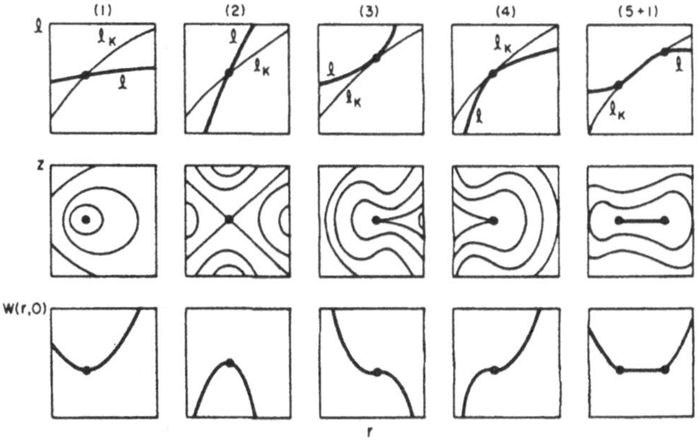


FIG. 10.—The four types of isolated neutral points. *First row*, the angular momentum inside the disk (*heavy lines*). *Second row*, the topology of equipressure surfaces; *third row*, the total potential in the equatorial plane. (1) This is an absolute maximum of the pressure and a minimum of the total potential in the horizontal direction. (2) Pressure has a maximum in the vertical direction and a minimum in the horizontal direction. (3), (4) Pressure has a maximum in the vertical direction and is monotonic in the horizontal direction. The total potential has a point of inflection. (5 + 1) If the curves $l = l(r)$ and $l = l_K(r)$ are tangent over a finite range of $r$, we also have four possibilities for neutral lines; only one of them is shown.

Let us consider (in a general, nonbarytropic case) two different types of rotation law: $l = l_1(r, z)$ and $l = l_2(r, z)$. Suppose that the first of them describes rotation in a region [1] and the second in a region [2] and that the boundary between these regions is given by $z = z(r)$. If angular momentum is continuous across the boundary, one has, of course, $l_1(r, z) = l_2(r, z)$ at the boundary. But this means that $dh/dr$ is continuous across the boundary:

$$\frac{dh}{dr} = -\frac{\partial\Phi/\partial r - l^2(r, z)r^{-3}}{\partial\Phi/\partial z}. \tag{A9}$$

One can first solve equation (A9) in the region [1] and then in the region [2]. Both solutions will fit perfectly through the boundary even if $\partial l/\partial r$ and $\partial l/\partial z$ are not continuous through it.

## APPENDIX B

Consider Figure 11. The line $APG$ is the Keplerian angular momentum. Suppose that the center of the disk is located on the circle $\xi = \xi_c$ (i.e., the disk's angular momentum distribution crosses the Keplerian one at the point $P$). This means that the curves describing the angular momentum of the disk must be placed between the lines $ANPRG$ and $APG$. (The lines $ANK$ and $DRG$ are for constant angular velocity.) In addition, such curves cannot have their slopes greater than $+2$ or less than 0. Consider only curves which are constructed from four straight line segments. Let the two free vertices of them lie somewhere on the lines $BJ$ and $EH$. (There are only *two* free vertices, as two are fixed at the points $A$, $G$, and another is fixed at the point $P$.) Where is the *highest* possible location of the vertex in the $BJ$ line? It has to be *below* the lines $MS$ and $AK$, so the highest location is at the point $O$. For the same reason the *lowest* possible location of the vertex in the line $HE$ is at the point $Q$.

Now check whether the condition (3.15) holds. Computation shows that the area $PQG$ is too big: the point $Q$ has to be shifted to $Q_0$. We see that if the angular momentum distribution in the ln $\lambda$ versus ln $\xi$ plane consists of four straight-line segments with vertices at vertical lines $\xi_A = 1$, $\xi_B$, $\xi_C$, $\xi_E$, and $\xi_F = 1/q$, then the vertices at $\xi = \xi_B$ and $\xi = \xi_E$ have to be located between $O, O_{10}$ and $Q_0, Q_{10}$, respectively. Now, let us divide the sector $Q_0 Q_{10}$ into 10 equal parts. For each $Q_i$ one can find the corresponding $O_i$ (using the area condition [3.15]), i.e., a distribution $AO_iQ_iG$ which corresponds to a model in mechanical equilibrium. For each such model (the example $i = 2$ is shown: see the heavy line in the figure) the corresponding luminosity can be calculated from equation (3.20). Therefore, for each $q$ and a given choice of $B, C, E$ we have 10 different disks. Because there are about $\frac{1}{3} \times 100^3$ possibilities for different choice of $B, C, E$, there will be about $3 \times 10^6$ different disk models for each value of $q$. ($AF$ is divided into 100 sections as indicated in Fig. 11. The example $B = A_5$, $C = A_{40}$, $E = A_{70}$ is given.) We have checked that for several ($\sim$ 100) different values of $q$, Jaroszyński's disk has a greater luminosity than all of these $\sim 3 \times 10^6$ disks. We have also checked in a few cases that an additional (6th) vertex does not change this conclusion.

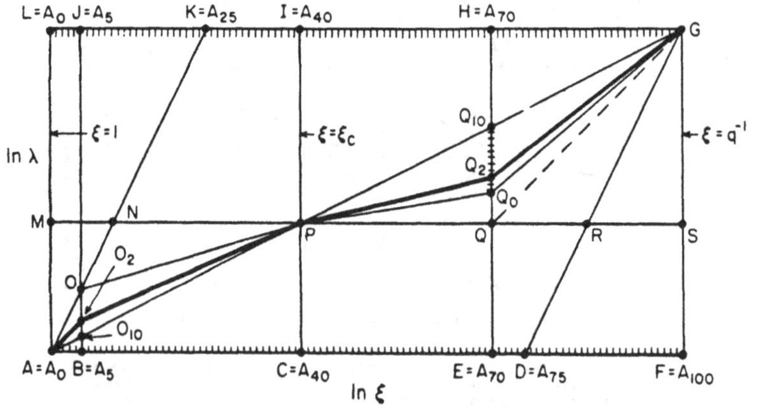


FIG. 11.—The variational problem connected with the maximal luminosity (Euler-Ritz approximation).

REFERENCES

Abramowicz, M. A., Jaroszyński, M., and Sikora, M. 1978, *Astr. Ap.*, **63**, 221.
Abramowicz, M. A., and Piran, T. 1980, *Ap. J.* (*Letters*), in press.
Bisnovatyi-Kogan, G. S., and Blinnikov, S. I. 1977, *Astr. Ap.*, **59**, 111.
Blandford, R. D., and Thorne, K. S. 1979, in *General Relativity, an Einstein Centenary Survey*, ed. S. W. Hawking and W. Israel (Cambridge: Cambridge University Press).
Bondi, H. 1952, *M.N.R.A.S.*, **112**, 195.
———. 1960, *Cosmology* (Cambridge: Cambridge University Press).
Carter, B., Gibbons, G. W., Lin, D. N. C., and Perry, M. J. 1976, *Astr. Ap.*, **52**, 427.
Damour, T. 1978, thesis, Observatoire de Paris.
Jaroszyński, M., Abramowicz, M. A., and Paczyński, B. 1980, *Acta Astr.*, **30**, 1 (JAP).
Kozłowski, M., Jaroszyński, M., and Abramowicz, M. A. 1978, *Astr. Ap.*, **63**, 209.
Landau, L. D., and Lifshitz, E. M. 1962, *The Classical Theory of Fields* (Oxford: Pergamon Press).
Liang, E. P. T., and Price, R. H. 1977, *Ap. J.*, **218**, 247.
Liang, E. P. T., and Thompson, K. A. 1980, *Ap. J.*, in press.
Lynden-Bell, D. 1978, *Phys. Script.*, **17**, 185.
Lynden-Bell, D., and Pringle, J. M. 1974, *M.N.R.A.S.*, **168**, 603.
Novikov, I. D., and Thorne, K. S. 1973, in *Black Holes*, ed. B. and C. DeWitt (New York: Gordon & Breach).
Paczyński, B. 1978, *Acta Astr.*, **28**, 91.
Paczyński, B., and Sienkiewicz, R. 1972, *Acta Astr.*, **22**, 73.
Paczyński, B., and Wiita, P. 1980, *Astr. Ap.*, in press (PW).
Pringle, J. M., and Rees, M. J. 1972, *Astr. Ap.*, **21**, 1.
Shakura, N. I., and Sunyaev, R. A. 1973, *Astr. Ap.*, **24**, 337.
Sikora, M. 1980, thesis, N. Copernicus Center, Warszawa (in Polish).
Tassoul, J. L. 1978, *Theory of Rotating Stars* (Princeton: Princeton University Press).

MAREK A. ABRAMOWICZ: Center for Relativity, Department of Physics, The University of Texas at Austin, Austin, TX 78712

MASSIMO CALVANI: Istituto di Astronomia, Università degli Studi di Padova, Vicollo del Osservatorio 5, 35100 Padova, Italia

LUCIANO NOBILI: Istituto di Fisica Galileo Galilei, Università degli Studi di Padova, via Marzolo 8, 35100 Padova, Italia

Astron. Astrophys. 88, 23–31 (1980)

ASTRONOMY AND ASTROPHYSICS

# Thick Accretion Disks and Supercritical Luminosities

B. Paczyńsky and P. J. Wiita*

N. Copernicus Astronomical Center, Polish Academy of Sciences, ul. Bartycka 18, PL-00-716 Warszawa, Poland

Received December 20, 1978; accepted November 12, 1979

**Summary.** When accretion rates exceed a critical value accretion disk around a black hole must become thick, vitiating an assumption necessary in the construction of most models. Even for subcritical rates, the inner regions of standard accretion disks are believed to be unstable and are expected to puff up. By replacing the usual assumption of local energy balance with a global conservation requirement, and by taking the local radiated flux to be critical, we construct families of consistent thick accretion disks. These have cusps at their inner edges, which can lie between the marginally bound and marginally stable orbits, and depend upon the angular momentum distribution specified but are not directly dependent on assumptions about the viscosity law. They can be matched onto relatively thin disks at a "transition radius". The accretion rates can become very large, and the total luminosities can exceed the nominal Eddington luminosity by substantial factors due to geometrical effects. Such luminosities produce huge radiative energy densities in the cusp region, so that the formation of well collimated beams is a distinct possibility. The calculations use a pseudo-Newtonian potential which reproduces many of the salient features of the Schwarzschild solution.

**Key words:** accretion – black holes – active galactic nuclei

## 1. Introduction

If one attempts to build a model of an accretion disk where the accretion rate becomes very large, it is clear that the disk must become thick, at least in the region close to the compact object. According to standard models, the maximum value of the ratio of the thickness of the disk to its radius is equal to the ratio of the accretion rate to a critical rate (Shakura and Sunyaev, 1973). Since these standard "α models" assume that the disk is relatively thin everywhere, they cannot treat this situation consistently. Even when the accretion is subcritical, it has been recognized that the inner portions of standard disk models around black holes are subject to various instabilities (e.g., Lightman and Eardley, 1974; Shakura and Sunyaev, 1976). These instabilities probably cause such disks to puff up and become geometrically thick, thus vitiating the assumptions of thinness and hydrostatic equilibrium which provide the foundations of these models. By making various assumptions, different authors have constructed models in which winds are driven off the disk's surface, or in which two streams of material are ejected along the axes of the disk (e.g., Shakura and Sunyaev, 1973; Shapiro et al., 1976; Callahan, 1977; Piran, 1977; Lynden-Bell, 1978; Meier, 1978) thus attempting to allow for the instabilities and the strong radiation pressure encountered in the central region of the disk. However Shakura and Sunyaev (1976) have shown that the faster growing instabilities are thermal, and are predominantly due to the assumption that $Q_-$, the rate at which energy is radiated per unit surface area, is equal to $Q_+$, the rate at which it is generated by friction within the disk. When the disk is both geometrically and optically thick, this equality cannot in general be true, for the energy produced at some point inside the bloated disk will not just diffuse vertically, but may emerge from essentially any part of the surface. We are thus led to consider how this assumption can be replaced by a more physical one, and if possible to construct models of thick inner regions of accretion disks that have a consistent time-averaged stationary behaviour.

The approach we have chosen was inspired by the recent general relativistic treatment of the structure of perfect fluid disks around a black hole given by Abramowicz et al. (1978) and Kozłowski et al. (1978); see also Fishbone and Moncrief (1976). These authors have shown that a cusp exists at the inner edge of such disks (or really rings) and that fat disks can be constructed whose inner edges extend down to the marginally bound circular orbit, $r_{mb}$ ($r_{mb}=2r_g=4GM/c^2$ for a Schwarzschild black hole). Thus, we investigate disks whose inner edge lies somewhere between $r_{mb}$ and the last stable circular orbit, $r_{ms}$ ($r_{ms}=3r_g$ for the Schwarzschild case); the portion of the disk inside $r_{ms}$ is supported by a non-Keplerian angular momentum distribution, and therefore does not immediately plummet into the black hole. However, unlike Abramowicz et al. (1978) we will allow for the generation of energy by viscous stresses and for its radiation. We will also match our thick disks onto thin disks at suitably large radii, thus giving the solutions more physical significance.

However our treatment will not be a correct general relativistic one, for we shall employ a pseudo-Newtonian potential, $\psi=-GM/(R-r_g)$, that correctly reproduces the positions of both $r_{ms}$ and $r_{mb}$, and yields efficiency factors in close agreement with the Schwarzschild solution. One great advantage of our approach is that it is not explicitly dependent upon the assumed form of the viscosity law, which is subject to considerable uncertainty; thus we only have to bring in the "α model" (Novikov and Thorne, 1973; Shakura and Sunyaev, 1973) to provide a consistency check on our results (Sect. 4c). There are however many assumptions and approximations in our model that make the quantitative conclusions inexact. We assume that: the equation of state is barytropic, so that the specific angular momentum is constant on

*Send offprint requests to:* B. Paczyński

* Permanent address: Department of Astronomy (E-1), University of Pennsylvania, Philadelphia, PA 19104, USA

cylinders; red-shift effects can be neglected; reabsorption of radiation and evaporation of the disk can be ignored (cf. Shakura and Sunyaev, 1973; Cunningham, 1975, 1976; Rees, 1978 and references therein). Because each of these approximations is crude our treatment is an idealized one. We hope to loosen some of these restrictions in subsequent work, and to perform a fully relativistic treatment. However, the removal of the barytropic assumptions would require solving a much more complicated set of differential equations.

Apart from being independent of assumptions about the viscosity, our approach has other advantages. One is simplicity, both conceptually and in the computations, allowing different angular momentum distributions to be investigated. Another is that the salient effects of general relativity are included via our pseudo-Newtonian potential. We are able to construct thick disks whose inner radii lie within $r_{ms}$ and the match onto more standard thin disks. The accretion rates can exceed the nominal critical rates, and because of the bloated shape of our disks, the total luminosities can exceed the nominal Eddington luminosity. We feel that these basic conclusions are likely to stand even if the details are inaccurate because of the nature of our idealized model, and we feel that this new approach does yield new insight.

In Sect. 2 we present the basic equations we use in studying stationary accretion disks of arbitrary thickness. The assumptions made and the specialized equations needed by our approach in obtaining the shape and luminosity of thick disks are given in Sect. 3. Specific cases are calculated and our results summarized in Sect. 4. In Sect. 5 we draw conclusions and point out modifications that should be included in more refined calculations.

## 2. Equations for Stationary Accretion Disks

We now summarize the important relations needed for our analysis. Naturally, most of these are standard and appear in earlier papers (Lynden-Bell, 1969; Pringle and Rees, 1973; Shakura and Sunyaev, 1973; Novikov and Thorne, 1973) but we derive them again here for two reasons. Firstly, we use a somewhat different approach in obtaining the equations, and also a slightly different notation. Secondly, we shall be using non-Newtonian potentials and non-Keplerian angular momentum distributions, and thus we must use a rather more general formulation of the equations.

We use cylindrical coordinates $(r, z, \varphi)$ centered on the black hole. Azimuthal symmetry is assumed and the spherical radius $R=(r^2+z^2)^{1/2}$. The generalized potential $\psi$ must satisfy the following conditions: $\psi(R)<0$; $\psi(R)\to 0$ as $R\to\infty$; $d\psi(R)/dR>0$.

The typical assumption of a thin disk has $z \ll r$ so that $r \approx R$. The circular velocity, angular velocity, and angular momentum per unit mass (which we henceforth refer to simply as angular momentum) are given, assuming Keplerian orbits, by

$$v=\left(r\frac{d\psi}{dr}\right)^{1/2}, \quad \Omega=\left(\frac{1}{r}\frac{d\psi}{dr}\right)^{1/2}, \quad l=\left(r^3\frac{d\psi}{dr}\right)^{1/2}. \tag{1}$$

The total mechanical energy per unit mass is given in general by

$$e=v^2/2+\psi, \tag{2}$$

as long as the radial and vertical velocities are much less than the azimuthal velocity; $e<0$ implies that a mass element is bound.

Let $\eta$ be the viscosity at some point in the disk, whose half thickness is given by $z_0$. The heat released due to friction can be expressed as (in erg cm$^{-3}$ s$^{-1}$)

$$\varepsilon=\left(r\frac{d\Omega}{dr}\right)^2\eta. \tag{3}$$

The torque applied by frictional forces at a given radius can be written as

$$g=2\pi r^3\left(\frac{-d\Omega}{dr}\right)\int_{-z_0}^{z_0}\eta dz. \tag{4}$$

Typical definitions of surface density and accretion rate are taken:

$$\Sigma=\int_{-z_0}^{z_0}\varrho dz, \quad \dot{M}=\int_{-z_0}^{z_0}2\pi r\varrho v_r dz. \tag{5}$$

The total couple gives us the angular momentum flux,

$$\dot{J}=\dot{M}l+g. \tag{6}$$

The radial energy flux can be expressed as

$$\dot{E}_r=\dot{M}\cdot e+g\Omega, \tag{7}$$

whilst the vertical energy flux is usually taken to be

$$\frac{\partial \dot{E}_z}{\partial r}=2\pi r\cdot 2F, \tag{8}$$

where $F$ is the energy flux from the surface.

In this notation the conservation of mass, angular momentum, and energy can be written as

$$\frac{\partial}{\partial t}(2\pi r\Sigma)+\frac{\partial \dot{M}}{dr}=0, \tag{9}$$

$$\frac{\partial}{\partial t}(2\pi r\Sigma l)+\frac{\partial \dot{J}}{\partial r}=0, \tag{10}$$

$$\frac{\partial}{\partial t}(2\pi r\Sigma e)+\frac{\partial \dot{E}_r}{\partial r}+\frac{\partial \dot{E}_z}{\partial r}=0. \tag{11}$$

We now employ the mass and angular momentum conservation laws directly, since they remain valid for thick disks as long as $\Omega$ is taken as a function of $r$, only i.e., if meridional circulation and radial velocities can be ignored. Substituting Eqs. (5) and (6) into Eqs. (9) and (10) immediately yields

$$\dot{M}\frac{\partial l}{\partial r}+\frac{\partial g}{\partial r}=0, \tag{12}$$

which is also true for thick disks (Loska, private communication). At this point we use the assumption of stationarity and take $\dot{M}=$ const. Eq. (12) can then be integrated to give

$$g=g_0+(-\dot{M})(l-l_0), \tag{13}$$

and we have accretion if $\dot{M}<0$. We now evaluate $g_0$ and $l_0$ at the inner edge of the disk, $r_0$, and make the following reasonable assumptions about the boundary conditions near a black hole: the torque at $r_0$, $g(r_0)=g_0=0$; the angular momentum at $r_0$, $l_0$, is given by the Keplerian value, and since the disk is thin at $r_0$ (it has a cusp there), Eq. (1) is adequate. For convenience we define

$$S=\int_{-z_0}^{z_0}\eta dz, \tag{14}$$

and then use Eq. (4) to write

$$S=\frac{g}{2\pi r^3}\left(\frac{-d\Omega}{dr}\right)^{-1},$$

which, using Eq. (13) with $g_0=0$ can be expressed as

$$S=(-\dot{M})\left(\frac{-d\Omega}{dr}\right)^{-1}(2\pi r^3)^{-1}(l-l_0). \tag{15}$$

The total energy generated within a given column of the disk is

$$Q_+=\int_{-z_0}^{z_0}\varepsilon dz=\left(r\frac{d\Omega}{dr}\right)^2\int_{-z_0}^{z_0}\eta dz=\left(r\frac{d\Omega}{dr}\right)^2 S, \tag{16}$$

where we have made use of Eqs. (3) and (14). We can eliminate the viscosity from our expression for the energy generation by inserting Eq. (15) into Eq. (16) to obtain a key relationship

$$Q_+=(-\dot{M})(2\pi r)^{-1}\left(-\frac{d\Omega}{dr}\right)(l-l_0). \tag{17}$$

If we make the usual assumption, valid for Newtonian disks, that $\psi=-GM/r$, and further assume a thin disk so that $Q_-\equiv 2F\equiv Q_+$, we immediately recover from Eq. (17), using Eq. (1), the well known expression for the flux,

$$F=(-\dot{M})\frac{3}{8\pi}\frac{GM}{r^3}\left[1-\left(\frac{r_0}{r}\right)^{1/2}\right].$$

However, we are interested in the more general situation where neither Eq. (1), nor $Q_-=Q_+$, is acceptable.

At this point we summarize the forms certain important quantities take on for the specific potential we use,

$$\psi=\frac{-GM}{(R-r_g)}. \tag{18}$$

In the regions of the disk where it is thin, and thus $r\approx R$:

$$v=\left(\frac{GM}{r}\right)^{1/2}\left[\frac{r}{r-r_g}\right],\quad \Omega=\left(\frac{GM}{r^3}\right)^{1/2}\left[\frac{r}{r-r_g}\right],$$
$$l=(GMr)^{1/2}\left[\frac{r}{r-r_g}\right]; \tag{19}$$

$$-\frac{d\Omega}{dr}=\tfrac{3}{2}\left(\frac{GM}{r^5}\right)^{1/2}\left[\frac{(r-\frac{1}{3}r_g)r}{(r-r_g)^2}\right],$$
$$\frac{dl}{dr}=\tfrac{1}{2}\left(\frac{GM}{r}\right)^{1/2}\left[\frac{(r-3_g)r}{(r-r_g)^2}\right]; \tag{20}$$

$$e=\left(\frac{-GM}{2r}\right)\left[\frac{(r-2r_g)r}{(r-r_g)^2}\right],$$
$$\frac{de}{dr}=\left(\frac{GM}{2r^2}\right)\left[\frac{r^2(r-3r_g)}{(r-r_g)^3}\right]. \tag{21}$$

In each of the above equations the correction due to the inclusion of the $r_g$ term is enclosed in brackets. Inspection of Eq. (21) shows that $e$, the binding energy, vanishes at $r=2r_g$, and thus we identify $r_{mb}=2r_g$, as for Schwarzschild geometry. Likewise, from Eq. (20), we see that $dl/dr=0$ and $de/dr=0$ when $r=3r_g$, and as an orbit can only be stable when $dl/dr\geq 0$ and $de/dr\geq 0$, we conclude that $r_{ms}=3r_g$, and no *Keplerian* disk can exist inside this radius. The efficiency of energy conversion is given by $\eta'=e/c^2$, and we note that at $r=r_{ms}$, Eq. (21) yields the result $\eta'=0.0625$, whilst the correct result for the Schwarzschild metric is 0.057. At smaller radii the relative agreement is even closer, so we expect that our estimation of luminosities for given mass fluxes will not be more than 10% too high because of the overestimation of $\eta'$, although the uncertainty is increased by other factors to be discussed later.

**3. An Approach to Thick Disks**

*a) Basic Assumptions and Specialized Equations*

The first important approximation, or idealization we make is to assume a barytropic equation of state, $P=P(\varrho)$. It then follows (von Zeipel's theorem) that $\Omega=\Omega(r)$, and this simplification is very helpful. However this approximation must eventually fail in the inner regions where the flow is almost spherical. In this barytropic situation we can define the enthalpy by $dH=dP/\varrho$, and in the equatorial plane of the disk ($z=0$), we denote the density, enthalpy and gravitational potential as

$$\varrho_c(r),\quad H_c(r),\quad \psi_c(r)=\psi(r). \tag{22}$$

We take the boundary conditions on the surface of the disk to be the vanishing of the enthalpy and density, so that on the surface, defined by $\pm z_0$, we have

$$\varrho_0(r)=0,\quad H_0(r)=0,\quad \psi_0(r)=\psi(r^2+z^2)^{1/2}. \tag{23}$$

Our next key assumption is that the equations of hydrostatic equilibrium hold. This certainly fails to be true to some extent in the very innermost region, close to the cusp, where both radial and vertical velocities will not actually be negligible. However, within our formulation the problems are not as severe as when an attempt is made to construct a consistent thin disk (e.g. Bisnovatyi-Kogan and Blinnikov, 1977), and we hope to lossen this restriction in a future paper. The equilibrium equations can be written as

$$\frac{1}{\varrho}\nabla P=\nabla H=\nabla\left(-\psi+\int_{r_0}^{r}\Omega^2 r dr\right). \tag{24}$$

Using the notation defined in Eqs. (22) and (23), the vertical component of Eq. (24) integrates to

$$H_c(r)=\psi_0(r)-\psi(r). \tag{25}$$

Differentiating Eq. (25) with respect to $r$ and comparing with the radial component of Eq. (24) yields the relation

$$r\Omega^2=\frac{dH_c}{dr}+\frac{d\psi}{dr}=\frac{d\psi_0}{dr}. \tag{26}$$

As we have already required that $\Omega=\Omega(r)$ we have the following basic equations that determine the angular quantities in terms of the potential at the surface, or *vice versa*:

$$\Omega=r^{-1/2}\left(\frac{d\psi_0}{dr}\right)^{1/2},\quad l=r^{3/2}\left(\frac{d\psi_0}{dr}\right)^{1/2},\quad v=r^{1/2}\left(\frac{d\psi_0}{dr}\right)^{1/2}. \tag{27}$$

With this interpretation, all the Eqs. (3–17) are just as valid for thick disks as for thin ones. When we make the further basic (and nearly universal) assumption that the self-gravity of the disk is negligible, we may choose Eq. (18) as the potential, with $M$ being taken as the mass of the black hole. [For investigations into accretion disks where self-gravity may be dominant see Paczyński (1978a, 1978b) and Kozłowski et al. (1979).

In order to progress further we must decide upon the best way of finding the flux radiated from the surface of the disk. It is important to recall that the typical "$\alpha$ disk" models should puff up in the inner regions, where radiation pressure dominates, electron scattering provides the bulk of the opacity, and where the disk is optically thick. Thus it is natural to assume that the disk is radiating critically, just as a stellar atmosphere under the same conditions would (Paczyński, 1978a). We thus take the power

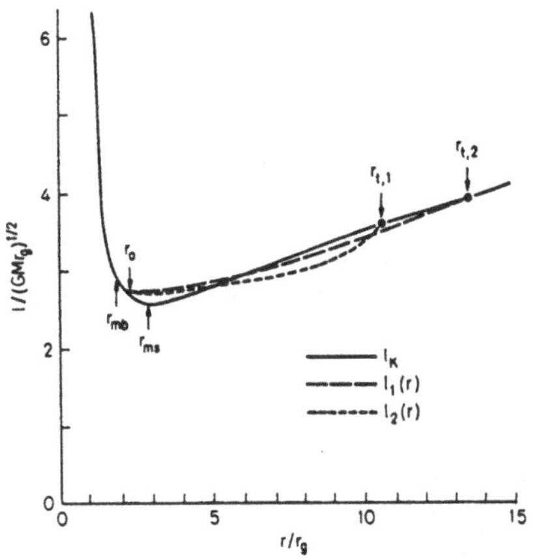


**Fig. 1.** The Keplerian specific angular momentum curve for the pseudo-Newtonian potential, $l_K(r)$, and two arbitrary angular momentum distributions that satisfy stability requirements. The marginally bound $r_{mb}$ and marginally stable $r_{ms}$ orbits are indicated, as are the inner radius $r_0$ and transition radii $r_{t,1}$, $r_{t,2}$ for the thick portions of the accretion disks

---

emitted per unit surface area as

$$F_{rad}=\frac{c}{K}g_{eff}, \tag{28}$$

where $K$ is the opacity per gram and $g_{eff}$ is the effective acceleration of gravity, including the centrifugal force. It is obvious that $g_{eff}$ is normal to the surface of the disk and that we have

$$g_z=g_{eff}\cos\theta, \tag{29}$$

where $g_z=\frac{\partial\psi_0}{\partial z}$ and $\theta$ is the instantaneous angle that the surface of the disk makes with the radial direction, i.e., $\cos\theta=\left[1+\left(\frac{dz_0}{dr}\right)^2\right]^{-1/2}$. Thus we may use Eq. (29) in Eq. (28) to obtain

$$F_{rad}=\frac{c}{K}\frac{\partial\psi(r,z_0)}{\partial z}\left[1+\left(\frac{dz_0}{dr}\right)^2\right]^{1/2}. \tag{30}$$

*b) Angular Momentum Distributions and the Form of the Disk*

The time has now come to specify how we construct our disk models, and the interplay of angular momentum distributions becomes crucial. In Fig. 1 we the pseudo-Keplerian $l_K(r)$ curve given by Eq. (19) as well as two other arbitrary angular momentum distributions that characterize possible disks, and which must have the following properties: (1) $l(r_0)=l_K(r_0)\equiv l_0$; (2) $dl/dr\geqq 0$, for $r\geqq r_0$; (3) $l(r)=l_K(r)$ at two values of $r>r_0$. The first condition is merely our boundary condition; the second is needed to ensure a stable disk (cf. Abramowicz et al., 1978). We impose the third so that the disk may again be thin at the second intersection, the transition radius $r_t$, in the sense that $z_0(r_t)\ll r_t$. At that point we can allow the disk to have once again a Keplerian distribution of angular momentum, for such a distribution will be stable for $r>r_{ms}$. If we have chosen $l(r)$ in such a way that $z_0(r_t)\ll r_t$ then the region $r>r_t$ can be adequately described by standard thin disk models because in that region $\psi_0(r)=\psi(R)\approx\psi(r)$. However, for $r_0\leqq r<r_t$ the ratio $z_0/r$ may not be negligible, and $\psi_0(r)\not\approx\psi(r)$, so the additional freedom allowed by a thick disk is necessary.

We can find the thickness of this thin disk at the transition radius if we assume that it is radiating critically there. It may not satisfy this assumption at larger radii and it will be necessary to check *a posteriori* that the physical preconditions for this assumption are valid (see Sect. 4c). If the disk is thin and radiating critically, then Eq. (30) can be replaced by

$$F_{rad}=\frac{c}{K}g_z=\frac{c}{K}\frac{\partial\psi}{\partial r}\frac{z_0}{r}=Q_-/2. \tag{31}$$

The energy generated within the disk is $Q_+$ which is given by Eq. (17). Since the disk *is* thin at this large radius we may take $Q_+=Q_-$, and for a thin disk under these conditions we have

$$z_0(r)=\left(\frac{-\dot M}{4\pi}\right)\frac{K}{c}\left(\frac{\partial\psi}{\partial r}\right)^{-1}\left(\frac{-d\Omega}{dr}\right)(l(r)-l_0). \tag{32}$$

When the derivatives and angular momentum are evaluated at $r_t$ we obtain a value of $z_t$ that depends only on $\dot M$, which in turn is totally determined by $l(r)$, as we shall see in the next subsection [Eq. (46)].

The basic constraint upon a consistent disk is the requirement that the thickness of the fat portion, given by Eq. (37) below, matches that of the thin portion, determined from Eq. (32, when both are evaluated at $r_t$. However we stress that the calculations in the next subsection are independent of this matching *via* Eqs. (31) and (32), so if a different method is proposed to obtain the thickness of the thin disk, the procedure we are about to describe could also be employed to satisfy it.

*c) The Shape of the Thick Portion*

It is not very difficult to obtain the differential equation that yields the shape of the disk as a function of the radius and the angular momentum distribution. If we define a total potential $\phi\equiv\psi-\int\Omega^2 r dr$, the surface of the disk is a generalized equipotential on which $\phi$ is constant. Thus we must have

$$d\phi=0=\left(\frac{\partial\psi}{\partial r}-\Omega^2 r\right)dr+\left(\frac{\partial\psi}{\partial z}\right)dz \tag{33}$$

on the surface, or

$$\frac{dz}{dr}=\left(\Omega^2 r-\frac{\partial\psi}{\partial r}\right)\left(\frac{\partial\psi}{\partial z}\right)^{-1}. \tag{34}$$

Substituting in Eqs. (18) and (23) this is transformed to

$$\frac{dz}{dr}=\frac{r}{z}\left\{\frac{\Omega^2}{GM}(r^2+z^2)^{1/2}[(r^2+z^2)^{1/2}-r_g^2]-1\right\}. \tag{35}$$

By a simple manipulation this takes the form

$$\frac{dR^2}{dr^2}=\frac{\Omega^2}{GM}R(R-r_g)^2,$$

whose solution is

$$R^2=\left[(r_0-r_g)\left(1-\frac{r_0-r_g}{GM}\int_{r_0}^{r}\Omega^2 r dr\right)^{-1}+r_g\right]^2, \tag{36}$$

where we have used the boundary condition $R(r_0)=r_0$. Thus the surface of the disk is given by

$$z_0=\left\{\left[\frac{r_0-r_g}{1-\frac{r_0-r_g}{GM}\int\limits_{r_0}^{r} l^2(r)r^{-3}dr}+r_g\right]^2-r^2\right\}^{1/2}, \tag{37}$$

where we have employed the relation $l=\Omega r^2$.

### *d) The Luminosity and Accretion Rate*

As we have presumed the disk to be radiating critically it is a simple matter to calculate the total energy radiated by the thick portion of the disk. When the disk is not nearly flat we must use the fact that an element of surface area will be given by the expression $d\sigma=(rd\varphi)(\sec\theta dr)$, so that

$$L_{rad}=2\int\limits_0^{2\pi}\int\limits_{r_0}^{r_t} F_{rad}d\sigma, \tag{38}$$

where both surfaces of the disk have been included. When we insert Eq. (30) into Eq. (38) and evaluate $\sec\theta$ in the expression for $d\sigma$ we obtain

$$L_{rad}=\frac{4\pi c}{K}\int\limits_{r_0}^{r_t}\frac{\partial\psi_0}{\partial z}\left[1+\left(\frac{dz_0}{dr}\right)^2\right]rdr. \tag{39}$$

But now that we have the solution for $z_0(r)$, this can be expressed as

$$L_{rad}=\frac{4\pi cGM}{K}\int\limits_{r_0}^{r_t}\left[\frac{l^4R(R-r_g)^2}{G^2M^2zr^5}-\frac{2l^2}{GMzr}+\frac{rR}{z(R-r_g)^2}\right]dr. \tag{40}$$

Let us now find the total amount of energy that is generated within the thick portion of the disk by intergrating the columnal energy generation over the appropriate region. We shall derive a general expression for the energy generated between any two radii and then specialize to our particular case.

$$L_{gen}(r_1,r_2)=\int\limits_{r_1}^{r_2} Q_+2\pi rdr. \tag{41}$$

Using Eq. (17) this becomes

$$\begin{aligned}L_{gen}(r_1,r_2)&=(-\dot M)\int\limits_{r_1}^{r_2}\left(-\frac{d\Omega}{dr}\right)(l-l_0)dr\\&=(-\dot M)\int\limits_{\Omega_2}(l-l_0)d\Omega\\&=(-\dot M)\left[\int\limits_{\Omega_2}^{\Omega_1} r^2\Omega d\Omega-l_0(\Omega_1-\Omega_2)\right],\end{aligned}$$

where we first change variables and then use the definition of $l$.

Integrating by parts we have

$$L_{gen}(r_1,r_2)=(-\dot M)\left[\tfrac{1}{2}r^2\Omega^2\Big|_{\Omega_2}^{\Omega_1}+\int\limits_{r_1}^{r_2}\Omega^2rdr-l_0(\Omega_1-\Omega_2)\right].$$

We now use the definition of $v$ and Eq. (27) to change this to

$$L_{gen}(r_1,r_2)=(-\dot M)[\tfrac{1}{2}v_1^2-\tfrac{1}{2}v_2^2+(\psi_0)_2-(\psi_0)_1-l_0(\Omega_1-\Omega_2)].$$

If the disk is thin enough *at both relevant radii*, then $\psi_0\approx\psi$, and with a little more manipulation we have

$$L_{gen}(r_1,r_2)=(-\dot M)[(e_2-e_1)+\Omega_1(l_1-l_0)-\Omega_2(l_2-l_0)], \tag{42}$$

where the definition of the binding energy, Eq. (2), has been employed.

The special case we are interested in is $r_1=r_0$ and $r_2=r_t$, but we first obtain the well known general relation that the total energy produced within the disk is

$$L_{gen,total}=(-\dot M)(-e_0), \tag{43}$$

which comes from Eq. (42) because $e_\infty=0$ and $\Omega_\infty\cdot l_\infty=0$. Returning to the total energy produced in the thick region of the disk:

$$L_{gen}(r_0,r_t)=(-\dot M)[(e_t-e_0)-\Omega_t(l_t-l_0)]. \tag{44}$$

At this point we must make one more critical assumption; that the material flowing into the disk at $r_t$ and flowing out of it at $r_0$ carries negligible internal energy in comparison with its mechanical energy. If this is true we can replace the local condition $Q_+=Q_-$ with the global one

$$L_{gen}=L_{rad}, \tag{45}$$

which is physically justifiable. A given distribution $l(r)$ will imply a radius $r_t$, which, along with $r_0$, enables us to integrate Eq. (40) to find $L_{rad}$. We then know $L_{gen}$ and everything on the right hand side of Eq. (44) except the accretion rate, which we solve for:

$$-\dot M=L_{rad}[(e_t-e_0)-\Omega_t(l_t-l_0)]^{-1}. \tag{46}$$

## 4. Calculations and Results

### *a) Procedure for Constructing Models*

We now draw together the threads of the previous section and show how we weave complete thick disk models. From now on we present most of the results in terms of non-dimensional units, since everything scales directly with the mass of the black hole. We take $GM=1$, and $r_g$ as our unit of length. As a first step we choose a potential and write Eq. (18) as

$$\psi=-1/(R-1). \tag{47}$$

The second key step is to choose a value of $r_0\in(2,3]$. Since angular momentum can be expressed in units of $(GMr_g)^{1/2}$ we have

$$l_K(r)=r^{3/2}/(r-1), \tag{48}$$

with $l_0=r_0^{3/2}/(r_0-1)$ as the point from which we start the assumed angular momentum distribution $l(r)$, chosen as step three in our procedure.

The first computational step is to find the two secondary intersections of $l(r)$ and $l_K(r)$. If they do not exist then that distribution of $l(r)$ is inadequate and a new one is chosen. This fourth step is followed by the calculation of the shape and luminosity of the disk between $r_0$ and $r_t$. The solution for $z_0$ is written

$$z_0(r)=\left\{\left[\frac{r_0-1}{1-(r_0-1)\int\limits_{r_0}^{r} l^2(r)r^{-3}dr}+1\right]^2-r^2\right\}^{1/2}. \tag{49}$$

Defining $L_{edd}\equiv4\pi cGM/K$ as the Eddington, or critical, luminosity (using a Newtonian potential and spherical symmetry) we may write

$$L_{rad}=L_{edd}I(r_0,r_t), \tag{50}$$

where $I(r_0,r_t)$ is the non-dimensional form of the integral appearing in Eq. (40). Although the effective gravity at any point on the

**Table 1.** Parameters of thick disks for $\beta=1.0$

| $r_0$ | $A_{1.0}$ | $r_t$ | $\frac{z(r_t)}{r_t}$ | $\left[\frac{z(r)}{r}\right]_{max}$ | $L$ | $-\dot{M}$ | $\eta^*$ | $\eta'$ | $L_{thin}$ |
|---|---|---|---|---|---|---|---|---|---|
| 2.05 | 0.015407 | 3847.6 | 0.135 | 4.406 | 1.19 (39) | 1.19 (20) | 0.0112 | 0.0113 | 2.07 (37) |
| 2.10 | 0.027969 | 1078.3 | 0.202 | 2.490 | 9.20 (38) | 5.12 (19) | 0.0200 | 0.0207 | 3.01 (37) |
| 2.20 | 0.047920 | 319.8 | 0.252 | 1.425 | 6.07 (38) | 2.07 (19) | 0.0326 | 0.0347 | 3.95 (37) |
| 2.30 | 0.063301 | 162.6 | 0.281 | 1.021 | 4.54 (38) | 1.25 (19) | 0.0404 | 0.0444 | 4.50 (37) |
| 2.40 | 0.075585 | 102.6 | 0.289 | 0.796 | 3.48 (38) | 8.62 (18) | 0.0449 | 0.0510 | 4.74 (37) |
| 2.50 | 0.085599 | 72.7 | 0.277 | 0.648 | 2.69 (38) | 6.35 (18) | 0.0472 | 0.0556 | 4.77 (37) |
| 2.60 | 0.093951 | 55.4 | 0.273 | 0.542 | 2.16 (38) | 5.01 (18) | 0.0480 | 0.0586 | 4.79 (37) |
| 2.70 | 0.100989 | 44.3 | 0.265 | 0.462 | 1.75 (38) | 4.10 (18) | 0.0476 | 0.0606 | 4.76 (37) |
| 2.80 | 0.106970 | 36.6 | 0.252 | 0.399 | 1.44 (38) | 3.43 (18) | 0.0465 | 0.0617 | 4.68 (37) |
| 2.90 | 0.112134 | 31.1 | 0.243 | 0.348 | 1.19 (38) | 2.94 (18) | 0.0448 | 0.0623 | 4.60 (37) |
| 3.00 | 0.116419 | 26.9 | 0.238 | 0.306 | 9.92 (37) | 2.58 (18) | 0.0428 | 0.0625 | 4.55 (37) |

surface of the disk will be less than the purely spherical attraction to the central mass at that distance, and thus the critical flux from the disk's surface would be less than that from a spherical surface at the same radius, we will find that our disks may have surface areas far in excess of such equivalent spheres. Thus there is no reason to require that $I(r_0, r_t) \leqq 1$, and it need not surprise us if the total emitted energy exceeds the Eddington luminosity. [Geometrical effects needed in computing the ratio of gravitational to radiative pressure forces were also considered by Bisnovatyi-Kogan and Blinnikov (1977), but in the framework of test particles near a thin disk.]

We can now proceed to the sixth step, which is the calculation of $\dot{M}$ by equating the total energy generated with the total radiated luminosity. We first define a critical accretion rate in the Newtonian fashion,

$$-\dot{M}_{cr} = \frac{2r_0}{GM} L_{edd} = 8\pi c r_0/K \tag{51}$$

and then define the reduced accretion rate as:

$$\dot{m} = (-\dot{M})/(-\dot{M}_{cr}). \tag{52}$$

We rewrite Eq. (44) for the energy generated

$$L_{gen}(r_0, r_t) = (-\dot{M})\frac{GM}{r_g}\left\{\frac{1}{2}\left[\frac{r_0-2}{(r_0-1)^2} - \frac{r_t-2}{(r_t-1)^2}\right] - \frac{l_t}{r_t^2}(l_t - l_0)\right\}$$

$$\equiv (-\dot{M})GMK(r_0, r_t)/r_g, \tag{53}$$

so that Eq. (46) is replaced by

$$\dot{m} = I(r_0, r_t)/(2r_0 K(r_0, r_t)). \tag{54}$$

Our final step, the seventh, is to compute $z_0(r_t)$ for the thin disk using Eq. (32) and see if it matches the value for the thick disk given by Eq. (49). If they agree to within 1% we consider that a consistent match has been found, although it must be confirmed that the errors caused by the finite value of $z(r_t)/r_t$ are small. On the other hand, if they do not agree, we must modify the angular momentum distribution (step three) and repeat stages three through seven until a suitable match is found.

*b) Specific Results*

For simplicity we chose to look at the most basic forms of angular momentum distributions that were likely to fit the various necessary conditions. We examined the simply two parameter family,

$$l(r) = l_0 + A_\beta (r - r_0)^\beta, \tag{55}$$

with $A_\beta \geqq 0$ so that $dl/dr \geqq 0$ and the disk is not unstable. The requirement that $S$ be non-negative implies that we must have $d\Omega/dr \leqq 0$ [cf. Eq. (15)] which imposes a complex constraint on the combination of $A_\beta$ and $\beta$, but certainly requires $\beta \geqq 1$, for otherwise (with $A_\beta \neq 0$) $d\Omega/dr \to +\infty$ as $r \to r_0$. When these simple relationships are chosen, the integral,

$$\int_{r_0}^{r} l^2(r) r^{-3}\, dr \tag{56}$$

which appears in the expression for $z_0(r)$ can be performed analytically. Thus, in this case, the only integral we must evaluate numerically is $I(r_0, r_t)$ of Eq. (40).

If $A_\beta$ is chosen too large there is no hope of a solution, as $l(r) > l_K(r)$ everywhere. As we reduce $A_\beta$ two intersections with the Keplerian curve are found, but the integral (56) grows too quickly and $z_0(r) \to \infty$ for some $r < r_t$. If $A_\beta$ is taken too small, $r_t$ becomes very large and we get $z_0(r) = 0$ at some $r < r_t$. We could actually find values of $A_\beta = A_\beta^0$ that yield $z_0(r_t) = 0$, but these accretion rings (like the solutions of Kozłowski et al., 1979) would not match onto exterior "thin" disks, and the value of $A_\beta$ must be taken somewhat larger than $A_\beta^0$ if we are to obtain an equality between Eqs. (49) and (32).

Let us consider linear angular momentum distributions first, where $\beta = 1.0$. Eleven values of $r_0$, ranging from 2.05 to 3.00 were considered, and for each the value of $A_1(r_0)$ that yielded an acceptable solution was obtained, using the iterative procedure we have just described. Our basic results are summarized in Table 1, where we have listed for each $r_0$ the corresponding $A_1(r_0)$, $r_t$, $z(r_t)/r_t$, and the maximum value of the ratio $z_0(r)/r$ in the first five columns. We also give the values of $L$ and $-\dot{M}$ (in cgs units) per solar mass of $\dot{M}$, where $K$ was taken as $0.34\,\text{cm}^2\,\text{g}^{-1}$, appropriate to solar envelope material. From these we calculate an efficiency factor $\eta^*$ using the relation

$$L = \eta^*(-\dot{M})c^2 \tag{57}$$

and compare it with the theoretical value, $\eta'$, found from Eq. (21). Because the total generated energy is given by Eq. (43), the luminosity radiated in the outer, thin, portion of the disk is

$$L_{thin} = (\eta' - \eta^*)(-\dot{M})c^2 \tag{58}$$

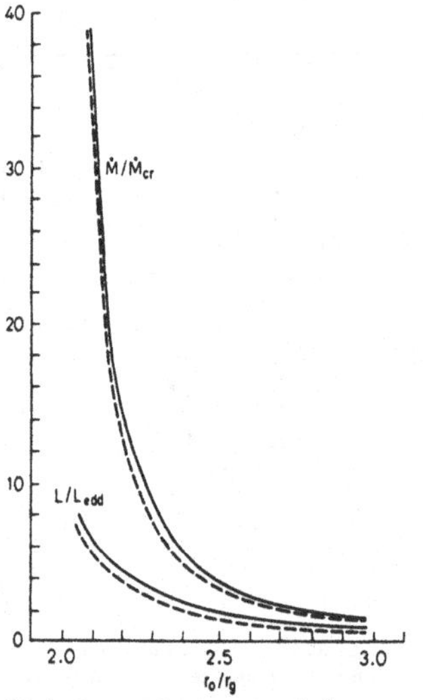


**Fig. 2.** Curves of $L/L_{edd}$ and $\dot{M}/\dot{M}_{cr}$ vs. $r_0$ for two families of angular momentum distributions, $\beta=1.0$ and $\beta=1.1$

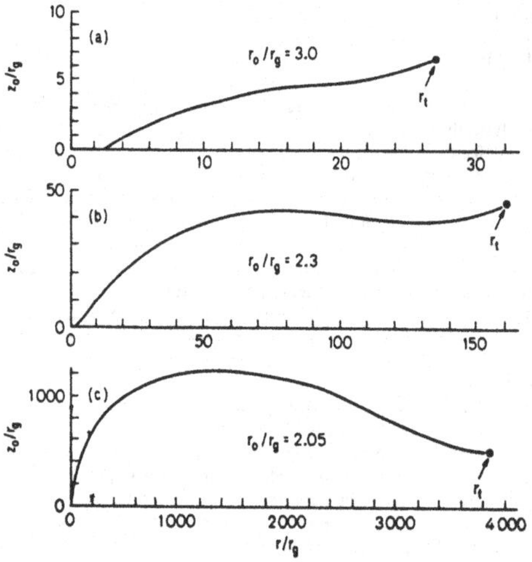


**Fig. 3. a** The shape of the thick portion of a disk (one quadrant only) constructed with $l(r)=l_0+A(r-r_0)$ and $r_0=3.0\,r_g$. **b** The same as **a** for $r_0=2.3\,r_g$. **c** The same as **a** for $r_0=2.05\,r_g$

and is tabulated in the last column of Table 1. We note that the fraction of the total luminosity contributed by the thick portion drops from 0.983 for $r_0=2.05$ to 0.686 for $r_0=3.00$. We also present the ratios $L/L_{edd}$ and $\dot{m}$ graphically in Fig. 2. The shapes of our thick disks are shown for three values of $r_0$ in Fig. 3; note the different scales in the sections of this figure.

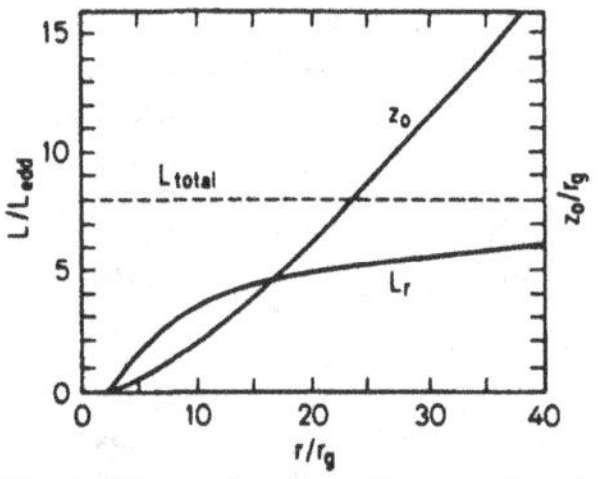


**Fig. 4.** The region near the cusp for the case $r_0=2.05\,r_g$ and $\beta=1.0$, expanded from Fig. 3c. Also shown is the luminosity emitted between $r_0$ and $r$, as well as the total luminosity of this thick disk model

One result is that $\dot{m}$ increases dramatically as $r_0$ decreases below 3, but this is to be expected as $\eta^*$ is bounded above by $\eta'$, which vanishes at $r_0=2$. A far more interesting result is that we find $L>L_{edd}$ for $r_0<2.8$ (when $\beta=1.0$); most of this excess is real and can be understood in terms of the geometrical effects mentioned in the previous subsection, other reasons for it are discussed below. We also note that as $r_0$ decreases, $r_t$ increases, so that the thick portions of the disk become very large indeed. Not only do these thick disks become longer, but they also become fatter as the maximum value of $z/r$ increases rapidly as $r_0$ approaches $r_{mb}$.

It is of interest to check from which portions of the disk the bulk of the energy emerges. As $r_0\rightarrow r_{mb}$ a larger fraction of the total luminosity is emitted from a relatively small fraction of the surface close to the cusp where the disk resembles a tunnel or funnel down towards the black hole. The shape of the innermost portion of our most extreme case, $r_0=2.05$ and $\beta=1.0$ is given in an expanded scale in Fig. 4. Also shown in Fig. 4 is the fraction of the luminosity $L(r)$, emitted between $r_0$ and $r$. Even though $r_t=3848$, and $L(r_t)/L_{edd}=8.07$, we see that 25% of the energy is emitted for $r<10$, 50% for $r<14$, 75% for $r<40$, and 90% for $r<290$.

We next considered a different family of disks by constructing models with $\beta=1.1$ in Eq. (55). Such disks have $dl/dr=0$ at $r_0$, and this may have some physical justification (Kozłowski et al., 1978). Our results for these models are summarized in Table 2, where the same quantities given for the linear angular momentum case in Table 1 are tabulated for these steeper disks. The $\beta=1.1$ case differs uniformly from the $\beta=1.0$ case in the following ways: $\dot{m}$, $L$, and $r_t$ are all reduced, but $z(r_t)/r_t$ is increased, thus implying less accuracy in these models. These values of $\dot{m}$ and $L/L_{edd}$ are also given in Fig. 2. Here the thin portion of the disk contributes a relatively greater part of the emission, and the fraction of the luminosity radiated from the thick part drops from 0.953 to 0.446 as $r_0$ goes from 2.05 to 3.00. All of these trends continue in the few cases we have calculated for higher values of $\beta$.

However, we found it impossible to construct disks for values of $\beta\gtrsim 1.5$, since the values of $r_t$ become quite small while $z(r_t)$ was greater than $r_t$. The nominally large values of $\dot{M}$ led to the contradictory situation where the thickness evaluated on the "thin" side of the transition radius was always greater than that evaluated on the "thick" side.

### *c) Tests and Caveats*

The first obvious check is to examine the validity of our assumption that $z/r\ll 1$ at $r_t$. For the $\beta=1.0$ case this ratio ranged between

**Table 2.** Parameters of thick disks for $\beta = 1.1$

| $r_0$ | $A_{1.1}$ | $r_t$ | $\frac{z(r_t)}{r_t}$ | $\left[\frac{z(r)}{r}\right]_{max}$ | $L$ | $-\dot{M}$ | $\eta^*$ | $\eta'$ | $L_{thin}$ |
|---|---|---|---|---|---|---|---|---|---|
| 2.05 | 0.012287 | 1344.8 | 0.354 | 3.315 | 1.09 (39) | 1.12 (20) | 0.0101 | 0.0113 | 5.36 (37) |
| 2.10 | 0.022334 | 453.5 | 0.409 | 1.980 | 8.02 (38) | 4.66 (19) | 0.0192 | 0.0206 | 6.34 (37) |
| 2.20 | 0.038649 | 156.5 | 0.429 | 1.171 | 5.16 (38) | 1.88 (19) | 0.0306 | 0.0347 | 6.95 (37) |
| 2.30 | 0.051643 | 86.1 | 0.431 | 0.843 | 3.70 (38) | 1.11 (19) | 0.0372 | 0.0444 | 7.18 (37) |
| 2.40 | 0.062313 | 55.7 | 0.409 | 0.654 | 2.73 (38) | 7.51 (18) | 0.0405 | 0.0510 | 7.11 (37) |
| 2.50 | 0.071312 | 40.3 | 0.387 | 0.527 | 2.09 (38) | 5.60 (18) | 0.0416 | 0.0556 | 7.04 (37) |
| 2.60 | 0.078977 | 31.0 | 0.357 | 0.435 | 1.61 (38) | 4.36 (18) | 0.0411 | 0.0586 | 6.84 (37) |
| 2.70 | 0.085781 | 24.7 | 0.341 | 0.366 | 1.27 (38) | 3.57 (18) | 0.0395 | 0.0606 | 6.75 (37) |
| 2.80 | 0.096250 | 20.3 | 0.311 | 0.311 | 9.91 (37) | 2.96 (18) | 0.0372 | 0.0617 | 6.53 (37) |
| 2.90 | 0.097000 | 17.0 | 0.290 | 0.290 | 7.81 (37) | 2.54 (18) | 0.0342 | 0.0623 | 6.44 (37) |
| 3.00 | 0.105000 | 12.2 | 0.332 | 0.332 | 6.30 (37) | 2.51 (18) | 0.0279 | 0.0625 | 7.82 (37) |

0.13 and 0.29; however, we can see that these values are sufficiently small that the approximation $\psi_0(r_t) = \psi(r_t)$ remains accurate to better than 4% and the error this implies for $\dot{M}$ is even smaller. The larger values of $z/r$ allowed when $\beta = 1.1$ mean that the difference between $\psi_0(r_t)$ and $\psi(r_t)$ can be as big as 9%, so that larger values of $\beta$ would yield very substantial errors.

The physical assumption that the disk is radiating critically [Eq. (28)] can also be partially verified *a posteriori*. In order for this to be true we must certainly have electron scattering as the dominant source of opacity and it is also essential that radiation pressure dominates, at least near the surface of the disk. However our models have made no detailed assumptions or predictions about the equation of state or viscosity mechanism, so that the only way to check this point is by indirect comparison with some other models. Here we consider the "$\alpha$ disk" models investigated by Shakura and Sunyaev (1973), where the boundary between "zone $a$" (radiation dominating pressure, electron scattering dominating opacity) and "zone $b$" (gas pressure dominating, still electron scattering) is given approximately by

$$r_{ab} = 89 r_0 (\alpha m)^{2/21} \dot{m}^{16/21}, \tag{59}$$

where $\alpha < 1$ is the canonical measure of viscosity, $m = M/M_{\odot}$, and we have renormalized Shakura and Sunyaev's expression using our definition of $\dot{m}$ [Eq. (52)]. Then $r_t < r_{ab}$ is a necessary condition for the consistency of our disks and this can be translated into a bound on the product $\alpha m$. We find that this becomes only slightly restrictive for small $r_0$ and $\beta$, for when $\beta = 1.0$ and $r_0 = 2.05$ we must demand $\alpha m > 0.023$, which is, however, satisfied for most values of $\alpha$. At larger $r_0$ values this condition is almost certainly met, for at $r_0 = 2.10$ it is $\alpha m > 2.8\,10^{-5}$ and at $r_0 = 3.0$ it is $\alpha m > 4.2\,10^{-12}$. Increasing $\beta$ decreases $r_t$, so even for the "worst" case that we have looked at $\beta = 1.1$, $r_0 = 2.05$, this condition is $\alpha m > 5.6\,10^{-7}$. Of course $m > 1$ is expected for all black holes, so the above bounds are even less strict when applied to $\alpha$ alone.

Another basic assumption, that the self-gravity of our disks is negligible, is connected with our choice of Eq. (18) or Eq. (47) as the potential. Because $\dot{m}$ does get very large it is not obvious that this is true for all our models, and it is difficult to estimate the total mass of the disk within our framework. It is likewise difficult to check that the flux of internal energy is negligible compared to mechanical energy, and thus that our global condition Eq. (45) is valid, without making further assumptions about the equation of state and viscosity. Preliminary estimates indicate that neither of these assumptions is bad for $r_0 \gtrsim 2.2$, but the very high accretion rates and fluxes predicted when $r_0 \approx r_{mb}$ will certainly be reduced by the latter condition breaking down.

It should also be mentioned that while we have deliberately built our models so that the thickness is continuous at the transition radius, the flux $F_{rad}$ need not be, since the slope $dz_0/dr$ is discontinuous across this boundary. If we went to the additional trouble of requiring that $\left.\frac{dl(r)}{dr}\right|_{r_t} = \left.\frac{dl_K(r)}{dr}\right|_{r_t}$ then this minor problem could be alleviated, but we would have to make the postulated angular momentum variation substantially more complex. We feel that little additional light would be thrown on the subject by adding this refinement.

Yet another source of inaccuracy is our neglect of the red-shift of the emitted radiation, for our pseudo-Newtonian potential introduces a gravitational shift that mimics that of the Schwarzschild metric. The basic effect would be to reduce the observed flux from the innermost portions of the disk, both by decreasing the frequency of the photons emitted and by lengthening the observed interval between them. This gravitational redshift will be coupled with a rotational one which will also be strong in the inner portions of the disk and may either increase or decrease the observed flux, depending upon the observer's orientation. One further effect that has been neglected, but is certainly not negligible for the central part of the disk, is the capture of radiation by the black hole, for a fair amount of the radiation emitted from the inward facing portion will be swallowed. Only a fully relativistic treatment can solve the question of what the disk will look like to an external observer (cf. Sikora, 1980).

### 5. Discussion and Conclusions

By making some simple physical assumptions we have been able to construct models of the bloated inner portions of accretion disks around black holes. These results are essentially independent of assumed viscosity laws and they satisfy several self-consistency tests. The procedure we have employed could be generalized to the Schwarzschild metric without much difficulty and it is possible that other extensions could be made.

Very large amounts of mass can be swallowed, and a nominally stable disk can be formed as we do not require $Q_+ = Q_-$. "Supercritical luminosities" are possible because we assume that the disk is radiating at the local equivalent of the Eddington flux, and even though the effective gravity is smaller, the bloated shape provides a very large surface area that allows the disk to dissipate much more energy than is possible under conditions of spherical symmetry. The reality of these "supercritical luminosities" will have to be verified in the future with fully general relativistic models.

We have only calculated the shapes and luminosities for two simple families of angular momentum distributions. However, within the power law form [Eq. (55)] we have come close to exhausting the valid possibilities, because $\beta < 1$ yields positive angular velocity gradients implying that angular momentum cannot be transferred outwards, while for $\beta \gtrsim 1.5$ no matching thin solution can be found. Thus, working within this framework, we have a fairly narrow relationship between $L$ or $\dot{M}$ and $r_0$. Because the case $\beta = 1.0$ is at an analytical limit and also yields the maximum values for $L$ and $\dot{M}$, we can be fairly confident that we have found approximate absolute maxima of these quantities. Previous calculations have assumed that the lowest allowed value for $r_0$ is $r_{ms}$ and in this case we do find $L \lesssim 0.66\, L_{edd}$, in agreement with earlier workers (e.g. Bisnovatyi-Kogan and Blinnikov, 1977).

The disk shapes we have discovered range from quite bloated ones that thin substantially before $r_t$ (for small $r_0$), to disks that increase monotonically in thickness until $r_t$ is reached, but which can still be considered to be relatively thin there (for large $r_0$). We did not find disks which matched onto external portions which became *physically* thin, i.e., $z/r_g$ was always greater than unity at $r_t$.

The calculations presented here are just an early stage in our understanding of the effects that cusps on the inner edges of accretion disks can produce. Only if other angular momentum distributions are considered will we be confident that $\beta = 1.0$ really does yield maximum values. In general, we expect the cusp to open up somewhat and the accretion will resemble material flowing through the inner Lagrangian point in a binary system. Including this effect will allow estimates to be made of the infall velocities and the amount of internal energy gobbled up by the hole. The simple approximation of Kozłowski et al. (1978) could be applied to our models to determine these quantities approximately. Related to this feature is the expectation that hydrostatic equilibrium in the inner region will be violated to some extent, and that this would tend to reduce our $\dot{M}$ and $L$ values.

As mentioned previously, red-shift effects and the trapping of radiation by the hole must be considered in a more accurate treatment. We should also note that the inward facing parts of the disk will be subjected to strong radiative bombardment from the opposite portion of the surface. This will evaporate part of the disk and cause it to puff up even more. The cavities produced on either side of the black hole will contain huge radiation energy densities and could provide collimated beams for quasars and radio galaxies, a prospect that has been suggested many times previously in one form or another (Lynden-Bell, 1978; Rees, 1978 and references therein). The extreme thickness of the disk outside the hole, and the sharp drop down towards the cusp, where the bulk of the energy is radiated, suggest a crude analogy with an ordinary star into which two tunnels have been bored, allowing us to see directly the high radiation densities in its core. This approach to the structure of the inner region may be very fruitful in the investigation of the details of such collimation. By combining a fully general relativistic approach to the tracing of photon trajectories (Sikora, 1980) with an iterative attack on the reflection of radiation in the disk, the appearance of such a disk to a distant observer could be found. Finally, more general equations of state and geometries ought to be investigated by a suitable generalization of our method.

All of the above problems deserve further attention and development, and we plan to report on such elaborations in the future.

*Acknowledgments.* This work has greatly benefited from conversations with M. Abramowicz, M. Jaroszyński, and Z. Loska. Special thanks are due to M. Ekiel who checked many of the calculations. P.J.W. was supported during his stay in Warsaw by Smithsonian Institution Grant FR6-50001. The computations reported here were made on a PDP 11/45 computer donated to the N. Copernicus Astronomical Center by the US National Academy of Sciences.

**References**

Abramowicz, M.A., Jaroszyński, M., Sikora, M.: 1978, *Astron. Astrophys.* **63**, 221
Bisnovatyi-Kogan, G.S., Blinnikov, S.I.: 1977, *Astron. Astrophys.* **59**, 111
Callahan, P.S.: 1977, *Astron. Astrophys.* **59**, 127
Cunnigham, C.T.: 1975, *Astrophys. J.* **202**, 788
Cunningham, C.T.: 1976, *Astrophys. J.* **208**, 534
Fishbone, L.G., Moncrief, V.: 1976, *Astrophys. J.* **207**, 962
Kozłowski, M., Jaroszyński, M., Abramowicz, M.A.: 1978, *Astron. Astrophys.* **63**, 209
Kozłowski, M., Wiita, P.J., Paczyński, B.: 1979, *Acta Astron.* **29**, 157
Lightman, A.P., Eardley, D.M.: 1974, *Astrophys. J.* **187**, L1
Lynden-Bell, D.: 1969, *Nature* **223**, 690
Lynden-Bell, D.: 1978, *Physica Scripta* **17**, 185
Lynden-Bell, D., Pringle, J.E.: 1974, *Monthly Notices Roy. Astron. Soc.* **168**, 603
Meier, D.L.: 1978 (preprint)
Novikov, I., Thorne, K.S.: 1973, in *Black Holes*, eds. B. DeWitt and C. DeWitt, Gordon and Breach, New York
Paczyński, B.: 1978a, *Acta Astron.* **28**, 91
Paczyński, B.: 1978b, *Acta Astron.* **28**, 241
Piran, T.: 1977, *Monthly Notices Roy. Astron. Soc.* **180**, 45
Pringle, J.E., Rees, M.J.: 1973, *Astron. Astrophys.* **21**, 1
Rees, M.J.: 1978, *Nature* **275**, 516
Shakura, N.I., Sunyaev, R.A.: 1973, *Astron. Astrophys.* **24**, 337
Shakura, N.I.: 1976, *Monthly Notices Roy. Astron. Soc.* **175**, 613
Shapiro, S.L., Lightman, A.P., Eardley, D.M.: 1976, *Astrophys. J.* **204**, 187
Sikora, M.: 1980, Ph. D. Thesis, N. Copernicus Astronomical Center

**Note added in proof.** A theory of non-barotropic thick disks in general relativity has been recently published by M. Jaroszyński, M. Abramowicz and B. Paczyński, 1980, *Acta Astron.* **30**, 1.

THE ASTROPHYSICAL JOURNAL, 204:187-199, 1976 February 15


# A TWO-TEMPERATURE ACCRETION DISK MODEL FOR CYGNUS X-1: STRUCTURE AND SPECTRUM

STUART L. SHAPIRO AND ALAN P. LIGHTMAN
Center for Radiophysics and Space Research, Cornell University*
AND
DOUGLAS M. EARDLEY
Physics Department, Yale University†
*Received 1975 May 5; revised 1975 August 4*

## ABSTRACT

We present a model for Cygnus X-1, involving an accretion disk around a black hole, which can explain the observed X-ray spectrum from 8 to 500 keV. In particular we construct a detailed model of the structure of an accretion disk whose inner region is considerably hotter and geometrically thicker than previous disk models. The inner region of the disk is optically thin to absorption, is gas-pressure dominated, and yields, from first principles, electron temperatures of $10^9$ K and ion temperatures 3–300 times hotter. The spectrum above 8 keV is produced by inverse Compton scattering of soft X-ray photons in the two-temperature inner region of the disk. This spectrum is computed by numerical integration of the Kompane'ets equation, modified to account for escape of photons from a region of finite (order unity) electron scattering optical depth.

*Subject headings:* radiative transfer — stars: black holes — X-rays: sources

## I. INTRODUCTION

Current estimates of the mass of the compact X-ray source Cyg X-1 give $M \gtrsim 10\ M_\odot$ (Paczynski 1974; Avni and Bahcall 1975), indicating that this object is the most likely candidate for a black hole. If Cyg X-1 contains a black hole, the observed X-rays would originate from the inner portion of an accretion disk surrounding the black hole. The disk would be formed from gas captured from the companion star of the binary system, HDE 226868. One of the most important requirements of this (or any other) model is that it reproduce the observed X-ray spectrum. In this paper we present detailed calculations of the structure and spectrum of an accretion disk around a black hole which yield a close fit to the observations of Cyg X-1. A summary of our results has previously appeared (Eardley *et al.* 1975).

Historically, the "cool disk model" proposed to explain accretion onto a black hole in a close binary system (Pringle and Rees 1972; Shakura and Sunyaev 1973; Novikov and Thorne 1973) was found to be incapable of producing the observed, hard X-rays ($\sim 100$ keV) from Cyg X-1 (Lightman and Shapiro 1975). Thorne and Price (1975) demonstrated that *if* the inner portion of an accretion disk consisted of optically thin, high-temperature gas ($\sim 10^9$ K) instead of the optically thick, low-temperature gas characterizing the cool disk model, *then* the observed hard component of the Cyg X-1 spectrum near 100 keV could be explained. They argued that the secular instability present in the cool disk inner region (Lightman and Eardley 1974) could swell this optically thick, radiation-pressure dominated region to a hot, gas-pressure dominated, optically thin region. Their model, though semiempirical, suggested how the observational data for Cyg X-1 and the accretion disk model might be reconciled.

The main result of this paper is that the cool disk model is not unique. There exists a second, much hotter, self-consistent solution for an accretion disk around a black hole, with the same boundary conditions. When applied to Cyg X-1, this solution yields from first principles the required thermal emission temperatures of $10^9$ K and is able to reproduce the observed X-ray spectrum above $\sim 8$ keV. The disk in Cyg X-1 would choose the hot rather than the cool disk structure, if the cool one is secularly unstable. In § II we discuss the physical conditions that must exist in the disk to explain the observed spectrum from Cyg X-1. In § III we describe the detailed structure of and spectrum from the hot, inner portion of the disk. In § IV the results of numerical studies of the proposed two-fluid accretion disk model are presented and analyzed.

## II. PHYSICAL CONDITIONS IN THE DISK AROUND CYGNUS X-1

It is useful to describe, independently of specific cooling mechanisms and spectral considerations, the physical conditions in an accretion disk around a black hole which are necessary to explain the observations of Cyg X-1. These conditions provide strong motivation for the model we shall construct.

* Supported in part by the National Science Foundation Grant MPS 05056-A02.
† Supported in part by the National Science Foundation (GP-36317).

### a) *Disk Temperature Range*

The maximum and minimum temperatures achievable in the disk are determined by the extreme limiting cases of (1) complete internalization of gravitational energy by an optically thin gas ($T \to T_{max}$) and (2) LTE in an optically thick gas ($T \to T_{min}$). From the flux, $F$, emitted by any accretion disk around a nonrotating black hole (independent of the details of vertical structure) at the radius $r$ (Shakura and Sunyaev 1973), we may write

$$\tfrac{1}{4}aT_{min}^4(r) = F(r) = (\tfrac{3}{8\pi})\dot{M}(GM/r^3)\mathscr{J}\,, \tag{1a}$$

or

$$T_{min} \sim 3 \times 10^6\ \mathrm{K}(M/10\ M_\odot)^{-1/2}(\dot{M}/6 \times 10^{17}\ \mathrm{g\ s^{-1}})^{1/4}\,, \tag{1b}$$

where $M$ and $\dot{M}$ are the mass of the black hole and mass accretion rate, respectively, $a$ is the radiation constant, and $\mathscr{J} \equiv 1 - (6GM/rc^2)^{1/2}$. In equation (1b), the flux has been evaluated at the point of maximum emission at $r_m \approx 10\ GM/c^2$, and $M$ and $\dot{M}$ have been normalized to values appropriate for Cyg X-1. The low value of $T_{min}$ explains why the optically thick, cool disk model fails for Cyg X-1. For mass accretion rates near the Eddington limit, the inner region of the cool disk becomes optically thin and approaches temperatures of $4 \times 10^8$ K. However, the predicted flux in the observable X-ray band is far too high (see Shakura and Sunyaev 1973, Lightman and Shapiro 1975, for details). For rapidly rotating holes, with the appropriate X-ray luminosity, disk models can have an emission temperature larger than $10^8$ K, but not as large as $10^9$ K (Thorne and Price 1975).

The temperature $T_{max}$ is determined by the gravitational binding energy of an ion at the inner edge of the disk at $r_l = 6GM/c^2$. Thus,

$$kT_{max} \sim GMm_p/r_l = \tfrac{1}{6}m_pc^2\,, \tag{2a}$$

or

$$T_{max} \sim 2 \times 10^{12}\ \mathrm{K}\,. \tag{2b}$$

Temperatures approaching $T_{max}$ characterize the fluid near the event horizon in the case of optically thin, spherically symmetric accretion (Shvartsman 1971; Shapiro 1974). The observations indicate that the hottest emission temperature from Cyg X-1, $T \sim 10^9$ K, lies between $T_{min}$ and $T_{max}$, assuming thermal emission. We of course cannot rule out alternative models in which a very different temperature, or no temperature at all, characterizes the emitting electrons; but we judge these possibilities to be less likely.

### b) *The Temperature-dependent Disk Structure*

Without considering cooling mechanisms, one can determine the disk structure in terms of the ion and electron temperatures, $T_i$ and $T_e$ (which may not be equal). From observations, one can infer an electron temperature, assuming the emission is thermal. Then, assuming $T_i \geq T_e$, we can determine the allowable range of the physical variables which describe the emitting region. Three of the principal equations which describe this hot, inner portion of a (Newtonian) disk composed of pure hydrogen are:

*hydrostatic equilibrium*:

$$P/h = \rho(GM/r^3)h\,; \tag{3}$$

*angular momentum conservation*:

(viscous stress)·(area)·$r$ = (angular momentum per unit mass)·(mass accretion rate), or

$$(\alpha P)(2\pi r \cdot 2h)r = (GMr)^{1/2}\dot{M}\mathscr{J}\,; \tag{4}$$

*equation of state*:

$$P = \rho k(T_i + T_e)/m_p\,. \tag{5}$$

In the above equations $P$, $\rho$, $T_i$, $T_e$, and $h$ are the pressure, density, ion and electron temperatures, and the disk half-thickness, respectively. In equation (4) the usual viscosity law has been assumed, i.e., shear stress = $\alpha P$ ($0.01 \leq \alpha \leq 1$ [see Shakura and Sunyaev 1973; Eardley and Lightman 1975]). The angular momentum per unit mass for gas in nearly circular Keplerian orbits is $(GMr)^{1/2}$, and $\mathscr{J}$ expresses the boundary condition that the viscous stress must vanish at the inner edge of the disk, $r = r_l$. The dominant pressure source is assumed to be gas pressure, which we verify below. Two additional equations, one which determines the coupling between ions and electrons in the plasma and the other which determines the electron temperature through radiative cooling, will be specified in § III. Without these additional equations, equations (3)–(5) can be solved as a function of $T_i$ and

$T_e$. Equations (3)–(5) may be found in Shakura and Sunyaev (1973) for the case $T_i = T_e$. Solving the equations, we obtain:

$$h = (6 \times 10^3\ \mathrm{cm}) M_* r_*^{3/2} \left(\frac{T_i + T_e}{10^9\ \mathrm{K}}\right)^{1/2}, \tag{6a}$$

$$\rho = (0.5\ \mathrm{g\ cm^{-3}}) M_*^{-2} r_*^{-3} \dot{M}_* \mathscr{J} \alpha^{-1} \left(\frac{T_i + T_e}{10^9\ \mathrm{K}}\right)^{-3/2}, \tag{6b}$$

$$P = (4 \times 10^{17}\ \mathrm{dyn\ cm^{-2}}) M_*^{-2} r_*^{-3} \dot{M}_* \mathscr{J} \alpha^{-1} \left(\frac{T_i + T_e}{10^9\ \mathrm{K}}\right)^{-1/2}, \tag{6c}$$

where $M_* \equiv M/3\ M_\odot$, $\dot{M}_* \equiv \dot{M}/10^{17}\ \mathrm{g\ s^{-1}}$, and $r_* \equiv r/(GM/c^2)$.

From equations (6) and (1a), we may compute three important quantities: the ratio of radiation to gas pressure in the disk, $P_R/P$; the optical depth to free-free absorption measured from the disk midplane to the surface, $\tau_*$; and the electron-scattering optical depth from the disk midplane to the surface, $\tau_{es}$. Since the dominant source of opacity is electron scattering, radiation pressure perpendicular to the disk plane is given by $P_R \sim F\tau_{es}/c$, where

$$\tau_{es} = \kappa_{es}\rho h = (1 \times 10^3) M_*^{-1} r_*^{-3/2} \dot{M}_* \mathscr{J} \alpha^{-1} \left(\frac{T_i + T_e}{10^9\ \mathrm{K}}\right)^{-1}, \tag{7a}$$

and where $\kappa_{es} = 0.38\ \mathrm{cm^2\ g^{-1}}$ is the scattering opacity. Thus, the ratio $P_R/P$ is given by

$$P_R/P \approx \frac{F\tau_{es}/c}{\rho\kappa(T_i + T_e)/m_p} = 30 M_*^{-1} r_*^{-3/2} \dot{M}_* \mathscr{J} \left(\frac{T_i + T_e}{10^9\ \mathrm{K}}\right)^{-1/2}. \tag{7b}$$

The absorption optical depth is given by

$$\tau_* = (\kappa_{es}\bar{\kappa}_{ff})^{1/2}\rho h = (7 \times 10^{-2}) M_*^{-2} r_*^{-3} \dot{M}_*^{3/2} \mathscr{J}^{3/2} \alpha^{-3/2} \left(\frac{T_e}{10^9\ \mathrm{K}}\right)^{-7/2} \left(1 + \frac{T_i}{T_e}\right)^{-7/4}, \tag{7c}$$

where $\bar{\kappa}_{ff} = 6 \times 10^{23} \rho T_e^{-7/2}$ is the Rosseland mean opacity for free-free absorption.

If the observed, hard X-rays from Cyg X-1 originate from the inner portion of an accretion disk with $T_e \gtrsim 10^9$ K, then equation (7b), evaluated at typical radii, implies that gas pressure dominates radiation pressure in that region. Equation (7c) indicates that this region is optically thin to absorption, since $\tau_* \ll 1$. Finally, equation (7a) shows that the inner portion of the disk may be marginally thick to electron scattering, since $\tau_{es} \lesssim 1$ for $T_i \gg T_e$. In general, therefore, the *observations of Cyg X-1 require the inner portion of an accretion disk about a central black hole to be optically thin to free-free absorption, and gas-pressure dominated.* The consequences of this conclusion are explored further in the following sections.

### *c) The Two-Temperature Requirement*

In steady-state the electron heating and cooling rates must be equal. Since the medium is optically thin to absorption, the electron cooling rate, $\Lambda_e$, can be expressed as:

$$\Lambda_e = (5.2 \times 10^{20} \rho^2 T_e^{1/2}) A_{ff}\ \mathrm{ergs\ s^{-1}\ cm^{-3}}, \tag{8}$$

where the term in parentheses is the bremsstrahlung cooling rate and the multiplying factor $A_{ff}$ allows for possible *additional* cooling processes, i.e., $A_{ff} \geq 1$.

The heating rate, arising from gravitational and viscous energy release, is $F/h$ (ergs $\mathrm{s^{-1}\ cm^{-3}}$). Equating the two rates and using equations (1a), (6a, b), and (8), we find

$$\left(1 + \frac{T_i}{T_e}\right)^{5/4} \left(\frac{T_e}{10^9\ \mathrm{K}}\right)^{3/4} = 20\, A_{ff}^{1/2} M_*^{-1/2} r_*^{-3/4} \dot{M}_*^{1/2} \mathscr{J}^{1/2} \alpha^{-1}. \tag{9}$$

If $T_e \sim 10^9$ K in the inner portion of the disk, we find from equation (9) that $T_i > T_e$. A one-temperature solution to the structure equations may exist for the inner region, but the required temperature $T_e = T_i$ exceeds $10^{10}$ K, as shown from equation (9), and is inconsistent with the simplest interpretation of the observations. (New radiative processes are important at $10^{10}$ K, such as pair production/annihilation, which we have not tried to compute. These processes increase the cooling rate, i.e., increase $A_{ff}$; so by equation (9), $T_e$ must be even greater, further increasing the cooling rate, and so on. Whether a one-temperature solution exists at all is an open question.) Thus, *the observations of Cyg X-1 indicate that in the inner portion of the accretion disk, the ion temperature exceeds the electron temperature.*

If the mechanism of viscous dissipation in the disk puts most of the heating into the ion population, then $T_i$ will naturally exceed $T_e$. Subsonic turbulent viscosity will probably do this; supersonic turbulence, or ohmic losses which accompany magnetic reconnection, probably will not. However, even if the electrons and ions are heated equally, the electrons can cool much more efficiently, so that for a given mass element, $T_i$ will substantially exceed $T_e$ most of the time. Our calculations below, which presume $10^9$ K $\sim T_e < T_i \sim 10^{10-11}$ K, will therefore be valid "most of the time" even for this case (see Shapiro and Salpeter 1975 for a rather similar situation involving accretion onto neutron stars). Then about half of the total luminosity would emerge from the disk as an inverse Compton spectrum at $T_e \sim 10^9$ K, as discussed below; whereas the other half would emerge as brief, intense "flashes" of radiation, as electrons quickly cool from $\sim T_i$ to $\sim 10^9$ K. These flashes might appear as $\gamma$-rays between 100 keV and 10 MeV if (relativistic) inverse Compton cooling and free-free emission dominate the cooling or at softer energies near $\lesssim 10$ keV if synchrotron radiation is significant. We will not attempt to calculate these effects.

## III. STRUCTURE OF THE TWO-TEMPERATURE DOMAIN

### *a) Basic Assumptions and Equations*

Applying the results reached above, we construct from first principles a modified accretion disk model which can explain the observed, hard X-ray spectrum of Cyg X-1. The key assumptions which determine the structure of the inner portion of the disk are:

1. The dominant pressure source is gas pressure ($P \gg P_R$).
2. The gas is optically thin to absorption ($\tau_* \ll 1$).
3. The ions and electrons are coupled by collisional energy exchange, and those plasma instabilities which could provide further coupling are absent.
4. The gas consists of pure hydrogen.
5. The black hole is nonrotating, and the disk is Newtonian.
6. A copious supply of soft photons is available, so that unsaturated Comptonization dominates the cooling.

The two additional structure equations which, together with equations (3)–(5), determine the disk structure in the hot, inner region, may be now specified. The first equation is a direct consequence of assumption (3):

*electron-ion energy exchange* (*ion energy balance*)

$$F/h = 3/2\nu_E \rho k(T_i - T_e)/m_p \,. \tag{10}$$

In the above equation, the electron-ion coupling rate, $\nu_E$, can be approximated by $\nu_E = 2.4 \times 10^{21} \ln \Lambda \rho T_e^{-3/2}$ (Spitzer 1962), where the Coulomb logarithm $\ln \Lambda \approx 15$ in the present case. Rapid thermalization by plasma instabilities leading to $T_e \sim T_i \gtrsim 10^{10}$ K is assumed to be absent. Collisionless shock heating, which can equilibrate electron and ion temperatures in a few ion plasma periods (Shapiro and Salpeter 1975) will not be significant when turbulent velocities are subsonic, i.e., $\alpha < 1$ (Shakura and Sunyaev 1973). It has been pointed out by Rees (1975) that if the viscosity is too high ($\alpha \sim 1$), then shock heating may be important.

The remaining structure equation expresses the fact that Comptonization is the dominant cooling mechanism in the inner disk:

*inverse* (*unsaturated*) *Compton cooling* (*electron energy balance*)

$$F/h = (4kT_e/m_e c^2)\rho\kappa_{es} U_r c \,, \qquad \text{i.e., } y = 1 \,, \tag{11}$$

where

$$y \equiv (4kT_e/m_e c^2) \,\text{Max}\,(\tau_{es}, \tau_{es}^2)$$

is the dimensionless parameter that characterizes Comptonization (see Zel'dovich and Shakura 1969). The radiation energy density, $U_r$, appearing above satisfies $U_r \approx (F/c)\,\text{Max}\,(1, \tau_{es})$. When bremsstrahlung is the main source of photons, the above expression for Compton cooling of the electrons is not appropriate. In that case our calculations indicate the process will *saturate*; i.e., the energy of a typical photon increases to $\sim 3kT_e$ by repeated scatterings, at which point further scatterings transfer no net energy from the electrons. *Saturated* Comptonization, characterized by $y \gg 1$, can be significant in the inner region of the cool disk model (Shakura and Sunyaev 1973; Lightman and Shapiro 1975). *Unsaturated* Comptonization, given by equation (11), gives $y \sim 1$ and is appropriate whenever there is a copious source of soft photons in the hot inner region; see § III*b*. The origin of this source in our model and the corresponding spectrum are discussed in § III*b*, *c*.

Equations (3)–(5), (10), and (11) now comprise a complete set of equations for the five unknown disk structure variables $\rho$, $h$, $P$, $T_i$, and $T_e$. The solution to the equations are:

$$T_e = (7 \times 10^8 \text{ K})(M_* \dot{M}_*^{-1} \alpha^{-1} \mathscr{J}^{-1})^{1/6} r_*^{1/4} \,, \tag{12a}$$

$$T_i = (5 \times 10^{11} \text{ K}) M_*^{-5/6} \dot{M}_*^{5/6} \alpha^{-7/6} \mathscr{J}^{5/6} r_*^{-5/4} \,, \tag{12b}$$

$$h = (1 \times 10^5 \text{ cm}) M_*^{7/12} \dot{M}_*^{5/12} \alpha^{-7/12} \mathscr{J}^{5/12} r_*^{7/8}, \quad (12c)$$

$$\rho = (5 \times 10^{-5} \text{ g cm}^{-3}) M_*^{-3/4} \dot{M}_*^{-1/4} \alpha^{3/4} \mathscr{J}^{-1/4} r_*^{-9/8}, \quad (12d)$$

$$P = (2 \times 10^{15} \text{ dyn cm}^{-2}) M_*^{-19/12} \dot{M}_*^{7/12} \alpha^{-5/12} \mathscr{J}^{7/12} r_*^{-19/8}. \quad (12e)$$

Important dimensionless quantities which may be computed from the above equations are

$$h/r \sim 0.2\, M_*^{-5/12} \dot{M}_*^{5/12} \alpha^{-7/12} \mathscr{J}^{5/12} r_*^{-1/8}, \quad (13a)$$

$$\tau_{es} \sim 2 M_*^{-1/6} \dot{M}_*^{1/6} \alpha^{1/6} \mathscr{J}^{1/6} r_*^{-1/4}, \quad (13b)$$

$$T_e/T_i \sim 1 \times 10^{-3} M_* \dot{M}_*^{-1} \alpha \mathscr{J}^{-1} r_*^{3/2}. \quad (13c)$$

We observe from equation (12a) that *the computed value of $T_e$ is near $\sim 10^9$ K and is nearly independent of the radius, $M$, $\dot{M}$, and $\alpha$ throughout the two-temperature domain.* Equation (13a) indicates that this region is geometrically thick, with $1 > h/r \gtrsim 0.2$, due to the high value of the ion temperature. The ratio, $A_{ff}$, of inverse Compton cooling to free-free emission in the two-temperature domain is

$$A_{ff} = \frac{F/h}{\Lambda_{ff}} = (8 \times 10^3) M_*^{-7/6} \dot{M}_*^{7/6} r_*^{-7/4} \alpha^{-5/6} \mathscr{J}^{7/6} \gg 1. \quad (14)$$

Thus, Comptonization is far more efficient than bremsstrahlung alone as a cooling mechanism.

The above equations apply to a Newtonian disk model. Since we assume that the black hole is nonrotating, general-relativistic corrections are not significant.

We have used the thin-disk approximation ($h/r \ll 1$) to derive this model; clearly, this approximation is no longer strictly valid. The theory of *thick* accretion disks (see Bardeen 1972) unfortunately has not yet been worked out. We believe that the thin-disk equations are still valid to a factor of 2 in the present model and that the conclusions we draw are *substantially* unaffected, *as long as* the disk is not so thick ($h/r \sim 1$?) that a substantial amount of matter is blown radially into the hole or vertically to infinity by pressure gradients; whether this is possible in our model is an open question.

### *b) The Emission Spectrum*

The formation of X-ray spectra by inverse Compton scattering has been discussed by Illarionov and Sunyaev (1972), and Felten and Rees (1972) (for the case of a thermal bremsstrahlung source); and by Zel'dovich and Shakura (1969) (for the case of a soft photon source). It is an essential assumption of our model that there exists for the two-temperature region a copious soft (X-ray) photon source, much stronger than bremsstrahlung, cyclotron, or double Compton scattering. In this situation cooling takes place under the condition $y \approx 1$ ("unsaturated Comptonization"; see eq. [11] above) and $A \gg 1$, where $A$ is the energy enhancement factor due to Comptonization for arbitrary emission processes. However, the method of Zel'dovich and Shakura is inapplicable to the case $y \approx 1$, $A \gg 1$ because of a subtle but crucial point: Most of the emitted energy is carried by those few photons which remain in the disk for many scatterings (even though $\tau_{es} \sim 1$) and which thereby become especially hard. Therefore, it is essential to model photon transport, not by a mean number of scatterings before escape, but by a mean probability of escape per scattering. The steady-state Kompane'ets equation (Kompane'ets 1956; Weymann 1965; Cooper 1971) governing the spectrum then takes the form

$$0 = \frac{1}{E^2} \frac{d}{dE} \alpha(E, T_e) \left[ n + kT_e \frac{dn}{dE} \right] + S(E) - \beta(E) n, \quad (15)$$

where

$n$ = photon occupation number $\left( I_\nu = \frac{2h\nu^3}{c^2} n \right)$, assumed isotropic ;

$E$ = photon energy (keV) ;

$kT_e$ = electron temperature (keV) ;

$S(E)$ = photon source (per state per Thomson scattering time) ;

$\beta(E)$ = mean rate of photon escape from the disk, per Thomson scattering time, per photon in the disk $\approx \text{Min}(\tau_{es}^{-1}, \tau_{es}^{-2})$; here $\tau_{es}$ is computed from the total Klein-Nishina cross section. Note that $\beta(0) \equiv (4kT_e/m_e c^2)/y$ ;

$$\alpha(E, T_e) = \frac{E^4}{m_e c^2} [1 + f(T_e)/(1 + 0.02E)][1 + 4.2 \times 10^{-3} E + 9 \times 10^{-6} E^2]^{-1},$$

where

$$f(T_e) = 2.5\theta + 1.875\theta^2(1 - \theta), \quad \text{and} \quad \theta \equiv kT_e/m_ec^2 .$$

In the expression for $\alpha$, the factors in square brackets are the special-relativistic corrections, due to Cooper (1971). These corrections are accurate for $E \lesssim 1$ MeV, $kT_e \lesssim 100$ keV. We have neglected the stimulated scattering term, proportional to $n^2$, in equation (15); this term is unimportant in the X-ray regime under present conditions.

When there exists a copious supply of soft photons [represented by a source term $S(E)$ in eq. (15)] which vanishes above some soft energy $E_s$, the shape of $n(E)$ for $E > E_s$ is completely independent of the details of $S(E)$. For $E > E_s$, $E \ll kT_e$, $n(E)$ is approximately self-similar:

$$n(E) \propto E^{m}, \quad \text{where} \quad m = -\frac{3}{2} - \left[\frac{9}{4} + \frac{4}{y(1+f)}\right]^{1/2} . \tag{16}$$

For $y \approx 1$, equation (16) gives an energy spectrum $I_\nu \propto E^3 n \propto E^{-1}$. For $E \gtrsim kT_e$, the spectrum falls roughly like $\exp(-E/kT_e)$. When the relativistic corrections to $\alpha$ and $\beta$ are omitted (valid for $kT_e \lesssim 25$ keV), equation (15) can be solved exactly for the special case $y = 1$:

$$I_\nu \propto E^{-1}e^{-x}\left(1 + x + \frac{1}{2}x^2 + \frac{1}{6}x^3 + \frac{1}{24}x^4\right) , \tag{17}$$

where $x \equiv E/kT_e$. This solution is not very accurate under present conditions, but it well illustrates the qualitative shape of the spectrum. In general, equation (15) must be solved numerically; we have used the scheme of Chang and Cooper (1970) for the numerical work.

Equation (15) may not strictly apply in the real world because (*a*) the Fokker-Planck approximation (diffusion approximation to photon scattering in $E$-space) may break down, (*b*) the simple, probabilistic rate $\beta$ may be an insufficient model of photon escape, (*c*) $n$ may deviate strongly from isotropy at high $E$. All these effects are difficult to estimate; we suspect (*c*) to be the most significant. If these effects are important, they will have strongest effects on the spectrum at $E > 100$ keV (deviation from exponential spectrum), or at $E_s < E \lesssim 2E_s$ (deviation from self-similar spectrum).

We can now understand the cooling condition $y \sim 1$ (eq. [11]) and examine its sensitivity to the soft photon luminosity $L_s$. The shape of $S(E)$ and the precise value of $T_e$ make little difference; let us choose a delta-function

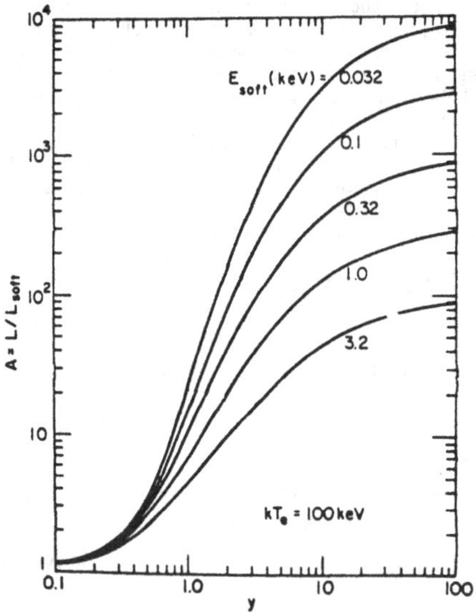


FIG. 1.—Energy enhancement fact $A$ for Comptonization of a soft photon source, as a function of the parameter $y \equiv (4kT_e/m_ec^2)\text{Max}(\tau_{es}, \tau_{es}^2)$ and of the characteristic energy $E_{soft}$ of the source. Electron temperature $kT_e \equiv 100$ keV ($T_e = 1.16 \times 10^9$ K) and source shape $S(E) \propto \delta(E - E_{soft})$, which are representative, are adopted.

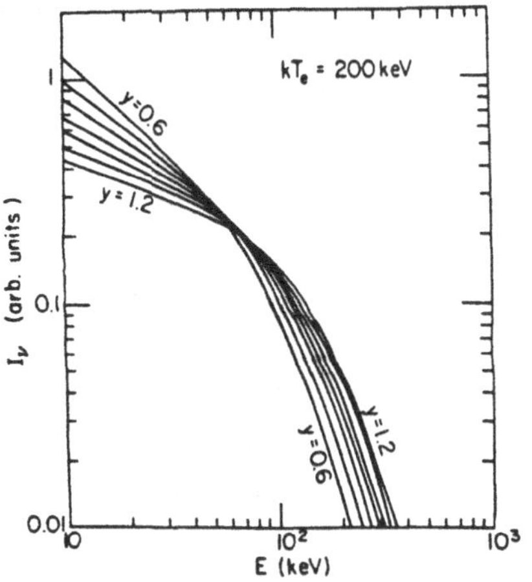


FIG. 2.—Spectral shapes of a Comptonized soft photon source, for photon energy $E > 10$ keV, and for $y$ in the range $0.6 \leq y \leq 1.2$ with steps $\Delta y = 0.1$. Each curve gives the spectral energy density $I_\nu$, normalized to the same total luminosity $L_{>10}$ above 10 keV. Shapes are not very dependent on adopted electron temperature $kT_e = 200$ keV ($T = 2.32 \times 10^9$ K) and are independent of soft source shape $S(E)$ provided $S(E)$ cuts off below 10 keV (and assuming the approximations that underlie equation 15 hold; see § IIIb). In our model, these changes in shape could be caused by changes in the luminosity $L_s$ of the soft source, $4.5 \lesssim L_{>10}/L_s \lesssim 41$, if $E_s = 0.1$ keV.

at $E_s$ for $S(E)$ and $kT_e = 100$ keV ($T_e = 1.16 \times 10^9$ K). Since the total luminosity $L$ is essentially fixed by observation, $L_s$ is determined by the energy enhancement factor $A \equiv L/L_s$.

Values of $A$ as a function of $E_s$ and $y$ are presented in Figure 1, from numerical integrations of equation (15). For small $E_s$, $A(y)$ rises very steeply in the region $y \sim 1$, much more steeply than $e^y$. This steep rise is due to the rare photons that remain in the disk for many scatterings and become hard. We can understand qualitatively the shape of the curves $A(y)$ through this derivation, which ignores the relativistic corrections to $\alpha$ and $\beta$: To express conservation of energy, we multiply equation (15) by $E^3$ and integrate over $E$. The result can be solved for $y$ to give the relation

$$y = \frac{1 - A^{-1}}{1 - \langle E \rangle / 4kT_e},$$

where $\langle E \rangle \equiv \int E^4 \, n dE / \int E^3 n dE$ is the intensity-weighted mean spectral energy. This relation explains the three regimes as $y$ ranges from 0 to $\infty$, which are apparent in Figure 1: (*a*) *Negligible Comptonization.* The emergent spectrum is essentially that of the input spectrum $S(E)$, $\langle E \rangle / 4kT_e \ll 1$; $A \approx 1$, so numerator $\ll 1$, hence $y \ll 1$. (*b*) *Unsaturated Comptonization.* As $y \to 1$, the spectrum remains soft, $\langle E \rangle / 4kT_e \ll 1$, so the denominator $\approx 1$. This gives $A^{-1} \ll 1$, i.e., $A$ becomes large. (*c*) *Saturated Comptonization.* Eventually $A$ increases to $\sim kT_e/E_s$, so most photons become hard, and $\langle E \rangle / 4kT_e$ becomes finite in the denominator, driving $y$ up. Finally spectrum approaches a Wien spectrum $n \propto e^{-E/kT}$, for which $\langle E \rangle = 4kT_e$, so $y \gg 1$, with $A \to 3kT_e/E_s$.

In nature, of course, the above calculation is reversed. The soft luminosity $L_s$ is determined by the thermal structure of the disk. As long as the soft photon source is sufficiently copious $E_s/kT_e \ll L_s/L \ll 1$, Figure 1 shows that $y \sim 1$, in the regime of unsaturated Comptonization. The value of $y$ is well buffered near 1 against large excursions in $L_s$; e.g., for $E_s = 0.1$ keV, $L_s$ can range over a factor of 100 while $y$ stays between 0.5 and 2.5. (In actuality, there may well be feedback from conditions in the disk to $L_s$ which further stiffens the buffering.) Therefore, the structure of the disk is very insensitive to changes in $L_s$ in the regime $y \sim 1$, as long as $E_s$ is small.

The predicted spectral shape, however, is sensitive to changes in $y$, and therefore is slightly sensitive to changes in $L_s$. It is the self-similar slope below $\sim kT_e$ that is most sensitive to changes in $y$; see equation (16). We present in Figure 2 computed spectral shapes for the range $0.6 \leq y \leq 1.2$, for adopted values $kT_e = 200$ keV ($T_e = 2.32 \times 10^9$ K) and $E_s = 0.1$ keV, with $S(E) = \delta(E - E_s)$. [The precise value of $T_e$ makes little difference; the value of $E_s$ or the shape of $S(E)$ makes no difference at all to the shape as long as $E_s < E$.] These spectra are all normalized to have the same total luminosity $L_{>10}$ above $E = 10$ keV (known from observations in the case of Cyg X-1). At $E_s = 0.1$ keV, this range of spectra corresponds to a range of $A$, $9.1 \leq A \leq 48.5$; or a range of the ratio of observed energy $L_{>10}$ to soft energy $L_s$, $4.5 \leq L_{>10}/L_s \leq 41.3$. These ranges would be smaller if $E_s$ were greater.

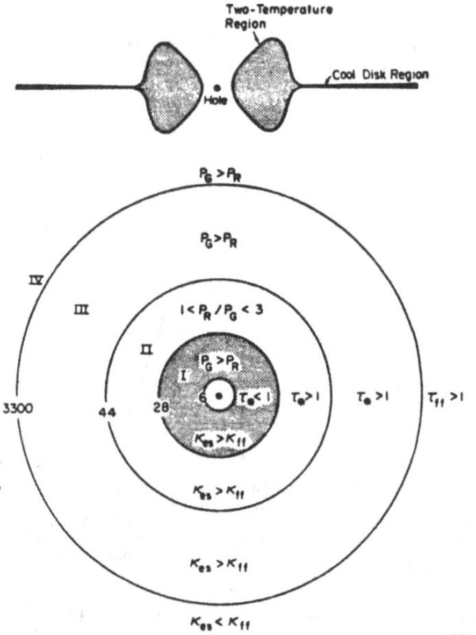


FIG. 3.—(*a*) (*top*) A side view (to scale) of the inner portions of an accretion disk around a nonrotating black hole. For this model, $M = 10\ M_{\odot}$, $\alpha = 0.1$, $\dot{M} = 7 \times 10^{17}$ g s$^{-1}$, $E_s = 0.5$ keV. The observed X-rays ($E > 1$ keV) are produced in the hot, bloated, two-temperature region, extending from 90 km $\leq r \leq$ 410 km and joining onto the thin cool disk region. (*b*) (*bottom*) A top view (not to scale) of the accretion disk for the same model parameters as Fig. 3*a*, illustrating the various disk regions. Optical depths, pressures, and opacities are denoted by $\tau$, $P$, $\kappa$, respectively (see text). Region IV is the cool disk "outer region," which generates in each radial annulus a blackbody spectrum. Regions III and II are the cool disk "middle" and "inner" regions, respectively, which produce modified (by scattering) blackbody spectra at surface temperatures $T_e \lesssim 3 \times 10^6$ K. Region I is the "two-temperature inner region," where $T_e \sim 2 \times 10^9$ K, $T_i \sim 1$–$2 \times 10^{11}$ K, and the spectra are produced by unsaturated Comptonization (see text). Radial boundaries of the regions are in units of $GM/c^2 = 14.7$ km.

Since we do not know $L_s$, we must make some assumption in building our model. We just choose $y \equiv 1$ throughout. In light of the preceding discussion, this choice is well justified for determining the disk structure, but it is to some extent arbitrary for determining the precise spectral shape. This point will be discussed in § IV*a*.

### *c) Origin of the Soft Photons*

The spectral energy curve given by equation (15) with $y \sim 1$ is valid in the hot region of the disk for $E > E_s$ whenever there exists a copious supply of soft photons with energy in the range 50 eV $\lesssim E_s \lesssim$ 5 keV. The lower limit to $E_s$ is determined by blackbody emission, since, whatever the soft photon source, it cannot produce more photons than a blackbody radiating at the electron temperature $T_e$ in the hot (nearly isothermal) zone. The intersection of a blackbody curve at $T_e \sim 10^9$ K with $I_\nu$ given by equation (15) (normalized to fit the observed spectrum shown in Fig. 3) occurs at 50 eV. Hence, Comptonization of photons softer than 50 eV cannot generate the observed hard X-ray flux. The upper limit to $E_s$ is also established by the observations, since, above $\sim$5 keV, the spectrum given by equation (15) with $y = 1$ fits the observed points quite well. Since the solution is only valid for $E > E_s$, the close fit requires $E_s < 5$ keV.

With these limits on $E_s$ we can eliminate thermal cyclotron emission as a significant soft photon source in the hot, inner region. The maximum magnetic field strength in the disk occurs if the magnetic viscosity is comparable to the turbulent viscosity, in which case $B \sim 10^7$ gauss in the inner regions (Shakura and Sunyaev 1973; Eardley and Lightman 1975). The maximum cyclotron frequency is then $h\nu_H \sim 0.1$ eV, which is far below the lower limit on $E_s$. High-harmonic cyclotron emission above 50 eV ($n \gtrsim 500$) is far too inefficient, even when $T_e \sim 10^9$ K.

Equation (14) eliminates thermal bremsstrahlung as a strong, soft photon source. Double Compton scattering can likewise be eliminated.

One source of soft X-rays is the cool ($T_e \sim 10^6$ K) region of the disk surrounding the hot, inner portion. A substantial flux of soft photons with energies near $\sim$0.5 keV impinge upon the hot, inner zone whenever $h/r \sim 1$

($\alpha \sim 0.1$). We have computed the fraction of the flux emitted in the cool outer regions of the disk which strikes the two-temperature inner region, assuming that the soft photon flux is emitted with a cosine limb-darkening law (approximately valid for a scattering atmosphere; see Chandrasekhar 1950). We find that for model parameters appropriate for Cyg X-1, the ratio $A$ of the total outgoing energy flux from the hot, inner region to the incident soft energy flux from the cool disk is, typically, $\sim 10$ for $\alpha \sim 0.1$. Figure 1 shows that $A \sim 10$ under these conditions is consistent with $y \sim 1$ and $E_s \sim 0.5$ keV; hence the model is self-consistent.

There may be additional soft photon sources in the inner region. One strong possibility includes emission from random high density, low temperature clumps in pressure and thermal equilibrium with the hot plasma, which form in the inner region due to the secular instability (Lightman 1974*b*). If the clumps have a size $d \leqslant r$, the energy of a typical soft photon should peak near $E \gtrsim kT_{\rm eff} \sim 0.5$ keV. We have not ruled out the possibility of soft photon illumination from the infalling gas *within* the region $2m < r < 6m(m = GM/c^2)$ prior to its capture by the black hole.

### *d*) *Formation of the Two-Temperature Domain*

We have demonstrated above that a new two-fluid solution to the disk structure equations can be found which differs considerably from the radiation-pressure dominated, optically thick inner region of the cool disk model. We now suggest that the secular instability (Lightman and Eardley 1974; Lightman 1974*b*) which probably occurs within the cool disk inner region will quickly drive the disk from a cool state ($T_e \sim 10^6$ K) to the hot, two-fluid state ($T_e \sim 10^9$ K) in this region.

The onset of the secular instability occurs at the radius where $\partial W(r, \Sigma)/\partial\Sigma$ changes sign; here $W$ is vertically integrated viscous stress and $\Sigma$ is surface density (Lightman and Eardley 1974; Lightman 1974*a*, *b*). To avoid the instability in the inner region, the derived quantity $\alpha(\Sigma, r) = W/2hP$ would have to obey $\partial \ln \alpha/\partial \ln \Sigma < -1$ when $P_R \gg P_G$(here $P_R$ = radiation pressure, $P_G$ = gas pressure), which does not seem likely. Consequently, the instability will set in at a radius where $P_R \approx P_G$ (near the outer boundary of the inner region). For instance, for the traditional law $\alpha$ = const., one has for the cool disk, the relation

$$\frac{\partial W}{\partial \Sigma} = \frac{5P_G + 2P_R}{3P_G - 2P_R} . \tag{18}$$

Hence the instability just occurs when

$$P_R = \tfrac{3}{2} P_G . \tag{19}$$

Using the "cool disk" values for $P_R$ and $P_G$, this condition occurs at the radius $r_0$ satisfying the relation

$$r_{0*}^{21/8} \mathscr{J}^{-2}(r_0) \approx 1 \times 10^4 \alpha^{1/4} M_*^{-7/4} \dot{M}_*^2 . \tag{20}$$

Thus, the hot, two-temperature domain will lie just within the cool disk inner region and will exist whenever the inequality

$$\alpha^{1/4} \dot{M}_*^2 M_*^{-7/4} \gtrsim 0.6$$

is satisfied. (This criterion applies to a nonrelativistic disk. For a fully relativistic disk, set $\beta = \frac{3}{2}$ in eq. [C1] of Eardley and Lightman 1975.)

The important feature of equation (18) is not this precise criterion for the location of the sign change, which depends on the special assumption $\alpha$ = const., but the fact that the sign changes via a *pole* rather than a zero, which seems to be a general feature. This suggests that the onset of instability is sudden and violent, and that the transition zone between the inner region of the cool disk and the hot, two-temperature region is a narrow region ($\Delta r/r \ll 1$) in which the disk effectively undergoes a "phase change" on the thermal time scale, as elements of the disk enter the "no-solution region" for cool-disk structure (Lightman 1974*a*).

Our predicted spectrum is quite insensitive to the precise criterion for onset of instability, because the two-temperature region is nearly isothermal (cf. eq. [12a]). In the numerical calculations of § IV, a slightly more conservative criterion for instability, namely, $P_R = 3P_G$, has been used.

We note that the hot, two-temperature solution is a valid solution for disk structure outside the inner region as well as within. A real disk might well choose to be in a hot state rather than a cool one, even outside the inner region, for reasons other than those considered here. This possibility was pointed out and discussed by Pringle *et al.* (1973) (although their hot state differs considerably from ours).

### *e*) *The Modified Accretion Disk Model*

We have incorporated the two-fluid disk domain in a modified disk model for Cyg X-1. The "outer" and "middle" regions of the cool disk (see Shakura and Sunyaev 1973; Novikov and Thorne 1973) describe the outer portions of the modified disk model. The inner region of the cool disk, whose outer boundary lies at the point where

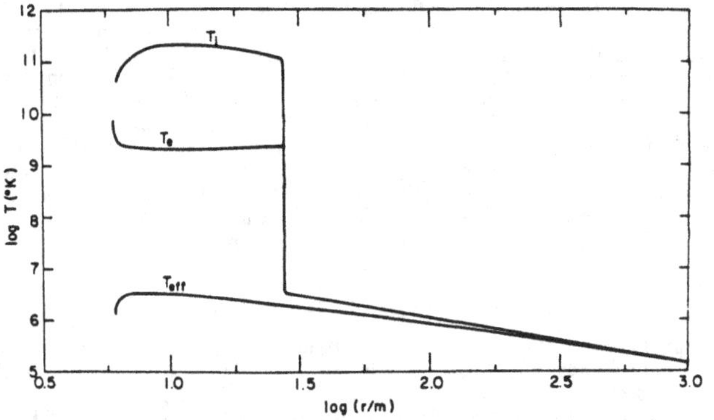


FIG. 4.—Temperature profile of the emitting layers of the identical disk illustrated in Fig. 3. The ion, electron, and effective temperatures ($T_i$, $T_e$, and $T_{eff}$, respectively) are plotted against the radial coordinate measuring distance from the central black hole. The "two-temperature inner region" extends from the inner edge of the disk at $r_I = 6m$ to $r = 28m$, where $m \equiv GM/c^2 = 14.8$ km.

$P_R = P_G$, extends inward to the radius $r_0$, where $P_R = 3P_G$. This radius now marks the outer boundary of the "two-temperature inner region," which then extends inward to the innermost stable circular orbit $r = r_I = 6GM/c^2$. The modified disk is therefore identical to the cool disk for $r \geq r_0$ but is described by the two-temperature structure equations for $r < r_0$.

Results of numerical calculations of the emergent continuous radiation spectrum from the entire accretion disk are discussed in the following section. The disk is divided into concentric rings, and the physical variables and emergent spectrum are determined for each ring. For $r \geq r_0$, we compute a model atmosphere in each ring from the known surface gravity, $g(r)$, and total energy flux $F(r)$ (see Lightman and Shapiro 1975 for details). For $r \leq r_0$, the spectrum in each ring is determined by equation (15) with $y = 1$. The total spectrum is then obtained by summing the contributions from each ring.

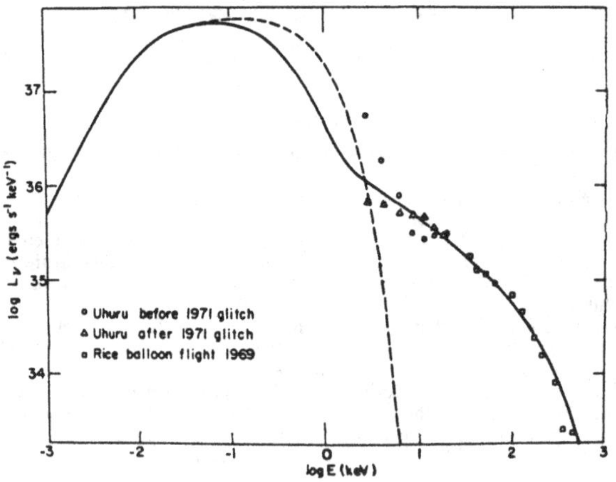


FIG. 5.—Comparison of observed and theoretical spectra for Cyg X-1. The solid curve gives the predicted spectrum of the two-temperature disk model for the same parameters as Figs. 3–4, with $y \equiv 1$. The dashed curve gives the predicted spectrum of the previous "cool disk" model for the same values of $M$, $\dot{M}$, $\alpha$. The 1969 measurements above 100 keV are uncertain to within factors of ~2, but error bars have been omitted in the figure for simplicity. A distance of 2.5 kpc to Cyg X-1 has been assumed in plotting the data.

TABLE 1

SPECTRAL PROPERTIES OF THE TWO-TEMPERATURE DISK MODEL

| $M$ ($M_\odot$) | $\dot{M}$ ($\dot{M}_E$)* | $\alpha$ | $E_s$ (keV) | $L$ ($10^{37}$ ergs s$^{-1}$) | $r_0$ (100 km) | $L_{r_0}/L$ | $T_{e,max}$ ($10^9$ K) | Luminosity ($10^{37}$ ergs s$^{-1}$) (keV range) (< 0.1) | (0.1–3.0) | (3.0–10) | (10–100) | (> 100) |
|---|---|---|---|---|---|---|---|---|---|---|---|---|
| Observations† of Cygnus X-1 | | | | | | | | ... | ... | 0.9 | 1.2 | 0.5 |
| 5....... | 0.065 | 0.1 | 0.5 | 4.1 | 3.1 | 0.68 | 2.3 | 0.23 | 1.5 | 0.41 | 1.2 | 0.54 |
| 10....... | 0.041 | 0.1 | 0.5 | 5.1 | 4.1 | 0.56 | 2.4 | 0.42 | 2.1 | 0.44 | 1.3 | 0.62 |
| 10....... | 0.041 | 1.0 | 0.5 | 5.1 | 5.6 | 0.65 | 1.7 | 0.41 | 1.6 | 0.46 | 1.8 | 0.84 |
| 10....... | 0.035 | 0.1 | 5.0 | 4.4 | 3.4 | 0.48 | 2.3 | 0.38 | 1.8 | 0.34 | 1.3 | 0.62 |
| 20....... | 0.028 | 0.1 | 0.5 | 7.0 | 5.6 | 0.41 | 2.3 | 0.81 | 3.6 | 0.43 | 1.3 | 0.65 |
| 20....... | 0.027 | 0.1 | 0.5 | 6.8 | 5.3 | 0.38 | 2.3 | 0.80 | 3.7 | 0.36 | 1.1 | 0.55 |

* $\dot{M}_E = 1.67 \times 10^{18} (M/M_\odot)$ g s$^{-1}$.

† Haymes and Harnden 1970; Schreier *et al.* 1971. Tabulated values give the pre–1971 glitch observations and assume a distance of 2.5 kpc.

## IV. NUMERICAL RESULTS AND DISCUSSION

Results of numerical calculations of the disk structure and emission spectrum are summarized in Figures 3–5 and Table 1. Figure 3 illustrates the geometrical profile and different physical regimes characterizing an accretion disk around Cyg X-1 for a typical model. Figure 5 gives the theoretical spectrum for the same model, together with some representative observed spectra of Cyg X-1 (Haymes and Harnden 1970; Schreier *et al.* 1971), and the predicted spectrum for the cool disk model. Figure 4 gives the ion, electron, and effective temperature profiles in the hot disk model for the same model depicted in Figure 3. The results for different model parameters, all chosen to fit the observations of Cyg X-1, are summarized in Table 1. It is clear from Figure 5 and Table 1 that excellent agreement between the theoretical spectrum and observations can be achieved for a suitable choice of model parameters.

The spectral shape for $E > E_s$ is specified by equation (15) with $y = 1$ and reproduces the observed self-similar shape, $I_\nu \propto \nu^{-1}$, for $E < kT_e$, independent of other model parameters. The two remaining features of the spectrum are (1) the location of the "knee" at $E \sim kT_e$ (observed near 125 keV) and (2) the total integrated luminosity above 3 keV ($\sim 3 \times 10^{37}$ ergs s$^{-1}$ for an assumed distance of 2.5 kpc [Bregman *et al.* 1973]). The first condition is satisfied whenever $T_e \sim 1$–$2 \times 10^9$ K, a result which follows from our model, almost independently of parameters (cf. eq. [12a]). The second condition determines the accretion rate, $\dot{M}$, for a given $\alpha$ and given black-hole mass $M$. The integrated luminosity from the inner edge of the disk to $r_0$ is

$$L_{r_0} = [1 + 2(\tfrac{1}{6}r_{0*})^{-3/2} - 18/r_{0*}]\dot{M}c^2/12 . \tag{21}$$

We assume that most of the radiation generated in the two-temperature inner region is observed ($E \geq 3$ keV). Employing equation (20) to find $r_{0*}$ as a function of $\dot{M}$ for a given $M$ and $\alpha$, the condition $L_{r_0} \sim 3 \times 10^{37}$ ergs s$^{-1}$ yields

$$\left[1 + 2\left(\frac{r_{0*}}{6}\right)^{-3/2} - \frac{18}{r_{0*}}\right] r_{0*}^{21/16}\left[1 - \left(\frac{r_{0*}}{6}\right)^{-1/2}\right]^{-1} = 2.8 \times 10^2\lambda , \tag{22}$$

where $\lambda \equiv (\alpha M_*^{-7})^{1/8}$. Since $\lambda$ is insensitive to $\alpha$, $r_{0*}$ and $\dot{M}$ are uniquely determined for a given $M$, by equations (20) and (22). These values are listed in Table 1. If the soft photons discussed in § III*b* have energies $E_s \ll 1$ keV and some of the Comptonized radiation of the two-temperature inner region comes out in the low-energy ($E <$ 3 keV) unobserved portion of the spectrum, then the coefficient in front of $\lambda$ in equation (22) must be increased. The last two entries in Table 1 indicate how sensitive the value of the emitted luminosity above 10 keV is to the mass accretion rate.

The parameter $\alpha$ is constrained by the model to lie within the range $0.05 < \alpha < 1$. The lower bound is set by the requirement that $h/r < 1$ for an accretion disk. The upper limit ensures that collisionless shock heating, which can bring about a rapid equipartition of ion and electron thermal energy, will not occur (see § III*a*). Since the shape of the disk surface in the hot, inner region depends on $\alpha$ (see eq. [12c]), a determination of the polarization properties of the hard X-rays originating from the region may constrain $\alpha$ further (See Lightman and Shapiro 1975).

The thermal instability discussed by Pringle *et al.* (1973), which operates when bremsstrahlung is the photon source in an optically thin gravitationally confined gas, probably is not present in our model. The source of soft photons, which subsequently cool the two-temperature inner region, is presumably thermally uncoupled from the hot optically thin gas.

### *a) Long-Term Variations in the Spectrum*

As indicated in Figure 3, a sudden change in the soft X-ray spectrum observed by *Uhuru* (2 keV $< E <$ 20 keV) occurred in 1971.[1] It is not clear whether the hard spectrum was also different after 1971. We have not attempted to fit the postglitch spectrum, because hard X-ray ($E \gtrsim$ 100 keV) observational data is more abundant before 1971.

Besides these major glitches, considerable variations in spectral shape have been reported by observers throughout the history of Cyg X-1 (for references and discussion, see Thorne and Price 1975). In particular, the slope $m$ of the energy spectrum, $I_\nu \propto E^{-m}$, for roughly the range 10 keV $< E <$ 100 keV, has been reported from $m =$ 0.5 to 1.0; and the variability seems to be real. Although we will not attempt a proper evaluation in this paper, we suggest that our model can accommodate these changes in $m$ by changes in $y$ in the region $y \sim 1$, the immediate cause of which would be changes in the soft photon source. To fit the data of Haymes and Harnden, we chose $y = 1$. Figure 2 presents a sequence of spectral shapes for $0.6 \leq y \leq 1.2$ that in fact roughly spans the reported range of slopes. The flatter slopes are more deficient in soft photons in our model. In this regard, we note that the observable soft component of the flux seems to have been small or absent during 1971–1975, and during this time the slope $m$ seems to have been nearly constant at the flat value $m \approx 0.5$ (Rothschild 1975). This coincidence suggests that there may be a connection between the observed soft flux and the theoretical one that we require.

### *b) Gamma Luminosity*

As is well known, an ion gas at $kT_i \gtrsim$ 100 MeV can develop a significant luminosity $L_\gamma$ for $\gamma$-rays, $10 \lesssim E \lesssim$ 100 MeV, through $\pi^0$ production. Dahlbacka *et al.* (1974) have pointed out that spherical accretion onto an isolated black hole might thereby produce $L_\gamma \approx 10^{22}$ ergs s$^{-1}$; this low luminosity is primarily a result of the notorious inefficiency of spherical accretion onto a hole. Our two-temperature disk offers the possibility of a much brighter $\gamma$-ray source. In the present model, with a nonrotating hole, $T_i$ is just not high enough ($kT_{i,\max} \sim$ 30 MeV), essentially because of the small binding energy of the innermost stable orbit ($\sim$60 MeV per baryon). We have made preliminary estimates for the case of a rapidly rotating black hole, for which the binding energy is greater ($\sim$300 MeV per baryon); we find some models with $L_\gamma \sim 10^{37}$ ergs s$^{-1}$ $\sim L_X$. We therefore suggest that rapidly rotating black holes, but not nonrotating ones, can be efficient, compact $\gamma$-ray sources, and that this effect may serve as a means to measure the spin of a black hole.

### *c) Future Investigations*

Future theoretical work on the accretion disk model for Cyg X-1 should correctly model the inner disk as *thick*, should deal with the origin and nature of the soft photon source required by the model (perhaps there are others in addition to the ones mentioned in § III) and should make the model fully relativistic. Our calculations have been for a Newtonian disk around a nonrotating black hole, for which general-relativistic corrections are not significant. General-relativistic effects for accretion onto rapidly rotating holes have been considered by Cunningham (1975) for very thin ($h/r \ll 1$) disk accretion and by Shapiro (1974) for spherical accretion, and will be important for the model considered here. Both general-relativistic effects (redshift, gravitational focusing, and black hole capture of photons) and special relativistic effects (Doppler shifts and forward beaming) will alter the disk structure and, more significantly, the observed radiation spectrum.

It is a pleasure to thank Kip S. Thorne and Richard H. Price for several useful discussions and suggestions and for critically reading the manuscript.

*Note added 1975 October.*—We have learned by personal communication and preprint that J. I. Katz (Institute for Advanced Study, Princeton) has investigated unsaturated Comptonization of a soft photon source in the context of QSO models and has reached conclusions similar to those that we present in § III*b*.

[1] The reverse transition seems to have occurred in 1975 April.

## REFERENCES

Avni, Y., and Bahcall, J. N. 1975, preprint.
Bardeen, J. M. 1973, in *Black Holes*, Les Houches, ed. C. DeWitt and B. DeWitt (New York: Gordon & Breach).
Bregman, J., Butler, D., Kemper, E., Koski, A., Kraft, R., and Stone, R. P. S. 1973, *Ap. J. (Letters)*, **185**, L117.
Chandrasekhar, S. 1950, *Radiative Transfer* (Oxford: Oxford University Press).
Chang, J., and Cooper, G. 1970, *J. Comp. Phys.*, **6**, 1.
Cooper, G. 1971, *Phys. Rev. D.*, **3**, 2312.
Cunningham, C. T. 1975, preprint.
Dahlbacka, G. H., Chapline, G. F., and Weaver, T. A. 1974, *Nature*, **250**, 36.
Eardley, D. M., and Lightman, A. P. 1975, *Ap. J.*, **200**, 187.
Eardley, D. M., Lightman, A. P., and Shapiro, S. L. 1975, *Ap. J. (Letters)*, **199**, L153.
Felten, J. E., and Rees, M. J. 1972, *Astr. and Ap.*, **17**, 226.
Haymes, R. C., and Harnden, F. R., Jr. 1970, *Ap. J.*, **159**, 1111.
Illarionov, A. F., and Sunyaev, R. A. 1972, *Soviet Astr.—AJ*, **16**, 45.
Kompane'ets, A. S. 1956, *Zh. Eksp. Teor. Fiz.*, **31**, 876 (English transl. in *Soviet Phys.—JETP*, **4**, 730 [1957]).
Lightman, A. P. 1974*a*, *Ap. J.*, **194**, 419.
———. 1974*b*, *ibid.*, p. 429.
Lightman, A. P., and Eardley, D. M. 1974, *Ap. J. (Letters)*, **187**, L1.
Lightman, A. P., and Shapiro, S. L. 1975, *Ap. J. (Letters)*, **198**, L73.
Novikov, I. D., and Thorne, K. S. 1973, in *Black Holes*, Les Houches 1973, ed. C. DeWitt and B. DeWitt (New York: Gordon & Breach).

Paczynski, B. 1974, *Astr. and Ap.*, **34**, 161.
Pringle, J. E., and Rees, M. J. 1972, *Astr. and Ap.*, **21**, 1.
Pringle, J. E., Rees, M. J., and Pacholczyk, A. G. 1973, *Astr. and Ap.*, **29**, 179.
Rees, M. J. 1975, private communication.
Rothschild, R. E. 1975, private communication.
Schreier, E., Gursky, H., Kellogg, E., Tananbaum, H., and Giacconi, R. 1971, *Ap. J.*, **170**, L21.
Shakura, N. I., and Sunyaev, R. A. 1973, *Astr. and Ap.*, **24**, 337.
Shapiro, S. L. 1974, *Ap. J.*, **189**, 343.
Shapiro, S. L., and Salpeter, E. E. 1975, *Ap. J.*, **198**, 671.
Shvartsman, V. F. 1971, *Astr. Zh.* (English transl. in *Soviet Astr.—AJ*, **15**, 377 [1971]).
Spitzer, L, Jr. 1962, *Physics of Fully Ionized Gases* (New York: Wiley).
Thorne, K. S., and Price, R. H. 1975, *Ap. J.* (*Letters*), **195**, L101.
Weymann, R. 1965, *Phys. Fluids*, **8**, 2112.
Zel'dovich, Ya. B., and Shakura, N. I. 1969, *Astr. Zh.*, **46**, 225 (English transl. in *Soviet Astr.—AJ*, **13**, 175 [1969]).

*Note added 1975 December.*—We have recently learned from K. S. Thorne that R. A. Sunyaev, N. I. Shakura, and A. F. Illarionov of the Soviet Union have constructed a hot disk model with $T_e \sim 10^{10}$ K very similar to the two-temperature model discussed here. Because of the decrease in the scattering cross section in the relativistic (Klein-Nishina) domain, a knee in the spectrum still occurs at ~100 keV from Comptonization of soft photons.

DOUGLAS M. EARDLEY: Physics Department, Yale University, New Haven, CT 06520

ALAN P. LIGHTMAN and STUART L. SHAPIRO: Center for Radiophysics and Space Research, Cornell University, Ithaca, NY 14853

*Nature* Vol. 295 7 January 1982 17

ARTICLES

# Ion-supported tori and the origin of radio jets

**M. J. Rees*, M. C. Begelman*‡, R. D. Blandford† & E. S. Phinney***

* Institute of Astronomy, Madingley Road, Cambridge CB3 0HA, UK
† Theoretical Astrophysics, California Institute of Technology, Pasadena, California 91125, USA
‡ Department of Astronomy, University of California, Berkeley, California 94720, USA

*While apparently supplying tremendous power to their extended radio-emitting regions, the nuclei of most radio galaxies emit little detectable radiation. It is proposed that at the centre of each is a spinning black hole surrounded by a torus of gas too hot and tenuous to radiate efficiently. The torus anchors magnetic fields which extract rotational energy from the hole in the form of two collimated beams of relativistic particles and fields. These in turn drive the observed radio jets and hot spots. A large supply of accreting gas is thus unnecessary and radio galaxies may be interpreted as starved quasars.*

ALL extended radio sources seem to be associated with elliptical galaxies or quasars, and the evidence[1] suggests that quasars too are located in ellipticals. As the present number-density of dead quasars is comparable with that of elliptical galaxies[2,3], it is reasonable to suppose that radio sources are powered by $\sim 10^8 M_\odot$ black holes[3] left over from a quasar phase. A long-standing problem in the theory of active galactic nuclei has been that of finding a supply of nuclear gas adequate to 'feed the monster'. Stars in the galactic nucleus cannot provide more than $\sim 10^{-3} M_\odot \mathrm{yr}^{-1}$, unless conditions are so extreme that stellar collisions dominate the evolution[4]. Gas originating further out must dispose of substantial angular momentum to enter the nucleus[5,6], and it must do so before succumbing to galactic winds or star formation.

The observational evidence is less ambiguous: radio sources almost certainly do have low fuelling rates. Low-luminosity double-jet radio sources of the sort being mapped with the VLA generally have no detectable nuclear emission in the optical or X-ray bands; of the more powerful double hot spot radio sources, only a few (such as 3C390.3 with $\sim 2\times 10^{44}$ erg s$^{-1}$ in both optical and X-ray bands[7], and Cygnus A with $\sim 2\times 10^{43}$ erg s$^{-1}$ in optical continuum[8]) have nuclear luminosities above $10^{42}$ erg s$^{-1}$ $= 10^{-4} M_8^{-1} L_{\mathrm{Edd}}$ (refs 9, 10). They cannot be produced by any process requiring super-Eddington luminosity, unless the hole mass is very small. This it cannot be because (even aside from the progenitor argument) some low-luminosity radio sources have giant ($\sim$Mpc) extensions with minimum energies[11] of up to $2\times 10^{60}$ erg s$^{-1}$ $= 10^6 M_\odot c^2$. Even if we make optimistic estimates of the efficiency of the central 'engine' and of the likely efficiency with which power emitted from the nucleus can accelerate relativistic electrons in the radio lobes, we must conclude that a mass of at least $10^7$–$10^8 M_\odot$ is involved (in computing the powers and energies above, we have used a Hubble constant of $H_0 = 50$ km s$^{-1}$ Mpc$^{-1}$).

Recent VLBI observations[12-17] have corroborated the view that the jets which power extragalactic sources are collimated on scales $\leq 1$ pc. These observations reveal small-scale structure which is generally aligned with larger-scale structure. Misalignments of up to $\sim 45°$, which have been found in the core-dominated superluminal sources, are probably caused by viewing a slightly curved jet along an almost parallel line of sight. In the case of the cores of extended radio sources, however, the alignment is fairly good. Any mechanism which purports to explain them must, therefore, be capable of producing collimated jets of at least trans-relativistic velocities while radiating as waste heat much less than an Eddington luminosity. Furthermore, it must, at least in some more extreme sources, put orders of magnitude more power into the jet than escapes as (detectable) radiation from the nucleus. These constraints rule out the class of models involving thick disks supported by radiation pressure[18-25].

## Sub-critical accretion

To understand radio sources, we should therefore study accretion at very low rates, $\dot M \ll M_{\mathrm{Edd}} \equiv L_{\mathrm{Edd}}/c^2$. Accreted gas should carry with it some magnetic flux, and shearing motions will maintain the magnetic field energy $B^2/8\pi$ at a significant fraction $\beta^{-1}$ of the pressure $P$. The Larmor radius is then, for the applications which interest us, about 10 orders of magnitude smaller than the scale of the flow. Ordinary 'molecular' viscosity will consequently be suppressed, and magnetic (or possibly turbulent) viscosity will probably govern the shear stress. The viscous stress may be written as $\alpha P$, where estimates of magnetic viscosity suggest $\alpha \geq 0.01$ (ref. 26). Although this estimate is uncertain, there is no reason why $\alpha$ should diminish as $M$ falls: if $\alpha$ is fixed, the torque per unit mass will be independent of $\dot M$ (except insofar as the disk structure depends on $\dot M$). Thus the inflow time in a disk of half-thickness $h(r)$ around a hole of gravitational radius $r_g = GM/c^2$,

$$t_i = \alpha^{-1}\left(\frac{r}{r_g}\right)^{3/2}\left(\frac{h}{r}\right)^{-2}\left(\frac{r_g}{c}\right) \quad (1)$$

does not depend explicitly on density. (See refs 27–29 for the standard theory of thin disks with $h \ll r$.) The cooling time, on the other hand, is inversely proportional to the density.

The thickness of the disk depends on how fast it can radiate (or otherwise dispose of, say by a wind) the binding energy liberated by viscous stresses. The major cooling mechanisms available to the accreting matter are illustrated in Fig. 1. Depending on initial conditions and external perturbations the matter may find itself in any of several temperature states. In fact, for $10^{-7} \leq (\dot M/\dot M_{\mathrm{Edd}})\alpha^{-2} \leq 50$ there are two self-consistent solutions to the fluid and radiation equations.

One is a thin, dense disk that remains at $\sim 10^4$ K. If matter with well defined angular momentum is fed in steadily (with high angular momentum) at a large radius, and there are no traumatic perturbations, this cold disk should be selected. These disks are, however, subject to instabilities[5,30,31], and if heated up and thickened, they will thereafter be unable to radiate the energy dissipated by viscous stresses, and will inflate to become pressure-supported tori with vortices along the rotation axis (see Fig. 2). The hot torus solution will also be selected if cool matter is preheated by high-frequency non-thermal radiation (for example, from a jet), or if the gas is supplied hot (for example by colliding stellar winds[6] or by tidal disruption of stars[32,33]). Both of these latter examples lead to flows which join naturally onto the ones described here.

0028-0836/82/010017—05$01.00

For an optically thin disk to radiate by bremsstrahlung, the energy dissipated between $r$ and $r/2$, the accretion rate must satisfy

$$\dot{m} = \dot{M}/\dot{M}_{\rm Edd} \gtrsim \alpha_f^{-1}\left(\frac{m_p}{m_e}\frac{r_g}{r}\right)^{1/2}\alpha^2\left(\frac{h}{r}\right)^2, \quad r > \left(\frac{h}{r}\right)^2\left(\frac{m_p}{m_e}\right)r_g \tag{2a}$$

$$\dot{m} \gtrsim \alpha_f^{-1}\alpha^2\left(\frac{h}{r}\right)^3(\log\gamma_{re})^{-1}, \quad r < \left(\frac{h}{r}\right)^2\left(\frac{m_p}{m_e}\right)r_g \tag{2b}$$

where $\alpha_f = 1/137$ is the fine structure constant and $\gamma_{re} = 3\,kT_e/m_e c^2$. Hence at the low accretion rates of interest ($\dot{m} \simeq 10^{-4}$), a disk once heated to $T \approx T_{\rm virial} = (m_p c^2/3k)(r_g/r)$ will be unable to cool through bremsstrahlung and must remain pressure-supported with $h \approx r$.

As the gas moves in, however, it will shear and amplify any seed magnetic field until $B^2/8\pi \simeq P_{\rm gas}$, and the fields begins to rise buoyantly out of the disk (moving perpendicular to the plane of a thin disk; towards the vortex funnel in a thick torus). We thus expect the ratio $\beta$ to be self-limiting at $\beta \simeq 1$, although $\beta$ may be $\ll 1$ in a corona if the latter is threaded by field lines anchored to interior (high-pressure) parts of the disk, as in the solar atmosphere. The corona may also join onto a wind (M.C.B. *et al.* in preparation, and ref. 34). This magnetic field provides the torus with a new cooling mechanism: for $r < (m_p/m_e)r_g$, electrons at the virial temperature of the ions would be relativistic, and each would radiate an energy

$$\Delta E_r \approx kT_i\left(\frac{50r_g}{r}\right)^{31/4}\dot{m}_{-4}^{1/2}\alpha_{-2}^{-3/2}\beta^{-3/2}M_8^{-1/2} \tag{3}$$

by optically thick synchrotron radiation as it spiralled in from $r$ to $r/2$ (where $\dot{m}_{-4} \equiv 10^4\,\dot{m}$ and $\alpha_{-2} = 10^2\alpha$). The actual cooling rate is considerably higher for

$$r < r_{\rm comp} \equiv \left(\frac{m_p}{m_e}\right)^{4/5}(\dot{m}\alpha^{-1})^{2/5} = 65\,\dot{m}_{-4}^{2/5}\,\alpha_{-2}^{-2/5} \tag{4}$$

since then the dominant cooling is due to Compton scattering of the synchrotron photons (Fig. 1). Thus if the electrons and ions are kept in energy equipartition, a thick torus will cool and deflate to a thin disk for $r \lesssim 50r_g$.

The torus will be 'collisionless' inside some large radius (typically $\sim$2,000 $r_g$, but defined more precisely below), in the sense that through two-body collisions alone protons at temperature $T_p$ give less than half their energy to electrons (with $T_e < T_p$) on the inflow time scale, if

$$\dot{m} < \left(\frac{m_p}{m_e}\right)(\ln\Lambda)^{-1}\alpha^2\left(\frac{m_p}{m_e}\cdot\frac{r_g}{r}\,\frac{T_e}{T_p}\right)^{3/2}$$

$$\approx 50\alpha^2\left(\left(\frac{2{,}000r_g}{r}\right)\left(\frac{T_e}{T_p}\right)\right)^{3/2} \quad kT_e < \tfrac{2}{3}m_e c^2 \tag{5a}$$

$$\dot{m} < \left(\frac{m_p}{m_e}\right)\left(\frac{\gamma_{re}}{\ln\Lambda}\right)\alpha^2 = 50\gamma_{re}\alpha^2 \quad kT_e > \tfrac{2}{3}m_e c^2 \tag{5b}$$

In these conditions, collective plasma phenomena will determine the particle distribution functions, the radiative efficiency, and thus the pressure and shape of the whole torus/disk. In most of what follows we assume that collective processes are unable to transfer more than half the proton's kinetic energy to the electrons [which would in turn quickly radiate it: compare with equation (3)]. Given the complexity of the plasma physics and the uncertainties in accretion theory, it may be a long time before this assumption can be checked. If the collective transfer can take place, the structure is similar to that already familiar from results at high accretion rates. If it cannot, one consequence, as already realized in the context of accretion disks around stellar-mass black holes[35,36], is a 'two-temperature' flow.

## Two-temperature tori

Consider an accretion flow where equation (5) is satisfied for $r < r_D$, and assume that the protons and electrons are then

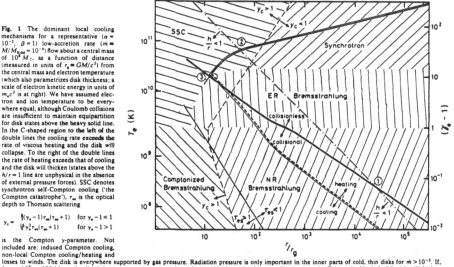

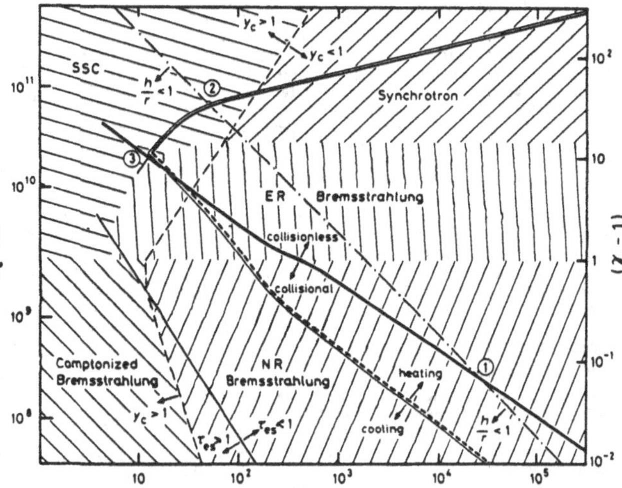


**Fig. 1** The dominant local cooling mechanisms for a representative ($\alpha = 10^{-2}$, $\beta = 1$) low-accretion rate ($\dot{m} \equiv \dot{M}/\dot{M}_{\rm Edd} = 10^{-4}$) flow about a central mass of $10^8 M_\odot$, as a function of distance (measured in units of $r_g \equiv GM/c^2$) from the central mass and electron temperature (which also parametrizes disk thickness; a scale of electron kinetic energy in units of $m_e c^2$ is at right). We have assumed electron and ion temperature to be everywhere equal, although Coulomb collisions are insufficient to maintain equipartition for disk states above the heavy solid line. In the C-shaped region to the left of the double lines the cooling rate exceeds the rate of viscous heating and the disk will collapse. To the right of the double lines the rate of heating exceeds that of cooling and the disk will thicken (states above the $h/r = 1$ line are unphysical in the absence of external pressure forces). SSC denotes synchrotron self-Compton cooling ('the Compton catastrophe'), $\tau_{es}$ is the optical depth to Thomson scattering

$$y_c = \begin{cases} \tfrac{8}{3}(\gamma_e - 1)\tau_{es}(\tau_{es}+1) & \text{for } \gamma_e - 1 \ll 1 \\ \tfrac{16}{9}\gamma_e^2\tau_{es}(\tau_{es}+1) & \text{for } \gamma_e - 1 > 1 \end{cases}$$

is the Compton $y$-parameter. Not included are: induced Compton cooling, non-local Compton cooling/heating and losses to winds. The disk is everywhere supported by gas pressure. Radiation pressure is only important in the inner parts of cold, thin disks for $\dot{m} > 10^{-3}$. If, for $r \lesssim 10^7 r_g \simeq 100\,M_8$ pc, the disk is heated to a temperature above that at which bremsstrahlung cooling becomes inefficient (double dashed line) it will thicken to a torus, $h \sim r$. For $r < r_D = 2 \times 10^4 r_g$ (point 1), Coulomb collisions cannot maintain equipartition. If plasma processes take over this job, the disk will begin to cool by SSC at $r \simeq 50r_g$ (point 2) and will gradually thin to $(h/r) \sim \frac{1}{3}$ at $r \simeq 10r_g$, at which point (3) bremsstrahlung will rapidly cool it to a new equilibrium at $T \sim 10^4$ K and $(h/r) \sim 10^{-4}$ where it radiates as a black body (optically thick bremsstrahlung and line cooling just balance viscous heating). If, as assumed here, equipartition cannot be maintained, the electrons continue to cool (Fig. 3) inside $r = r_D$ (point 1) but the ions remain at the virial temperature. The torus will thus remain thick, with the structure shown in Fig. 2.

indeed thermally decoupled from each other. Equation (2) shows that, for $\dot{m}_{-4} \simeq 1$, the thick ($h \simeq r$) torus is a self-consistent solution even for $r \gg r_R = 2{,}000 r_g$ (see Fig. 1). At $r_D \gtrsim r_R$, electrons and ions will decouple; but as neither can cool, they may remain at the same temperature. For $r < r_R$, the electrons become relativistic. If the heating of both species is adiabatic, then $T_p \propto n^{2/3}$ while $T_e \propto n^{1/3}$, so the ions are preferentially heated. Shock heating may likewise favour the ions, as isotropization of the bulk flow velocities will give to each species a thermal energy proportional to its mass. Fermi acceleration processes will amplify any temperature difference once it exists. Thus the electrons are granted little of the ions' energy, despite their ability to radiate it [equation (3)]. The ions remain at the virial temperature, and the two-temperature torus remains thick even inside $r \simeq 50 r_g$ [equations (3) and (4)].

If the torus surrounds a spinning black hole, the Lense–Thirring effect can enforce axisymmetry with respect to the hole's spin axis[37] out to a radius $\sim \alpha^{-2/3}(J/J_{max})^{2/3} r_g$, where $J$ is the hole's angular momentum and $J_{max} = M r_g c$ is the maximum value allowed by the Kerr metric. We would expect the holes to have been spun up to $J \approx J_{max}$ during their previous quasar phase[38,39]. There may be misalignment outside this radius if the gas is not supplied in the equatorial plane of the hole. The isotherms within the torus can then be calculated in the same way as for radiation-supported tori[18,19,22]. If the pressure support is due primarily to non-relativistic ions, then the effective specific heat ratio will be about 5/3. If $\alpha$ is independent of radius, then $n \propto r^{-3/2}$, and most of the mass of the torus resides at large radii. If both these conditions are fulfilled, then the ions are roughly isentropic.

The contours of Fig. 2 have been computed by this procedure. The results are subject to the same uncertainties as those for radiation-supported tori: problems with self-consistency in viscosity (R.D.B. and M. Jarozyński, in preparation), and the absence of a formalism for properly matching one of these inner solutions onto a flow field specified at large radii. Magnetic fields will modify the simple structure shown. However, essential features, such as the characteristic vortex funnel, which becomes parabolic in the limit of a flow with constant specific angular momentum $l$ and is less collimated than parabolic for a flow with $l$ increasing outwards, are unlikely to be changed. (Flows for which $l$ decreases outwards are argued to be unstable; note, however, that $l$ must be constant along the surface if there is no poloidal circulation and we insist that the surfaces have zero energy-at-infinity.) In this 'zone of non-stationarity', matter cannot have enough angular momentum to achieve near-equilibrium between centrifugal force, pressure and gravity. A fluid flow in the funnel must either fall straight into the hole, or have enough enthalpy to leave the region rapidly. Hydromagnetic effects[40] or flares (compare with previous remarks about buoyant fields) on the funnel walls will inject into it particles of positive energy-at-infinity. The liberated power will escape along the two funnels, leading to a pair of inertially collimated jets carrying some fraction of $\dot{M}c^2 = 10^{42}\, \dot{m}_{-4} M_8\ \mathrm{erg\ s^{-1}}$. This may be enough to explain the low-luminosity double-jet radio sources.

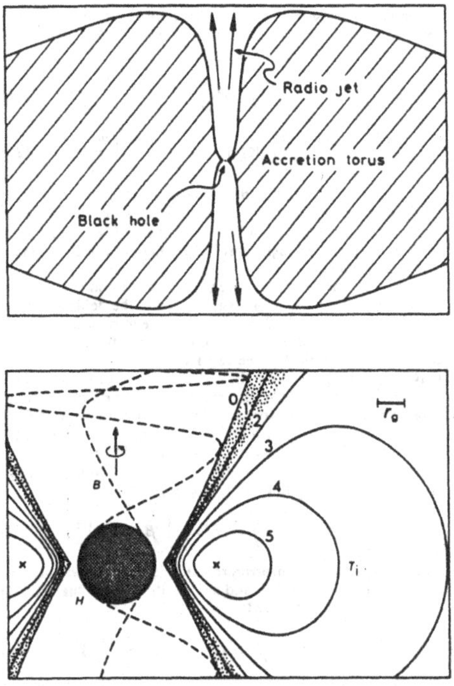


**Fig. 2** An accretion torus supported by ion pressure surrounds a spinning black hole. The torus defines a pair of funnels capable of collimating a relativistic hydromagnetic flow which ultimately forms the observed radio jets. The enlargement shows the inner regions of the torus and displays contours of ion temperature in units of $10^{11}$ K (solid lines). These have been calculated assuming that the ions are an isentropic gas with constant specific angular momentum and that the hole (horizon H) has $J = 0.9\, J_{max}$ (see ref. 19). The outer isotherm is also the surface of zero binding energy. The dashed lines represent the magnetic field lines in the configuration discussed in the text. A poloidal current flows in the funnel. Large toroidal currents and the return poloidal currents flow in a thin surface layer (stippled).

## Electromagnetic extraction of the hole's rotational energy

If the central black hole is near-maximally rotating ($J \simeq J_{max}$), a much greater supply of free energy is available: the rotational energy of the hole (up to 29% of $\dot{M}c^2$ for $J = J_{max}$). If an ordered electromagnetic field can be supported around the hole, then it can put particles (which may in fact be inside the horizon if the field threads the latter) systematically on negative energy orbits. The negative energy swallowed by the hole appears as positive energy in the external fields and thus drives away from the hole a Poynting flux of luminosity[41–43]

$$L_{EM} \lesssim B_{pH}^2 r_g^2 (J/J_{max})^2 c \tag{6}$$

where $B_{pH}$ is the poloidal field in the flux tube that intersects the hole's horizon. If $J \simeq J_{max}$, then this power $\lesssim 10^{44} M_8 \alpha_{-2}^{-1} \dot{m}_{-4} \beta_{pH}^{-1}\ \mathrm{erg\ s^{-1}}$ (where $\beta_{pH}$ = (maximum gas pressure in torus)/($B_{pH}^2/8\pi$)) can exceed by $\gtrsim \alpha^{-1}$ the rate of liberation of binding energy by the accreting matter. The torus will be able to confine and initially collimate the extracted power, as the associated radiation (or ram) pressure will be less than the gas pressure in the torus if

$$\dot{M}c^2/4\alpha > L_{EM} \lesssim \dot{M}c^2/(20\alpha\beta_{pH}) \cdot (J/J_{max})^2 \tag{7}$$

Both these inequalities are nearly always satisfied since $\beta_{pH} \gtrsim 1$.

To extract power predominantly from the hole, we must have $\beta_{pH} \simeq 1$ and $\alpha \ll 1$. Is this consistent with a physically attainable disk structure? One might naively suppose that the poloidal field could be generated by currents distributed through the torus, and thus be an extension of an ordered field structure within it. But if this were the case, the internal fields would transport energy and angular momentum outwards so efficiently that $\alpha \simeq 1$ and $L_{EM}$ would be $\lesssim \dot{M}c^2$. Furthermore, in the presence of a magnetosphere any current loop in keplerian rotation around a central mass loses energy from its outer light cylinder at a rate (~vacuum magnetic dipole power) $\simeq (J_{max}/J)^2$ times the rate at which it could extract power were the central mass a black hole of angular momentum $J$. If the current loop runs in an accretion disk, the missing power must be provided by the liberation of binding energy, which again implies $L_{EM} \lesssim \dot{M}c^2$ (compare with the thin-disk solution of Blandford and Znajek[42] in which $L_{EM} \simeq 0.3\ (J/J_{max})^2\ L_{disk}$).

To understand how a torus could avoid all this energy transport and loss, it is helpful to consider the time evolution of a perfectly conducting torus with a very weak ($\beta_p \gg 1$) magnetic field piercing it at large radii. (It may also have internally closed field loops of arbitrary strength.) This weak field is convected in with the flow, and by applying Faraday's law to a sufficiently large loop encircling the rotation axis, we find that the magnetic flux through the funnel increases as we carry field in, even though the field lines are causally disconnected from the matter in which they were frozen after the latter has crossed the horizon[43]. If the torus behaves like a perfect conductor, it will exclude this increasing external flux by developing surface currents. The field and surface currents will increase (perhaps shutting off the accretion altogether) until either some massive plasma-field interchange instability leads to an explosive energy release or, more likely, dissipation in the surface layer destroys piercing field as fast as it is convected in. The structure of such a torus is rather different from that of the field-free torus: the $\mathbf{J}_T \times \mathbf{B}_p$ force across the surface layer (funnel boundary) must be balanced by a gradient of gas pressure, so that at its base $P_{gas} \simeq B_{pH}^2/8\pi$. If we want $\beta_{pH} \simeq 1$, most of the drop of gas pressure must occur in the surface layer, the inner parts of the torus being at roughly constant pressure.

We can imagine the following self-consistent solution for a torus with a jet (Fig. 2): toroidal currents in the thin surface layer of the vortex funnel produce a poloidal field which is torqued by the hole, driving an outgoing Poynting flux. This will soon become plasma-loaded; toroidal fields will begin to dominate, causing self-collimation[34,44,45]. The return poloidal current must also flow in the thin surface layer, pulling material from it into the funnel, but leaving the bulk of the torus unaffected. The only forces from the jet transmitted to the body of the torus are those due to the poloidal field (transmitted by the gas pressure in the subcutis), whose pressure drops as (area of funnel)$^{-2}$ and which thus has less effect on the structure of the torus the further out we go. Simple solutions ignoring the jet can therefore correctly model the torus, while the jet may collimate itself on the paraboloidal field lines thus defined without exerting much modifying force.

The above discussion shows how a torus supported by ion pressure, with a low accretion rate $\dot{M}$, can catalyse the extraction of a hole's spin energy at a rate $\gtrsim \dot{M}c^2$. Moreover, we show below that the luminosity from the torus itself may be $\ll \dot{M}c^2$. Thus if the meagre losses from the torus can be replaced by energy and angular momentum extracted from the hole, there need be no accretion: only the inertia of the torus, and the currents in it are needed to confine the magnetic fields that tap the hole's energy.

## Losses and radiation from the torus

The total power dissipated in the surface layer of thickness $s$ is $\sim (s/r_g)L_{EM} \ll L_{EM}$. Moreover, as only a very weak field need penetrate the torus, transport and losses from the outer light cylinder may also be negligible. The radiation emitted by the body of an ion-supported torus will be primarily non-thermal; its magnitude depends sensitively on the efficiency of electron acceleration. The torus will generally be optically thin to Thomson scattering, with a total optical depth along a line of sight in the equatorial plane, from the hole to infinity, of $\sim \dot{m}\alpha^{-1}$. Incoherent cyclotron and synchrotron radiation will be self-absorbed. (The conditions envisaged resemble those of a so-called 'plasma turbulent reactor'[46], so that a power law spectrum may be set up). The torus will, however, be optically thin to $\gamma$ radiation produced by relativistic electron bremsstrahlung, and to inverse Compton radiation at frequencies above $10^{13}$ Hz. The relevant relaxation times that go into fixing the electron temperature are plotted in Fig. 3, which shows that the electrons will be able to radiate efficiently from the inner part of the torus only if their temperatures exceed $10^{10}$ K. At higher temperatures the power radiated increases very rapidly with temperature: the optically thick synchrotron losses vary as $T_e^7$, and the dependence on $T_e$ is steepened further because repeated Compton scattering of synchrotron photons up to the $\gamma$-ray band dominates when $\gamma_e^2\tau_{es} > 1$. However, as long as equation (5) is satisfied, the electrons can radiate only the small fraction of the total energy which they can drain from the ions (or derive from adiabatic heating) and the ion-supported torus will not collapse. For the parameter range of interest, the radiation will be in the far IR or the $\gamma$-ray region of the spectrum, according as $\dot{m} \lesseqgtr 10^{-5}\alpha_{-2}^{3/5}\beta^{-3/5}M_8^{-1/5}$. We cannot exclude the possibility that coherent radiation from fields $\sim 100$ G may be important. But, although this might seem relevant to the phenomenon of low-frequency radio variability[47] it would be subject to electron synchrotron absorption and induced Compton scattering[48,49].

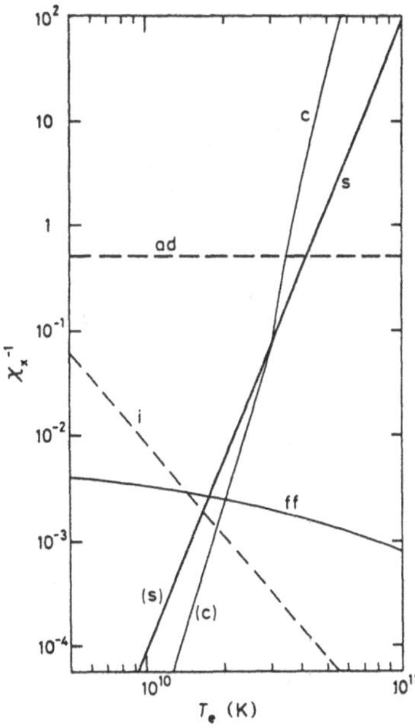


**Fig. 3** Dimensionless electron relaxation rates $\chi_x^{-1}$ as a function of the electron temperature $T_e \cdot \chi_x^{-1}$ is related to the heating or cooling rate per unit volume $\varepsilon_x$ by

$$\chi_x^{-1} = \frac{\varepsilon_x}{3n^2kT_e\sigma_T c}.$$

The lines are drawn for $\alpha = 10^{-2}$, $\beta = 8\pi nkT_i/B^2 = 1$, $\dot{m} \equiv \dot{M}/\dot{M}_{Edd} = 10^{-4}$, $r_* = r/r_g = 5$, and $M = 10^8 M_\odot$. The relaxation rates with appropriate scalings are: compressional heating, $\chi_{ad}^{-1} \propto \alpha^2\dot{m}^{-1}$; collisional heating by ions, $\chi_i^{-1} \propto r_*^{-1}$, optically thick synchrotron cooling, $\chi_s^{-1} \propto \alpha^{1/2}\beta^{-3/2}r_*^{-7/4}\dot{m}^{-1/2}M^{-1/2}$; optically thin Compton cooling, $\chi_c^{-1} \propto \alpha^{-2}\beta^{-1}r_*^{-3.25}\dot{m}^2$, and relativistic electron bremsstrahlung $\chi_{ff}^{-1}$ (no scaling). Self-absorption of the Compton radiation, double-Compton and pion cooling and pair production are unimportant for the range of densities and temperatures under consideration. The virial temperature of the ions is $7\times10^{11}$ K $(r_*/5)^{-1}$ (see also Fig. 2), while in the absence of losses, compressional heating of electrons transrelativistic at $r_* \simeq 1{,}000$ would give $T_e \simeq 3\times10^{10}(r_*/5)^{-1/2}$ K.

The $\gamma$-ray flux from the inner regions of the torus (which may include a redshifted positron annihilation line) would have only a small probability of being reabsorbed near the hole. Nevertheless, the small number of interactions that will occur can produce sufficient electron–positron pairs within the funnel to complete the circuit and allow electric current to flow through the hole. For example, to support an electromagnetic power of $10^{44}$ erg s$^{-1}$ in the funnel, a current of $3\times10^{17}$ A must flow, which requires a minimum pair creation rate of $\sim10^{36}$ s$^{-1}$. This can be maintained by $\gamma$ rays from the torus.

If, as assumed above, the electrons and ions are indeed thermally decoupled when equation (5) is satisfied, then the inner part of the torus will radiate much less than $0.1\,\dot{M}c^2$, and this power is in the far IR or the $\gamma$-ray band. But if, on the other hand, collective plasma effects couple the ions and electrons at all radii, then (see Fig. 1) there will be a thin disk for $r\leqslant 50\,r_g$, radiating $\sim0.1\,\dot{M}c^2$ with a spectrum peaking in the optical. For $\dot{m}_{-4}>50\,\alpha_{-2}^2$, even Coulomb coupling of electrons and ions is efficient enough to cool the ions, and one would expect a thin disk radiating in the near UV. When $\dot{m}$ is low, the predicted emission from galactic nuclei depends so drastically on whether or not the electrons and ions are coupled by plasma effects, that this issue can perhaps be settled observationally more readily than by the efforts of plasma theorists.

The maximum power that can be self-consistently extracted from a black hole by a thick ion-supported torus, $\dot{M}$ being given, is achieved when $\alpha$ has the lowest value consistent with the electron–ion decoupling condition $\dot{m}<50\,\alpha^2$. Even if $\dot{M}$ is a free parameter, the maximum power than can be self-consistently extracted from a black hole by a thick torus supported by ion pressure is $\sim10^{45}\,\alpha_{-2}\,M_8$ erg s$^{-1}$. However, if strong magnetic fields are anchored in a dense radiation–pressure-supported torus[18–21], whose thermal luminosity is $\geqslant L_{\rm Edd}$, there is no clearcut reason why the power extracted electromagnetically from the hole should not be orders of magnitude larger. A model along these lines may be relevant to objects such as 3C273 where (in contrast to the radio galaxies) there is a nuclear thermal luminosity $\sim L_{\rm Edd}$, but where there is evidence also for a powerful relativistic jet.

## Astrophysical context

We have argued that, provided the viscosity parameter $\alpha$ is not too low and the electron–ion coupling is sufficiently weak, a two-temperature, ion-supported torus can be maintained around a massive black hole accreting gas at a very subcritical rate in a galactic nucleus. The inner region of such a torus can collimate a pair of relativistic jets comprising electrons, positrons and electromagnetic fields. If radio jets originate in the funnels of accretion tori, the latter are likely to be supported by ion pressure rather than radiation pressure in most observed cases. The luminosity of the torus would then be far less than the kinetic energy flux in the jets. (The scale of the observed jets is, of course, many orders of magnitude larger than the torus itself, so some further collimation process may also be operative).

If most of the power is extracted from a spinning black hole, a fuelling rate of $\sim10^{-3}\,M_\odot$ yr$^{-1}$ suffices for most radio sources. A 'baseline' value for $\dot{M}$ in galactic nuclei would be supplied by debris from stars which pass so close to the hole that they are tidally disrupted[4,32,33]. This debris would have an angular momentum appropriate to an orbit at a few times $r_g$ (for $M_8\simeq1$) but a specific binding energy of only $\sim10^{-5}\,c^2$ (ref. 33). This debris would then form a torus extending out to $\sim10^5\,r_g$; a single disrupted star could then maintain $\dot{m}\approx10^{-4}$ for $\sim10^4$ yr.

If the energy in radio source jets derives mainly from the spin of the central black hole, then the hole probably underwent a phase of rapid mass accretion at an earlier epoch. One can speculate on a possible evolutionary sequence governed by accretion rate. If $\dot{m}>10$, a radiation-supported torus can be formed. For $50\alpha^2\leqslant\dot{m}\leqslant10$ the situation is less clear, but an ion-supported torus could possibly result, with most of the mass being contained within cool clouds or a thin disk.

Disks with $\dot{m}>50\alpha^2$ will be strong sources of ionizing radiation, and should be associated with strong line emission from the nucleus. Tori with $\dot{m}<50\alpha^2$ will be essentially invisible except as $\gamma$-ray sources. Both may drive the strongest of extended radio sources. Recent quasar observations have shown that the majority are radio-quiet, have small percentage polarization, and relatively low X-ray luminosity[50–52]. Presumably these do not create large fluxes of relativistic particles, and are perhaps $\sim10^8\,M_\odot$ black holes surrounded by radiation-supported tori without dominant magnetic fields. Seyfert galaxies may be similar, but with $\sim10^6\,M_\odot$ black holes. The radio loud and (usually) X-ray luminous objects may then be associated with electromagnetically powered tori, the BL Lac and superluminal radio sources perhaps being viewed at small angles to the jet axis.

Finally, the physics we have been discussing, with the exception of the detailed radiative properties of the torus, involves no intrinsic mass scale. Indeed, there may be an ion-supported torus around a $\leqslant10^6\,M_\odot$ hole in our galactic centre, and relativistic jets collimated by tori around stellar-mass black holes may exist within our Galaxy.

M.C.B. acknowledges partial support under NSF grant AST 79-23243. R.D.B. acknowledges hospitality at the Institute of Astronomy and support under NSF grant AST 80-11752 and from the Alfred P. Sloan foundation. E.S.P. thanks the Marshall Commission for financial support.

Received 28 July; accepted 9 November 1981.

1. Wyckoff, S., Wehinger, P. & Gehren, T. *Astrophys. J.* **247**, 750 (1981).
2. Schmidt, M. *Phys. Scr.* **17**, 135 (1978).
3. Soltan, A. *Mon. Not. R. astr. Soc.* (in the press).
4. Frank, J. *Mon. Not. R. astr. Soc.* **184**, 87 (1978).
5. Gunn, J. E. *Active Galactic Nuclei* (eds Hazard, C. & Mitton, S.) (Cambridge University Press, 1978).
6. Sparke, L. S. & Shu, F. H. *Astrophys. J. Lett.* **241**, L65 (1980).
7. Barr, P. *et al. Mon. Not. R. astr. Soc.*, **193**, 549 (1980).
8. Osterbrock, D. E. & Miller, J. S. *Astrophys. J.* **197**, 535 (1975).
9. Koski, A. T. *Astrophys. J.* **223**, 56 (1978).
10. Costero, R. & Osterbrock, D. E. *Astrophys. J.* **211**, 675 (1977).
11. Waggett, P. C., Warner, P. J. & Baldwin, J. E. *Mon. Not. R. astr. Soc.* **181**, 465 (1977).
12. Cohen, M. H. & Readhead, A. C. S. *Astrophys. J. Lett.* **233**, L101 (1979).
13. Readhead, A. C. S. *et al. Nature* **276**, 768 (1978).
14. Linfield, R. P. *Astrophys. J.* **244**, 436 (1981).
15. Pearson, T. J. *et al. Nature* **290**, 365 (1981).
16. Cotton, W. D., Shapiro, I. I. & Wittels, J. J. *Astrophys. J. Lett.* **244**, L57 (1981).
17. Jones, D. L., Sramek, R. A. & Terzian, V. *Astrophys. J. Lett.* **247**, L57 (1981).
18. Abramowicz, M. A., Jaroszyński, M. & Sikora, M. *Astr. Astrophys.* **63**, 221 (1978).
19. Kozlowski, M., Jaroszyński, M. & Abramowicz, M. A. *Astr. Astrophys.* **63**, 209 (1978).
20. Abramowicz, M. A., Calvani, M. & Nobili, L. *Astrophys. J.* **242**, 772 (1980).
21. Paczyński, B. & Wiita, P. *Astr. Astrophys.* **88**, 23 (1980).
22. Jaroszyński, M., Abramowicz, M. A. & Paczyński, B. *Acta Astr.* **30**, 1 (1980).
23. Abramowicz, M. & Piran, T. *Astrophys. J. Lett* **241**, L7 (1980).
24. Sikora, M. *Mon. Not. R. astr. Soc.* **196**, 257 (1981).
25. Sikora, M. & Wilson, D. B. *Mon. Not. R. astr. Soc.* (in the press).
26. Eardley, D. M. & Lightman, A. P. *Astrophys. J.* **200**, 187 (1975).
27. Shakura, N. I. & Sunyaev, R. A. *Astr. Astrophys.* **24**, 337 (1973).
28. Novikov, I. D. & Thorne, K. S. in *Black Holes* (eds DeWitt, C. & B.S.) (Gordon & Breach, New York, 1973).
29. Pringle, J. E. *Ann. Rev. Astr. Astrophys.* **19**, 137 (1981).
30. Pringle, J. E. *Mon. Not. R. astr. Soc.* **177**, 65 (1976).
31. Shakura, N. I. & Sunyaev, R. A. *Mon. Not. R. astr. Soc.* **175**, 613 (1976).
32. Hills, J. G. *Mon. Not. R. astr. Soc.* **182**, 517 (1978).
33. Frank, J. *Mon. Not. R. astr. Soc.* **187**, 883 (1979).
34. Blandford, R. D. & Payne, D. *Mon. Not. R. astr. Soc.* (in the press).
35. Shapiro, S. L., Lightman, A. P. & Eardley, D. M. *Astrophys. J.* **204**, 187 (1976).
36. Eardley, D. M., Lightman, A. P., Payne, D. G. & Shapiro, S. L. *Astrophys. J.* **224**, 53 (1978).
37. Bardeen, J. M. & Petterson, J. A. *Astrophys J. Lett.* **195**, L65 (1975).
38. Thorne, K. S. *Astrophys. J.* **191**, 507 (1974).
39. Abramowicz, M. A. & Lasota, J. P. *Acta Astr.* **30**, 35 (1980).
40. Galeev, A. A., Rosner, R. & Vaiana, G. S. *Astrophys. J.* **229**, 318 (1979).
41. Znajek, R. L. thesis. Cambridge Univ. (1976).
42. Blandford, R. D. & Znajek, R. L. *Mon. Not. R. astr. Soc.* **179**, 433 (1977).
43. Macdonald, D. & Thorne, K. S. *Mon Not. R. astr. Soc.* (in the press).
44. Blandford, R. D. *Mon. Not. R. astr. Soc.* **176**, 465 (1976).
45. Chan, K. L. & Henriksen, R. N. *Astrophys. J.* **241**, 534 (1980).
46. Norman, C. A. & ter Haar, D. *Phys. Rep.* **17**, 307–358 (1975).
47. Condon, J. J. *et al. Astr. J.* **84**, 1 (1979).
48. Levich, E. V. & Sunyaev, R. A. *Astrophys. Lett.* **7**, 69 (1970).
49. Wilson, D. B. & Rees, M. J. *Mon. Not. R. astr. Soc.* **185**, 297 (1978).
50. Condon, J. J. *et al. Astrophys. J.* **244**, 5 (1981).
51. Ku, W. H. M., Helfand, D. J. & Lucy, L. B. *Nature* **288**, 323 (1980).
52. Angel, J. R. P. & Stockman, H. S. *A. Rev. Astr. Astrophys.* **18**, 321 (1980).

THE ASTROPHYSICAL JOURNAL, 273:L59–L63, 1983 October 15


# SHOCKS IN SPHERICALLY ACCRETING BLACK HOLES: A MODEL FOR CLASSICAL QUASARS

P. MÉSZÁROS[1]
Harvard-Smithsonian Center for Astrophysics
AND
J. P. OSTRIKER
Princeton University Observatory
*Received 1983 March 18; accepted 1983 June 15*

## ABSTRACT

Spherically symmetric accretion onto black holes is thought to be very inefficient if the flow is nondissipative because of the inadequacy of compression heating and the short time available to radiate. We show here that standing shock solutions exist where the downstream protons are heated to temperatures at which the fluid becomes almost collisionless. The material then moves inwards at the much slower diffusion rate given by the *p-e* energy exchange time scale. For a given accretion rate, a self-consistent shock radius is found by matching the flow equations above and below that value, showing that such shocks can be self-sustaining. Efficiencies reaching up to 0.1–0.3 can be achieved, with most of the luminosity in the hard X-ray and γ-ray, as well as optical–IR ranges. This provides a new model for the classical (radio-quiet) quasars and active galactic nuclei and possibly for galactic sources such as Cyg X-1.

*Subject headings:* black holes — galaxies: nuclei — quasars

## I. INTRODUCTION

Black hole models of active galactic nuclei may not have accretion disks (see Rees 1977), as when the accreting gas is injected near enough to the hole so that the intrinsic angular momentum is less than that in a Keplerian orbit near the Schwarzschild radius. The accretion flow is then quasi-spherical, and for dissipationless free-fall, it is well known that the radiation efficiency of subcritical accretion would typically be less than $10^{-4}$ (Shvartsman 1971; Shapiro 1973). More realistic flows may involve entropy generation, if the flow is nonlaminar, which could raise the efficiency up to about $10^{-1}$ (Mészáros 1975). These calculations, dependent on magnetic fields, turbulence, etc., are quite uncertain. They also suffer from not having any distinguishing length scale, which would enable them to predict specific variability time scales. Such preferred length scales have been introduced by Ostriker *et al.* (1976) (see also Cowie *et al.* 1978; Grindlay 1978), who pointed out the importance of inverse Compton heating for limiting the region of *stable* free-fall inflow. The length scales introduced by this effect are appropriate for the long-term variability of extragalactic sources, but there remains the problem of the short time scales, which require length scales not much larger than the Schwarzschild radius. In this *Letter* we show that shocks can exist near the horizon, which provide short time variability and also a high efficiency of energy conversion.

The possibility of a standing shock arising in a flow around a black hole has been in the past considered unlikely, because of the absence of a solid surface exerting a back pressure which would uphold the shock. This view, however, is based on an overly hasty extrapolation of the usual experience with one-temperature fluids. A nonfluid, two-temperature description may be more appropriate (e.g., Begelman 1977; Rees *et al.* 1982), and we show below that a self-consistent shock solution can then be found. We also know, in a sense, that accretion can proceed smoothly after a shock transition, since successful subsonic flow solutions have been constructed (cf. Flammang 1982) for accretion onto black holes embedded within normal stars (with boundary condition $r = c$ at the horizon). In the above work and in the related paper by Thorne and Żytkow (1977), it was assumed that the plasma could be treated as an ordinary hydrodynamic fluid incorporating radiation pressure. Here we take the opposite physical model and show that, in this case also, there exist well-behaved subsonic solutions.

The situation we envisage is that of a spherical flow, which in the outer parts is in free fall, the proton random energy being small compared with its radial bulk energy motion. Such cold, supersonic infall is highly unlikely to remain laminar, since small disturbances in

[1]Visiting Scientist, Smithsonian Astrophysical Observatory, supported partly by NASA grant NAGW-246; on leave from Max-Planck-Institut für Physik und Astrophysik MPA, Garching.

the density at the accretion radius grow large by the time they reach the horizon (Petterson *et al.* 1980), and this could lead to a shock. Such a shock converts the radial motion energy of the upstream protons into *random* energy, below the shock. However, once the proton random energy (or "temperature") is of the order of the gravitational temperature $T_{gr} \sim GMm_p k^{-1} r^{-1}$, one can verify that the protons will be essentially collisionless, since the Coulomb time $t_{pp} \gg t_{ff}$, the free-fall time. This holds also for $t_{pe}$, the proton-electron energy exchange time by Coulomb encounters. Furthermore, the protons below the shock will now have large orbital eccentricities and, as in the stellar dynamical case around a black hole, will only be able to accrete into the hole on a diffusion (Coulomb) time scale $\gg t_{ff}$ (cf. Bahcall and Wolff 1977), since this is the time in which they are able to dissipate their excess orbital angular momentum. (For a different view of a shock model, see Protheroe and Kazanas 1983.) To maintain a steady state, this requires a larger density in the postshock (diffusion) region than in the upstream free-fall region. It is easy to see that, once established, such a situation can perfectly well persist also in steady state.

## II. ACCRETION PICTURE

A schematic diagram of the flow is shown in Figure 1. We assume that the upstream material below the accretion radius $r_a$ is in free fall. The temperature of the gas in the region above the shock is determined by the balance of Compton heat gains (due to illumination by hard X-ray photons of average energy $\langle h\nu_\gamma \rangle$ coming from below the shock) and bremsstrahlung cooling. Upstream of the shock, we have $n_p(\tilde{r}) = 5 \times 10^{10} m_8^{-1} \dot{m} \tilde{r}^{-3/2}$ cm$^{-3}$ and $T_e(r) \approx T_1(\tilde{r}/\tilde{r}_s)^{-1}$ K, where $T_1 = (1/4k)\langle h\nu_\gamma \rangle$; $\tilde{r}$ is in Schwarzschild units of $2GM/c^2$, $m_8 = M_{BH}/10^8\ M_\odot$, and $\dot{m} = \dot{M}/\dot{M}_{crit} = \dot{M}c^2/(2L_E)$, with $L_E = 4\pi GMm_p c/\sigma_T = 1.3 \times 10^{46}\ m_8$ ergs s$^{-1}$.

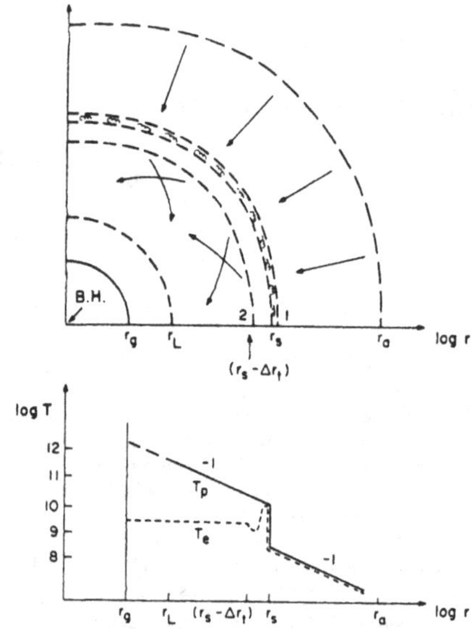


FIG. 1.—(*upper*) Schematic view (not to scale) of the spherical accretion geometry. Indicated are the accretion radius $r_a$, shock radius $r_s$, relaxation layer $\Delta r_f$, the lower radius at which the diffusion treatment is valid $r_L$, and the Schwarzschild radius $r_g$; The collisionless shock is in the thin, marked region above $r_s$. The jump conditions are applied between 1 and 2, on either side of $\Delta r_f$. Between $r_a$ and $r_s$, the matter is in free fall, while between $r_s - \Delta r_f$ and $r_L$, the protons and electrons behave as individual particles, much as a star cluster would, slowly diffusing inwards on the *p-e* Coulomb scattering time scale. (*lower*) The proton temperature $T_p$ and electron temperature $T_e$ corresponding to the radial coordinate of Fig. 1 (*upper*).

A collisionless shock, arising randomly in the flow, would rapidly acquire spherical symmetry, as a result of the high effective sound speed downstream. The structure of such shocks without large-scale quasi-perpendicular magnetic fields has been explored at low Mach numbers, but at high Mach numbers the situation remains fairly complex (cf. Max *et al.* 1982). The order of magnitude of the shock thickness is expected to be approximately $l_s \sim V_p(m_p/4\pi e^2 n_p)^{1/2} \sim 2 \times 10^1 m_8^{1/2} \dot{m}^{-1/2} \tilde{r}_s^{1/4}$ cm, where $\tilde{r}_s$ is the shock standoff distance in Schwarzschild units. The turbulent electric and magnetic fields created in the shock are of magnitude

$$E \sim (c/v_e)B \sim \frac{\delta\phi}{l_s} \sim 3 \times 10^4 m_8^{-1/2} \dot{m}^{1/2} r_s^{-5/4} \text{ gauss} \tag{1}$$

(cf. Zel'dovich and Raizer 1967). These turbulent fields will survive downstream for only a short time, an absolute upper limit on which is given by the magnetic decay time $t_B \sim 4\pi\sigma_c l_s^2/c^2$, where $\sigma_c$ is the conductivity; a conservative estimate is $t_B \lesssim 10^2 \omega_{op}^{-1} \sim 10^{-7} m_8^{1/2} \dot{m}^{-1/4} \tilde{r}_s^{-1/4}$. The turbulent fields will therefore be dissipated very shortly behind the shock transition layer. What about any possible large-scale magnetic fields that are present in the upstream flow? These would have to be significantly below equipartition for the upstream region to be subsonic and freely falling. As long as $B_1 < [(\gamma - 1)/(\gamma + 1)]B_{eq} \sim (1/4)B_{eq}$, these large-scale fields will be below equipartition also after adiabatic compression in the shock transition. However, the chaotic $E$ and $B$ fields produced in the shock itself are by definition at equipartition, since $B \sim E \sim kT/el_s$ implies $B^2/8\pi \sim (1/2)\rho V_p^2$. The large-scale fields will be dominated by the chaotic ones, which vary over the extremely small length scale $l_s$ (eq. [3]), and the chaotic currents will cause wiggles of scale $l_s$ to appear in the large-scale field loops. Having acquired this small-scale structure, they are dissipated on the same short time scale $t_B$ as the chaotic fields, provided magnetic reconnection occurs on this scale. If it does not, the large-scale radial field component persists. Thus, as long as the upstream magnetic energy is significantly sub-equipartition and magnetic reconnection occurs in the chaotic layer below the shock, and as long as one may neglect any possible anomalous scattering mechanisms, our approximation of a nonfluid, stellar dynamic type of regime in the downstream region is justified.

In the transition layer $\Delta r_t$, the electrons cool quickly and thus half of the dissipated turbulent field energy is transformed into photons via the synchrotron and inverse Compton mechanisms. An equilibrium is reached with $T_e \ll T_p$ behind the turbulent field zone. The electrons are heated by Coulomb encounters with protons at $T_p \sim T_{gr}$ and cooled by bremsstrahlung and inverse Compton. For $\tilde{r}_s \lesssim 4 \times 10^2$, they reach an equilibrium temperature at $\theta_e = (kT_e/m_e c^2) \lesssim 1$, above which the losses increase rapidly due to relativistic processes. Downstream of $\Delta r_t$, we have $T_p(\tilde{r}) \sim T_{gr} \sim 2 \times 10^{12}\tilde{r}^{-1}$ K and $T_e(\tilde{r}) \sim \theta_e(m_e c^2/k) = \theta_e 6 \times 10^9$ K. The relaxation shell is of relative width $\Delta r_t/r_s \sim T_e/T_p \sim 2.7 \times 10^{-3}\theta_e \tilde{r}_s \ll 1$.

Below the relaxation region, we have a quasi-collisionless flow, where matter flows inwards on a diffusion time scale. For the solutions found here, the proton-electron Coulomb time $t_{pe} \ll t_{pp}$, the proton-proton time, so we neglect proton heat conduction effects. Both the diffusion of matter inwards and the energy exchange occur therefore on the time scale $t_{pe} \approx (m_p c^2 \theta_e)/n_e \sigma_T m_e c^3 \ln \Lambda \approx 1.5 \times 10^{-8}\theta_e \rho^{-1}$, valid for relativistic electrons if $T_e \ll T_p$ (Gould 1981).

We apply jump conditions to the flow on both sides of the relaxation region (see Fig. 1). In the downstream (2) side, we assume that the matter moves inwards with a diffusion velocity $u_2 = \delta r t_{pe}^{-1}$, where $\delta \sim 1$ is a constant. This means that for $r < r_2 = r_s - \Delta r_t$, one has $\rho \propto r^{-3/2}$, where the exponent is fortuitously the same as in free fall, although the constant in front is different. The downstream one-dimensional velocity dispersion $V_2$ is linked to the "pressure" via $p_2 = \rho_2 V_2^2$ and to the "temperature" via $T_2 = k^{-1} m_p V_2^2$. This is taken to be the gravitational value corresponding to a $\rho \propto r^{-3/2}$ law, $(V_2/c)^2 = (2/5)GM/rc^2 = (1/5)\tilde{r}^{-1}$. Upstream, we may take $u_1 = (2GM/r)^{1/2}$, and $p_1 = T_1 = 0$ is taken for simplicity. The density jump is, therefore,

$$\frac{\rho_2}{\rho_1} = \frac{u_1}{u_2} = 1.67 \times 10^1 \dot{m}^{-1/2}\delta^{-1/2}\theta_2^{1/2}\eta_2^{-1/2}, \quad (2)$$

where $\eta_2$ is the ratio of electrons (and positrons, if any) to protons. We do not expect a large pair density, i.e., $\eta_2 \sim 1$ (see below). The density jump is higher than for an adiabatic (fluid) shock, since $u_2$ is not a hydrodynamic velocity but a diffusion velocity and because the relaxation zone is not adiabatic. Momentum conservation can be written as

$$\rho_2 u_2^2 + \rho_2 V_2^2 - \frac{4}{3}\gamma \frac{r^2 \rho_2}{2t_{pe}} \frac{u_1}{\Delta r} = \rho_1 u_1^2. \quad (3)$$

The third term on the left is the viscous stress tensor $\sigma \approx (4/3)(\lambda^2 \rho/t)(du/dr)$; for the mean free path $\lambda$, we take $\lambda^2 \approx \gamma r^2$, with $\gamma = \text{const} \sim (1/2)$; for the viscous (forward momentum) exchange time, we take $2t_{pe}$, this being the appropriate value for protons on electrons (e.g., Spitzer 1962); and for $du/dr$, we replace $\sim u_1/\Delta r$, since $u_2 \ll u_1$. We take $\Delta r \sim \Delta r_t$. Condition (3) implies that the shock standoff distance is

$$\tilde{r}_s \approx 50 \dot{m}^{1/2}\theta_2^{-3/2}\eta_2^{1/2}\gamma\delta^{-1/2}. \quad (4)$$

As $\dot{m}$ decreases from unity, the shock radius approaches the Schwarzschild radius $\tilde{r} = 1$, reaching it for $\dot{m}_s = 4 \times 10^{-4}\theta_2^{1/3}\eta_2^{-1}\gamma^{-2}\delta$. Since at $\tilde{r}_s$ we satisfy $t_{pe} \ll t_{pp}$, we do not normally expect a region dominated by proton conduction to exist.

The electron temperature in the inner region is almost uniform because the electron exchange time is very short compared with the *e-p* time. Behind the shock, photons are produced by *e-p* and *e-e* bremsstrahlung. Comptonization plays a role in limiting the temperature below about $\theta_e \lesssim 1$. The temperature adjusts itself to a value where the Comptonization is in the unsaturated regime. Pair production will also tend to stabilize the temperature at $\theta_e \lesssim 1$, although mostly $n_\pm < n_p$, $\eta \sim 1$. The proton density downstream of the shock is given by $n_p(\tilde{r}) = 8.3 \times 10^{11}\dot{m}^{1/2}m_8^{-1}\theta_2^{1/2}\eta^{-1/2}\delta^{-1/2}\tilde{r}^{-3/2}$ cm$^3$, which is a factor of approximately $17\dot{m}^{-1/2}$ times the upstream free-fall density. This is also the ratio of *p-e* scattering time to free-fall time. Below $\tilde{r} \lesssim 3.5$, the protons, at the temperature given by equation (4), have enough energy to make pions via *p-p* collisions. Using the rates of Kolykhalov and Sunyaev (1979), we have $t_\pi/t_{ff} \approx 10^2\dot{m}^{-1}\tilde{r}^{0.86}$. Thus for $\tilde{r} < \tilde{r}_l \sim 3$, this is the principal energy loss mechanism. The dynamics is further complicated by the presence of general relativistic and loss cone effects, both of which become important below $\tilde{r}_l \approx 3$. These tend to accelerate the flow in this region, so that the lower boundary provides no new constraints on the inflow rate.

### III. IMPLICATIONS

The fact that the protons diffuse inward by transferring their energy to the electrons, which then radiate it away, implies a very high efficiency, of order near unity. It does not reach unity since the simple diffusion breaks down as one approaches the horizon because of new effects that start playing a role near the horizon. These, however, should still leave the efficiency above 0.1–0.3.

The X-ray spectrum produced by the upstream gas is a superposition of optically thin bremsstrahlung spectra with varying $\rho(r)$ and $T(r)$, giving a power-law $F_\nu \sim \nu^{-b}$, with $b \sim 0.6$–$0.7$ (cf. Mészáros 1983). At 2 keV ($4.83 \times 10^{17}$ Hz), the spectral flux density is $l_x \sim 10^{27} m_8 \dot{m}^{7/4} (\bar{r}_l/3)^{-1}$ ergs s$^{-1}$ Hz$^{-1}$, where $\bar{r}_l$ is the lowest radius entering equation (5). The corresponding luminosity is a fraction $\tau$(upstream) of the value given by equation (5).

There will be another spectral component in the IR–optical range, resulting from Comptonized synchrotron photons produced in the turbulent fields near the collisionless shock itself. The synchrotron peak, in the chaotic fields of the strength given by equation (1), depends on the electron Lorentz factor $\Gamma$, $\nu_m \sim 1.5 \sin\alpha[(eB)/(2\pi m_e c)]\Gamma^2 \gtrsim 10^{12} m_8^{-1/2} \dot{m}^{9/8}$, where we took the adiabatic shock temperature, $\Gamma_{th} \sim 10^3 \bar{r}^{-1}$. Additionally, a fraction of the electrons may acquire a superthermal power-law distribution (Blandford and Ostriker 1978; Eichler 1979) so that $\Gamma_{st} \geq \Gamma_{th}$, and the spectrum will be a power law extending blueward of the peak $\nu_m$. A blueward power law is also expected because half of the synchrotron photons, initially emitted downwards, encounter the hot ($\theta_e \lesssim 1$, $\tau_s \gtrsim 1$, see below) downstream diffusion region, which leads to a spectrum $F_\nu \propto \nu^{-\alpha}$ (Shapiro *et al.* 1976; see also Katz 1976), with $\alpha \gtrsim 1$, typical of observed optical spectra. The total IR–optical luminosity due to the shock is $L_s \approx G M \dot{M}/(2 r_s)^{-1} \sim 1.5 \times 10^{44} m_8 \dot{m}^{1/2} \theta_2^{3/2}$ ergs s$^{-1}$. We estimate for the optical flux density at 3000 Å ($10^{15}$ Hz) $l_o \approx 10^{30} (a_0/5) \dot{m}^{1/2} \theta_2^{3/2}$ ergs s$^{-1}$ Hz$^{-1}$, where $a_o$ is the amplification factor from $\nu_m$ to $10^{15}$, which itself also depends on $\dot{m}$ in a complicated manner through $\tau_s(\dot{m})$ and $\theta(\dot{m})$.

The downstream diffusion region will emit mainly in the poorly observed, hard X-ray region, $\gtrsim 200$ keV. The typical scattering opacity seen by a photon is approximately a few. The spectrum will be isothermal bremsstrahlung, moderately Comptonized, i.e., with a moderate hump at $\langle h\nu_\gamma \rangle \sim (1$–$3)kT_e \sim 0.5$–$1.5$ MeV and fewer photons below. The luminosity in this component is close to the total one,

$$L_t \sim L_d \approx \frac{GM\dot{M}}{r_l} \approx 4 \times 10^{45} m_8 \dot{m} (\bar{r}_l/3)^{-1} \text{ ergs s}^{-1}, \tag{5}$$

where $\bar{r}_l \gtrsim 3$ is the lowest radius to which our classical $p$-$e$ diffusion treatment can be extended. The pair recombination luminosity in the region $\bar{r}_l < \bar{r} < \bar{r}_s - \Delta\bar{r}_s$ will be significant, contributing to the hump above approximately 0.5 MeV. For $\bar{r} \lesssim 3$, general relativistic effects and pion production by $p$-$p$ collisions are important, and there may be loss cone effects as well.

The total emission in the form of $\gamma$-rays depends on the dynamic inflow time in the region $\bar{r} \lesssim 3.5$, where the main energy loss mechanism is pion production by $p$-$p$ collisions. A dynamic time $t_\pi$ would lead to a luminosity $L_\gamma \sim 10^{37} m\dot{m}(t_{ff}/10^2 t_{dy}) \sim 10^{45} m_8 \dot{m}(t_{ff}/10^2 t_{dy})$, where we took the upper limit given by $t_{dy} \sim t_\pi$ and included gravitational trapping of photons. The actual $\gamma$-ray luminosity may be lower, due to general relativistic effects on the particle orbits, which would drive $t_{dy}$ toward $t_{ff}$, which could also lead to increasing photon trapping by convection. The neutrino luminosity will be of the same order. The intrinsic $\gamma$-ray photon spectrum due to this process would peak at about 20 MeV, with about 10% of the photons at $E_\gamma \gtrsim 100$ MeV. The actual spectrum that escapes, however, may not extend beyond $E_\gamma \lesssim$ MeV, due to photon-photon absorption with the X-ray photons, at the higher $\dot{m}$ values. This opacity decreases with $\dot{m}$, $\tau_{\gamma\gamma} \sim 1.3 \times 10^1 \dot{m}(\varepsilon/0.1)^{-1}(\bar{r}/\bar{r}_l)^{-1}$, so that for $\dot{m} \lesssim 10^{-1}$ one may see the full $\gamma$-ray spectrum extending to more than $10^2$ MeV. For $\dot{m} \gtrsim 10^{-1}$, the degradation of the high-energy $\gamma$-ray photons will add to the approximately MeV hump.

The short-term variability of the optical emission will be, for large-amplitude variations ($\Delta L/L \sim 1$), $t_o \sim t_{ff}(\bar{r}_s) = 4 m_8 \dot{m}^{3/4} \theta_2^{-9/4}$ days. For $m \sim 10$, corresponding to stellar black holes, this time is about 30 milliseconds. In terms of the optical luminosity, one has $t_o \propto l_o^{1.5}$, and for $\dot{m}$ approaching unity, one expects large-amplitude relaxation oscillations to set in. The X-ray emission can have short-period, $\Delta L/L \sim 1$ variations on the time scale $t_x \sim t_{pe}(\bar{r}_s) = 4 \times 10^1 m_8 \dot{m}^{1/4} \theta_2^{-3/2}$ days, or $t_x \propto l_x^{1/7}$. The small-amplitude variations can occur on the light-travel time $t'_x \sim t_L(r_s) \sim 1.4 \times 10^1 m_8 \dot{m}^{1/2}$ hr, with $t'_x \propto l_x^{2/7}$. The hard X-rays will vary on $t_{HX} \sim t_{pe}(\bar{r}_s) = 40 m_8 \dot{m}^{1/4}$ days, or $t_{HX} \propto L_{HX}^{1/4}$. The very high energy (> 100 MeV) $\gamma$-rays, when $\dot{m}$ is low enough for their escape, should have large-amplitude variations on a time scale of $t_{dyn}$ at $\bar{r} \sim 3$, which lies between $t_\gamma \sim$ (3) $\sim 1.5 m_8$ hr and $t'_\gamma \sim t_{pe}(3) \sim 2 \times 10^1 m_8 \dot{m}^{-1/2} \theta_2^{\frac{1}{2}}$ hr, $t'_\gamma \propto L_\gamma^{-1/2}$. For $\dot{m} \gtrsim 10^{-1}$, when the $\gamma$-ray spectrum cuts off at about 1 MeV due to photon-photon absorption on the X-rays, the variability is the same as that of the hard X-rays.

Our model as presented here applies mainly to radio-quiet QSOs, Seyfert 1 galaxies, and X-ray selected active galactic nuclei. Scaled-down versions ($m \sim 10$) would apply to galactic black hole sources such as Cyg X-1, Cir X-1, and LMC X-3. Notice that, for $m \sim 10$, the $\nu_m$ of the shock luminosity is in the UV. It is worth noting that, since Seyfert galaxies and quasars contribute significantly to the diffuse X-ray and $\gamma$-ray background, the $\gamma$-ray feature at $E \sim 20$ MeV seen in our models may account for the well-known feature in the $\gamma$-ray background at 4–6 MeV (e.g., Fabian 1980). A test of our spherical models may be possible through the time variabilities that we predict as a function of luminosity. A specific prediction is that, in the γ-ray domain, we expect a broad peak around 0.5–1.5 MeV, due to the hot downstream electrons, and another broad peak around 20 MeV, from *p-p* reactions near the horizon.

This research has been supported partly by NASA grant NAGW-246 (P. M.) and NSF grant AST 80-22785 (J. P. O.). We are grateful to R. Kulsrud, J. Krolik, D. Q. Lamb, A. P. Lightman, R. Lovelace, and C. Max for comments.

REFERENCES

Bahcall, J. N., and Wolff, R. A. 1977, *Ap. J.*, **214**, 90.
Begelman, M. C. 1977, *M.N.R.A.S.*, **181**, 347.
Blandford, R. D., and Ostriker, J. P. 1978, *Ap. J.* (*Letters*), **221**, L29.
Cowie, L. L., *et al.* 1978, *Ap. J.*, **226**, 1041.
Eichler, D. 1979, *Ap. J.*, **229**, 419.
Fabian, A. C. 1980, *Ann. NY Acad. Sci.*, **375**, 235.
Flammang, R. A. 1982, *M.N.R.A.S.*, **199**, 833.
Gould, R. J. 1981, *Phys. Fluids*, **24**, 102.
Grindlay, J. E. 1978, *Ap. J.*, **221**, 234.
Katz, J. I. 1976, *Ap. J.*, **206**, 910.
Kolykhalov, P. I., and Sunyaev, R. A. 1979, *Soviet Astr.—AJ*, **23**, 189.
Max, C., *et al.* 1982, *Bull. AAS*, **14**, 937.
Mészáros, P. 1975, *Astr. Ap.*, **49**, 59.
Mészáros, P. 1983, *Ap. J.* (*Letters*), in press.
Ostriker, J. P., *et al.* 1976, *Ap. J.* (*Letters*), **208**, L61.
Petterson, J., *et al.* 1980, *M.N.R.A.S.*, **191**, 571.
Protheroe, R. J., and Kazanas, D. 1983, *Ap. J.*, **265**, 620.
Rees, M. J., *et al.* 1982, *Nature*, **295**, 17.
Shapiro, S. L. 1973, *Ap. J.*, **180**, 531.
Shapiro, S. L., *et al.* 1976, *Ap. J.*, **204**, 187.
Shvartsman, V. F. 1971, *Soviet Astr.—AJ*, **15**, 377.
Spitzer, L., Jr. 1962, *Physics of Fully Ionized Gases* (New York: Wiley Interscience).
Thorne, K. S., and Żytkow, A. N. 1977, *Ap. J.*, **212**, 832.
Zel'dovich, Ya. B., and Raizer, Yu. P. 1967, *Physics of Shock Waves and High-Temperature Hydrodynamic Phenomena* (New York: Academic).

P. MÉSZÁROS: Pennsylvania State University, Astronomy Department, University Park, PA 16802

J. P. OSTRIKER: Princeton University Observatory, Peyton Hall, Princeton, NJ 08540

THE ASTROPHYSICAL JOURNAL, 234:296–316, 1979 November 15


# ACCRETION BY ROTATING MAGNETIC NEUTRON STARS. III. ACCRETION TORQUES AND PERIOD CHANGES IN PULSATING X-RAY SOURCES[1]

P. GHOSH[2]
Department of Physics, University of Illinois at Urbana-Champaign; and Astrophysics Branch, Space Science Laboratory, NASA Marshall Space Flight Center

AND

F. K. LAMB[3]
Department of Physics, University of Illinois at Urbana-Champaign; and California Institute of Technology
*Received 1978 September 25; accepted 1979 June 4*

## ABSTRACT

We use the solutions of the two-dimensional hydromagnetic equations obtained previously to calculate the torque on a magnetic neutron star accreting from a Keplerian disk. We find that the magnetic coupling between the star and the plasma outside the inner edge of the disk is appreciable. As a result of this coupling the spin-up torque on fast rotators is substantially less than that on slow rotators; for sufficiently high stellar angular velocities or sufficiently low accretion rates this coupling dominates that due to the plasma and the magnetic field at the inner edge of the disk, braking the star's rotation even while accretion, and hence X-ray emission, continues.

We apply these results to pulsating X-ray sources, and show that the observed secular spin-up rates of all the sources in which this rate has been measured can be accounted for quantitatively if one assumes that these sources are accreting from Keplerian disks and have magnetic moments $\sim 10^{29}$–$10^{32}$ gauss cm$^3$. The reduction of the torque on fast rotators provides a natural explanation of the spin-up rate of Her X-1, which is much below that expected for slow rotators. We show further that a simple relation between the secular spin-up rate $-\dot{P}$ and the quantity $PL^{3/7}$ adequately represents almost all the observational data, $P$ and $L$ being the pulse period and the luminosity of the source, respectively. This "universal" relation enables one to estimate any one of the parameters $P$, $\dot{P}$, and $L$ for a given source if the other two are known. We show that the short-term period fluctuations observed in Her X-1, Cen X-3, Vela X-1, and X Per can be accounted for quite naturally as consequences of torque variations caused by fluctuations in the mass transfer rate. We also indicate how the spin-down torque at low luminosities found here may account for the paradoxical existence of a large number of long-period sources with short spin-up time scales. Finally, we stress the need for a sequence of simultaneous period and luminosity measurements of each source. Such measurements would provide a direct check on our theory, as well as valuable information about both the spin evolution of pulsating sources and temporal variations in the mass transfer process in accreting X-ray binaries.

*Subject headings:* hydromagnetics — stars: accretion — stars: magnetic — stars: neutron — X-rays: binaries

## I. INTRODUCTION

The interpretation of most pulsating X-ray sources as accreting neutron stars, based on the qualitative features of their spectra and their secular spin-up rates, is now relatively secure. The period changes observed in these sources are of some importance because they offer the possibility of a direct, quantitative comparison of theoretical predictions with accurate observations. Furthermore, an understanding of this phenomenon would provide an important tool for exploring other outstanding problems, such as the characteristics of mass transfer in binaries and the properties of neutron stars. Thus, for example, observations of secular period changes probe the average circulation of the accreting plasma at the magnetospheric boundary, the strength of the star's dipole field, and the size of its moment of inertia, while measurements of short-term period fluctuations probe the stability of the accretion flow, the relative inertial moments of the crust and superfluid neutron core, and the frequencies of internal collective modes (Lamb 1977; Lamb, Pines, and Shaham 1978*a*, *b*).

This is the third in a series of papers in which we are developing a quantitative theory of accretion flows and period changes in pulsating X-ray sources. In the first paper (Ghosh, Lamb, and Pethick 1977, hereafter Paper I) we investigated the flow of accreting plasma and the configuration of the magnetic field inside the

[1] Research supported in part by NSF grant PHY78-04404 at the University of Illinois and NASA contract NAS5-23315 at Caltech.

[2] NAS-NRC Resident Research Associate.

[3] Alfred P. Sloan Foundation Research Fellow.

magnetosphere. There we showed that the early dimensional estimate (Pringle and Rees 1972; Lamb, Pethick, and Pines 1973) of the sign and magnitude of the torque on slowly rotating neutron stars is correct if the transition zone at the magnetospheric boundary is narrow and that in this case this same estimate is also approximately correct even for fast rotators. This result argued strongly that the transition zone is in fact broad, since otherwise the behavior of Her X-1, which almost certainly is accreting from a disk but has a secular spin-up rate ~40 times smaller than that predicted by the slow rotator estimate (Elsner and Lamb 1976), would be difficult to understand. Thus the results of Paper I underlined the importance of calculating the size and structure of the transition zone.

In a second paper (Ghosh and Lamb 1979, hereafter Paper II) we investigated the interaction between the stellar magnetic field and the disk plasma at the magnetospheric boundary. We found that, due to growth of the Kelvin-Helmholtz instability, turbulent diffusion, and magnetic field reconnection, the magnetospheric field readily invades the disk over a broad region near its inner edge. Assuming a stationary axisymmetric flow and treating the slippage of the stellar field lines through the disk plasma by an effective conductivity, we obtained solutions to the two-dimensional hydromagnetic equations which describe the radial and vertical structure of the transition zone. These solutions show that the inner radius of the transition zone is located where the integrated stress of the stellar magnetic field becomes comparable to the integrated material stress of the disk plasma, while the outer radius is located where the electrical currents flowing in the transition zone screen the stellar magnetic field to zero. The transition zone itself is composed of two qualitatively different parts, a broad outer part where the angular velocity is Keplerian, and a narrow inner part where it departs significantly from the Keplerian value.

In the present paper we use the accretion flow solutions obtained in Paper II to calculate the torque on the neutron star and discuss the implications for pulsating X-ray sources. In § II we describe the method we use to determine the accretion torque, discuss the results of this calculation, and compare our results with those of other workers, including Scharlemann (1978) and Ichimaru (1978). We discuss the general problem of interpreting period changes in pulsating X-ray sources in § III. In § IV we consider the limited evidence regarding accretion flow patterns which is provided by the relatively sparse period and X-ray flux measurements that have been made thus far, while in § V we estimate the dipole magnetic moments of nine pulsating X-ray sources by fitting the theoretical spin-up equation to estimates of the average luminosity and spin-up rate of each source. Accretion theory predicts that fluctuations in the mass accretion rate will cause fluctuations in both the accretion luminosity and the accretion torque. In § VI we show that torque variations that arise in this way can easily be large enough to account for the period wandering observed in the well-studied pulsating sources, if these sources are fed by disks. In § VII we consider the numerous long-period sources with relatively short spin-up times and describe how the existence of many such sources may be understood, given the braking torque found in the present calculations. Finally, in § VIII we summarize our results and point to a number of critical observations. A brief account of our results has been given previously in Ghosh and Lamb (1978*a*).

## II. THE ACCRETION TORQUE

In this section, we use the two-dimensional flow solutions found in Paper II to calculate the accretion torque $N$ acting on a magnetic neutron star accreting matter from a disk. First, we outline the method of calculation. Next, we explain the results in physical terms, describe the behavior of the torque as a function of the rotation period $P$, and the mass accretion rate $\dot{M}$, and discuss the generality of this behavior. Finally we compare our results with other work.

### *a) Method of Calculation*

As in Paper II we assume that the stellar magnetic field is dipolar with moment $\mu$, and that the flow is steady and has axial symmetry everywhere. We generally use cylindrical coordinates ($\varpi$, $\phi$, $z$) centered on the neutron star and aligned with the stellar rotation axis, but sometimes also refer to the distance $r = (\varpi^2 + z^2)^{1/2}$ from the center of the star (note that in the disk plane, $r = \varpi$).

Figure 1 shows schematically the character of the flow solutions obtained in Paper II. Between the unperturbed accretion disk and the magnetosphere there is a broad transition zone where the stellar magnetic field threads the disk. This zone divides into two parts, a broad outer part stretching from the screening radius $r_s$ inward to $r_0$, and a narrow inner part or boundary layer between $r_0$ and the corotation radius $r_{co}$, inside which plasma is forced to corotate with the star. The angular velocity is Keplerian outside $r_0$ but then falls sharply to the stellar angular velocity at $r_{co}$. Matter flows from the disk plane toward the star along the bundle of magnetic field lines that thread the boundary layer.

For steady accretion, the torque on the star is given by the integral of the angular momentum flux across any surface, $S$, enclosing the star. If the flow is also axisymmetric, this integral may be written in the form (Lamb 1977)

$$N = \int_S \left(-\rho v_p \varpi^2 \Omega + \varpi \frac{B_p B_\phi}{4\pi} + \eta \varpi^2 \nabla \Omega\right) \cdot \hat{n} dS\,, \quad (1)$$

which displays explicitly the various stresses that contribute to the total torque. Here $\rho$ is the mass density, $v_p$ the poloidal velocity, and $\Omega$ the angular velocity of the plasma, $B_p$ and $B_\phi$ are the poloidal and azimuthal components of the magnetic field, $\eta$ is the effective dynamic viscosity, and $\hat{n}$ is a unit outward normal. The three terms on the right-hand side of equation (1) represent, in turn, the contributions of

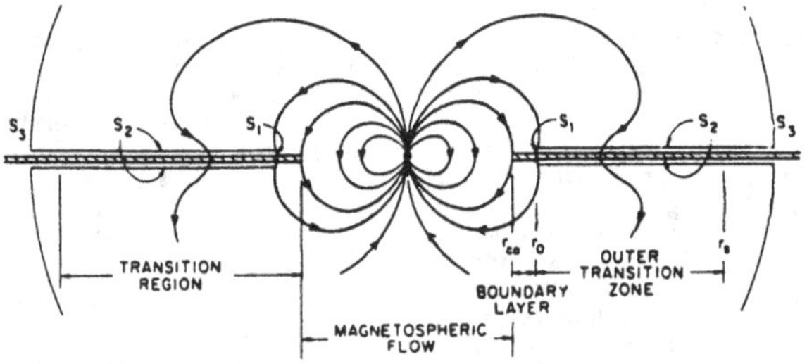


FIG. 1.—Side view of the accretion flow and the surfaces used to evaluate eq. (1), showing the transition region, composed of a broad outer zone where the angular velocity is Keplerian and a narrow boundary layer where it departs significantly from the Keplerian value, and the region of magnetospheric flow. The width $\delta = r_0 - r_{co}$ of the boundary layer is typically $\sim 0.04 r_0$ (see text). $S_1$ is a cylindrical surface of radius $r_0$ and height $2h$ while $S_2$ is composed of two plane surfaces just above and below the disk and $S_3$ comprises two hemispherical surfaces located at infinity.

the material, magnetic, and viscous stresses to the accretion torque. We note that the relative sizes of these three contributions depend on the surface used to evaluate the integral. Ultimately, however, the angular momentum flux is carried entirely by the magnetic stress in the sense that the other two contributions to the torque are completely negligible compared to the magnetic torque if the integral is evaluated on a surface that lies close to the surface of the neutron star.

In evaluating equation (1), it is convenient to choose the surface shown in Figure 1, which is composed of three parts: (1) a cylindrical surface $S_1$ of height $2h$ located at the radius $r_0$ that separates the boundary layer from the outer transition zone, (2) a surface $S_2$ consisting of two sheets running just above and below the disk from $r_0$ to infinity, and (3) two hemispherical surfaces at infinity. Here $h$ is the semi-thickness of the disk. The integral over $S_1$ gives the torque $N_{in}$ that is eventually communicated to the star by the magnetic field lines that thread the inner transition zone, while the integral over $S_2$ gives the torque $N_{out}$ communicated by the magnetic field lines that thread the outer transition zone. The integral over $S_3$ vanishes.

To an excellent approximation the torque $N_{in}$ is given by the material stress on $S_1$, since the viscous stress on $S_1$ is negligible by comparison (see § V of Paper II) while the magnetic stress has no component perpendicular to $S_1$. Now the angular velocity of the plasma at $r_0$ is closely Keplerian, by definition, so that

$$N_{in} \approx -\rho v_r r_0^2 \Omega_K(r_0) \cdot 2\pi r_0 \cdot 2h$$

$$= \dot{M}(GMr_0)^{1/2} \equiv N_0 \,, \qquad (2)$$

where $\Omega_K(r) = (GM/r^3)^{1/2}$ is the Keplerian angular velocity at $r$ in terms of the mass $M$ of the neutron star. The torque $N_{out}$, on the other hand, is given by the magnetic stress on $S_2$, since the material stress on $S_2$ is negligible (no matter crosses it) while the viscous stress has no component perpendicular to $S_2$. Thus

$$N_{out} = \int_{S_2} (rB_z B_\phi / 4\pi) dS \,. \qquad (3)$$

On combining contributions (2) and (3), one finds for the total torque on the star the result

$$N = N_0 + \int_{r_0}^{r_s} \gamma_\phi(r) B_z^2(r) r^2 dr \,, \qquad (4)$$

where

$$\gamma_\phi \equiv -(B_\phi/B_z)_{z=h} = (B_\phi/B_z)_{z=-h} \qquad (5)$$

is the average azimuthal pitch of the stellar magnetic field at the upper and lower surfaces of the disk and $r_s$ is the outer radius of the transition zone, beyond which the stellar field is screened to zero. Equation (4) can be evaluated by using the poloidal magnetic field $B_z(r)$ given by equations (39)–(41) of Paper II and the azimuthal pitch, $\gamma_\phi(r)$, given by equation (37) of that paper. The result is

$$N = n(\omega_s) N_0 \,, \qquad (6)$$

where the dimensionless accretion torque,

$$n(\omega_s) = 1 + \tfrac{1}{3}(1 - \omega_s)^{-1} \times \int_1^{y_s} b_{out}(y)(y^{-3/2} - \omega_s) y^{-31/40} dy \,, \qquad (7)$$

depends only on the fastness parameter (Elsner and Lamb 1977),

$$\omega_s \equiv \Omega_s / \Omega_K(r_0) \,, \qquad (8)$$

and the dimensionless outer radius of the transition zone, $y_s = r_s/r_0$. The function $b_{out}(y)$ is the dimensionless poloidal magnetic field in the outer transition zone given by equation (40) of Paper II.

### *b*) *Results*

The dimensionless accretion torque $n$ is primarily a function of the fastness parameter $\omega_s$, a fact we have emphasized by explicitly displaying this dependence in equation (7). For slow rotators ($\omega_s \ll 1$), one has $n(\omega_s) \approx 1.4$. For faster rotators spinning in the same direction as the disk flow, $n$ decreases with increasing $\omega_s$ and vanishes at a certain critical fastness $\omega_c$; for $\omega_s > \omega_c$, $n$ is negative and becomes increasingly so with increasing $\omega_s$. Finally, for $\omega_s$ greater than a certain maximum fastness $\omega_{max}$ (typically $\approx 0.95$) there are no stationary solutions to the two-dimensional flow equations of Paper II, and the torque on the star cannot be calculated in the manner described here.

The behavior of $n$ as a function of $\omega_s$ can be understood as follows. The total accretion torque is the sum of the torque $N_{in}$ eventually communicated to the star by the field lines that thread the inner transition zone and the torque $N_{out}$ due to the twisted field lines threading the outer transition zone. $N_{in}$ always acts to spin up a star rotating in the same sense as the disk flow, whereas $N_{out}$ can have either sign. This is because the azimuthal pitch of the field lines threading the outer transition zone changes sign at the radius

$$r_c = (GM/\Omega_s^2)^{1/3}, \tag{9}$$

where the angular velocity $\Omega_K$ of the disk plasma is the same as that of the star, as shown in Figure 2*a*. The contribution to the torque from the field lines threading the disk between $r_0$ and $r_c$ is positive whereas the contribution from the field lines threading the disk between $r_c$ and $r_s$ is negative. For slow rotators, $r_0 \ll r_c$, the positive part dominates the negative, and $N_{out}$ adds a further spin-up torque $\approx 0.4N_0$ to the torque $N_{in}$ which is equal to $N_0$. In contrast, for fast rotators $r_0 \sim r_c$, the negative part dominates, and $N_{out}$ contributes a spin-down torque which partly cancels $N_{in}$. For sufficiently fast rotators, the spin-down torque contributed by $N_{out}$ dominates the spin-up torque contributed by $N_{in}$ and there is a net spin-down torque on the star. The contribution to $N_{out}$ made by the field lines threading the disk interior to a given radius is shown in Figure 2*b* for three values of $\omega_s$.

For $\omega_s > \omega_{max}$, the equations describing steady, axisymmetric inflow within the boundary layer have no solution (see § V of Paper II). The reason for this is that the centrifugal force and the force due to the magnetic pressure gradient, which is also outward, are so large that radial inflow is halted at $r_0$ in a distance small compared to the boundary layer width $\delta$. We speculate that unsteady accretion may occur for values of $\omega_s$ larger than but comparable to $\omega_{max}$, while for $\omega_s \gg \omega_{max}$, disk accretion may cease altogether (compare Davidson and Ostriker 1973 and Lamb, Pethick, and Pines 1973).

In addition to its strong dependence on $\omega_s$, the accretion torque also depends weakly on the four boundary layer constants $C_b$, $C_\omega$, $C_p$, and $\gamma_0$ introduced in Paper II, through the weak dependence of $y_s$ on these constants.[4] Although these constants are not determined by our model, appropriate values are expected to be of order unity. In order to give the reader a feeling for the uncertainty in the torque which stems from our lack of knowledge of the precise structure of the boundary layer, we show in Figures 3 and 4 the dependence of the torque curve $n(\omega_s)$ and the critical fastness $\omega_c$ on the value of $\gamma_0$, the constant to which they are most sensitive. Figure 3 shows that $n$ is essentially independent of $\gamma_0$ for low angular velocities ($\omega_s \approx 0$) but becomes more sensitive to $\gamma_0$ as the angular velocity increases; however, even at the highest angular velocity for which there are stationary

[4] Actually, $\omega_s$ is also weakly dependent on the boundary layer constants, only three of which are linearly independent (see Paper II).

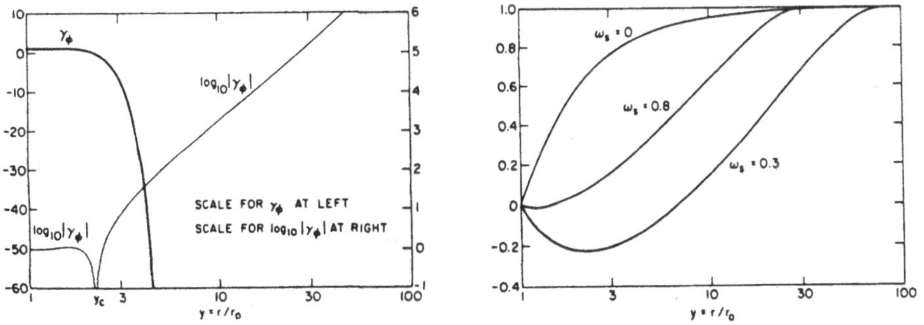


FIG. 2*a* FIG. 2*b*

FIG. 2.—Magnetic coupling between the outer transition zone and the star. (*a*) Azimuthal magnetic pitch in the outer transition zone as a function of the dimensionless radius $y = r/r_0$, for a star of fastness $\omega_s = 0.3$. The corotation point is at $y_c$. (*b*) Contribution to the torque $N_{out}$ made by that part of the outer transition zone which is interior to radius $y$, in units of the total torque $N_{out}$, for three values of $\omega_s$.

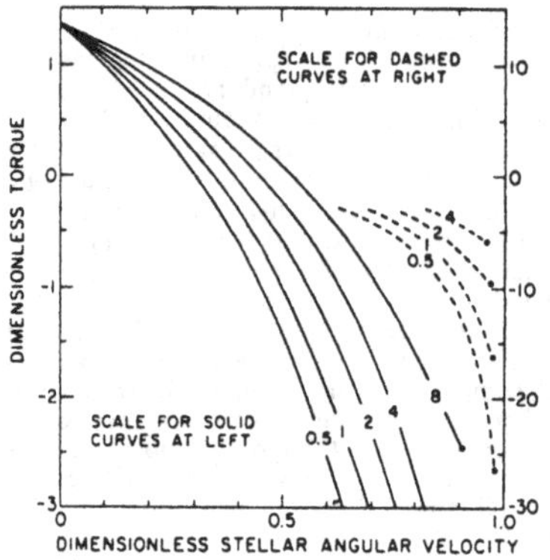


FIG. 3.—The dimensionless torque $n$ as a function of the dimensionless stellar angular velocity or fastness $\omega_s$, for five values of the magnetic pitch in the boundary layer, $\gamma_0$. Each curve is labeled with the corresponding value of the pitch. Those parts of the curves which continue off the bottom of the figure for large $\omega_s$ are shown as dashed curves on the reduced scale at right. The termination of each curve at the maximum fastness $\omega_{max}$ for which a steady flow is possible, is indicated by a dot.

flow solutions ($\omega_s = \omega_{max}$), $n$ varies only by a factor $\sim 2.7$ for values of $\gamma_0$ in the range 0.5–2. The critical fastness $\omega_c$ is even less sensitive to $\gamma_0$, varying by only 20–30% for values of $\gamma_0$ in the range 0.5–2, as shown in Figure 4.[5] The boundary layer structure, and hence $n$, is significantly less sensitive to the other boundary

[5] A simple analytic approximation to $\omega_c$, accurate to 1% for values of $\omega_c$ in the interval 0–0.8, is provided by the root of the equation

$$\alpha^{0.0986}(0.25 + 0.22\omega_c)\gamma_0^{0.123} = \omega_c(1 - \omega_c)^{0.173},$$

where $\alpha$ is the viscosity parameter of the disk.

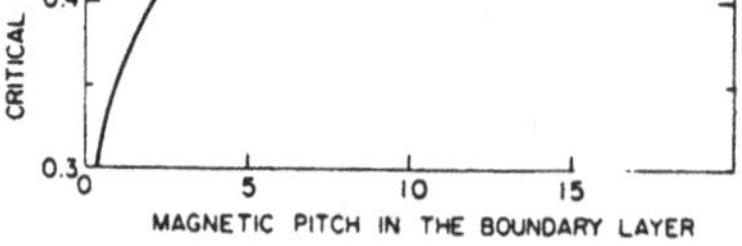

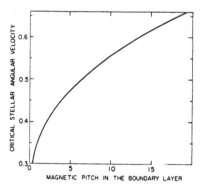

FIG. 4.—The critical value $\omega_c$ of the dimensionless stellar angular velocity or fastness parameter, as a function of the magnetic pitch in the boundary layer, $\gamma_0$.

layer constants (see Fig. 1 of Paper II). Thus the lack of precise knowledge of the boundary layer structure introduces a negligible uncertainty in the torque on slow rotators and only a moderate uncertainty in the torque on fast rotators. Throughout the remainder of this paper we shall assume $C_b = 2.5$, $C_p = 2$, and $\gamma_0 = 1$, the same values adopted for the bulk of the calculations presented in Paper II.

Once the boundary layer constants are fixed, $y_s$ is a function only of $\omega_s$ (see Paper II) and hence the accretion torque $N$ depends only on $\omega_s$ and $N_0$. A useful approximate expression for the dimensionless torque is

$$n(\omega_s) \approx 1.39\{1 - \omega_s[4.03(1 - \omega_s)^{0.173} - 0.878]\}(1 - \omega_s)^{-1}, \quad (10)$$

which is accurate to within 5% for $0 \le \omega_s \le 0.9$.

The fastness parameter $\omega_s$ can be expressed in terms of $\mu$, $M$, $\Omega_s$, and $\dot{M}$ by using expression (30) of Paper II for $r_0$, namely,

$$r_0 \approx 0.52 r_A^{(0)} = 0.52\mu^{4/7}(2GM)^{-1/7}\dot{M}^{-2/7}. \quad (11)$$

Here $r_A^{(0)}$ is the characteristic Alfvén radius for spherical accretion (Elsner and Lamb 1977). On substituting this expression into equation (8) one obtains

$$\omega_s = 1.19 P^{-1}\dot{M}_{17}^{-3/7}\mu_{30}^{6/7}(M/M_\odot)^{-5/7}, \quad (12)$$

where $P$ is the spin period in seconds, $\dot{M}_{17}$ is $\dot{M}$ in units of $10^{17}$ g s$^{-1}$, and $\mu_{30}$ is $\mu$ in units of $10^{30}$ gauss cm$^3$. In equations (11) and (12) and those that follow, one should use a value of $\mu$ slightly larger than the unscreened dipole moment of the neutron star since the screening currents flowing in the boundary layer enhance the magnetic field within the magnetosphere. An accurate determination of the appropriate correction factor, which is expected to be of order unity (see Paper II), must await a detailed calculation of the accretion flow between $r_0$ and the flow-alignment radius $r_f$ (again see Paper II).

### c) *Discussion*

Equations (2) and (10)–(12) show that the accretion torque on a neutron star of given mass and magnetic moment depends on both the mass accretion rate, $\dot{M}$, and the star's spin period, $P$, through the dependence of $N_0$, $r_0$, and $\omega_s$ on these quantities. However, the behavior of the accretion torque can be simply described in the two following situations of some astrophysical interest.

Consider first a star rotating *in* the same direction as the disk and accreting at a constant rate. Such a star can experience either spin-up or spin-down, depending on its spin period. If $P$ is sufficiently long, $\omega_s$ is small compared to unity, and the star experiences a strong spin-up torque $\sim 1.4 N_0$. As $P$ decreases, $\omega_s$ increases, and the spin-up torque falls, vanishing at the critical spin period $P_c$ at which $\omega_s = \omega_c$. If, on the other hand,

$P$ is less than $P_c$, the star experiences a spin-down torque. As $P$ increases, $\omega_s$ decreases, and the spin-down torque diminishes, vanishing at $P_c$. Thus the spin period of such a star will approach the critical period $P_c$ that corresponds to its accretion rate, and will then remain there. For $P$ less than the period $P_{min}$ at which $\omega_s = \omega_{max}$, accretion, if it occurs, is not steady.

In contrast, the accretion torque on a star of constant spin period varies with the accretion rate as shown in Figure 5. If $\dot{M}$ is sufficiently large, $\omega_s$ is small compared to unity, defining the star as a slow rotator, and the star experiences a strong spin-up torque $\sim 1.4 N_0$. As $\dot{M}$ decreases, the fastness $\omega_s$ increases, and the spin-up torque falls, vanishing at the critical accretion rate $\dot{M}_c$ at which $\omega_s = \omega_c$. For accretion rates less than $\dot{M}_c$, the star experiences a spin-down torque, the magnitude of which increases steadily until $\dot{M}$ reaches the minimum accretion rate $\dot{M}_{min}$ consistent with steady accretion. At this accretion rate $\omega_s = \omega_{max}$. If accretion continues at mass flow rates less than $\dot{M}_{min}$, it is not steady.

To what extent does this behavior depend on the approximations inherent in the present model and to what extent is it likely to be a general feature of disk accretion? We consider the generality of our results first in the context of stationary axisymmetric flows and then in the wider context of more general flows.

As discussed in Paper II, the structure of the transition zone that forms the basis for our calculation of

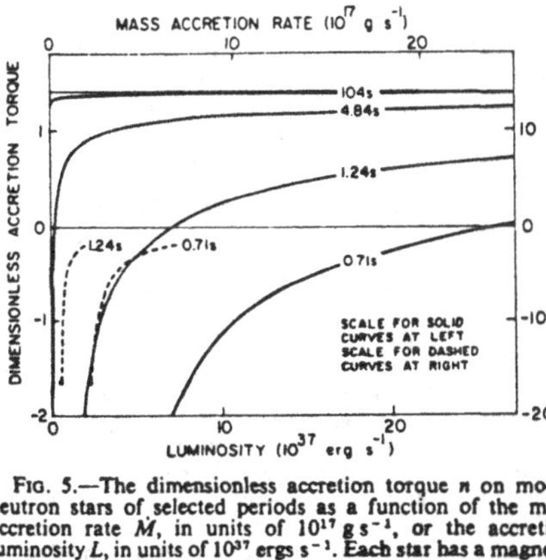


FIG. 5.—The dimensionless accretion torque $n$ on model neutron stars of selected periods as a function of the mass accretion rate $\dot{M}$, in units of $10^{17}$ g s$^{-1}$, or the accretion luminosity $L$, in units of $10^{37}$ ergs s$^{-1}$. Each star has a magnetic moment $\mu = 10^{30}$ gauss cm$^3$ and a mass $M = 1.3\ M_\odot$, and has been constructed using the TI equation of state of Pandharipande, Pines, and Smith (1976). The period of each star is written on the corresponding curve. The periods are those of the sources SMC X-1, Her X-1, Cen X-3, and A0535+26. Those parts of the solid curves that go off scale for low accretion rates are shown as dashed curves on the reduced scale at right. The termination of each curve at the minimum accretion rate consistent with steady inflow is marked by a dot. The light solid line above the curve for 104 s is the asymptotic behavior obtained in the slow rotation limit, $P \rightarrow \infty$.

the torque, namely a narrow inner zone where most of the screening occurs together with a broad outer zone where the residual stellar flux threads the disk, appears to be a general feature of steady axisymmetric disk accretion. Moreover, the radius of the inner edge of the disk and the width of the inner transition zone do not depend on the details of the dissipative process in the boundary layer, but only on the approximate isotropy of the effective conductivity and the reasonable assumption that $\gamma_\phi \sim 1$ at the radius $r_0$ where the magnetic field begins to control the flow. Thus the contribution of $N_{in}$ to the total torque is accurately given by equations (2) and (11). In contrast, $N_{out}$ depends on the configuration of the magnetic field in the outer transition zone. Although the fraction ($\sim 0.2$) of the total stellar flux that threads the outer transition zone appears to be relatively insensitive to the structure of the boundary layer, the azimuthal magnetic pitch in the outer zone is sensitive to the details of the magnetic field dissipation process there. Furthermore, according to the present model of the outer transition zone the azimuthal pitch increases steeply with radius beyond the disk corotation point. As a result, for some values of $\omega_s$ a substantial contribution to $N_{out}$ comes from radii as large as $10$–$20 r_0$, where the pitch is very large. Therefore, should improvements in the model lead to a smaller outer transition zone (see Paper II), $N_{out}$ would be somewhat reduced and $\omega_c$ somewhat increased.

These considerations suggest that for slow rotators, where $N_{in}$ and $N_{out}$ are additive and $N_{out}$ is small, the torque given by the present model is likely to be fairly accurate. For fast rotators, on the other hand, $N_{in}$ and $N_{out}$ have opposite signs, $N_{out}$ is large, and the present calculation is likely to be less accurate. Nevertheless, the qualitative behavior of the torque found here, including the braking torque on fast rotators, appears to be a general feature of steady axisymmetric disk accretion.

Finally, consider briefly the torque produced by more general accretion flows. In reality, disk accretion by aligned rotators is probably unsteady, at least on sufficiently small spatial and temporal scales. Even so, stationary flow models may provide an adequate description of the average accretion torque. In the case of oblique rotators, the flow is of necessity time-dependent and the coupling between the star and the disk altered from that of the aligned case (see Paper II). Nevertheless, we expect the qualitative structure of the flow to be similar to that of the present model, with a relatively narrow shear boundary layer and a more extended region where the disk and the star are magnetically coupled. If so, the accretion torque will be qualitatively similar to that found here.

### *d) Comparison with Other Work*

The accretion torque on magnetic neutron stars was considered in the very first work on such stars. Lamb, Pethick, and Pines (1973; see also Pringle and Rees 1972) gave arguments which showed that the torque on slow rotators accreting from a disk is $\sim N_0$. In comparing observed changes in the rotation rates of accreting neutron stars with this theoretical estimate, Elsner and Lamb (1976) drew attention to the fact that Her X-1, which almost certainly is accreting from a disk, has a spin-up rate $\sim 40$ times smaller than that which would correspond to the torque $N_0$. As a possible explanation of this discrepancy they noted that if Her X-1 were a fast rotator ($\omega_s \sim 1$), the magnetic and viscous stresses at the inner edge of the disk would tend to cancel the material stress there even if matter were not ejected, so that the total torque would be less than the material torque $N_0$. They argued further that as the fast rotation limit is approached ($\omega_s \rightarrow 1$), the magnetic and viscous stresses might more than offset the material stress, causing a net braking torque on the star even while accretion continues.

Motivated in part by this discrepancy, in Paper I we sought to make the torque argument of Lamb, Pethick, and Pines more precise. There we showed that for a steady axisymmetric flow with a general dependence of $\Omega$ on $r$ outside the Alfvén surface, the total torque is $\sim \dot{M} r_A{}^2 \Omega(r_A)$ if (1) the effective viscosity at $r_A$ is less than a certain reference value or (2) the transition zone has a width less than or equal to its radius, magnetic stresses within it are negligible, and the stellar angular velocity is less than $\Omega(r_A)$. For stars accreting from a disk we showed further that if the transition zone were narrow compared to its radius, then the total accretion torque would be bounded above by the torque

$$N_{max} = N_0 \tag{13}$$

and below by the torque

$$N_{min} = \tfrac{1}{2}(1 + \omega_s)N_0 + O(h/r_0)^2 \, . \tag{14}$$

Thus, for disks rotating in the same sense as the star, a narrow transition zone would imply a spin-up torque $\sim N_0$ even on fast rotators. This result argued strongly that the transition zone in disk accretion must be broad, since otherwise the Her X-1 spin-up rate would be extremely difficult to understand.

As an alternative way to resolve the apparent discrepancy between the torque $N_0$ expected to act on slow rotators and the much smaller torque apparently experienced by Her X-1 and some other accreting neutron stars, Scharlemann (1978) suggested that the transition zone in disk accretion is narrow (width $\sim$ height) but that the torque is always comparable to

$$N_1 \equiv \dot{M} r_0{}^2 \Omega_s = \omega_s N_0 \, .$$

On this basis Scharlemann argued that all seven of the stars that he considered, including Her X-1, were slow rotators with weak dipole magnetic fields. Unfortunately, this suggestion cannot be correct, for the following reason. Scharlemann showed that if one makes the ad hoc assumption that the magnetic field lines threading the neutron star all have positive pitch ($\gamma_\phi \geq 0$), then the accretion torque cannot be *less* than $N_1$, but did not show that the torque $N_1$ could ever be achieved. In fact, for a narrow boundary layer and slow rotation the accretion torque can never be as small as $N_1$, since it is bounded below by the much larger torque $N_{min}$, as shown in Paper I (a torque as small as $N_1$ would require a substantial violation of energy conservation).

In Ghosh and Lamb (1978*a*) and Paper II we showed that there need not be any discrepancy between the predicted and observed spin-up rates of sources like Her X-1, since the transition zone in disk accretion generally is *not* narrow, but is instead rather broad. Hence the bounds (13) and (14) do not apply. Here we have used the two-dimensional hydromagnetic flow solutions obtained in Paper II to calculate the accretion torque on the star. These calculations show that the magnetic coupling between the disk and the star due to the stellar field lines that thread the disk in the outer transition zone is appreciable and can be the dominant component of the accretion torque when the stellar angular velocity is high. The coupling via these field lines increases the spin-up torque on slow rotators, but reduces the torque on fast rotators and may even cause the total torque to become negative (braking the star's rotation) if the angular velocity is sufficiently high.

Recently Ichimaru (1978) attempted a calculation of the torque on a neutron star accreting from a disk. Unfortunately, Ichimaru's model is defective in important respects, some of which were noted in Paper II. Among those which are the most serious for his calculation of the accretion torque are the following. First, Ichimaru's equation (29) assumes that the flux of angular momentum toward the star is exactly zero. This assumption is far too restrictive; it implies that there is never any change in the angular momentum of the star and hence that the neutron star generally spins down as it accretes, since the added matter increases the star's moment of inertia. Second, Ichimaru assumes that the transition zone is narrow but that the torque on the star can be much less than the lower bound $N_{min}$ given above. As noted above, this would require a substantial violation of energy conservation.

In closing this comparison with previous work, we note that the spin-down torque on fast rotators found here is quite distinct from possible spin-down torques associated with ejection of matter by very fast rotators which have been discussed previously by several authors (Davidson and Ostriker 1973; Illarionov and Sunyaev 1975; Fabian 1975; Shakura 1975; Kundt 1976; Savonije and van den Heuvel 1977). The spin-down torque found here operates on stars with angular velocities which lie between the critical angular velocity and the maximum allowable angular velocity consistent with steady accretion, does not involve any mass ejection from the vicinity of the neutron star, and acts even while steady accretion continues. On the other hand, both the magnetic spin-down torque suggested by Davidson and Ostriker (1973) and the "propeller" spin-down torques suggested by Illarionov and Sunyaev (1975) and Shakura (1975) were assumed to operate only on very fast rotators ($\omega_s > 1$), and both involve mass ejection from the vicinity of the neutron star in an essential way. Nevertheless, at the maximum allowable fastness, $\omega_{max}$, the torque calculated here roughly agrees with the spin-down torques conjectured by Davidson and Ostriker, Illarionov and Sunyaev, and Shakura, evaluated at $\omega_s = 1$ (at this fastness all the latter torques are equal). Thus it may be possible to develop a consistent description of the torque on rotating stars by using the present model for $\omega_s < 1$ and models involving some mass ejection for $\omega_s > 1$.

### III. INTERPRETING PERIOD CHANGES IN PULSATING X-RAY SOURCES

Most, and perhaps all, pulsating X-ray sources other than pulsars are accreting neutron stars in which the pulsation period is the rotation period of the neutron star crust (see Lamb 1977). In all sources that have been studied carefully, the pulsation period has been found to change with time (see Schreier 1977). These changes are thought to be due to changes in the angular momentum, and hence the rotation period, of the crust. According to current ideas, the change in period over a sufficiently long time is due to the action of the external accretion torque (Pringle and Rees 1972; Lamb, Pethick, and Pines 1973), while period fluctuations on shorter time scales are caused either by fluctuations in the accretion torque (Elsner and Lamb 1976) or by variations in the torque exerted on the crust by the liquid interior (Lamb, Pines, and Shaham 1978*a*).

The external torque depends on the inflow rate and flow pattern of the accreting plasma, while the internal torque depends on the strength and nature of the coupling between the liquid interior and the crust. The response of the neutron star to a torque acting on the crust, whether external or internal, is expected to depend on the dynamical properties of the star: the "applied signal" represented by the torque is, in effect, "filtered" by the coupled crust-core system to produce an "output" represented by changes in the pulsation period. Thus, a detailed study of period changes in a given X-ray source can provide valuable information both about the properties of the accretion flow onto the star and about the properties of the star itself.

In the following subsections we first describe the nature of the interpretational problem in more detail, outline a possible solution, and discuss the comparison of theory with observation. We then consider the pulsation period changes predicted by current models of Keplerian and non-Keplerian accretion flows. Finally, we summarize how the goals listed above can be accomplished.

#### *a) The Nature of the Problem and a Possible Solution*

Period measurements made at intervals ranging from days to years have been reported for nine of the 17 presently known pulsating X-ray sources (see Giacconi 1974; Fabbiano and Schreier 1977; Mason 1977; Ögelman *et al.* 1977; Rappaport and Joss 1977; Schreier 1977; White, Mason, and Sanford 1977; Becker *et al.* 1978; Charles *et al.* 1978; Rappaport *et al.* 1978; Boynton and Deeter 1979; Jernigan and Nugent 1979). The sources whose period derivatives have been measured in this way have so far been found to be spinning up, over the long term. In addition, the four sources that have been studied most carefully (Her X-1, Cen X-3, Vela X-1, and X Per) show substantial short-term period fluctuations with occasional episodes of spin down. Figure 6 shows the behavior of Cen X-3, which is typical. Figure 7 shows recent data on Vela X-1 which suggest that its period behavior is similar to that of Her X-1, Cen X-3, and X Per.

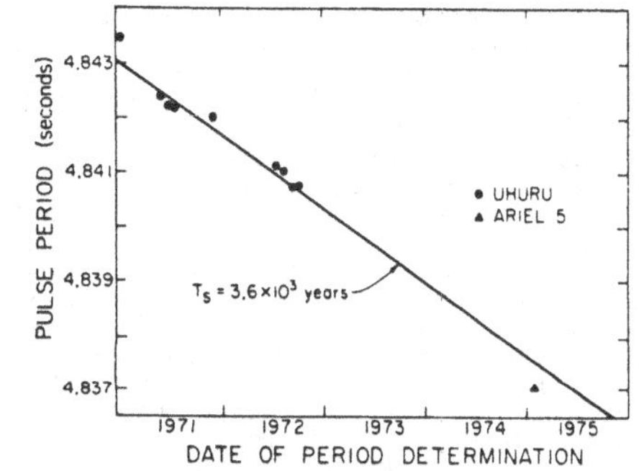


FIG. 6.—The behavior of the pulsation period of Cen X-3, showing the short-term period fluctuations and secular period decrease. Note the period increase of 1972 September-October. After Fabbiano and Schreier (1977).

The interpretation of these period changes is complicated by the fact that the response of the neutron star interior is not known *a priori*. Hence, such changes might be due either to the internal and external torques acting at the time, or to the response of the star to previous values of these torques. If the change in the rotation rate were smooth and the total observing interval were long, the likelihood that the star was

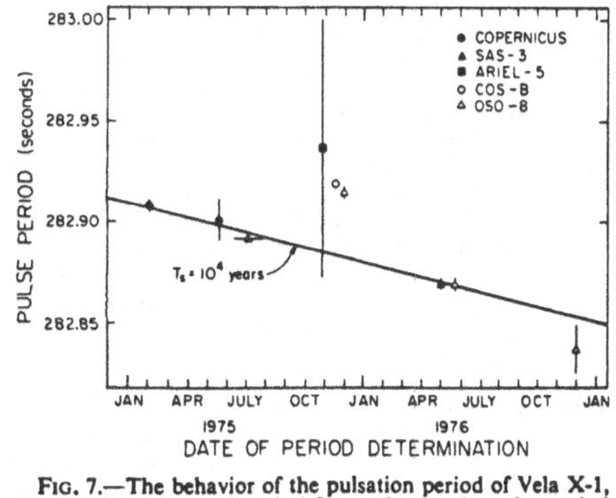


FIG. 7.—The behavior of the pulsation period of Vela X-1, showing the short-term period fluctuations and secular period decrease. Note the period increase of 1975 November. After Becker *et al.* (1978).

still responding to earlier values of these torques would be less, although the interpretation would still be ambiguous. In fact, the period changes in those sources that have been studied carefully are observed to be highly irregular.

Recently, Lamb, Pines, and Shaham (1978*a*) pointed out that if the torque fluctuations that cause the short-term period variations can be described by a simple noise process, then these interpretational difficulties can be overcome by using the period variations themselves to determine the dynamical response of the neutron star. These authors further suggested that either white or red torque fluctuations are physically plausible models and showed that the available data on Her X-1 and Cen X-3 are consistent with either type of torque noise. Motivated by these studies, Boynton and Deeter (1979) used the *Uhuru* data on Her X-1 to compute the power density spectrum of the pulse phase fluctuations in this source. They find that the torque fluctuations in Her X-1 are describable as white noise over a wide range of frequencies and derive significant constraints on the dynamical response of the neutron star from their analysis. Thus, this method of determining the neutron star response appears quite promising. Use of this method requires a dense, regular sequence of pulse phase measurements.

Measurements of the short-term behavior of the pulse phase and X-ray flux also provide the best type of data for probing the accretion flow pattern and the large-scale structure of the magnetosphere. First, accretion theory predicts that the change in spin period between two observations depends sensitively on the behavior of the accretion luminosity $L$ during the interval. If $L$ is highly variable, as it often is, spot checks at widely spaced times may lead to very large errors in the inferred luminosity history $L(t)$. Such errors can be minimized by measuring the period and flux at frequent intervals. Second, theoretical models necessarily include a number of poorly known parameters, such as the mass, radius, effective moment of inertia, and magnetic dipole moment of the star. Although the uncertainty in these parameters does not permit one to fit every possible value of $P$, $\dot{P}$, and $L$, in practice the possibility of adjusting some of these parameters introduces a substantial degree of freedom. If only the average values of $\dot{P}$ and $L$ are known, there is often not enough information to test the theory. However, once these parameters are determined, there is no more freedom, so that additional measurements of $P$, $\dot{P}$, and $L$ provide a quantitative test of the theory. Finally, since the effective inertial moment of the star generally is not constant, one must combine the theoretical relation for the accretion torque as a function of X-ray luminosity with the observed luminosity behavior $L(t)$ and a dynamical model of the neutron star crust and core in order to solve for the predicted period behavior $P(t)$ of the neutron star crust. Only then can a comparison be made with the observed period behavior. This procedure can be carried out only if a dense, regular sequence of period and flux measurements is available.

In summary, only studies of the short-term behavior of the pulse phase and flux of neutron star X-ray sources seem likely to provide the type of data required to test current theoretical models of their magnetospheres and interiors.

### *b) Comparing Theory and Observation*

Although measurement of pulse period fluctuations appears to be the most promising method for obtaining information about the magnetospheric structure and accretion flow pattern of a given source, the extraction of this information from such measurements clearly requires some care. First, the period fluctuations must be shown to be due to fluctuations in the accretion torque rather than internal torques. Second, the response of the neutron star must be considered in interpreting the data, since the response time of the liquid interior may be comparable to the time scale for changes in the accretion torque.

Accretion torque fluctuations can potentially be identified by searching for correlated changes in the pulse period $P$ and the X-ray flux $F$ at Earth, since accretion theory predicts that fluctuations in the mass accretion rate will cause fluctuations in both the accretion luminosity and the accretion torque; correlated changes in $P$ and $F$ are much less likely if the period changes are caused by internal torque fluctuations.[6] Fluctuations in the accretion rate appear quite natural and, as we show in § VI, can produce torque variations large enough to account for the period wandering observed in the well-studied pulsating sources. We shall therefore focus on torque variations that arise in this way.

To the extent that the dominant torque variations can be described by a simple noise process, one can disentangle the variations in the torque from the time-dependent response of the star in the manner described by Lamb, Pines, and Shaham (1978*a*). Once this is accomplished, a sequence of pulse period and X-ray flux measurements can potentially be used to (1) confirm that the X-ray source is indeed a neutron star, (2) determine whether the source is fed by a Keplerian accretion disk or by some other accretion flow pattern, (3) test quantitatively the theory of disk accretion, (4) determine accurately the dipole moment of the X-ray star, and (5) establish the *nature* of the accretion torque fluctuations.

The first step in achieving these goals is to construct a theoretical relation between the X-ray flux $F(t)$ at Earth and the pulse period $P(t)$. Such a relation can be constructed if one has available (1) a relation between $F$ and the accretion luminosity $L$, (2) a relation between $L$ and the mass accretion rate $\dot{M}$, (3) a relation between $\dot{M}$ and the torque $N$, and (4) a model for the change in the rotation of the neutron star crust caused by $N$. Assuming that the X-ray flux at Earth accurately reflects the X-ray luminosity and that the latter is

[6] We emphasize that the *absence* of correlated changes would not necessarily imply that accretion torque fluctuations are absent, since accretion flow variations different from those considered here could conceivably produce a change in the torque with little or no change in the mass accretion rate.

essentially the accretion luminosity $L$, then $F = L/4\pi D^2$, where $D$ is the distance to the source, while $L = \dot{M}(GM/R)$. One can then turn to accretion theory for a relation between $\dot{M}$ and $N$, such as equation (6). Finally, the stellar properties required to determine the change in the rotation period $P$ of the neutron star crust produced by the torque $N$ are fixed by the power density spectrum of pulse phase fluctuations.

Once a theoretical relation between $F(t)$ and $P(t)$ has been constructed, it can be tested by comparison with a sequence of X-ray flux and pulse period measurements. To illustrate how this approach can be applied, we shall discuss briefly the spin-up equations given by accretion theory (*a*) for disk-fed sources and (*b*) for sources that accrete matter which has insufficient angular momentum to form a disk, which we refer to as "wind-fed" sources.[7] For simplicity we shall assume that the neutron star responds like a rigid body with an effective moment of inertia $I_{eff}$, but we note that such a simple model can only be justified by a measurement of the dynamical response of the star over the time scales of interest.

### *c) Accretion from a Disk*

To the extent that the effective inertial moment of the neutron star is constant, the model of disk accretion described in Paper II predicts a simple relation between the spin-up rate, $-\dot{P}$, and the quantity $PL^{3/7}$, for a given source. This relation follows from equation (2) and the equation for the change in the stellar angular velocity $\Omega_s$ produced by the torque $N$. If we neglect the typically small effect of the change in the effective moment of inertia due to accretion (see Paper I), the latter becomes $\dot{\Omega}_s = N/I_{eff}$. Combining these two equations yields

$$-\dot{P} = 5.0 \times 10^{-5}\mu_{30}^{2/7}n(\omega_s)S_1(M)(PL_{37}^{3/7})^2 \text{ s yr}^{-1}, \tag{15}$$

where the function $n(\omega_s)$ is given by equation (7) or, approximately, by equation (10), and

$$\omega_s \approx 1.35\mu_{30}^{6/7}S_2(M)(PL_{37}^{3/7})^{-1}. \tag{16}$$

Here

$$S_1(M) = R_6^{6/7}(M/M_\odot)^{-3/7}I_{45}^{-1} \tag{17}$$

and

$$S_2(M) = R_6^{-3/7}(M/M_\odot)^{-2/7} \tag{18}$$

are structure functions that depend on the mass, equation of state, and dynamical response of the neutron star. $L_{37}$ is the accretion luminosity in units of $10^{37}$ ergs s$^{-1}$, $R_6$ is the stellar radius $R$ in units of $10^6$ cm, and $I_{45}$ is $I_{eff}$ in units of $10^{45}$ g cm$^2$. Since $n$ depends only on $\omega_s$ (see § II) and $\omega_s$ scales as $(PL^{3/7})^{-1}$, as shown by equation (16), *the value of $\dot{P}$ for a star of given mass and magnetic moment is a function only of $PL^{3/7}$.*

[7] Note that a neutron star whose binary companion is losing matter via a wind need not be wind-fed, since the matter that is accreted may still have sufficient angular momentum to form a disk (see, for example, Petterson 1978; Savonije 1978).

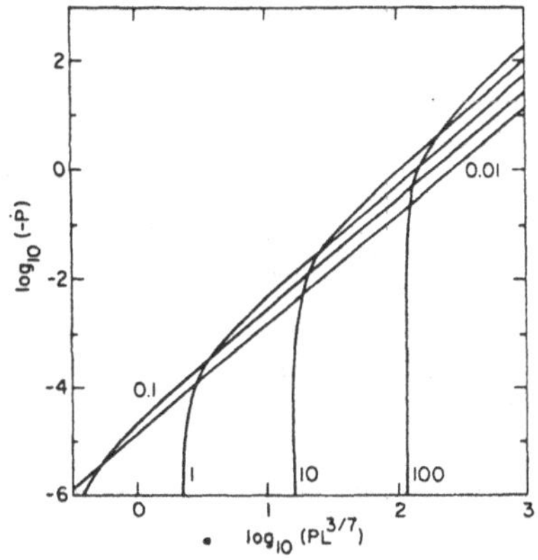


FIG. 8.—The theoretical relation between the spin-up rate $-\dot{P}$ of a 1.3 $M_\odot$ PPS neutron star and the quantity $PL^{3/7}$, for five values of the stellar dipole magnetic moment, $\mu$. Each curve is labeled with the corresponding value of the magnetic moment in units of $10^{30}$ gauss cm$^3$. The units of $-\dot{P}$, $P$, and $L$ are s yr$^{-1}$, s, and $10^{37}$ ergs s$^{-1}$, respectively.

The character of the relation between $-\dot{P}$ and $PL^{3/7}$ is shown in Figure 8. For large values of $PL^{3/7}$, the star is a slow rotator ($\omega_s \ll 1$), $n(\omega_s)$ is approximately constant (see eq. [10]), and $-\dot{P}$ scales as $(PL^{3/7})^2$. Thus, in a plot of $\log(-\dot{P})$ versus $\log(PL^{3/7})$, equation (15) predicts a straight line of slope 2 in the region of slow rotation. As $PL^{3/7}$ decreases, the fastness parameter $\omega_s$ becomes larger, $n(\omega_s)$ decreases, and $\log(-\dot{P})$ falls below the extrapolation of this line. Finally, at the value of $PL^{3/7}$ for which $\omega_s = \omega_c$, $\dot{P}$ vanishes and $\log(-\dot{P})$ diverges. The value of $PL^{3/7}$ at which the spin-up curve begins to fall below the extrapolated straight line depends on the magnetic moment of the star, and scales as $\mu^{6/7}$.

### *d) Accretion from a Wind*

When an accretion disk does not form, as may happen if the neutron star accretes from a wind, the radial velocity of the accreting plasma near the accretion capture radius $r_a$ is expected to be large, its shear is expected to be small, and the stellar magnetic field is expected to be confined to the interior of a magnetospheric cavity whose radius is much smaller than $r_a$ (Lamb, Pethick, and Pines 1973; Arons and Lea 1976; Elsner and Lamb 1977). As a result, if the surface in equation (1) is placed at $r_a$, the material stress completely dominates and the torque is just

$$N = \dot{M}l_a, \tag{19}$$

where $\dot{M}$ is the mass capture rate and $l_a$ is the specific angular momentum of the accreting plasma at $r_a$. The capture rate is given by

$$\dot{M} = \pi \rho_W v_0 r_a^2 , \quad (20)$$

where $\rho_W$ is the wind density and $v_0^2 \approx v_{orb}^2 + v_W^2 + c_s^2$ is the square of the capture velocity. Here $v_{orb}$ is the orbital velocity of the neutron star and $v_W$ and $c_s$ are the wind velocity with respect to the companion star and the local sound speed, evaluated at $r_a$. The specific angular momentum $l_a$ can be expressed as (Shapiro and Lightman 1976)

$$l_a \approx \tfrac{1}{4} a v_{orb} (r_a/a)^2 , \quad (21)$$

where $a$ is the binary separation. Finally, the capture radius is given by

$$r_a = \xi (2GM/v_0^2) , \quad (22)$$

where $\xi$ is a parameter of order unity. Thus equation (19) can be rewritten as

$$N = \pi^2 \xi^2 (2GM)^4 \rho_W v_0^{-7} P_{orb}^{-1} , \quad (23)$$

where $P_{orb}$ is the orbital period of the binary system. An equation for wind accretion analogous to equation (15) for disk accretion is

$$-\dot{P} = 3.8 \times 10^{-5} R_6 (M/M_\odot)^{-1} I_{45}^{-1} \times (l_a/10^{17}\ \mathrm{cm^2\ s^{-1}}) P^2 L_{37}\ \mathrm{s\ yr^{-1}} . \quad (24)$$

$\dot{P}$ is independent of the stellar magnetic moment because the magnetosphere plays no role in determining either $\dot{M}$ or $l_a$.

As the physical properties of the wind change in time, both $\dot{P}$ and $L$ will vary. The behavior of $\dot{P}$ as a function of $L$ depends on how $\rho_W$ and $v_0$ vary with changes in the wind:

1. *The wind density $\rho_W$ varies but $v_0 = const.$* This will be the case if $v_{orb}^2 \gg v_W^2 + c_s^2$. In this case $r_a =$ const., $l_a =$ const., $\dot{M} \propto \rho_W$, and $-\dot{P} \propto L$.

2. *The velocity $v_0$ varies but $\rho_W = const.$* In this case $r_a \propto v_0^{-2}$, $l_a \propto v_0^{-4}$, $\dot{M} \propto v_0^{-3}$, and $-\dot{P} \propto L^{7/3}$.

3. *Both $\rho_W$ and $v_0$ vary, but $\rho_W v_0 = const.$* In this case $r_a \propto \rho_W^2$, $l_a \propto \rho_W^4$, $\dot{M} \propto \rho_W^4$, and $-\dot{P} \propto L^2$.

The theory of winds in X-ray binaries is not sufficiently well developed to determine which of these (or other) possibilities for the variation of $\rho_W$ and $v_0$ is correct (Conti 1978; Vitello 1979). Thus, at present wind theory cannot predict the behavior of $\dot{P}$ as a function of $L$ for a given source.

### *e) Summary*

These results point to the importance of obtaining a frequent, regular sequence of period and flux measurements in developing an understanding of a given source. Such a sequence can provide information about the accretion flow pattern and the X-ray star that is impossible to obtain from average values or from any single measurement.

If the effective moment of inertia of the neutron star is essentially constant over the time scales of interest, it is appropriate to plot the observed values of $P$, $\dot{P}$, and $L$ on a graph of $\log(-\dot{P})$ vs. $\log(PL^{3/7})$. If the result can be described qualitatively by equation (15), this would constitute strong evidence that the X-ray source is a neutron star which is disk-fed and that the period variations are due largely to variations in the mass accretion rate. The shape of the curve would then give the size of the dipole magnetic moment and the stellar structure functions $S_1(M)$ and $S_2(M)$, while a detailed comparison of the data with the theoretical curve would provide a quantitative test of the theory.

If the inertial moment is *not* constant over the time scales of interest, one must instead combine the more fundamental equation for the accretion torque as a function of the mass accretion rate, equation (6), with the relation between $\dot{M}$ and $L$, the observed luminosity behavior $L(t)$ of the star, and the equations of motion of the neutron star crust and core, in order to solve for the rotation rate of the crust. The solution can then be compared with the observed period behavior $P(t)$ of the crust. Once again, if there is qualitative agreement between the predicted and observed period behavior, this would be strong evidence that the source is a disk-fed neutron star and that the period variations are due largely to variations in the mass accretion rate, and would allow a determination of the magnetic moment and structure of the star, as well as a quantitative test of the theory.

Finally, a search of pulse arrival times for relatively rare, very large pulse phase changes might allow one to resolve the torque noise process. If the noise process can be at least partially resolved, a detailed comparison of the time history of the pulse period and X-ray flux with the theoretical model can confirm that the largest torque fluctuations are caused by fluctuations in the mass accretion rate and determine the sign distribution of these torque excursions, their characteristic rise and fall times, and their mean rate of occurrence as a function of size. This information can then be compared with the expected properties of accretion torque fluctuations.

In the following sections we examine in turn the evidence furnished by the currently available data regarding accretion flow patterns, neutron star magnetic fields, and the origin of pulse period fluctuations. Where the value of $I_{eff}$ is needed, we take it to be the inertial moment of the whole star.

## IV. ACCRETION FROM DISKS OR WINDS?

In the previous section we have seen that a sequence of accurate period and luminosity measurements can establish unambiguously whether any given source is accreting from a disk. Unfortunately, such sequences are not yet available. Nevertheless, some evidence on accretion flow patterns can be obtained by comparing the currently available *time-average* values of $P$, $\dot{P}$, and $L$ with theoretical predictions for a collection of sources. In the present section we consider this evidence.

### *a) The Disk Hypothesis*

If we consider a *collection* of pulsating X-ray sources, equation (15) predicts that they would all lie

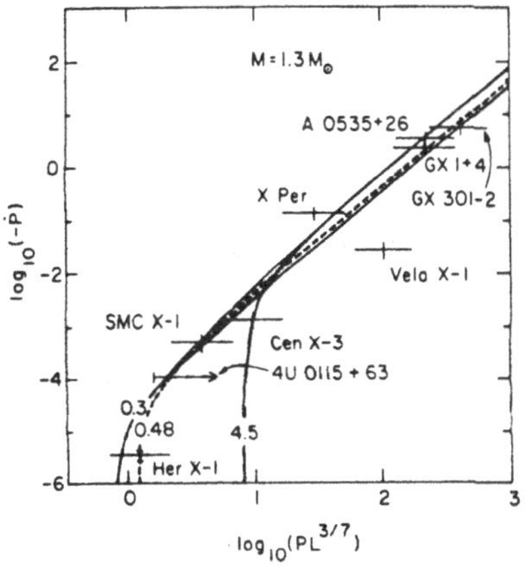


FIG. 9.—The theoretical relation between the spin-up rate, $-\dot{P}$, and the quantity $PL^{3/7}$, superposed on the data for nine pulsating X-ray sources. The units of $-\dot{P}$, $P$, and $L$ are s $yr^{-1}$, s, and $10^{37}$ ergs $s^{-1}$, respectively. Shown are the theoretical curves for three values of the stellar magnetic moment, assuming a 1.3 $M_{\odot}$ PPS model for all neutron stars. Each curve is labeled with the corresponding value of the magnetic moment, in units of $10^{30}$ gauss $cm^3$. *Shaded area*, region spanned by the theoretical curves for $0.3 \leq \mu_{30} \leq 4.5$. *Dashed line*, theoretical curve for $\mu_{30} = 0.48$, which gives a rough best fit to the data. Because the curves corresponding to different magnetic moments cross (see text), the upper boundary of the shaded region is defined by the envelope of the curves (*light solid line*).

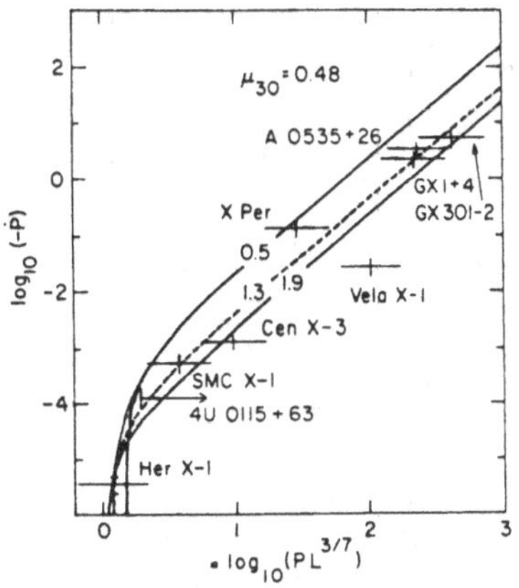


FIG. 10.—Same as Fig. 9, but showing the effect of varying the neutron star mass, assuming a stellar magnetic moment $\mu_{30} = 0.48$ and the PPS equation of state for all stars. Each curve is labeled with the corresponding value of $M/M_{\odot}$. *Shaded area*, region spanned by the theoretical curves for $0.5 \leq M/M_{\odot} \leq 1.9$. *Dashed line*, theoretical curve for $M/M_{\odot} = 1.3$, which gives a rough best fit to the data. Because the curves corresponding to different stellar masses cross, the upper boundary of the shaded region is defined by the envelope of the curves (*light solid line*).

on the same curve $-\dot{P} = f(PL^{3/7})$ if they all had (1) the same mass and (2) the same magnetic moment. Although all pulsating X-ray sources are not expected to have identical masses and magnetic moments, there should still be a correlation between $\dot{P}$ and $PL^{3/7}$ if the spread in masses and magnetic moments is not too large. It is therefore interesting to plot the logarithms of the observed values of $-\dot{P}$ against the logarithms of the observed values of $PL^{3/7}$, as in Figures 9 and 10. Such a plot tends to order the sources according to their fastness, since for fixed $M$ and $\mu$ the fastness parameter $\omega_s$ is a function only of $PL^{3/7}$. The data used in constructing these figures are given in Table 1.

The effect of varying the stellar magnetic moment is shown in Figure 9, which displays the function $f(PL^{3/7})$ for $M = 1.3\ M_{\odot}$, assuming the tensor interaction (TI) equation of state of Pandharipande, Pines, and Smith (1976, hereafter PPS), and various values of the stellar magnetic moment. Also shown are the observed data on nine pulsating X-ray sources. The curve for $\mu_{30} = 0.48$ is a rough best fit to the data. Except for Vela X-1 = 4U 0900−40, all the sources lie in the region spanned by the curves corresponding to values of $\mu_{30}$ in the range 0.3–4.5. Note that the curves for different values of $\mu$ cross one another, so that the upper edge of this region is given by the envelope of the various curves, which is indicated by the light solid line in the figure. The crossing of the curves for different values of $\mu$ occurs because the higher moment gives the larger torque and hence the larger spin-up rate in the slow rotator region (large values of $PL^{3/7}$), but also gives the larger critical value of $PL^{3/7}$ (see eq. [16]), so that the spin-up rate in the fast rotator region falls more rapidly for the higher moment than for the lower one.

The effect of varying the stellar mass is illustrated in Figure 10, which shows the function $f(PL^{3/7})$ for $\mu_{30} = 0.48$, the best-fit value of Figure 9, and various values of the stellar mass, again assuming the TI equation of state. Once again the curves for different masses cross one another. All sources except Vela X-1 lie in the region spanned by stellar masses in the range 0.5–1.9 $M_{\odot}$ (the latter is the maximum stable mass for stars obeying the TI equation of state).

### *b) The Wind Hypothesis*

Even though wind theory is not sufficiently advanced to predict the behavior of $\dot{P}$ as a function of the luminosity of any given source (see § III), nevertheless the wind hypothesis does, under some conditions, predict a correlation between $\dot{P}$ and other observable quantities when one compares a collection of sources, and it is therefore interesting to see if there is any evidence for such a correlation in the currently available data. In particular, the terminal velocities of winds from early-type stars like the massive companions of possibly wind-fed neutron stars are usually comparable to the escape velocities from these stars (Castor, Abbott, and Klein 1975; Abbott 1978). Moreover, the companion star must be inside its

Roche lobe, and the wind velocity near its terminal value at the accretion capture radius, if the flow is not to form an accretion disk around the neutron star (see Petterson 1978). Hence one expects $v_W \sim v_{esc}$, where $v_{esc} = (2GM_c/R_c)^{1/2}$ in terms of the mass $M_c$ and radius $R_c$ of the companion star. In addition, in most of these systems the binary separation $a$ is comparable to $R_c$, so that the neutron star orbital velocity $v_{orb}$ is also comparable to $v_{esc}$. Thus, if the local sound speed $c_s \lesssim (v_W^2 + v_{orb}^2)^{1/2}$, then the characteristic capture velocity $v_0$ (see § III) will scale from system to system as $v_{esc}$ (Shapiro and Lightman 1976). And equation (24) can be rewritten in terms of the spin-up time scale $T_s \equiv -P/\dot{P}$ as

$$PLT_s = fP_{orb}\,, \tag{25}$$

where the function $f$ is given by

$$f = (2GMI_{eff}/R)q^2R_c^{-2}\gamma^{-4}\,, \tag{26}$$

with $q = M_c/M$ and

$$\gamma \approx \xi^{1/2}[(v_W/v_{esc})^2 + (v_{orb}/v_{esc})^2]^{-1/2}\,. \tag{27}$$

Since the above scaling yields $\gamma \approx$ const., equation (25) predicts a correlation between the X-ray source properties $P$, $L$, $T_s$, and $M$, and the binary system properties $P_{orb}$, $M_c$, $R_c$, and $\gamma$. Given the current observational uncertainties in the quantities that enter $f$, one cannot hope to evaluate it for each individual source. However, to the extent that it also is roughly constant, there should be a correlation between $PLT_s$ and $P_{orb}$.

Figure 11 shows the currently available data for seven well-studied X-ray sources on a plot of $\log(PLT_s)$ against $\log(P_{orb})$. The data used in constructing this figure are given in Table 1. Also shown in the figure is the straight line

$$\log[P(s)L_{37}T_s(y)] = \log[P_{orb}(d)] + 3.82\,, \tag{28}$$

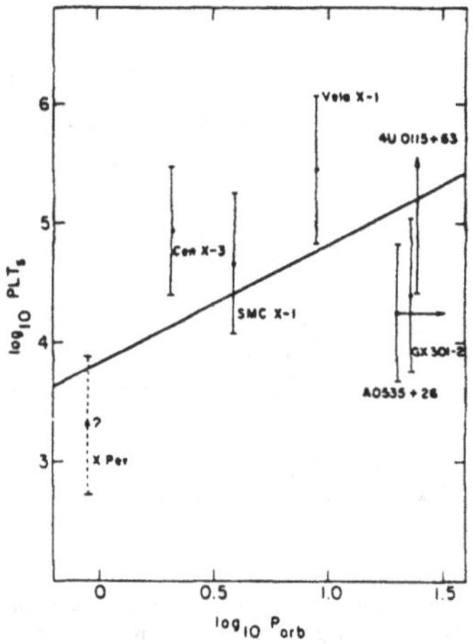


FIG. 11.—Observed values of the quantity $PLT_s$ plotted against observed values of the binary orbital period $P_{orb}$, for seven pulsating X-ray sources. The units of $P$, $L$, $T_s$, and $P_{orb}$ are seconds, $10^{37}$ ergs s$^{-1}$, years, and days, respectively. The straight line is the theoretical relation expected for wind-fed sources if $q = 15$, $R_c = 25\,R_\odot$, and $\gamma = 1$ (see text). The 22[h] X-ray period is shown for X Per, although the evidence suggests that it is not the binary period (see Table 1).

which corresponds to $q = 15$ (see Cowley 1977), $R_c = 25\,R_\odot$ (see Ziółkowski 1977), and $\gamma = 1$. Although the observed values of $PLT_s$ lie within an order of magnitude of the values expected theoretically, there is little evidence for any correlation with

TABLE 1

OBSERVED PROPERTIES OF NINE PULSATING X-RAY SOURCES[a]

| Source | $P$ (s) | Mean $L$ ($10^{37}$ ergs s$^{-1}$) | $T_s$ (years) | Binary Period (days) |
|---|---|---|---|---|
| 4U 0115−73 = SMC X-1 | 0.71 | 50 | $(1.3 \pm 0.4) \times 10^3$ | 3.89 |
| 4U 1653+35 = Her X-1 | 1.24 | 1 | $(3.3 \pm 0.6) \times 10^5$ | 1.70 |
| 4U 0115+63 | 3.6[b] | $\geq 0.9$[c,d] | $\sim 3.1 \times 10^4$[e] | 24.3[d] |
| 4U 1118−60 = Cen X-3 | 4.84 | 5 | $(3.6 \pm 0.6) \times 10^3$ | 2.09 |
| A0535+26 | 104 | 6 | $29 \pm 8$ | $\geq 20$[e] |
| 4U 1728−24 = GX 1+4 | 121 | 4 | $50 \pm 13$ | ... |
| 4U 0900−40 = Vela X-1 | 283 | 0.1 | $(1.0 \pm 0.4) \times 10^4$[ef] | 8.97 |
| 4U 1223−62 = GX 301-2 | 700 | 0.3 | $120 \pm 60$ | 23[e] |
| 4U 0352+30 = X Per | 836 | $4 \times 10^{-4}$ | $(5.9 \pm 1.5) \times 10^3$[g] | 0.9[h] |

[a] From Rappaport and Joss 1977, unless otherwise noted.
[b] Cominsky *et al.* 1978.
[c] Johnston *et al.* 1978.
[d] Rappaport *et al.* 1978.
[e] Bradt, Doxsey, and Jernigan 1979.
[f] Becker *et al.* 1978.
[g] White, Mason, and Sanford 1977; Jernigan and Nugent 1979.
[h] This period corresponds to a variation sometimes seen in the X-ray flux (see White *et al.* 1975 and Culhane *et al.* 1976), although current evidence suggests that it is not the binary period.

$P_{orb}$ (the $22^h$ X-ray variation sometimes observed in X Per is shown in the figure, although present evidence suggests that it is not the binary period of the source; if this point is given a low weight, there is no evidence for any increase of $PLT_s$ with $P_{orb}$).

### *c) Discussion*

Figures 9 and 10 show that the present theory of disk accretion taken together with the available observational data is consistent with a relatively narrow range ($\sim 1$ decade) of stellar magnetic moments for a given stellar mass or, alternatively, a narrow range ($\sim$ a factor of 4) of stellar masses for a given magnetic moment. The only source for which this is not the case is Vela X-1 which either has a much larger magnetic moment than the other sources or is not accreting from a disk.

Observational support for the existence of a significant region where the disk is magnetically coupled to the star and, ipso facto, support for the disk hypothesis comes from the fact that the full torque given by the present model (eq. [6]), which predicts a steep fall in $-\dot{P}$ for fast rotators, fits the data significantly better than the torque $N_0$ (eq. [2]).[8] If we exclude 4U 0900−40, then for a fixed mass and neutron star equation of state the present torque model can achieve agreement with the remaining sources with a factor of 15 variation in $\mu$, while for fixed $\mu$ the model can achieve agreement with a factor of 3.8 variation in $M$. For comparison, the torque $N_0$, which neglects the disk-star magnetic coupling, would require, for a fixed mass and equation of state, a factor of 200 (if Her X-1 is excluded) or of $10^5$ (if Her X-1 is included) variation in $\mu$ in order to achieve a similar agreement, while for fixed $\mu$ a factor of 20 (Her X-1 excluded) or of $10^3$ (Her X-1 included) variation in $M$ would be required, were this possible. The fall in $-\dot{P}$ at small values of $PL^{3/7}$, which is responsible for the better agreement of the present torque model with observation, is not expected in wind-fed sources. Conversely, the current observational data do not show the correlation with $P_{orb}$ that might be expected for wind-fed sources, although the observational uncertainties are large and may mask such a correlation.

Considered as a group, the currently measured pulsating X-ray sources are remarkably well described by a single "universal" relation between $P$, $\dot{P}$, and $L$, obtained from equation (15) by assuming that all are 1.3 $M_\odot$ neutron stars with $5 \times 10^{29}$ gauss cm$^3$ dipole moments accreting from disks. Although all pulsating sources are not actually expected to have identical masses and dipole moments, the good agreement of this relation with the data collected so far suggests that it holds approximately for almost all sources. If so, then given the values of two of the three quantities, $P$, $\dot{P}$, and $L$ for a particular source, one can use this relation to estimate the value of the third parameter. The good agreement of this relation with the data also argues strongly that the nine sources considered here are indeed neutron stars, since the corresponding relation for magnetic white dwarfs predicts values of $\dot{P}$ which are several orders of magnitude smaller than the values observed in these sources (see Lamb 1977).

In summary, the striking agreement between the present torque model and the average spin-up rates of eight of the nine currently measured sources that is indicated in Figure 10 and the apparent absence of any similar correlation in Figure 11 suggests that these eight sources are disk-fed. In the remaining sections of the present paper we shall therefore adopt the disk hypothesis. However, given our limited knowledge of the magnetic moments, wind conditions, and binary system parameters of these sources, this evidence is far from decisive, a situation which underscores the importance of carrying out the potentially conclusive studies outlined in § III.

[8] For this reason, plotting $\log(-\dot{P}/P)$ against $\log(PL^{6/7})$, which tends to intermix fast rotators with slow ones, produces a markedly larger scatter in the data (see, for example, Rappaport and Joss 1977) than does a plot of $\log(-\dot{P})$ against $\log(PL^{3/7})$, which orders sources according to fastness.

## V. NEUTRON STAR MAGNETIC FIELDS

In § III we showed that a sequence of accurate period and luminosity measurements can establish unambiguously that a given X-ray star is accreting from a disk and that such a sequence can then be used to determine accurately the magnetic moment of the star. Even without data of this quality, a first estimate of the magnetic moment can be obtained by fitting the theoretical spin-up equation to the *average* luminosity and spin-up rate of the source derived from the relatively sparse period determinations and X-ray flux measurements currently available (note, however, the cautionary remarks in § III*d*). In the present section we assume that all the observed sources are disk-fed and use their average spin-up time scales to infer their dipole moments.

### *a) Spin-up Time Scale*

According to equation (15), the spin-up time scale $T_s \equiv -P/\dot{P}$ predicted by the present disk accretion model is

$$T_s = 2.0 \times 10^4 \mu_{30}^{-2/7} n^{-1} S_1^{-1} L_{37}^{-6/7} \text{ yr}, \quad (29)$$

where $n$ is given by equation (7), and $S_1$ by equation (17). These equations show that for a given source (specified by a spin period $P$ and luminosity $L$) and a given neutron star model (specified by a mass $M$, radius $R$, and effective inertial moment $I_{eff}$), the predicted value of $T_s$ depends only on the magnetic moment $\mu$ of the star.

The behavior of $T_s$ as a function of $\mu$ for a given source and neutron star model is illustrated in Figures 12 and 13, which show the functions $T_s(\mu)$ that correspond to the observed parameters of the sources Her X-1, GX 1+4, and X Per, for the TI neutron star models of PPS. For small values of $\mu$, one has $T_s \propto \mu^{-2/7}$ and hence $T_s$ decreases with increasing $\mu$; this is the region of slow rotation. As $\mu$ is further

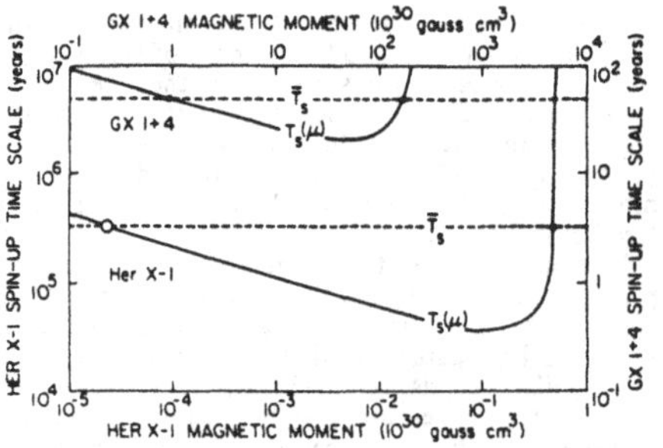


FIG. 12.—Spin-up time scale for GX 1+4 = 4U 1728−24 (*top curves*) and Her X-1 (*bottom curves*). *Solid curves*, theoretical spin-up time scales $T_s(\mu)$ for a 1.3 $M_\odot$ PPS neutron star. *Dashed lines*, observed average spin-up time scales $\bar{T}_s$. Theoretical and observed spin-up time scales agree at the points marked by circles. A filled circle indicates the physically acceptable solution for Her X-1. For GX 1+4, both solutions are physically acceptable.

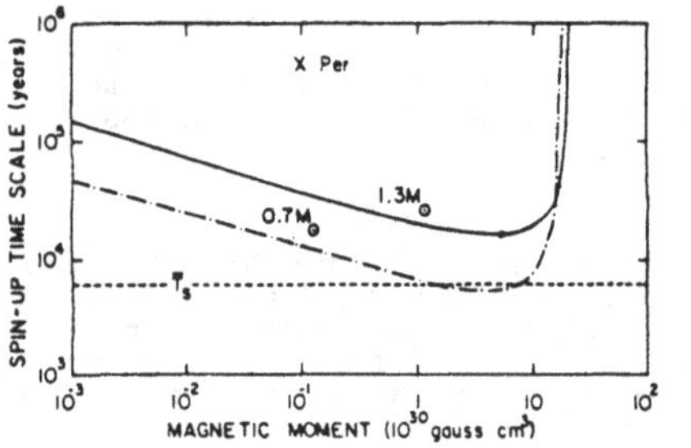


FIG. 13.—Same as Fig. 12, but for 4U 0352+30 = X Per. *Solid curve*, theoretical spin-up time scale $T_s(\mu)$ for a 1.3 $M_\odot$ PPS neutron star. *Dot-dashed curve*, theoretical spin-up time scale $T_s(\mu)$ for a 0.7 $M_\odot$ PPS neutron star. *Dashed line*, observed average spin-up time scale $\bar{T}_s$. A filled circle indicates the spin-up time scale and magnetic moment given in Table 2.

increased, $T_s$ passes through a minimum and then begins to rise; the region of rising $T_s$ is that of fast rotation. Finally, as $\mu$ approaches the critical value $\mu_c$ at which $\omega_s = \omega_c$, the accretion torque vanishes and hence $T_s$ diverges. Thus, given a particular X-ray source with an observed average spin-up time scale $\bar{T}_s$ and a particular neutron star model, either there are two values of $\mu$ for which the theory gives a spin-up time scale in agreement with the observed value, if $\bar{T}_s$ lies above the minimum of the theoretical curve $T_s(\mu)$, or there are none, if $\bar{T}_s$ lies below the minimum.[9] When there are two solutions for $\mu$, one is a "slow rotator" solution whereas the other corresponds to a "fast rotator." In practice, one of these two solutions can often be ruled out on other grounds; for example, a value of $\mu$ which corresponds to an inner disk radius $r_0$ less than or equal to the stellar radius $R$ can be

[9] A single value of $\mu$ can also give agreement in the special case where $\bar{T}_s$ exactly equals the minimum value of $T_s(\mu)$. This is the case for SMC X-1.

TABLE 2

DERIVED PARAMETERS OF NINE PULSATING X-RAY SOURCES[a]

| SOURCE | SLOW ROTATOR SOLUTION | | | FAST ROTATOR SOLUTION | | |
|---|---|---|---|---|---|---|
| | $\mu_{30}$[b] | $r_0$($10^8$ cm) | $\omega_s$ | $\mu_{30}$[b] | $r_0$ ($10^8$ cm) | $\omega_s$ |
| SMC X-1 | 0.50 | 0.36 | 0.11 | 0.50 | 0.36 | 0.11 |
| Her X-1 | ($2.4 \times 10^{-5}$) | ($3.8 \times 10^{-3}$) | ($6.9 \times 10^{-5}$) | 0.47 | 1.1 | 0.35 |
| 4U 0115+63 | $2.9 \times 10^{-3}$ (?) | $6.0 \times 10^{-2}$ (?) | $1.5 \times 10^{-3}$ (?) | 1.4 | 2.0 | 0.30 |
| Cen X-3 | $1.0 \times 10^{-2}$ (?) | $7.4 \times 10^{-2}$ (?) | $1.6 \times 10^{-3}$ (?) | 4.5 | 2.4 | 0.29 |
| A0535+26 | 3.3 | 1.9 | $9.7 \times$ [illegible]$^{-3}$ | 148 | 17 | 0.25 |
| GX 1+4 | 0.93 | 1.1 | $3.4 \times$ [illegible]$^{-3}$ | 170 | 21 | 0.29 |
| Vela X-1 | ($2.7 \times 10^{-5}$) | ($7.8 \times 10^{-3}$) | ($9.0 \times$ [illegible]$^{-7}$) | 86 | 40 | 0.35 |
| GX 301-2 | 0.3 | 1.2 | $6.7 \times 10^{-4}$ | 394 | 70 | 0.31 |
| X Per | 4.8[c] | 38 | 0.10 | 4.8[c] | 38 | 0.10 |

[a] Observed properties are taken from Table 1.

[b] For each source, the stellar magnetic moment was adjusted to obtain the best possible agreement between the observed value of $\bar{T}_s$ and the theoretical value of $T_s$ for a 1.3 $M_\odot$ PPS neutron star. In general, two values of the moment give agreement for those sources for which an exact agreement is possible. For SMC X-1 these two values of $\mu$ coincide, as the observed value of $\bar{T}_s$ exactly equals the minimum theoretical value. The slow rotator solutions for Her X-1 and Vela X-1 (enclosed in parentheses) are ruled out because the corresponding inner disk radii would lie inside the neutron star. The slow rotator solutions for Cen X-3 and 4U 0115+63 are suspect because the corresponding inner disk radii would lie close to the stellar surface. If one chooses the underlined solutions for A0535+26, GX 1+4, and GX 301−2, then the magnetic moments of all sources except Vela X-1 would lie within an order of magnitude of one another. The inferred magnetic moments listed here differ slightly from the preliminary values reported in Ghosh and Lamb (1978*a*) because our preliminary calculations did not incorporate the self-consistency condition (eq. [34] of Paper II) between the four constants $C_b$, $C_\omega$, $C_p$, and $\gamma_0$.

[c] For X Per, exact agreement between the observed and calculated values of $T_s$ was not possible for a 1.3 $M_\odot$ PPS neutron star, since the minimum theoretical value, $1.5 \times 10^4$ years, was somewhat greater than the observed value, $5.9 \times 10^3$ years. The agreement is, however, well within the observational uncertainties (see text). The value of $\mu$ given for this source corresponds to the minimum theoretical value of $T_s$.

rejected (in Fig. 12 the "slow rotator" solution for Her X-1 is such a case).

### b) *Inferred Dipole Moments*

As noted in § III, period measurements over relatively long time scales (~ months-years) have been made for nine of the 16 currently known pulsating X-ray sources. For each of these nine sources we have attempted to match the theoretical spin-up time scale given by equations (7), (17), and (29) to the observed average spin-up time scale by assuming that the neutron star has a mass of 1.3 $M_\odot$ and obeys the TI equation of state of PPS, and then adjusting $\mu$. The results are summarized in Table 2. Two solutions for $\mu$, corresponding to slow and fast rotation, are possible for all sources except SMC X-1, for which only a single value of $\mu$ fits the observations [$\bar{T}_s$ happens to equal the minimum value of $T_s(\mu)$], and X Per, for which no solution is possible given the current estimate of the mean source luminosity and the neutron star model chosen here [$\bar{T}_s$ lies below the minimum value of the curve $T_s(\mu)$]. The case of X Per is discussed further below.

Among those sources for which two solutions for $\mu$ are possible, the slow rotator solution is definitely ruled out for the two sources Her X-1 and Vela X-1 since it would imply that the inner radius of the disk lies inside the stellar surface. The rejected solutions are enclosed in parentheses in Table 2. In addition, the slow rotator solutions for Cen X-3 and 4U 0115+63 are very suspect, though not completely ruled out, since they correspond to inner disk radii only a few times larger than the stellar radius, and channelling of matter by the stellar magnetic field may be marginal in such a situation. For the other sources that admit of two solutions, there is as yet no compelling reason for rejecting either of them.

### c) *Discussion*

First, consider the sources Vela X-1 and X Per, which are somewhat exceptional and deserve further comment.

In order to fit the relatively long spin-up time scale of Vela X-1, a magnetic moment ~$10$–$10^2$ times larger than that inferred for the other sources is required. While such a large magnetic moment is certainly allowed by our current understanding of neutron star magnetic fields, the fact that Vela X-1 alone requires such a large value of $\mu$ suggests the alternative possibility that Vela X-1 is wind-fed rather than disk-fed (Lamb 1977), in which case it could have a relatively long spin-up time scale even if its magnetic moment were similar to those of the other sources.

For X Per, on the other hand, the observed value of $\bar{T}_s$, $5.9 \times 10^3$ yr, lies somewhat below the minimum of the theoretical curve, which corresponds to $1.5 \times 10^4$ yr, and hence no solution is possible for current estimates of the mean source luminosity and the neutron star model chosen here (see Fig. 13; the value of $\mu$ given in Table 2 corresponds to the minimum of the theoretical curve). However, the present lack of intersection is not significant, since there are major uncertainties in several of the input parameters upon which the theoretical curve depends. First, the mean luminosity of the source over several years, which is required in the calculation, is very poorly known at present. The uncertainty in the current estimate (~$4 \times 10^{33}$ ergs $s^{-1}$) of the mean luminosity of X Per is not known. For the sake of definiteness, Rappaport and Joss (1977) *assumed* a plausible but arbitrary uncertainty of a factor of 3 in the mean luminosities of all sources, including X Per. If this uncertainty in luminosity is accepted, then solutions for X Per exist for values of the mean luminosity within this range but a factor of ~2.5 less than the estimate adopted by Rappaport and Joss. Second, for the sake of definiteness we *assumed* that *all* sources, including X Per, have a mass of 1.3 $M_\odot$. This is the approximate value inferred for Her X-1, the only source for which a fairly precise determination of this quantity is at present available (Middleditch and Nelson 1976; Bahcall and Chester 1977). However, the present observational bounds on the neutron star masses in other X-ray sources are very wide (Joss and Rappaport 1976; Avni 1978; Bahcall 1978). For a large range of values of the stellar mass within these bounds, the minimum theoretical value of $T_s$ for X Per is below the observed value, giving two solutions. An example is presented in Figure 13, which shows the theoretical curves $T_s(\mu)$ corresponding to both 0.7 $M_\odot$ and 1.3 $M_\odot$ neutron stars. The former gives two solutions. Third, we *assumed* that neutron star matter follows the TI equation of state of PPS. However, the range of equations of state given by current conventional theories of many-body nuclear physics (see Baym and Pethick 1975, 1979) is wide enough to bring the minimum theoretical value of $T_s$ well below the observed value for somewhat softer, but quite acceptable, equations of state, even for the values of luminosity and mass adopted here.

These remarks show that an acceptable fit is possible for all nine measured sources for magnetic moments in the range $3 \times 10^{29}$–$4 \times 10^{32}$ gauss cm$^3$. Such magnetic moments are consistent with our present meager knowledge of neutron star formation and evolution, and we therefore conclude that the observational data on these sources are *consistent* with their being disk-fed, although the secular change in period and the average luminosity are not by themselves sufficient to demonstrate that they are accreting from disks. If they *are* disk-fed, the possible value or values of $\mu$ for these sources are those given in Table 2. We note, however, that the inferred value of $\mu$ depends sensitively on the *mean* X-ray luminosity, averaged over the months or years involved in the secular spin-up time scale determination and that for many sources this important observational quantity is poorly known.

Furthermore, Table 2 shows that by adopting the acceptable solution for those sources in which only one solution is acceptable and then choosing one of the solutions for those sources in which two solutions are acceptable, one can find a set of solutions with magnetic moments all lying within the relatively narrow range $\sim 5 \times 10^{29}$–$5 \times 10^{30}$ gauss cm$^3$, that is consistent with the data on all the measured sources except Vela X-1. Where two are acceptable, the solution that corresponds to this set is underlined in Table 2. The fact that such a set of stellar magnetic moments can be found explains why the "universal" relation between the parameters $\dot{P}$ and $PL^{3/7}$ given in § IV, which assumes a single value of $\mu$ for all sources, describes the observational data so well.

Finally, we remark that the comparatively large dipole magnetic moments inferred for these accreting neutron stars (including Her X-1, which may be as old as $\sim 10^8$ years) argues against the universality of rapid magnetic field decay, a hypothesis that has been widely discussed as a possible explanation for some observed properties of pulsars (Gunn and Ostriker 1970; Lyne, Ritchings, and Smith 1975; Flowers and Ruderman 1977; Fujimura and Kennel 1979). The fact that the dipole moment inferred for Her X-1 is, nevertheless, substantially weaker than would be deduced by assuming that the stellar surface field inferred from apparent cyclotron features in the X-ray spectrum (Trümper *et al.* 1978) is dipolar in character, suggests that the surface field strength is largely in higher multipole moments (Lamb 1978). Elsner and Lamb (1976) have argued on different grounds that the surface magnetic fields of some X-ray stars are complex. The presence of higher multipole moments, which tend to decay faster than the dipole moment, would further constrain the magnetic field decay time scales in these sources.

These preliminary conclusions can be tested by a detailed comparison of the theoretical spin-up equation with sequences of accurate period and X-ray flux measurements, as discussed in § III.

## VI. THE ORIGIN OF PULSE PERIOD FLUCTUATIONS

The well-studied pulsating X-ray sources have all been shown to exhibit short-term fluctuations or irregularities in their pulsation periods in addition to a secular spin-up trend. One possible explanation of this phenomenon is that it is caused by fluctuations in the accretion torque (Lamb, Pines, and Shaham 1976, 1978*a*, *b*). Here we show that if the present model applies, then torque fluctuations of the required size can be caused by fluctuations in the mass accretion rate. Since fluctuations in the accretion rate cause similar fluctuations in the luminosity of the source, this possibility can be checked directly by observation.

In order to relate the fluctuations in $\dot{M}$ to period and luminosity fluctuations, one needs a relation between $\dot{M}$ and $L$, a relation between $\dot{M}$ and $N$, and a model for the response of the neutron star to $N$. If we assume disk accretion and restrict ourselves to fluctuations on time scales longer than the inflow time through the transition zone ($\lesssim 3$ s) and the free-fall time from the inner radius of the disk ($\lesssim 0.1$ s), the stationary flow solutions presented in Paper II apply and the torque $N$ as a function of $\dot{M}$ is given by equations (6) and (7). Thus, for a source of given $P$, $L$, $M$, and $\mu$, these two equations, plus the accretion luminosity equation $L = (GM/R)\dot{M}$ and the neutron star equation of state, relate the relative fluctuation in the torque, $\delta N/N$, to the relative fluctuation in the accretion luminosity, $\delta L/L$. At the same time, the stellar model relates the relative fluctuation in the spin change rate, $\delta\dot{P}/\dot{P}$, to the relative fluctuation in the torque, $\delta N/N$. Hence one has a relation between $\delta\dot{P}/\dot{P}$ and $\delta L/L$. In Table 3 we give the ratio of $\delta\dot{P}/\dot{P}$ to $\delta L/L$ for the nine pulsating X-ray sources listed in Table 2, based on the choice of parameters given there and the assumption that the star responds as a rigid body. The behavior of fast rotators differs greatly from that of slow rotators. For slow rotators, one has $\delta\dot{P}/\dot{P} \sim \delta L/L$, whereas for fast rotators $\delta\dot{P}/\dot{P}$ may be orders of magnitude larger than $\delta L/L$. As examples, the chosen parameters give $\delta\dot{P}/\dot{P} = 2.8\delta L/L$ for Cen X-3, whereas for Her X-1, whose angular velocity is very close to the critical value, fluctuations in the accretion luminosity correspond to fluctuations in $\dot{P}$ which are 70 times larger.

Adopting the statistical description of torque fluctuations developed by Lamb, Pines, and Shaham (1978*a*) and the two-component model of the stellar response (Baym *et al.* 1969), and assuming a crustal moment of inertia comparable to that of the core, the condition for spin-down episodes to occur from time to time is that the root-mean-square relative torque fluctuation $\langle(\delta N/N)^2\rangle^{1/2}$ should exceed $(T/T_1)(RT)^{-1/2}$, for type 1 torque fluctuations (which correspond to white noise in $N$), or $(RT)^{-1/2}$, for type 2 fluctuations (which correspond to white noise in the time derivative of the torque, $\dot{N}$). Here $R$ is the average rate of torque fluctuation events, $T_1$ is the duration of a type 1 event, and $T$ (typically $\sim 3 \times 10^7$ to $10^8$ s) is the length of the observing period.

In Her X-1, for example, the observational data show two spin-down episodes over a period $\sim 14$ months (Giacconi 1974). According to the above condition and the results in Table 3, type 1 fluctuations in $\dot{M}$ of relative size $\delta\dot{M}/\dot{M} \sim 1$ occurring about once every $10^3$ s and lasting $\sim 2 \times 10^3$ s would account for these episodes. Alternatively, type 2 fluctuations at the same rate but of relative size $\sim 10^{-4}$ would suffice. In Cen X-3, on the other hand, Table 3 indicates that substantially larger fluctuations in $\dot{M}$ are needed to

TABLE 3

RATIOS OF FLUCTUATIONS IN $\dot{P}$ TO FLUCTUATIONS IN LUMINOSITY PREDICTED BY THE PRESENT MODEL

| Source | $(\delta\dot{P}/\dot{P})/(\delta L/L)$ |
|---|---|
| SMC X-1 | 1.0 |
| Her X-1 | 70 |
| 4U 0115+63 | 3.2 |
| Cen X-3 | 2.8 |
| A0535+26 | 0.86 |
| 4U 1728−24 | 0.86 |
| 4U 0900−40 | 60 |
| 4U 1223−62 | 0.86 |
| 4U 0352+30 | 1.0 |

account for the spin-down episode observed in 1972 September-October (Fabbiano and Schreier 1977). There the observations lasted for a period $\sim 21$ months. Thus a type 1 fluctuation of relative size $\delta \dot{M}/\dot{M} \sim 1$ occurring once every $10^5$ s and lasting $\sim 8 \times 10^5$ s would suffice, as would type 2 fluctuations of relative size $\sim 10^{-2}$ occurring at the same rate. Of course, fluctuations on shorter and longer time scales are also possible. If the observed period variations *are* caused by fluctuations in $\dot{M}$, the accretion luminosity should display similar fluctuations, with periods of reduced luminosity associated with periods of slower spin-up or, if the fluctuation is large enough, spin-down. We note that Fabbiano and Schreier have reported a possible decrease of relative size $\delta L/L \sim 0.5$ in the X-ray luminosity of Cen X-3 during the spin-down episode of 1972 September-October.

## VII. THE NATURE OF THE LONG-PERIOD SOURCES

A large fraction of the known pulsating X-ray sources have long periods ($\geq 10^2$ s) but are observed to be spinning up on time scales ($\sim$ 50–100 years) which are extremely short compared to their expected lifetimes as bright X-ray sources ($\sim 10^3$–$10^6$ years; compare van den Heuvel 1977; Amnuel and Guseinov 1976; Ziółkowski 1977; and Savonije 1978). Examples include A0535+26, GX 1+4, and GX 301-2. This apparent paradox has been noted and discussed previously by numerous authors (see Illarionov and Sunyaev 1975; Lamb, Lamb, and Arnett 1975; Fabian 1975; Wickramasinghe and Whelan 1975; Kundt 1976; Lea 1976; Lipunov and Shakura 1976; Savonije and van den Heuvel 1977; Paper I; Holloway, Kundt and Wang 1978; Davies, Fabian, and Pringle 1979; Wang 1979). Almost all of these discussions have focused on mechanisms for producing a long period at the onset of accretion, when X-ray emission first begins. However, given the very short observed time scales noted above and the significant fraction of potential X-ray binaries which are actually emitting X-rays (see van den Heuvel 1977; Ziółkowski 1977), it appears that the slow rotation of many pulsating X-ray sources must somehow be maintained over evolutionary time scales, even though they are acted upon by strong spin-up torques. If so, the principal issue is how the sources *maintain* their long periods, rather than how these are produced initially; indeed, the solution of the former problem may partially solve the latter problem as well. As discussed in § IV, the existence of a universal relation between $\dot{P}$ and $PL^{3/7}$ which adequately reproduces almost all the available observational data suggests that most of the observed sources are accreting from disks. Thus, whatever the mechanism by which slow rotation is maintained, it should probably function for accreting sources with orbital rather than radial inflow.

The braking torque on fast rotators found in the present calculation suggests one way that slow spin rates can be maintained (Ghosh and Lamb 1978*a*). Thus, if a pulsating X-ray source has recurrent "low" states during which the accretion rate is much reduced, it would be subjected to strong, recurrent spin-down torques. Although the published data on the luminosity behavior of the long-period sources is extremely scant at present, there is some evidence for variations by factors $\sim 30$ or more (S. Holt, private communication). Depending on the luminosities in alternating high and low states and the relative durations of such states, the spin-down that occurs during the low states can, on average, equal or even exceed the spin-up that occurs during the high states. If the spin-down on average *equals* the spin-up, a neutron star born with a relatively long spin period or spun down to that period just after birth by electromagnetic and particle emission prior to the start of accretion will be maintained with a long spin period during much of its life as an accreting X-ray source. If the spin-down actually *exceeds* the spin-up, on average, even a neutron star born with a relatively short rotation period will, during the course of its life as an X-ray source, be spun down to a long period. In either case, the source would be much more readily observed during high states, when it is spinning up, than during low states, when it is spinning down. This would account for the fact that all nine sources with measured $\dot{P}$'s are usually observed to be spinning up.

The mechanism for maintaining slow spin rates suggested here would produce cyclic variations in the spin period of a given source together with a slow drift in the mean period on a time scale related to the time scale of the binary evolution. The lengths of the cycles in $P(t)$ would be determined by the durations of successive high and low states. Although in a certain sense this proposed mechanism simply transforms the problem of understanding the persistence of long spin periods into a problem of understanding the pattern in time of mass transfer in X-ray binaries, it can be directly tested independently of the development of the latter understanding. Thus, on a plot of observed luminosity versus spin period, such a source should trace out a nearly closed curve whose center drifts on an evolutionary time scale (Ghosh and Lamb 1978*b*). Theoretical evolutionary paths of this type which are suitable for comparison with the behavior of observed sources have been computed and will be published elsewhere (Elsner, Ghosh, and Lamb 1979). Some evidence for this picture is provided by Vela X-1, which shows variations by a factor $\gtrsim 30$ in luminosity together with alternating episodes of strong spin-up and spin-down (see Fig. 7), although as noted in § IV, this particular source may not be disk-fed.

In closing this discussion, we remark that the spin-down torque during low states may involve both the braking torque on fast rotators found here and the braking torque on very fast rotators due to mass ejection which has been suggested by other authors (Davidson and Ostriker 1973; Illarionov and Sunyaev 1975; Shakura 1975), if the mass accretion rate becomes sufficiently low. Our braking torque, which operates for values of $\omega_s$ between the critical fastness, $\omega_c$, and the maximum allowable fastness for steady accretion, $\omega_{max} \approx 1$, and involves no mass ejection from the vicinity of the magnetosphere, would act at luminosities between those corresponding to these two values of the fastness, whereas the braking torque suggested by the above authors, which is assumed to operate for $\omega_s > \omega_{max}$ and to involve mass ejection from the vicinity of the magnetosphere, would act at luminosities below that corresponding to $\omega_s = \omega_{max}$.

## VIII. CONCLUDING REMARKS

In Paper I we showed that no model of axisymmetric stationary disk accretion by magnetic stars which involves a thin transition zone between the disk and the magnetosphere can produce an accretion torque significantly less than $\dot{M}(GMr_0)^{1/2}$ and hence that such models cannot account for the much smaller accretion torque acting on Her X-1. In Paper II we showed that in fact the transition zone in disk accretion is broad, owing to invasion of the disk plasma by the stellar magnetic field via the Kelvin-Helmholtz instability, turbulent diffusion, and magnetic flux reconnection. In the present paper we have used the solutions to the two-dimensional hydromagnetic equations obtained in Paper II to calculate the torque on the star. We find (1) that the magnetic coupling between the star and the plasma in the outer transition zone is appreciable; (2) that as a result of this coupling the spin-up torque on fast rotators is substantially less than that on slow rotators; and (3) that for sufficiently high stellar angular velocities or sufficiently low mass accretion rates the rotation of the star can be braked while accretion continues.

Applying these results to pulsating X-ray sources, we have shown that a star of given spin period rotating in the same direction as the disk can experience either spin-up or spin-down, depending on its luminosity. At high luminosities the star is a "slow rotator" and experiences a strong spin-up torque whereas at low luminosities it becomes a "fast rotator" and can experience a strong spin-down torque. If the accretion luminosity remains approximately constant, the spin period will eventually approach and then maintain the critical spin period at which the accretion torque vanishes.

We have discussed the general problem of interpreting period changes in pulsating X-ray sources and have described how one can disentangle the effect of the accretion torque from the effects of possible internal torques and the dynamical response of the neutron star utilizing a dense sequence of pulse period and X-ray flux measurements of each source. We have shown that the model of disk accretion developed in Paper II gives an equation for the accretion torque that can potentially be used to (1) establish the accretion flow pattern, (2) determine the properties of the X-ray star, and (3) establish the cause of pulse period fluctuations.

We have also considered the limited evidence regarding accretion flow patterns that is provided by the period and flux measurements that are currently available. Assuming that the effective inertial moments of the X-ray sources are roughly constant on the time scales of interest, disk-fed sources should show a particular correlation between $-\dot{P}$ and the quantity $PL^{3/7}$, and such a correlation is in fact indicated by the data. On the other hand wind-fed sources should, given certain assumptions, show a linear correlation between $PLT_s$ and $P_{orb}$, the orbital period of the binary system. The current observational data do not show such a correlation, although the observational uncertainties are large and may mask it at present. The striking agreement of the observational data with the theoretical spin-up equation for disk accretion and the apparent absence of any correlation of $PLT_s$ with $P_{orb}$ suggests that most of the currently measured sources are disk-fed. The current evidence is, however, far from conclusive, a situation which underlines the importance of obtaining a regular sequence of accurate period and flux measurements for each source.

Assuming that a given source is disk-fed, its average luminosity and spin-up rate can be used to estimate its dipole magnetic moment. In most cases a single value of the magnetic moment is indicated, but in a few cases two solutions are possible. Acceptable solutions are possible for all the sources whose spin-up rates have so far been measured, including Her X-1, and yield magnetic moments in the range $3 \times 10^{29}$–$4 \times 10^{32}$ gauss cm$^3$. Solutions that correspond to magnetic moments in the much narrower range $5 \times 10^{29}$–$5 \times 10^{30}$ gauss cm$^3$ can be found for all sources except Vela X-1. The much larger magnetic moment required to fit Vela X-1 may indicate that this source is not disk-fed. The substantial values of the inferred magnetic moments argue against rapid, universal magnetic field decay in neutron stars. These preliminary conclusions can be tested by comparing the theoretical spin-up equation with regular sequences of accurate period and X-ray flux measurements.

The short-term period fluctuations and spin-down episodes observed in Her X-1, Cen X-3, Vela X-1, and X Per follow naturally from the present model of disk accretion as consequences of fluctuations in the mass accretion rate. In particular, the model predicts a braking torque sufficient to account for the observed spin-down episodes without ejection of mass from the vicinity of the neutron star, if there is a reduction in the accretion rate. Fast rotators differ dramatically from slow rotators in the size of the relative $\dot{P}$ fluctuation, $\delta\dot{P}/\dot{P}$, that accompanies a given relative fluctuation $\delta L/L$ in the accretion luminosity. For slow rotators one has $\delta\dot{P}/\dot{P} \sim \delta L/L$ whereas for fast rotators $\delta\dot{P}/\dot{P}$ may be many orders of magnitude larger than $\delta L/L$. Again, these predictions can be checked by comparison with a dense sequence of period and X-ray flux measurements.

Finally, we have pointed out that an understanding of the statistics of the long period pulsating sources requires not only mechanisms leading to slow rotation at the onset of accretion, when X-ray emission first begins, but also a means for maintaining long periods in the face of strong spin-up torques. We have noted that the braking torque at low accretion rates found in the present calculations could provide an explanation, if these sources have recurrent periods of low luminosity. The evolutionary tracks in the period-luminosity plane that are predicted by this hypothesis can be tested directly by comparison with a sequence of period and luminosity measurements.

The results presented here point clearly to the critical need for a regular sequence of accurate period and X-ray flux measurements of a number of sources. At a minimum, such sequences would make possible a better estimate of the mean luminosity of these sources and hence allow a more accurate comparison with the secular period changes predicted by theory. Far more interestingly, a regular sequence of measurements of each source would allow a determination of the dynamical response of the neutron star and a direct comparison with the quantitative theory of the accretion torque presented here. Such a comparison would provide a check on the theory while at the same time establishing which of these sources are accreting from disks and which are not. Assuming that the theory is confirmed, the observations could be used to accurately determine the dipole magnetic field of the neutron star and to probe the nature of short-term fluctuations in the accretion flow.

It is a pleasure to thank Professors P. Boynton, D. Q. Lamb, C. J. Pethick, and D. Pines for stimulating discussions. One of us (F. K. L.) thanks Professor G. Garmire and the Department of Physics at Caltech for their warm hospitality.

## REFERENCES


Abbott, D. C. 1978, *Ap. J.*, **225**, 893.
Amnuel, P. R., and Guseinov, O. H. 1976, *Ap. Space Sci.*, **45**, 283.
Arons, J., and Lea, S. 1976, *Ap. J.*, **207**, 914.
Avni, Y. 1978, talk presented at 16th General Assembly of the IAU, in *Highlights of Astronomy*, Vol. **4**, in press.
Bahcall, J. N. 1978, *Ann. Rev. Astr. Ap.*, **16**, 241.
Bahcall, J. N., and Chester, T. J. 1977, *Ap. J. (Letters)*, **215**, L21.
Baym, G., and Pethick, C. J. 1975, *Ann. Rev. Nucl. Sci.*, **25**, 27.
———. 1979, *Ann. Rev. Astr. Ap.*, **17**, in press.
Baym, G., Pethick, C. J., Pines, D., and Ruderman, M. 1969, *Nature*, **224**, 872.
Becker, R. H., Rothschild, R. E., Boldt, E. A., Holt, S. S., Pravdo, S. H., Serlemitsos, P. J., and Swank, J. H. 1978, *Ap. J.*, **179**, 585.
Boynton, P. E., and Deeter, J. E. 1979, in preparation.
Bradt, H., Doxsey, R., and Jernigan, G. 1979, preprint.
Castor, J. L., Abbott, D. C., and Klein, R. I. 1975, *Ap. J.*, **195**, 157.
Charles, P. A., Mason, K. O., White, N. E., Culhane, J. L., Sanford, P. W., and Moffat, A. F. J. 1978, *M.N.R.A.S.*, **183**, 813.
Cominsky, L., Clark, G. W., Li, F., Mayer, W., and Rappaport, S. 1978, *Nature*, **273**, 367.
Conti, P. S. 1978, *Ann. Rev. Astr. Ap.*, **16**, 371.
Cowley, A. P. 1977, in *Proc. 8th Texas Symposium on Relativistic. Astrophysics* (*Ann. NY Acad. Sci.*, **302**, 1).
Culhane, J. L., Mason, K. O., Sanford, P. W., and White, N. E. 1976, in *X-Ray Binaries, Proceedings of the Goddard Conference*, ed. E. Boldt and Y. Kondo (NASA SP-389), p. 1.
Davidson, K., and Ostriker, J. P. 1973, *Ap. J.*, **179**, 585.
Davies, R., Fabian, A., and Pringle, J. 1979, *M.N.R.A.S.*, **186**, 779.
Elsner, R. F., Ghosh, P., and Lamb, F. K. 1979, in preparation.
Elsner, R. F., and Lamb, F. K. 1976, *Nature*, **262**, 356.
———. 1977, *Ap. J.*, **215**, 897.
Fabbiano, G., and Schreier, E. J. 1977, *Ap. J.*, **214**, 235.
Fabian, A. C. 1975, *M.N.R.A.S.*, **173**, 161.
Flowers, E., and Ruderman, M. A. 1977, *Ap. J.*, **215**, 302.
Fujimura, F. S., and Kennel, C. F. 1979, preprint.
Ghosh, P., and Lamb, F. K. 1978*a*, *Ap. J. (Letters)*, **223**, L83.
———. 1978*b*, *Bull. AAS*, **10**, 507.
———. 1979, *Ap. J.*, **232**, 259 (Paper II).
Ghosh, P., Lamb, F. K., and Pethick, C. J. 1977, *Ap. J.*, **217**, 578 (Paper I).
Giacconi, R. 1974, in *Astrophysics and Gravitation, Proceedings of the 16th International Solvay Conference* (Brussels: Université de Bruxelles), p. 27.
Gunn, J. E., and Ostriker, J. P. 1970, *Ap. J.*, **160**, 979.
Holloway, N., Kundt, W., and Wang, Y.-M. 1978, *Astr. Ap.*, **70**, L23.
Ichimaru, S. 1978, *Ap. J.*, **224**, 198.
Illarionov, A. F., and Sunyaev, R. A. 1975, *Astr. Ap.*, **39**, 185.
Jernigan, J. G., and Nugent, J. J. 1979, in preparation.
Johnston, M., Bradt, H., Doxsey, R., Gursky, H., Schwartz, D., and Schwartz, J. 1978, *Ap. J. (Letters)*, **223**, L71.
Joss, P. C., and Rappaport, S. 1976, *Nature*, **264**, 219.
Kundt, W. 1976, *Phys. Letters*, **57A**, 195.
Lamb, D. Q., Lamb, F. K., and Arnett, W. D. 1975, *Bull. AAS*, **7**, 545.
Lamb, F. K. 1977, in *Proc. 8th Texas Symposium on Relativistic Astrophysics* (*Ann. NY Acad. Sci.*, **302**, 482).
———. 1978, review presented at the International Conference on Cyclotron Lines in Accreting Neutron Stars, Munich, 1978 May.
Lamb, F. K., Pethick, C. J., and Pines, D. 1973, *Ap. J.*, **184**, 271.
Lamb, F. K., Pines, D., and Shaham, J. 1976, in *X-Ray Binaries, Proceedings of the Goddard Conference*, ed. E. Boldt and Y. Kondo (NASA SP-389), p. 141.
———. 1978*a*, *Ap. J.*, **224**, 969.
———. 1978*b*, *Ap. J.*, **225**, 582.
Lea, S. M. 1976, *Ap. J. (Letters)*, **209**, L69.
Lipunov, V. M., and Shakura, N. I. 1976, *Soviet Astr. Letters*, **2**, 343.
Lyne, A. G., Ritchings, R. T., and Smith, F. G. 1975, *M.N.R.A.S.*, **171**, 579.
Mason, K. O. 1977, *M.N.R.A.S.*, **178**, 81 P.
Middleditch, J., and Nelson, J. 1976, *Ap. J.*, **208**, 567.
Ögelman, H., Beuermann, K. P., Kanbach, G., Mayer-Hasselwander, H. A., Capozzi, D., Fiordilino, E., and Molteni, D. 1977, *Astr. Ap.*, **58**, 385.
Pandharipande, V. R., Pines, D., and Smith, R. A. 1976, *Ap. J.*, **208**, 550 (PPS).
Petterson, J. A. 1978, *Ap. J.*, **224**, 625.
Pringle, J. E., and Rees, M. J. 1972, *Astr. Ap.*, **21**, 1.
Rappaport, S., Clark, G. W., Cominsky, L., Joss, P. C., and Li, F. 1978, *Ap. J. (Letters)*, **224**, L1.
Rappaport, S., and Joss, P. C. 1977, *Nature*, **266**, 683.
Savonije, G. J. 1978, *Astr. Ap.*, **62**, 317.
Savonije, G. J., and van den Heuvel, E. P. J. 1977, *Ap. J. (Letters)*, **214**, L19.
Scharlemann, E. 1978, *Ap. J.*, **219**, 617.
Schreier, E. 1977, in *Proc. 8th Texas Symposium on Relativistic Astrophysics* (*Ann. NY Acad. Sci.*, **302**, 445).
Shakura, N. I. 1975, *Pis'ma Astr. Zh.*, **1**, 23 (English transl. in *Soviet Astr. Letters*, **1**, 223).
Shapiro, S. L., and Lightman, A. P. 1976, *Ap. J.*, **204**, 555.
Trümper, J., Pietsch, W., Reppin, C., Voges, W., Staubert, R., and Kendziorra, E. 1978, *Ap. J. (Letters)*, **219**, L105.
van den Heuvel, E. P. J. 1977, in *Proc. 8th Texas Symposium on Relativistic Astrophysics* (*Ann. NY Acad. Sci.*, **302**, 14).
Vitello, P. 1979, in *X-Ray Astronomy, Proc. 21st COSPAR Symposium*, ed. W. A. Baity and L. E. Peterson (Oxford: Pergamon), in press.
Wang, Y.-M. 1979, preprint.

White, N. E., Mason, K. O., and Sanford, P. W. 1977, *Nature*, **267**, 229.
White, N. E., Mason, K. O., Sanford, P. W., and Murdin, P. 1975, *M.N.R.A.S.*, **176**, 201.
Wickramasinghe, D. T., and Whelan, J. A. J. 1975, *Nature*, **258**, 502.
Ziółkowski, J. 1977, in *Proc. 8th Texas Symposium on Relativistic Astrophysics* (*Ann. NY Acad. Sci.*, **302**, 47).

P. GHOSH: Code ES62, George C. Marshall Space Flight Center, Marshall Space Flight Center, AL 35812
F. K. LAMB: Department of Physics, University of Illinois at Urbana-Champaign, Urbana, IL 61801

# Chapter 5 LIST OF REFERENCES

**References** *

Abbott D.C., 1978, "The Terminal Velocity of Stellar Winds from Early-Type Stars", *Astrophys. J.*, **225**, 893. [29].

Abramowicz M.A. and Lasota J.P., 1980, "Spin-up Black Holes by Thick Accretion Disks", *Acta Astr.*, **30**, 35. [27]

Abramowicz M.A. and Marsi C., 1987, "Accretion Research: How It Started", *Observatory*, **107**, 245. [1]

Abramowicz M.A. and Piran T., 1980, "On Collimation of Relativistic Jets from Quasars", *Astrophys. J. Lett.*, **241**, L7. [24, 27]

Abramowicz M.A., Calvani M. and Nobili L., 1980, "Thick Accretion Disks with Super-Eddington Luminosities", *Astrophys. J.*, **242**, 772. [1, 27]

Abramowicz M.A., Jaroszynski M. and Sikora M., 1978, "Relativistic, Accreting Disks", *Astron. Astrophys.*, **63**, 221. [24, 25, 27]

Abramowicz M.A., Czerny B., Lasota J.P. and Szuszkiewicz E., 1988, "Slim Accretion Disks", *Astrophys. J.*, **332**, 646. [1]

Adams W.S. and Joy A.H., 1922, *Pop. Astr.*, **30**, 103. [6]

Agrawal P.C., Biswas S., Gokhale G.S., Iyengar V.S., Kunte P..K., Manchanda R.K. and Sreekantan B.V., 1970, "Energy Spectra of Several Discrete X-Ray Sources in the 20-120 keV Range", in *Non-solar X- and $\gamma$-Ray Astronomy, I.A.U. Symposium No. 37*, ed. L.Gratton (Dordrecht: Reidel). [18]

Aizu K., 1973, "X-Ray Emission Region of a White Dwarf with Accretion", *Progr. Theor. Phys.*, **49**, 1184. [1, 18, 21]

Allen C.W., 1973, *Astrophysical Quantities*, 3rd edition, Athlone Press, London. [21]

Alme M.L. and Wilson J.R., 1973, "X-Ray Emission from a Neutron Star Accreting Material", *Astrophys. J.*, **186**, 1015. [1, 16, 18]

Ambartsumjan V.A. and Saakyan G.S., 1960, "The Degenerate Superdense Gas of Elementary Particle", *Astron. Zh.*, **37**, 193 [*Sov. Astron.-AJ*, **4**, 187]. [9]

* This list contains references reported in the 29 papers of this collection. In parenthesis we give the number of the paper which refers to the article.

Ambartsumjan V.A. and Saakyan G.S., 1963, *Voprosy Kosmog.*, **9**, 91. [9]

Amnuehl' P.R. and Gusejnov O.Kh., 1968, *Izv. Akad. Nauk Azerbaidzhan SSSR, Ser. Fiz. Tekh. Mat.*, **3**, 70. [12]

Amnuehl' P.R. and Gusejnov O.Kh., 1976, "On the Lifetime of Bright X-Ray Sources", *Astroph. Space Sci.*, **45**, 283. [29]

Angel J.R.P, 1969, "X-Ray Line Emission from Sco X-1", *Nature*, **224**, 160. [13]

Angel J.R.P and Stockman H.S., 1980, "Optical and Infrared Polarization of Active Extragalactic Objects", *Ann. Rev. A. A.*, **18**, 321. [27]

Arnett W.D., 1966, "Gravitational Collapse and Weak Interactions", *Canad. J. Phys.*, **44**, 2553. [9]

Arnett W.D., 1967, "Mass Dependence in Gravitational Collapse of Stellar Cores", *Canad. J. Phys.*, **45**, 1621. [9]

Arons J. and Lea S.M., 1976, "Accretion onto Magnetized Neutron Stars: Structure and Interchange Instability of a Model Magnetosphere", *Astrophys. J.*, **207**, 914. [21, 29]

Arp H.C., 1961, "U–B and B–V Colors of Black Bodies", *Astrophys. J.*, **133**, 874. [6]

Avni Y., 1978, "Masses of Compact X-Ray Sources", talk presented at 16th General Assembly of the I.A.U., in *Highlights of Astronomy*, Vol. 4. [29]

Baan W.A. and Treves A., 1973, "On the Pulsation of X–Ray Sources", *Astron. Astrophys.*, **22**, 421. [1]

Bahcall J.N. and Wolf R.A., 1977, "The Star Distribution around a Massive Black Hole in a Globular Cluster. II. Unequal Star Masses", *Astrophys. J.*, **216**, 883. [28]

Bahcall J.N., 1978, "Masses of Neutron Stars and Black Holes in X-Ray Binaries", *Ann. Rev. A. A.*, **16**, 241. [29]

Bahcall J.N. and Chester T.J., 1977, "On the Mass Determination of Hercules X-1", *Astrophys. J. Lett.*, **215**, L21. [29]

Bardeen J.M., 1970, "Kerr Metric Black Holes", *Nature*, **226**, 64. [13]

Bardeen J.M., 1973, "Timelike and Null Geodesic in the Kerr Metric", in *Black Holes*, Les Houches, ed. C. DeWitt and B. DeWitt (New York: Gordon and Breach). [26]

Bardeen J.M. and Petterson J.A., 1975, "The Lense-Thirring Effect and Accretion Disks around Kerr Black Holes", *Astrophys. J. Lett.*, **195**, L65. [27]

Bardeen J.M., Press W.H. and Teukolsky S.A., 1972, "Rotating Black Holes: Locally Nonrotating Frames, Energy Extraction, and Scalar Synchrotron Radiation", *Astrophys. J.*, **178**, 347. [15]

Barr P., Pollard G., Sanford P.W., Ives J.C., Ward M., Hine R.G., Longaire M.S., Penston M.V., Boksenberg A., Lloyd C., 1980, "The Variability of 3C 390.3", *Mon. Not. R. astr. Soc.*, **193**, 549. [27]

Barr P., 1986, "Rapid X-Ray Variability in Radio-Quiet AGN: A Probe of the Innermost Regions of the Active Nucleus", in *The Physics of Accretion onto Compact Objects*, ed. K.O. Mason, M.G. Watson and N.E. White (Berlin: Springer). [1]

Basko M.A. and Sunyaev R.A., 1973, "Interaction of the X-Ray Source Radiation with the Atmosphere of the Normal Star in Close Binary Systems", *Astroph. Space Sci.*, **23**, 71. [14]

Bath G.T., 1969, "Dynamical Instabilities in Semidetached Close Binary System with Possible Applications to Novae and Novalike Variables", *Astrophys. J.*, **158**, 571. [11]

Bath G.T., 1972, "Time-dependent Studies of Dynamical Instabilities in Semidetached Binary Systems", *Astrophys. J.*, **173**, 121. [11]

Bath G.T., 1973, "Periodicities and Disks in Dwarf Novae", *Nature Phys. Sci.*, **246**, 84. [11]

Bath G.T., 1975, "An Oblique Rotator Model for HD Herculis", *Mon. Not. R. astr. Soc.*, **171**, 311. [22]

Bath G.T. and Pringle J.E., 1981, "The Evolution of Viscous Discs. I. Mass Transfer Variations", *Mon. Not. R. astr. Soc.*, **194**, 967. [1, 22]

Bath G.T. and Pringle J.E., 1982, "The Evolution of Viscous Discs. II. Viscous Variations", *Mon. Not. R. astr. Soc.*, **199**, 267. [1]

Bath G.T., Evans W.D. and Pringle J.E., 1974, "Dynamical Instabilities and Mass Exchange in Binary Systems", *Mon. Not. R. astr. Soc.*, **166**, 113. [21]

Bath G.T., Evans W.D., Papaloizou J. and Pringle J.E., 1974, "The Accretion Model of Dwarf Novae with Application to Z Chamaleontis", *Mon. Not. R. astr. Soc.*,

**169**, 447. [21]

Baym G. and Pethick C.J., 1975, "Neutron Stars", *Ann. Rev. Nucl. Sci.*, **25**, 27. [29]

Baym G. and Pethick C.J., 1979, "Physics of Neutron Stars", *Ann. Rev. A. A.*, **17**, 415. [29]

Baym G., Pethick C.J., Pines D. and Ruderman M., 1969, "Spin Up in Neutron Stars: the Future in the Vela Pulsar", *Nature*, **224**, 872. [29]

Becker R.H., Rothschild R.E., Boldt E.A., Holt S.S., Pravdo S.H., Serlemitsos P.J. and Swank J.H., 1978, "Extended Observations of Vela X-1 by *OSO*", *Astrophys. J.*, **221**, 912. [29]

Begelman M.C., 1977, "Nearly Collisionless Spherical Accretion", *Mon. Not. R. astr. Soc.*, **181**, 347. [28]

Begelman M.C., 1978, "Black Holes in Radiation–dominated Gas: an Analogue of the Bondi Accretion Problem", *Mon. Not. R. astr. Soc.*, **184**, 53. [1]

Begelman M.C., 1979, "Can a Spherically Accreting Black Hole Radiate very near the Eddington Limit ?", *Mon. Not. R. astr. Soc.*, **187**, 237. [1]

Begelman M.C. and Chuieh T., 1988, "Thermal Coupling of Ions and Electrons by Collective Effects in Two–Temperature Accretion Flows", *Astrophys. J.*, **332**, 872. [1]

Begelman M.C., Blandford R.D. and Rees M.J., 1984, "Theory of Extragalactic Radio Sources", *Rev. Mod. Phys.*, **56**, 255. [1]

Bekefi G., 1966, *Radiation Processes in Plasmas* (New York: Wiley). [20]

Bisnovatyi-Kogan G.S., 1966, "The Critical Mass of a Hot Isothermal White Dwarf with General Relativistic Effects Taken into Account", *Astron. Zh.*, **43**, 89 [*Sov. Astron.-AJ*, **10**, 69 (1966)]. [9]

Bisnovatyi-Kogan G.S., 1979, "Magnetohydrodynamical Processes near Compact Objects", *Riv. Nuovo Cimento*, **2**, 1. [1]

Bisnovatyi-Kogan G.S. and Blinnikov S.I., 1977, "Disk Accretion onto a Black Hole at Subcritical Luminosity", *Astron. Astrophys.*, **59**, 111. [24, 25]

Bisnovatyi-Kogan G.S. and Fridman A.M., 1969, "A Mechanism for Emission of X–Rays by a Neutron Star", *Astron. Zh.*, **46**, 721 [*Sov. Astron.-AJ*, **13**, 566 (1970)]. [12, 16]

Bisnovatyi-Kogan G.S. and Sunyaev R.A., 1971, "Galaxy Nuclei and Quasars as Infrared Emission Sources", *Astron. Zh.*, **48**, 881 [*Sov. Astron.-AJ*, **15**, 697 (1972)]. [14, 16]

Blaes O. M., 1987, "Stabilization of Non–axisymmetric Instabilities in a Rotating Flow by Accretion on to a Central Black Hole", *Mon. Not. R. astr. Soc.*, **227**, 975. [1]

Blandford R.D., 1976, "Accretion Disk Electrodynamics. A Model for Double Radio Sources", *Mon. Not. R. astr. Soc.*, **176**, 465. [27]

Blandford R.D. and Ostriker J.P., 1978, "Particle Acceleration by Astrophysical Shocks", *Astrophys. J. Lett.*, **221**, L29. [28]

Blandford R.D. and Payne D., "Compton Scattering in a Converging Fluid Flow. I. The Transfer Equation", *Mon. Not. R. astr. Soc.*, **194**, 1033. [27]

Blandford R.D. and Payne D., "Compton Scattering in a Converging Fluid Flow. III. Spherical Supercritical Accretion", *Mon. Not. R. astr. Soc.*, **196**, 781. [27]

Blandford R.D. and Thorne K.S., 1979, "Black Hole Astrophysics", in *General Relativity. An Einstein Centenary Survey*, ed. S.W. Hawking and W. Israel (Cambridge: Cambridge University Press). [24]

Blandford R.D. and Znajek R., 1977, "Electromagnetic Extraction of Energy from Kerr Black Holes", *Mon. Not. R. astr. Soc.*, **179**, 433. [27]

Blumenthal G.R., 1974, "The Poynting–Robertson Effect and Eddington Limit for Electron Scattering with Hard Photons", *Astrophys. J.*, **188**, 121. [18]

Blumenthal G.R. and Gould R.J., 1970, "Bremsstrahlung, Synchrotron Radiation , and Compton Scattering of High–Energy Electrons Traversing Dilute Gases", *Rev. Mod. Phys.*, **42**, 237. [18]

Bohm D., 1949, *The Characteristic of Electrical Discharges in Magnetic Fields*, ed. A. Guthrie and R.K. Wakerling (New York: McGraw–Hill). [16]

Bolton C.T., 1972, "Identification of Cygnus X-1 with HDE 226868", *Nature*, **235**, 271. [13]

Bondi H., 1952, "On Spherical Symmetrical Accretion", *Mon. Not. R. astr. Soc.*, **112**, 195. [1, 4, 12, 13, 17, 24]

Bondi H., 1960, *Cosmology* (Cambridge: Cambridge University Press). [24]

Bondi H. and Hoyle F., 1944, "On the Mechanism of Accretion By Stars", *Mon. Not. R. astr. Soc.*, **104**, 273. [3, 4, 12, 16]

Boynton P.E., Groth E.J., Hutchinson D.P., Nanos G.P., Partridge R.B. and Wilkinson D.T., 1972, "Optical Timing of the Crab Pulsar NP 0532", *Astrophys. J.*, **175**, 217. [16]

Bradt H.V. and McClintock J.E., 1983, "The Optical Counterparts of Compact Galactic X-Ray Sources", *Ann. Rev. A. A.*, **21**, 13. [1]

Brandt J.C. and Rosen R.G., 1969, "Messier 87: The Galaxy of Greatest Known Mass", *Astrophys. J. Lett.*, **156**, L59. [10]

Bregman J., Butler D., Kemper E., Koski A., Kraft R. and Stone R.P.S., 1973, "On the Distance to Cygnus X-1 (HDE 226868)", *Astrophys. J. Lett.*, **185**, L117. [26]

Brown R.L. and Gould R.J., 1970, "Interstellar Absorption of Cosmic X–Ray", *Phys Rev. D.*, **1**, 2252. [21]

Buff J.M. and McCray R., 1974, "Accretion Flows in Galactic X-Ray Sources. I. Optically Thin Spherically Symmetric Model", *Astrophys. J.*, **189**, 147. [21]

Burbridge E.M., 1967, "Theoretical Ideas Concerning X–Ray Sources", in *Radio Astronomy and Galactic System, I.A.U. Symposium No. 31*, (London and New York: Academic Press). [8]

Burbridge E.M., Lynds C.R. and Stockton A.N., 1967, "On the Binary Nature of Cyg X-2", *Astrophys. J. Lett.*, **150**, L95. [8, 12]

Burger H.L. and Katz J.I., 1983, "The Eddington Limit and Supercritical Accretion. II. Time–dependent Calculations", *Astrophys. J.*, **265**, 393. [1]

Byram E.T., Chubb T.A. and Freidman H., 1966, "Cosmic X-Ray Sources, Galactic and Extragalactic", *Science*, **152**, 66. [10]

Callahan P.S., 1977, "On the Accretion Disk Model of QSOs", *Astron. Astrophys.*, **59**, 127. [25]

Cameron A.G.W. and Mock M., 1967, "Stellar Accretion and X–Ray Emission", *Nature*, **215**, 464. [8, 9, 12]

Canuto V. and Chiu H.Y., 1971, "Intense Magnetic Fields in Astrophysics", *Space Sci. Rev.*, **12**, 3. [16]

Carter B., Gibbons G.W., Lin D.N.C. and Perry M.J., 1976, "Black Hole Emission Process in the High Energy Limit", *Astron. Astrophys.*, **52**, 427. [24]

Cassinelli J.P. and Castor J.I., 1973, "Optically Thin Stellar Winds in Early-Type Stars", *Astrophys. J.*, **179**, 189. [18]

Castor J.L., Abbott D.C. and Klein R.I., 1975, "Radiation-driven Winds in Of Stars", *Astrophys. J.*, **195**, 157. [29]

Chan K.L. and Henriksen R.N., 1980, "On the Supersonic Dynamics of Magnetized Jets of Thermal Gas in Radio Galaxies", *Astrophys. J. Lett.*, **241**, 534. [27]

Chandrasekhar S., 1950, *Radiative Transfer* (Oxford: Oxford University Press). [26]

Chau Wai-Yin, 1967, "Gravitational Radiation from Neutron Stars", *Astrophys. J.*, **147**, 664. [9]

Charles P.A., Mason K.O., White N.E., Culhane J.L., Sanford P.W. and Moffat A.F.J., 1978, "X-Ray and Optical Observations of 3U0090−40 (Vela X-1)", *Mon. Not. R. astr. Soc.*, **183**, 813. [29]

Chiu H.Y., 1964, "Supernovae, Neutrinos and Neutron Stars", *Ann. Phys.*, **26**, 364. [9]

Cline Y.L., Desai U.D., Klebesadel R.W. and Strong I.B., 1973, "Energy Spectra of Cosmic Gamma-Ray Bursts", *Astrophys. J. Lett.*, **185**, L1. [18]

Cohen M.H. and Readhead A.C.S., 1979, "Misalignment in the Radio Jets of NGC 6251", *Astrophys. J. Lett.*, **233**, L101. [27]

Colpi M., 1988, "Multiple Compton Scattering by Thermal Electrons in a Spherical Inflow: the Effects of Bulk Motion", *Astrophys. J.*, **326**, 223. [1]

Colpi M., Maraschi L. and Treves A., 1984, "Two-Temperature Model of Spherical Accretion onto a Black Hole", *Astrophys. J.*, **280**, 319. [1]

Colpi M., Maraschi L. and Treves A., 1986, "Gamma-Ray Emission from Accretion onto a Rotating Black Hole", *Astrophys. J.*, **311**, 150. [1]

Cominski L., Clark G.W., Li F., Mayer W. and Rappaport S., 1978, "Discovery of 3.6-s X-Ray Pulsations from 4U 0115+63", *Nature*, **273**, 367. [29]

Condon J.J., Ladder J.E., O'Dell S.L. and Dennison B., 1979, "318-Mhz Variability of Complete Samples of Extragalactic Radio Sources", *Astron. J.*, **84**, 1. [27]

Condon J.J., Condon M.A., Jauncey D.L., Smith M.G., Turtle A.J. and Wright A.E., 1981, "A Multifrequency Radio Observations of Optically Selected Quasars",

*Astrophys. J.*, **244**, 5. [27]

Conti P.S., 1978, "Mass Loss in Early–Type Stars", *Ann. Rev. A. A.*, **16**, 371. [29]

Cooper J., 1971, "Compton Fokker–Planck Equation for Hot Plasmas", *Phys. Rev. D.*, **3**, 2312. [26]

Costero R. and Osterbrok D.E., 1977, "The Optical Spectra of Narrow–Line Radio Galaxies", *Astrophys. J.*, **211**, 675. [27]

Cotton W.D., Shapiro I.I. and Wittels J.J., 1981, "Observations of the Jet near the Core of M87", *Astrophys. J. Lett.*, **244**, L57. [27]

Cowie L.L., Ostriker J.P. and Stark A.A., 1978, "Time–dependent Spherically Symmetric Accretion onto Compact X–Ray Sources", *Astrophys. J.*, **226**, 1041. [1, 28]

Cowley A.P., 1977, "Low–Mass X–Ray Binaries and their Relation to the Non–X–Ray Sources", *Proc. 8th Texas Symposium on Relativistic Astrophysics (Ann. NY Acad. Sci.*, **302**, 1). [29]

Cox D.P. and Tucker W.H., 1969, "Ionization Equilibrium and Radiative Cooling of a Low–Density Plasma", *Astrophys. J.*, **157**, 1157. [13, 17]

Crawford J.A. and Kraft R.P., 1956, "An Interpretation of AN Aquarii", *Astrophys. J.*, **123**, 44. [11]

Cunningham C., 1975, "The Effects of Redshifts and Focusing on the Spectrum of an Accretion Disk around a Black Hole", *Astrophys. J.*, **202**, 788. [25, 26]

Cunningham C., 1976, "Returning Radiation in Accretion Disks around Black Holes", *Astrophys. J.*, **208**, 534. [25]

Czerny B. and Elvis M., 1987, "Constraints on Quasar Accretion Disks from the Optical/ Ultraviolet/Soft X–Ray Big Bump", *Astrophys. J.*, **321**, 305. [1]

Dahlbacka G.H., Chapline G.F. and Weaver T.A., 1974, "Gamma Rays from Black Holes", *Nature*, **250**, 36. [20, 26]

Davidsen A., Henry J.P., Middleditch J. and Smith H.E., 1972, "Identification of the X-Ray Pulsar in Hercules: A New Optical Pulsar", *Astrophys. J. Lett.*, **177**, L97. [16]

Davidson K., 1973, "Model for Sources Like Cen X-3 and Her X-2", *Nature Phys. Sci.*, **246**, 1. [18]

Davidson K. and Ostriker J.P., 1973, "Neutron-Star Accretion in a Stellar Wind: Model for a Pulsed X-Ray Source", *Astrophys. J.*, **179**, 585. [1, 11, 16, 18, 21, 29]

Davies R., Fabian A. and Pringle J., 1979, "Spindown of Neutron Stars in Close Binary Systems", *Mon. Not. R. astr. Soc.*, **186**, 779. [29]

Dean A., Gerardi G., De Martinis C., Monastero G.F., Russo A. and Scarsi L., 1973, "Scan of the Cygnus Region in the (1–10) MeV $\gamma$–Ray Energy Range", *Astron. Astrophys.*, **28**, 131. [18]

DeGregoria A.J., 1974, "An Investigation of Accretion of Matter onto White Dwarfs as a Possible X-Ray Mechanism", *Astrophys. J.*, **189**, 555. [21]

Dessler A.J., 1968, "Solar Wind Interactions and the Magnetosphere", in *Physics of the Magnetosphere*, ed. R.F. Carovillano, J.F. McClay and H.R. Radoski (Dordrecht: Reidel). [16]

Dilworth C., Maraschi L. and Reina C., 1974, "On the Luminosity Function of Galactic X–Ray Sources", *Astron. Astrophys.*, **28**, 71. [23]

Dolan J.F., 1971, "Eclipsing Binary Model of Cygnus XR-1", *Nature*, **233**, 109. [13]

Doroshkevich A.G., Zel'dovich Ya.B. and Novikov D., 1967, "The Origin of Galaxies in an Expanding Universe", *Astron. Zh.*, **44**, 295 [*Sov. Astron.–AJ*, **11**, 233]. [12]

Doxey R., Bradt H.V., Levine A., Murthy G.T., Rappaport S. and Spada G., 1973, "X-Ray Pulse Profile and Celestial Position of Hercules X-1", *Astrophys. J. Lett.*, **182**, L25. [16]

Dungey J.W., 1958, *Cosmic Electrodynamics* (Cambridge: Cambridge University Press). [16]

Eardley D.L. and Lightman A.P., 1975, "Magnetic Viscosity in Relativistic Accretion Disks", *Astrophys. J.*, **200**, 187. [26, 27]

Eardley D.L., Lightman A.P. and Shapiro S.L., 1975, "Cygnus X-1: A Two–Temperature Accretion Disk Model which Explains the Observed Hard X-Ray Spectrum", *Astrophys. J. Lett.*, **199**, L153. [20, 26]

Eardley D.L., Lightman A.P., Payne D.G. and Shapiro S.L., 1978, "Accretion Disks around Massive Black Holes: Persistent Emission Spectra", *Astrophys. J.*, **224**, 53. [20, 27]

Edelson P.A. and Malkan M.A., 1986, "Spectral Energy Distributions of Active Galactic Nuclei between 0.1 and 100 Microns", *Astrophys. J.*, **308**, 59. [1]

Eichler D., 1979, "Particle Acceleration in Collisionless Shocks: Regulated Injection and High Efficiency", *Astrophys. J.*, **229**, 419. [28]

Elsner R.F. and Lamb F.K., 1976, "Accretion Flows in the Magnetospheres of Vela X-1, A0535+26 and Her X-1", *Nature*, **262**, 356. [29]

Elsner R.F. and Lamb F.K., 1977, "Accretion by Magnetic Neutron Stars. I. Magnetospheric Structure and Stability", *Astrophys. J.*, **215**, 897. [29]

Elsner R.F., Ghosh P. and Lamb F.K., 1980, "On the Origin and Persistence of Long–Period Pulsating X–Ray Sources", *Astrophys. J. Lett.*, **241**, L155. [29]

Elvey C.T. and Babcock H.W., 1943, "The Spectra of U Geminorum Type Variable Stars", *Astrophys. J.*, **97**, 412. [6]

Elvis M., Czerny B. and Wilkes B. J., 1986, "Continuum Features in Quasars", in *The Physics of Accretion onto Compact Objects*, ed. K.O. Mason, M.G. Watson and N.E. White (Berlin: Springer). [1]

Evans W.D., Belian R.D., Conner J.P., 1970, "Observations of the Development and Disappearance of the X-Ray Source Centaurus XR-4", *Astrophys. J. Lett.*, **159**, L57. [12]

Fabbiano G. and Scherier E.J., 1977, "Further Studies of the Pulsation Period and Orbital Elements of Centaurus X-3", *Astrophys. J.*, **214**, 235. [29]

Fabian A.C., 1975, "Slowly Rotating Neutron Stars and Transient X–Ray Sources", *Mon. Not. R. astr. Soc.*, **173**, 161. [21, 29]

Fabian A.C., 1980, "The X– and Gamma–Ray Backgrounds", *Proc. 10th Texas Symposium on Relativistic Astrophysics* ( *Ann. NY Acad. Sci.*, **375**, 235). [28]

Fabian A.C., Pringle J.E. and Webbink R.F., 1975, "Ariel 1118–61. A Very Close Binary System or a Slowly Rotating Neutron Star ?", *Astroph. Space Sci.*, **42**, 161. [21]

Fabian A.C., Pringle J.E. and Webbink R.F., 1975, "Possible Identification of Ariel 1118−61", *Nature*, **255**, 208. [21]

Fabian A.C., Pringle J.A. and Rees M.J., 1976, "X–Ray Emission from Accretion onto White Dwarfs", *Mon. Not. R. astr. Soc.*, **175**, 43. [1]

Feast M.W., 1975, "SY Fornacis and the Mira Ceti B Phenomenon", *Observatory*, **95**, 19. [21]

Felten J.E., and Rees M.J., 1972, "Continuum Radiative Transfer in a Hot Plasma with

Application to Scorpius X-1", *Astron. Astrophys.*, **17**, 226. [13, 14, 16, 18, 19, 26]

Felten J.E., Rees M.J. and Adams T.F., 1972, "Transfer Effects on X–Ray Lines in Optically Thick Celestial Sources", *Astron. Astrophys.*, **21**, 139. [13]

Fishbone L.G. and Moncrief V., 1976, "Relativistic Fluid Disks in Orbit around Kerr Black Holes", *Astrophys. J.*, **207**, 962. [25]

Flammang R.A., 1982, "Stationary Spherical Accretion into Black Holes. II. Theory of Optically Thick Accretion", *Mon. Not. R. astr. Soc.*, **199**, 833. [1, 28]

Flammang R.A., 1984, "Stationary Spherical Accretion into Black Holes. III. Optically Thick Accretion in Particular Cases", *Mon. Not. R. astr. Soc.*, **206**, 589. [1]

Flowers E. and Ruderman M.A., 1977, "Evolution of Pulsar Magnetic Fields", *Astrophys. J.*, **215**, 302. [29]

Francey R.J., Fenton A.G., Harries J.R. and McCracken K.G., 1967, "Variability of Centaurus XR-2", *Nature*, **216**, 773. [8]

Frank J., 1978, "Tidal Disruption by a Massive Black Hole and Collisions in Galactic Nuclei", *Mon. Not. R. astr. Soc.*, **184**, 87. [27]

Frank J., 1979, "The Fate of the Debris of Tidal Disruption by a Massive Black Hole in a Dense Star Cluster", *Mon. Not. R. astr. Soc.*, **187**, 883. [27]

Frank J., King A. R. and Raine D. J., 1986, *Accretion Power in Astrophysics* (Cambridge: Cambridge University Press). [1]

Frye G.M., Staib J.A., Zych A.D., Hopper W.R., Rawlinson W.R. and Thomas J.A., 1969, "Evidence for a Point Source of High Energy Cosmic Gamma Rays", *Nature*, **223**, 1320. [12]

Fusco–Femiano R. and Massaro E., 1973, "On Gamma–Ray Emission from Cygnus X-2", *Astroph. Space Sci.*, **25**, 239. [18]

Galeev A.A., Rosner R. and Vaiana G.S., 1979, "Structured Coronae of Accretion Disks", *Astrophys. J.*, **229**, 318. [27]

Ghosh P. and Lamb F.K., 1978, "Disk Accretion by Magnetic Neutron Stars", *Astrophys. J. Lett.*, **223**, L83. [29]

Ghosh P. and Lamb F.K., 1979, "Accretion by Rotating Magnetic Neutron Stars. II. Radial and Vertical Structure of the Transition Zone in Disk Accretion",

*Astrophys. J.*, **232**, 259. [29]

Ghosh P., Lamb F.K. and Pethick C.J., 1977, "Accretion by Rotating Magnetic Neutron Stars. I. Flow of Matter inside the Magnetosphere and Its Implications for Spin–up and Spin–down of the Star", *Astrophys. J.*, **217**, 578. [29]

Giacconi R., Gorenstein P., Gursky H., Husher P.D., Waters J.R., Sandage A., Osmer P. and Peach J.V., 1967, "On the Optical Search for the X-Ray Sources Cyg X-1 and Cyg X-2", *Astrophys. J. Lett.*, **148**, L129. [8]

Giacconi R., Gursky H., Kellog E., Levinson R., Scherier E. and Tananbaun H., 1973, "Further X-Ray Observations of Hercules X-1 from *UHURU*", *Astrophys. J.*, **184**, 227. [11]

Giacconi R., Gursky H., Kellog E., Scherier E. and Tananbaun H., 1971, "Discovery of Periodic X-Ray Pulsations in Centaurus X-3 from *UHURU*", *Astrophys. J. Lett.*, **167**, L67. [13]

Ginzburg V.L. and Syrovatskii S.I., 1964, *The Origin of Cosmic Rays* (New York: MacMillan). [20]

Ginzburg V.L. and Syrovatskii S.I., 1964, "Gamma Rays and Cycloton Radiation X–Rays of Galactic and Metagalactic Origin", *J.E.P.T.*, **45**, 353 [*Sov. Astron.-JEPT*, **18**, 245 (1963)]. [20]

Ginzburg V.L. and Syrovatskii S.I., 1965, "Cosmic Magnetobremsstrahlung (Synchrotron Radiation)", *Ann. Rev. A. A.*, **3**, 297. [10]

Ginzburg V.L. and Syrovatskii S.I., 1969, *Elementary Processes for Cosmic Ray Astrophysics* (New York: Gordon and Breach). [20]

Glasby J.S., 1970, *The Dwarf Novae* (London: Constable & Co.). [1]

Gnedin Yu.N. and Sunyaev R.A., 1973, "Luminosity of Thermal X–Ray Sources with a Strong Magnetic Field", *Mon. Not. R. astr. Soc.*, **162**, 53. [14]

Gorenstein P., Giacconi R. and Gursky H., 1967, "The Spectra of Several X-Ray Sources of Cygnus and Scorpio", *Astrophys. J. Lett.*, **150**, L85. [8]

Gould R.J., 1981, "Kinetic Theory of Relativistic Plasma", *Phys. Fluids*, **24**, 102. [28]

Grad H., 1961, "Boundary Layer between a Plasma and a Magnetic Field", *Phys. Fluids*, **4**, 1366. [16]

Grant G., 1955, "Short-Period Fluctuations in the Intensity of SS Cygni", *Astrophys. J.*, **122**, 556. [6]

Grant G. and Abt H.A., 1959, "Photoelectric Photometry of an Outburst of SS Cygni", *Astrophys. J.*, **129**, 323. [6]

Greenstein J.L. and Kraft R.P., 1959, "The Binary System Nova DQ Herculis. I. The Spectrum and Radial Velocity during the Eclipse Cycle", *Astrophys. J.*, **130**, 99. [6]

Greenstein J.L. and Schmidt M., 1964, "The Quasi-stellar Radio Sources 3C 48 and 3C 273", *Astrophys. J.*, **140**, 1. [10]

Greisen K., 1966, "End to the Cosmic-Ray Spectrum ?", *Phys. Rev. (Letters)*, **16**, 748. [20]

Grindlay J.E., 1978, "Thermal Limit for Spherical Accretion and X-Ray Bursts", *Astrophys. J.*, **221**, 234. [28]

Gunn J.E., 1978, *Active Galactic Nuclei*, ed. C. Hazard and S. Mitton (Cambridge: Cambridge University Press). [27]

Gunn J.E. and Ostriker J.P., 1970, "On the Nature of Pulsars. III. Analysis of Observations", *Astrophys. J.*, **160**, 979. [29]

Gusejnov O.Kh. and Zel'dovich Ya.B., 1966, "Collapsed Stars in Binary Systems", *Astron. Zh.*, **43**, 313 [*Sov. Astron.-AJ*, **10**, 251]. [12]

Haffner H., 1937, "Ein W Ursae Majoris-Stern in der Praesepe", *Z. f. Ap.*, **14**, 285. [6]

Harwitt M. and Salpeter E.E., 1973, "Radiation from Comets near Neutron Stars", *Astrophys. J. Lett.*, **186**, L37. [18]

Hasinger G., 1987, "A Classification of Fast Quasi-Periodic X-Ray Oscillators: Is 6 Hz a Fundamental Frequency ?", *Astron. Astrophys.*, **186**, 153. [1]

Hatchett S., Buff J. and McCray R., 1976, "Transfer of X-Rays through a Spherically Symmetric Gas Cloud", *Astrophys. J.*, **206**, 847. [21]

Hayakawa S., 1974, "Circumstellar Matter in the Accretion Model of Cosmic X-Ray Sources", *Prog. Theor. Phys.*, **50**, 459. [21]

Hayakawa S. and Matsuoka M., 1964, "Origin of Cosmic X-Rays", *Prog. Theor. Phys. Suppl.*, **30**, 204. [8]

Haymes R.C. and Harnden F.N.Jr., 1970, "Low–Energy Gamma Radiation from Cygnus", *Astrophys. J.*, **159**, 1111. [26]

Haymes R.C., Harnden F.N., Johnson W.N., Prichard H.M. and Bosch H.E., 1972, "The Low–Energy Gamma–Ray Spectrum of Scorpius X-1", *Astrophys. J. Lett.*, **172**, L47. [13, 18]

Herbig G.H., 1944, "The Variable Star UZ Serpentis", *Publ. Astron. Soc. Pac.*, **56**, 230. [6]

Herbig G.H., 1960, "Observations and an Interpretation of VV Puppis", *Astrophys. J.*, **132**, 76. [6]

Herbst W., Hesser J.E. and Ostriker J.P., 1974, "The 71-Second Variation of DQ Herculis", *Astrophys. J.*, **193**, 679. [21]

Herterich K., 1974, "Absorption of Gamma Rays in Intense X–Ray Sources", *Nature*, **250**, 311. [20]

Hills J.G., 1972, "Stellar Debris Clouds in Quasars and Related Objects", *Mon. Not. R. astr. Soc.*, **182**, 517. [27]

Hirschfield J.L., Baldwin D.E., and Brown S.C., 1961, "Cyclotron Radiation from a Hot Plasma", *Phys. Fluids*, **4**, 198. [20]

Holloway N., Kundt W. and Wang Y.M., 1978, "Propeller Spindown of Rotating Magnets", *Astron. Astrophys.*, **70**, L23. [29]

Holmes A., 1937, *The age of the Earth* (Nelson Classic). [2]

Holzer T. E. and Axford W. I., 1970, "The Theory of Stellar Winds and Related Flows", *Ann. Rev. A. A.*, **8**, 31. [1]

Hoshi R., 1973, "X-Ray Emission from White Dwarfs in Close Binary Systems", *Progr. Theor. Phys.*, **49**, 776. [1, 18, 21]

Hoyle F. and Fowler W., 1963, "Nature of Strong Radio Sources", *Nature*, **197**, 533. [5]

Hoyle F. and Lyttleton, 1939, "The Effect of Interstellar Matter in Climatic Variation", *Proc. Cam. Phil. Soc.*, **35**, 405. [1, 3, 4, 12, 16]

Hoyle F. and Lyttleton, 1940, *Proc. Camb. Phil. Soc.*, **36**, 325. [3]

Hoyle F. and Lyttleton, 1940, "On the Physical Aspect of Accretion by Stars", *Proc. Camb. Phil. Soc.*, **36**, 424. [3]

Hoyle F. and Narlikar J.V., 1963, "Mach's Principle and the Creation of Matter", *Proc. Roy. Soc.*, **273**, No. 1352, 1. [5]

Hoyle F., Fowler W., Burbridge G. and Burbridge M., 1964, "On Relativistic Astrophysics", *Astrophys. J.*, **139**, 909. [5, 10]

Huang S.S., 1956, "A Dynamical Problem in Binary Systems and its Bearing on Stellar Evolution", *Astron. Astrophys.*, **61**, 49. [6]

Hunt R., 1971, "A Fluid Dynamical Study of the Accretion Processes", *Mon. Not. R. astr. Soc.*, **154**, 141. [16, 17]

Ichimaru S., 1978, "Interaction of the Accretion Disk with the Rotating Stellar Magnetic Field: Plasma Diffusion and Angular Momentum Transfer across the Alfvén Surface", *Astrophys. J.*, **224**, 198. [29]

Ichimaru S. and Nakano T., 1973, "Plasma Theory of Scorpius X-1", *Prog. Theor. Phys.*, **50**, 1867. [21]

Illarionov A.F. and Sunyaev R.A., 1972, "Compton Scattering by Thermal Electrons in X-Ray Sources", *Astron. Zh*, **49**, 58 [*Soviet Astr.-AJ*, **16**, 45, (1972)]. [14, 16, 19, 26]

Illarionov A.F. and Sunyaev R.A., 1975, "Why the Number of Galactic X-Ray Stars is so small ?", *Astron. Astrophys.*, **39**, 185. [20, 21, 29]

Inoué H. and Hoshi R., 1975, "X-Ray Emission from a White Dwarf with a Strong Magnetic Dipole Field", *Prog. Theor. Phys.*, **54**, 415. [21]

Ipser J. R. and Price R. H., 1977, "Accretion onto Pregalactic Black Holes", *Astrophys. J.*, **216**, 578. [1]

Ipser J. R. and Price R. H., 1982, "Synchrotron Radiation from Spherically Accreting Black Holes", *Astrophys. J.*, **255**, 654. [1]

Ipser J. R. and Price R. H., 1983, "Comptonization Effects in Spherical Accretion onto Black Holes", *Astrophys. J.*, **267**, 371. [1]

Jackson J.C., 1962, *Classical Electrodynamics* (New York: Wiley). [20]

Jackson J.C., 1972, "Model of Cygnus X-1", *Nature Phys. Sci.*, **236**, 39. [13]

Jaroszynski M., Abramowicz M.A. and Paczynski B., 1980, "Supercritical Accretion Disks around Black Holes", *Acta. Astr.*, **30**, 1. [1, 24, 27]

Jeffreys H., 1929, *The Earth* (Cambridge: Cambridge University Press). [2]

Jones D.L., Sramek R.A. and Terzian V., 1981, "Extended Radio Emission Aligned with Compact Nuclear Sources in Normal Galaxies", *Astrophys. J. Lett.*, **247**, L57. [27]

Johnston M., Bradt H., Doxsey R., Gursky H., Schwartz D. and Schwartz J., 1978, "Position and Pulse Profile of the X-Ray transient 4U 0115+63", *Astrophys. J. Lett.*, **223**, L71. [29]

Joss P.C. and Rappaport S.A., 1976, "Observational Constraints on the Masses of Neutron Stars", *Nature*, **264**, 219. [29]

Joss P.C. and Rappaport S.A., 1984, "Neutron Stars in Interacting Binary Systems", *Ann. Rev. A. A.*, **22**, 537. [1]

Joy A.H., 1956, "Radial–Velocity Measures of SS Cygni at Minimum Light", *Astrophys. J.*, **124**, 317. [6]

Kaplan S.A., 1949, *J.E.P.T.*, **19**, 951. [12, 14]

Kaplan S.A., 1954, *Doklady Acad. Nauk SSSR*, **94**, 33. [12]

Kaplan S.A. and Pikel'ner S.B., 1970, *The Interstellar Medium* (Harvard University Press). [12]

Kaplan S.A. and Tsytovich V.N., 1973, *Plasma Astrophysics* (Oxford: Pergamon). [20]

Kardashev N.S., 1964, "Magnetic Collapse and the Nature of Intense Sources of Cosmic Radio–frequency Emission", *Astron. Zh.*, **41**, 807 [*Sov. Astron.-AJ.*, **8**, 643 (1965)]. [12]

Karzas W.J. and Latter R., 1961, "Electron Radiative Transitions in a Coulomb Field", *Astrophys. J. Suppl.*, **6**, 167. [15]

Katz J.I., 1975, "The Structure of DQ Herculis", *Astrophys. J.*, **200**, 298. [21]

Katz J.I., 1976, "Nonrelativistic Compton Scattering and Models of Quasars", *Astrophys. J.*, **206**, 910. [28]

Katz J.I. and Salpeter E.E., 1974, "X-Ray Emission from Vibrating White Dwarfs", *Astrophys. J.*, **193**, 429. [21]

Kemp J.C., Swedlund J.B. and Wolstencroft R.D., 1974, "DQ Herculis: Periodic Linear Polarization Synchronous with the Rapid Light Variations", *Astrophys. J. Lett.*, **193**, L15. [21]

Kihara T. and Aono O., 1963, "Unified Theory of Relaxations in Plasmas. I. Basic Theorem", *J. Phys. Soc. Japan*, **18**, 837. [15]

Kitamura M., 1959, *Publ. Astr. Soc. Japan*, **11**, 216. [6]

Klebesadel R.W., Strong I.B. and Olson R.A., 1973, "Observations of Gamma-Ray Bursts of Cosmic Origin", *Astrophys. J. Lett.*, **182**, L85. [18]

Klein R. I., Stockman H. S. and Chevalier R. A., 1980, "Supercritical Time-Dependent Accretion onto Compact Objects. I. Neutron Stars", *Astrophys. J.*, **237**, 912. [1]

Kolykhalov P.I. and Sunyaev R.A., 1979, "Gamma Emission during Spherically Symmetric Accretion onto Black Holes in Binary Stellar Systems", *Astron. Zh.*, **56**, 338 [*Soviet Astr.-AJ*, **23**, 189]. [28]

Kompaneets A.S, 1956, "The Establishment of Thermal Equilibrium between Quanta and Electrons", *J.E.P.T.*, **31**, 876 [*Sov. Phys.-J.E.P.T.*, **4**, 730 (1957)]. [9, 14, 18, 26]

Koski A.T., 1978, "Spectrophotometry of Seyfert 2 Narrow Galaxies and Narrow–Line Radio Galaxies", *Astrophys. J.*, **223**, 56. [27]

Kozlowski M., Jaroszynski M. and Abramowicz M.A., 1978, "The Analytic Theory of Fluid Disks Orbiting the Kerr Black Hole", *Astron. Astrophys.*, **63**, 209. [24, 25, 27]

Kozlowski M., Wiita P.J. and Paczynski B., 1979, "Self–Gravitating Accretion Disks Models with Realistic Equations of State and Opacities", *Acta Astron.*, **29**, 157. [25]

Kraft R.P., 1958, "The Binary System Nova T Coronae Borealis", *Astrophys. J.*, **127**, 625. [6]

Kraft R.P., 1959, "The Binary System Nova DQ Herculis. II. An Interpretation of the Spectrum during the Eclipse Cycle", *Astrophys. J.*, **130**, 110. [6]

Kraft R.P., 1962, "Binary Stars among Cataclysmic Variables. I. U Geminorum Stars (Dwarf Novae)", *Astrophys. J.*, **135**, 408. [11]

Kraft R.P., 1963, "Cataclysmic Variables as Binary Stars", *Adv. Astron. Astroph.*, **2**, 43, ed. Z. Kopal. [11, 14, 21]

Kraft R.P. and Demoulin M.H., 1967, "On the Remarkable Spectroscopic Complexities of Cyg X-2", *Astrophys. J. Lett.*, **150**, L183. [8, 12]

Kraft R.P. and Luyten W.J., 1965, "Binary Stars among Cataclysmic Variables. VI. On the Mean Absolute Magnitude of U Geminorum Variables", *Astrophys. J.*, **142**, 1041. [11]

Kristian J., Sandage A.R. and Westphal J., 1967, "Rapid Photometric and Spectroscopic Variations of the X-Ray Source Cyg X-2", *Astrophys. J. Lett.*, **150**, L99. [8, 12]

Krzeminski W., 1965, "The Eclipsing Binary U Geminorum", *Astrophys. J.*, **142**, 1051. [11]

Ku W.H.M., Helfand D.J. and Lucy L.B., 1980, "X–Ray Properties of Quasars", *Nature*, **288**, 323. [27]

Kuan P. and Kuhi L.V., 1975, "Cygni Stars and Mass Loss", *Astrophys. J.*, **199**, 148. [21]

Kuiper J.P., 1941, "On the Interpretation of B Lyrae and Other Close Binaries", *Astrophys. J.*, **93**, 133. [6]

Kuiper J.P. and Johnson J.R., 1956, "Dimensions of Contact Surfaces in Close Binaries", *Astrophys. J.*, **123**, 90. [6]

Kukarkin B., Parenago P., Efremov Y. and Kholopov P., 1958, *General Catalogue of Variable Stars*, **1** (Moscow: Akademia Nauk U.S.S.R.). [6]

Kundt W., 1976, "Spinning Neutron Stars and Cosmic Rays", *Phys. Letters*, **57A**, 195. [29]

Lamb D.Q., 1974, "DQ Herculis: Weak Sister to HZ Herculis", *Astrophys. J. Lett.*, **192**, L129. [21]

Lamb F.K., 1977, "Knowledge of Neutron Stars from X–Ray Observations", *Proc. 8th Texas Symposium on Relativistic Astrophysics (Ann. NY Acad. Sci.*, **302**, 482). [29]

Lamb D.Q., 1985, "Recent Developments in the Theory of AM Her and DQ Her Stars", in *Cataclysmic Variables and Low Mass X–ray Binaries*, ed. D.Q. Lamb and J. Patterson. [1]

Lamb D.Q., Lamb F.K. and Pines D., 1973, "Soft Gamma–Ray Bursts from Accreting Compact Objects", *Nature Phys. Sci.*, **246**, 52. [18]

Lamb F.K., Pethick C. J. and Pines D., 1973, "A Model for Compact X-Ray Sources: Accretion by Rotating Magnetic Stars", *Astrophys. J.*, **184**, 271. [1, 11, 18, 21, 29]

Lamb F.K., Pines D. and Shaham J., 1978, "Period Variations in Pulsating X-Ray Sources. I. Accretion Flows Parameters and Neutron Star Structure from Timing Observations", *Astrophys. J.*, **224**, 969. [29]

Lamb F.K., Pines D. and Shaham J., 1978, "Period Variations in Pulsating X-Ray Sources. II. Torque Variations and Stellar Response", *Astrophys. J.*, **225**, 582. [29]

Landau L. D. and Lifshitz, 1959, *The Classical Theory of Fields* (Cambridge, Mass.: Addison-Wesley Press). [8, 24]

Landau L. D. and Lifshitz, 1959, *Fluid Mechanics* (New York: Pergamon). [1]

Landau L. D. and Lifshitz, 1969, *Statistical Physics* (Addison-Wesley, Reading). [5]

Lawrence A., 1987, "Classification of Active Galaxies and the Prospect of a Unified Phenomenology", *Publ. Astron. Soc. Pac.*, **99**, 309. [1]

Lea S.M., 1976, "Pulsating X-Ray Sources: Slowly Rotating Neutron Stars ?", *Astrophys. J. Lett.*, **209**, L69. [29]

Ledoux P. and Walraven Th., 1958, "Variable Stars", *Handbuch der Physik*, **51** (Berlin: Springer-Verlag). [11]

Levich E.V. and Sunyaev R.A., 1970, "The Heating of Gas in the Vicinity of Quasars, Nuclei of Seyfert Galaxies and Pulsars by the Induced Compton Effects", *Astrophys. Lett.*, **7**, 69. [27]

Levich E.V. and Sunyaev R.A., 1971, "Heating of Gas near Quasars, Seyfert-Galaxy Nuclei, and Pulsars by Low-Frequency Radiation", *Astron. Zh.*, **48**, 461 [*Sov. Astron.-AJ*, **15**, 363]. [14]

Lewin W.H.G., McClintock J.E., Ryckman S.G. and Smith W.B., 1971, "Detection of a High-Energy X-Ray Flare from a Source in Crux", *Astrophys. J. Lett.*, **166**, L69. [13]

Lewin W.H.G., McClintock J.E. and Smith W.B., 1970, "Decrease in the High-Energy X-Ray Flux from Centaurus XR-2", *Astrophys. J. Lett.*, **159**, L193. [12]

Liang E.P.T. and Price R.H., 1977, "Accretion Disk Coronae and Cygnus X-1", *Astrophys. J.*, **218**, 247. [24]

Liang E.P.T. and Thompson K.A., 1980, "Transonic Disk Accretion onto Black Holes", *Astrophys. J.*, **240**, 271. [24]

Liebert J. and Stockman H. S., 1985, "The AM Herculis Magnetic Variables", in *Cataclysmic Variables and Low Mass X-ray Binaries*, ed. Patterson J. and Lamb D.Q. (Dordrecht: Reidel). [1]

Lightman A.P., 1974, "Time-dependent Accretion Disks around Compact Objects. I. Theory and Basic Equations", *Astrophys. J.*, **194**, 419. [23, 26]

Lightman A.P., 1974, "Time-dependent Accretion Disks around Compact Objects. II. Numerical Models and Instability of Inner Region", *Astrophys. J.*, **194**, 429. [1, 23, 26]

Lightman A.P. and Eardley D.M., 1974, "Black Holes in Binary Systems: Instability of Disk Accretion", *Astrophys. J. Lett.*, **187**, L1. [1, 11, 22, 25, 26]

Lightman A.P. and Shapiro S.L., 1975, "Spectrum and Polarization of X-Ray from Accretion Disks around Black Holes", *Astrophys. J. Lett.*, **198**, L73. [26]

Lightman A.P., Zdziarski A.A. and Rees M.J., 1987, "Effects of Electron-Positron Pair Opacity for Spherical Accretion onto Black Holes", *Astrophys. J. Lett.*, **315**, L113. [1]

Linfield R.P., 1981, "VLBI Observations of Jets in Double Radio Galaxies", *Astrophys. J.*, **244**, 436. [27]

Lipunov V.M. and Shakura N.I., 1976, "On the Nature of Binary-System X-Ray Pulsars", *Soviet Astr. Letters*, **2**, 133. [29]

Lucy L.B. and Solomon P.M., 1970, "Mass Loss by Hot Stars", *Astrophys. J.*, **159**, 879. [21]

Lyndell-Bell D., 1969, "Galactic Nuclei as Collapsed Old Quasars", *Nature*, **223**, 690. [1, 12, 13, 14, 17, 25]

Lyndell-Bell D., 1978, "Gravity Power", *Phys. Scripta*, **17**, 185. [24, 25]

Lyndell-Bell D. and Pringle J.E., 1974, "The Evolution of Viscous Discs and the Origin of the Nebular Variables", *Mon. Not. R. astr. Soc.*, **168**, 603. [21, 24, 25]

Lyndell-Bell D. and Rees M.J., 1971, "On Quasars, Dust and the Galactic Center", *Mon. Not. R. astr. Soc.*, **152**, 461. [13]

Lynds C.R., 1967, "Spectroscopic Observations of Cyg X-2", *Astrophys. J. Lett.*, **149**, L41. [8]

Lyne A.G., Ritchings R.T. and Smith F.G., 1975, "The Period Derivatives of Pulsars", *Mon. Not. R. astr. Soc.*, **171**, 579. [29]

Lyutyi V.M., Sunyaev R.A. and Cherepashchuk A.M., 1972, "Nature of the Optical Variability of HZ Herculis (Her X-1) and BD+34°3815 (Cyg X-1)", *Astron. Zh.*, **50**, 3 [*Sov. Astron.-AJ*, **17**, 1 (1973)]. [14]

Macdonald D. and Thorne K.S., 1982, "Black-Hole Electrodynamics: an Absolute-Space/Universal-Time Formulation", *Mon. Not. R. astr. Soc.*, **198**, 345. [27]

Madau P., 1988, "Thick Accretion Disks around Black Holes and the UV/Soft X-Ray Excess in Quasars", *Astrophys. J.*, **327**, 116. [1]

Madej J. and Paczynski B., 1977, *Nonstationary Evolution of Close Binaries, IInd Symp. Physics and Evolution of Stars*, ed. A. Zytkow (Warsaw: Polish Scientific Publisher). [22]

Malkan M.A. and Sargent W.L., 1982, "The Ultraviolet Excess of Seyfert 1 Galaxies and Quasars", *Astrophys. J.*, **254**, 22. [1]

Manley O., 1966, "X-Ray Emission from Sco X-1", *Astrophys. J.*, **144**, 1253. [8]

Maraschi L., Reina C. and Treves A., 1974, "On Spherical Accretion near Eddington Luminosity", *Astrophys. J.*, **35**, 389. [1]

Maraschi L., Reina C. and Treves A., 1978, "The Effect of Radiation Pressure on Spherical Accretion", *Astron. Astrophys.*, **66**, 99. [1]

Maraschi L., Roasio R. and Treves A., 1982, "The Effect of Multiple Compton Scattering on the Temperature and Emission Spectra of Accreting Black Holes", *Astrophys. J.*, **253**, 312. [1]

Maraschi L., Perola G. C., Reina C. and Treves A., 1979, "Turbolent Accretion onto Massive Black Holes", *Astrophys. J.*, **230**, 243. [1]

Margon B. and Ostriker J.P., 1973, "The Luminosity Function of Galactic X-Ray Sources: A Cutoff and a 'Standard Candle' ?", *Astrophys. J.*, **186**, 9. [23]

Martynov D.Ya., 1971, *Course of General Astrophysics* (Moscow: Nauka). [14]

Mason K.O., 1977, "Secular Period Changes in X-Ray Pulsator", *Mon. Not. R. astr. Soc.*, **178**, 81P. [29]

Mattews T. and Sandage A., 1962, "3C 196 as a Second Radio Star", *Publ. Astron. Soc. Pac.*, **74**, 406. [5]

Maxon S., 1972, "Bremsstrahlung Rate and Spectra from a Hot Gas (Z=1)", *Phys. Rev.*, **5**, 1630. [18, 20]

McCrea W.H., 1953, "The Rate of Accretion of Matter by Stars", *Mon. Not. R. astr. Soc.*, **113**, 162. [4, 12]

McKee C.F., 1970, "Simulation of Counterstreaming Plasmas with Application to Collisionless Electrostatic Shocks", *Phys. Rev. (Letters)*, **24**, L990. [18]

Mestel L., 1968, "Magnetic Braking by a Stellar Wind", *Mon. Not. R. astr. Soc.*, **138**, 359. [16]

Mestel L., 1971, "Pulsar Magnetosphere", *Nature Phys. Sci.*, **233**, 149. [16]

Meszaros P., 1975, "Primeval Black Holes and Galaxy Formation", *Astron. Astrophys.*, **38**, 5. [20]

Meszaros P., 1975, "Radiation from Spherical Accretion onto Black Holes", *Astron. Astrophys.*, **44**, 59. [1, 28]

Meszaros P., 1983, "A Thermal Interpretation of the X–ray Spectra of Quasars, Active Galactic Nuclei and Cygnus X-1", *Astrophys. J. Lett.*, **274**, L13. [28]

Meszaros P. and Nagel W., 1985, "X–Ray Pulsar Models. I. Angle–dependent Cyclotron Line Formation and Comptonization", *Astrophys. J.*, **298**, 147. [1]

Meszaros P. and Nagel W., 1985, "X–Ray Pulsar Models. II. Comptonized Spectra and Pulse Shape", *Astrophys. J.*, **299**, 138. [1]

Meyer F. and Meyer-Hofmeister E., 1981, "On the Elusive Cause of Cataclysmic Variable Outbursts", *Astron. Astrophys.*, **104**, L10. [1]

Middleditch J. and Nelson J., 1976, "Studies of Optical Pulsations from HZ Herculis/Hercules X-1: A Determination of the Mass of the Neutron Star", *Astrophys. J.*, **208**, 567. [29]

Morrison P., 1967, "Extrasolar X–Ray Sources", *Ann. Rev. A. A.*, **5**, 325. [9]

Morton D.C., 1960, "Evolutionary Mass Exchange in Close Binary Systems", *Astrophys. J.*, **132**, 146. [6]

Neugebauer G., Oke J.B., Becklin E. and Garmire G., 1969, "A Study of Visual and Infrared Observations of Sco XR-1", *Astrophys. J.*, **155**, 1. [13]

Noerdlinger P.D., 1960, "Concerning Certain Collisionless Plasma–Shock Waves Models", *Astrophys. J.*, **133**, 1034. [18]

Norman C.A. and ter Haar D., 1973, "On the Black-Hole Model of Galactic Nuclei", *Astron. Astrophys.,* **24**, 121. [17]

Norman C.A. and ter Haar D., 1975, "Plasma Turbolent Reactors: an Astrophysical Paradigm", *Phys. Rep.*, **17**, 307. [27]

Novikov I.D. and Thorne K.S., 1973, "Black Hole Astrophysics", in *Black Holes*, ed. C. DeWitt and B. DeWitt (London and New York: Gordon and Breach). [1, 14, 17, 19, 23, 24, 25, 26, 27]

Novikov I.D. and Zel'dovich Ya.B., 1966, "Physics of Relativistic Collapse", *Suppl. Nuovo Cimento*, **4**, 810. [9]

Oegelman H., Beuermann K.P., Kanbach G., Mayer-Hasselwander H.A., Capozzi D., Fiordilino E. and Molteni D., 1977, "Increase in the Pulsational Period of 3U0900−40", *Astron. Astrophys.,* **58**, 385. [29]

Oppenheimer J. and Snyder H., 1939, "On Continued Gravitational Contraction", *Phys. Rev.*, **56**, 455. [5]

Oppenheimer J. and Volkoff G., 1938, "On Massive Neutron Cores", *Phys. Rev.*, **55**, 374. [5]

Osaki Y., 1970, "A Mechanism for the Outbursts of U Geminorum Stars", *Astrophys. J.*, **162**, 621. [11]

Osaki Y., 1974, "An Accretion Model for the Outbursts of U Geminorum Stars", *Publ. Astr. Soc. Japan*, **26**, 429. [22]

Osterbrock D.E. and Miller J.S., 1975, "The Optical Emission-Line Spectrum of Cygnus A", *Astrophys. J.,* **197**, 535. [27]

Ostriker J.P., Rees M.J. and Silk J., 1970, "Some Observable Consequences of Accretion by Defunct Pulsars", *Astrophys. Lett.*, **6**, 179. [16]

Ostriker J.P., McCray R., Weaver R. and Yahil A., 1976, "A New Luminosity Limit for Spherical Accretion onto Compact X-Ray Sources", *Astrophys. J. Lett.,* **208**, L61. [28]

Pacholczyk A.G., 1970, *Radio Astrophysics* (San Francisco:Freeman). [20]

Paczynski B., 1965, *Acta Astron.*, **15**, 89. [11]

Paczynski B., 1974, "Mass of Cygnus X-1", *Astron. Astrophys.,* **34**, 161. [26]

Paczynski B., 1977, *Nonstationary Evolution of Close Binaries, IInd Symposium Physics and Evolution of the Stars*, ed. A. Zytkow (Warsaw: Polish Scientific Publishers). [22]

Paczynski B., 1978, "A Model of Selfgravitating Accretion Disk", *Acta Astron.*, **28**, 91. [24, 25]

Paczynski B., 1978, "A Model of Selfgravitating Disk with a Hot Corona", *Acta Astron.*, **28**, 241. [25]

Paczynski B. and Sienkiewicz R., 1972, "Evolution of Close Binaries.VIII. Mass Exchange on the Dynamical Time Scale", *Acta Astron.*, **22**, 73. [24]

Paczynski B. and Wiita P., 1980, "Thick Accretion Disks and Supercritical Luminosities", *Astron. Astrophys.*, **88**, 23. [1, 24, 27]

Pandharipande V.R., Pines D. and Smith R.A., 1976, "Neutron Star Structure: Theory, Observation and Speculation", *Astrophys. J.*, **208**, 550. [29]

Papaloizou J.C.B. and Bath G.T., 1975, "Stellar Stability in Close Binary Systems", *Mon. Not. R. astr. Soc.*, **172**, 339. [22]

Papaloizou J.C.B. and Pringle J.E., 1977, "Tidal Torques on Accretion Discs in Close Binary Systems", *Mon. Not. R. astr. Soc.*, **181**, 441. [22]

Papaloizou J. C. and Pringle J. E., 1984, "The Dynamical Stability of Differentially Rotating Discs with Constant Specific Angular Momentum", *Mon. Not. R. astr. Soc.*, **208**, 721. [1]

Papaloizou J. C. and Pringle J. E., 1985, "The Dynamical Stability of Differentially Rotating Discs. II.", *Mon. Not. R. astr. Soc.*, **213**, 799. [1]

Papaloizou J., Faulkner J. and Lin D. N., 1983, "On the Evolution of Accretion Disc Flow in Cataclysmic Variables. II. The Existence and Nature of Collective Relaxation Oscillations in Dwarf Nova Systems", *Mon. Not. R. astr. Soc.*, **205**, 487. [1]

Parker E. N., 1963, *Interplanetary Dynamical Processes* (New York: Interscience). [1]

Parker E.N., 1967, "The Dynamical State of the Interstellar Gas and Field. II. Non-linear Growth of Clouds and Forces in Three Dimensions", *Astrophys. J.*, **149**, 517. [10]

Parker E.N., 1972, "Topological Dissipation and the Small-Scale Fields in Turbolent Gases", *Astrophys. J.*, **174**, 499. [16]

Parker E.N. and Ferraro V.C.A., 1971, "Theoretical Aspects of the Worldwide Magnetic Storm Phenomenon", *Handbuch der Physik*, **49**/3 (Berlin: Springer-Verlag). [16]

Pearson T.J., Uwin S.C., Cohen M.H., Linfield R.P., Readhead A.C.S., Seielstad G.A., Simon R.S. and Walker R.C., 1981, "Superluminal Expansion of Quasar 3C 273", *Nature*, **290**, 365. [27]

Peebles P.J.E. and Dicke R.H., 1968, "Origin of the Globular Star Clusters", *Astrophys. J.*, **154**, 891. [12]

Petterson J.A., 1978, "On the Occurrence of Streams and Disks in Massive X-Ray Binary Systems", *Astrophys. J.*, **224**, 625. [29]

Petterson J.A., Silk J. and Ostriker J.P., 1980, "Variations on a Spherically Symmetrical Accretion Flow", *Mon. Not. R. astr. Soc.*, **191**, 571. [28]

Pines D., Pethick C.J. and Lamb F.K., 1972, "Models for Compact X-Ray Sources", *Proc. 6th Texas Symposium on Relativistic Astrophysics (Ann. NY Acad. Sci.*, **224**, 237). [11]

Piran T., 1977, "Secondary Winds and Evaporation from Accretion Discs", *Mon. Not. R. astr. Soc.*, **180**, 45. [25]

Piran T., 1978, "The Role of Viscosity and Cooling Mechanisms in the Stability of Accretion Disks", *Astrophys. J.*, **221**, 652. [1]

Podurets M.A., 1964, "The Collapse of a Star with Back Pressure Taken into Account", *Doklady Acad. Nauk SSSR*, **154**, 300 [*Sov. Phys.-Doklady*, **9**, 1]. [5]

Poolley G.G., 1969, "5C3: A Radio Continuum Survey of M31 and its Neighbourhood", *Mon. Not. R. astr. Soc.*, **144**, 101. [10]

Pottasch S.R., 1970, "Mass Loss from Stars", in *Interstellar Gas Dynamics, I.A.U. Symposium No. 39*, (Dordrecht: Reidel). [14]

Pozdnyakov L. A., Sobol' I. M. and Sunyaev R. A., 1983, "Comptonization and the Shaping of X-Ray Source Spectra: Monte Carlo Calculations", *Astroph. Space Sci. Rev.*, **2**, 189. [1]

Prendergast K.H., 1960, "The Motion of Gas Streams in Close Binary Systems", *Astrophys. J.*, **132**, 162. [14]

Prendergast K.H. and Burbridge G.R., 1968, "On the Nature of Some Galactic X-Ray Sources", *Astrophys. J. Lett.*, **151**, L83. [1, 12, 13, 14, 16]

Pringle J.E., 1975, "Period Changes in Eruptive Binaries", *Mon. Not. R. astr. Soc.*, **170**, 633. [21]

Pringle J.E., 1976, "Thermal Instabilities in Accretion Discs", *Mon. Not. R. astr. Soc.*, **177**, 65. [27]

Pringle J.E., 1981, "Accretion Discs in Astrophysics", *Ann. Rev. A. A.*, **19**, 137. [1, 22, 27]

Pringle J.E. and Rees M.J., 1972, "Accretion Disc Models for Compact X-Ray Sources", *Astron. Astrophys.*, **21**, 1. [1, 11, 14, 16, 17, 18, 19, 20, 23, 24, 25, 26, 29]

Pringle J.E., Rees M.J. and Pacholczyk A.G., 1973, "Accretion onto Massive Black Holes", *Astron. Astrophys.*, **29**, 179. [1, 19, 26]

Protheroe R.J. and Kazanas D., 1983, "On the Origin of Relativistic Particles and Gamma-Rays in Quasars", *Astrophys. J.*, **265**, 620. [28]

Rappaport S. and Joss P.C., 1977, "Accretion Torque in X-Ray Pulsars", *Nature*, **266**, 683. [29]

Rappaport S., Cash W., Doxsey R., McClintock J. and Moore G., 1974, "Possible Detection of Very Soft X-Rays from SS Cygni", *Astrophys. J. Lett.*, **187**, L5. [11, 21]

Rappaport S., Clark G.W., Cominsky L., Joss P.C. and Li F., 1978, "Orbital Elements of 4U 0115+63 and the Nature of the Hard X-Ray Transients", *Astrophys. J. Lett.*, **224**, L1. [29]

Readhead A.C.S., Cohen M.H., Pearson T.J. and Wilkinson P.N., 1978, "Bent Beams and the Overall Size of Extragalactic Radio Sources", *Nature*, **276**, 768. [27]

Rees M.J., 1978, "Relativistic Jets and Beam in Radio Galaxies", *Nature*, **275**, 516. [25]

Rees M.J., 1978, "Accretion and the Quasar Phenomenon", *Phys. Scripta*, **17**, 193. [1]

Rees M.J., 1984, "Black Hole Models for Active Galactic Nuclei", *Ann. Rev. A. A.*, **22**, 471. [1]

Rees M.J., Begelman M.C., Blanford R.D. and Phinney E.S., 1982, "Ion-supported Tori and the Origin of Radio Jets", *Nature*, **295**, 17. [1, 28]

Roberts D.H. and Sturrock P.A., 1973, "Pulsar Magnetospheres, Braking Index, Polar Caps and Period-Pulse Width Distribution", *Astrophys. J.*, **181**, 161. [16]

Robinson E.L., 1973, "High-Speed Photometry of Z Camelopardalis", *Astrophys. J.*, **180**, 121. [11]

Robinson E.L., 1973, "Detection of Mass Loss from the Dwarf Nova Z Camelopardalis", *Astrophys. J.*, **186**, 347. [11]

Robinson E.L., 1976, "The Masses of Cataclysmic Variables", *Astrophys. J.*, **203**, 485. [21]

Russel H.N., Dugan R.S. and Stewart J.Q., 1938, *Astronomy* (Ginn & Co.). [2]

Rybicki G.B. and Lightman A., 1979, *Radiative Processes in Astrophysics* (New York: Wiley). [1]

Ryle M., 1968, *Highlights in Astronomy*, ed. L. Perek and D. Reidel. [10]

Sagdeev R.Z., 1966, *Rev. Plasma Physics*, **4**, 23, ed. M.A. Leontovich (New York: Consultants Bureau). [16]

Sahade J., 1959, *Liége Symposium: Modeles d'etoiles et evolution stellaire*, p. 76. [6]

Salpeter E.E., 1964, "Accretion of Interstellar Matter by Massive Objects", *Astrophys. J.*, **140**, 796. [1, 10, 12, 14, 17, 18]

Salpeter E.E., 1973, "Models for Compact X-Ray Sources", in *X- and Gamma-Ray Astronomy, I.A.U. Symposium No. 55*, ed. H. Bradt and R. Giacconi (Dordrecht: Reidel). [23]

Sandage A., Osmer P., Giacconi R., Gorenstein P., Gursky H., Waters J., Bradt H., Garmire G., Sreekantan B.V., Oda M., Osawa K. and Jugaku J., 1966, "On the Optical Identification of Scorpius X-1", *Astrophys. J.*, **146**, 316. [8]

Saslaw W.C., 1968, "The Effects of Accretion on White Dwarf Stars", *Mon. Not. R. astr. Soc.*, **138**, 337. [9]

Savonije G.J., 1978, "Roche-lobe Overflow in X-Ray Binaries", *Astron. Astrophys.*, **62**, 317. [29]

Savonije G.J. and Van den Heuvel E.P.J., 1977, "On the Rotational History of the Pulsars in Massive X-Ray Binaries", *Astrophys. J. Lett.*, **214**, L19. [29]

Scharlermann E., 1978, "The Fate of Matter and Angular Momentum in Disk Accretion onto a Magnetized Neutron Star", *Astrophys. J.*, **219**, 617. [29]

Schatzman E., 1955, in *Gas Dynamics of Cosmic Clouds, I.A.U. Symposium No. 2* (Amsterdam: North Holland Publishing Co.). [4]

Schreier E., 1977, "Timing Effects in Rotating Neutron Stars", *Proc. 8th Symposium on Relativistic Astrophysics (Ann. NY Acad. Sci.*, **302**, 445 ). [29]

Schreier E., Gursky H., Kellogg E., Tananbaum H. and Giacconi R., 1971, "Further Observations of the Pulsating X-Ray Source Cygnus X-1 from *UHURU*", *Astrophys. J. Lett.*, **170**, L21. [13, 19, 26]

Scherier E., Levinson R., Gursky H., Kellogg E., Tananbaum H. and Giacconi R., 1972, "Evidence for Binary Nature of Centaurus X-3 from *UHURU* X-Ray Observations", *Astrophys. J. Lett.*, **172**, L79. [13, 16]

Schmidt M., 1962, "Spectrum of a Stellar Object Identified with the Radio Source 3C 286", *Astrophys. J.*, **136**, 684. [5]

Schmidt M., 1963, "3C 273: A Star-like Object with Large Red-shift", *Nature*, **197**, 1040. [5]

Schmidt M., 1978, "The Local Space Density of Quasars and Active Nuclei", *Phys. Scripta*, **17**, 135. [27]

Schwarzchild M., 1958, *Structure and Evolution of the Stars* (Princeton: Princeton University Press). [6]

Shakura N.I., 1972, "Disk Model of Gas Accretion on a Relativistic Star in a Close Binary System", *Astron. Zh.*, **49**, 921 [*Sov. Astron.-AJ*, **16**, 756]. [1, 14]

Shakura N.I., 1972, "Effects of Thomson Scattering on the Emission Spectrum of an Optically Semiopaque Plasma", *Astron. Zh.*, **49**, 652 [*Sov. Astron.-AJ*, **16**, 532]. [14]

Shakura N.I., 1972, "The Long Period X-Ray Pulsar 3U0900−40 as a Neutron Star with an Abnormally Strong Magnetic Field", *Pis'ma Astron. Zh.*, **1**, 23 [*Sov. Astron. Letters*, **1**, 223). [29]

Shakura N.I. and Sunyaev R.A., 1973, "Black Holes in Binary Systems. Observational Appearance", *Astron. Astrophys.*, **24**, 337. [1, 16, 17, 18, 19, 20, 22, 23, 24, 25, 26, 27]

Shakura N.I. and Sunyaev R.A., 1976, "A Theory of the Instability of Disk Accretion on to Black Holes and the Variability of Binary X-Ray Sources, Galactic Nuclei and Quasars", *Mon. Not. R. astr. Soc.*, **175**, 613. [1, 25, 27]

Shapiro S.L., 1973, "Accretion onto Black Holes: The Emergent Radiation Spectrum", *Astrophys. J.*, **180**, 531. [1, 17, 18, 20, 28]

Shapiro S.L., 1973, "Accretion onto Black Holes: The Emergent Radiation Spectrum. II. Magnetic Effects", *Astrophys. J.*, **185**, 69. [1, 17, 20]

Shapiro S.L., 1974, "Accretion onto Black Holes: The Emergent Radiation Spectrum. III. Rotating (Kerr) Black Holes", *Astrophys. J.*, **189**, 343. [26]

Shapiro S.L. and Lightman A.P., 1976, "Black Holes in X-Ray Binaries: Marginal Existence and Rotation Reversals of Accretion Disks", *Astrophys. J.*, **204**, 555. [21, 29]

Shapiro S.L. and Salpeter E.E., 1975, "Accretion onto Neutron Stars under Adiabatic Shock Conditions", *Astrophys. J.*, **198**, 671. [1, 21, 26]

Shapiro S.L. and Teukolsky S.A., 1983, *Black Holes, White Dwarfs and Neutron Stars* (New York N.Y.: Wiley). [1]

Shapiro S.L., Lightman A.P. and Eardley D.M., 1976, "A Two-Temperature Accretion Disk Model for Cygnus X-1: Structure and Spectrum", *Astrophys. J.*, **204**, 187. [1, 25, 27, 28]

Shields G., 1978, "Thermal Continuum from Accretion Disks in Quasars", *Nature*, **272**, 706. [1]

Shkarofsky I.P., Johnston T,W. and Bachynski M.P., 1966, *The Particle Kinetics of Plasmas* (Reading: Addison Wesley). [15]

Shklovskii I.S., 1962, *Astron. Zh.*, **39**, 591 [*Sov. Astron.-AJ*, **6**, 465]. [5]

Shklovskii I.S., 1967, "The Nature of the X-Ray Source Sco X-1", *Astron. Zh.*, **44**, 930 [*Sov. Astron.-AJ*, **11**, 749 (1968)]. [9, 12, 14]

Shklovskii I.S., 1967, "On the Nature of the Source of X-Ray Emission of Sco XR-1", *Astrophys. J. Lett.*, **148**, L1. [1, 8, 9, 12, 13, 14]

Shonfelder V. and Lichti G., 1974, "Upper Limits to Soft X-Ray Flux from Seven X-Ray Sources and from the Galactic Plane", *Astrophys. J. Lett.*, **192**, L1. [18]

Shvartsman V.F., 1970, *Radiofisika*, **13**, 1852. [12]

Shvartsman V.F., 1970, "Ionization Zones around Neutron Stars: $H_\alpha$ Emission, Heating of the Interstellar Medium and the Influence on Accretion", *Astron. Zh.*, **47**, 824 [*Sov. Astron.-AJ*, **14**, 662 (1971)]. [12]

Shvartsman V.F., 1971, "Halos Around Black Holes", *Astron Zh.*, **48**, 479 [ *Sov. Astron.-AJ*, **15**, 377]. [1, 13, 14, 16, 17, 20, 26, 28]

Shvartsman V.F., 1971, "Neutron Stars in Binary Systems should not be Pulsars", *Astron. Zh.*, **48**, 438 [*Sov. Astron.-AJ*, **15**, 342]. [13, 14, 16]

Sikora M., 1981, "Superluminous Accretion Discs", *Mon. Not. R. astr. Soc.*, **196**, 257. [27]

Sikora M. and Wilson D.B., 1981, "The Collimation of Particle Beams from Thick Accretion Discs", *Mon. Not. R. astr. Soc.*, **197**, 529. [27]

Smak J.I., 1971, "Eruptive Binaries. II. U Geminorum", *Acta Astron.*, **21**, 15. [11]

Smak J.I., 1984, "Outbursts of Dwarf Novae", *Publ. Astron. Soc. Pac.*, **96**, 5. [1]

Smart W.M., 1938, *Stellar Dynamics* (Cambridge: Cambridge University Press). [6]

Soltan A., 1982, "Masses of Quasars", *Mon. Not. R. astr. Soc.*, **200**, 115. [27]

Sparke L.S. and Shu F.H., 1980, "Extended Radio Doubles are Found in Elliptical Galaxies", *Astrophys. J. Lett.*, **241**, L65. [27]

Spitzer L., 1955, *Physics of Fully Ionized Gases* (New York: Interscience). [10, 15, 16, 18, 26, 28]

Spreiter J.R. and Summers A.L., 1967, "On Conditions near the Neutral Points on the Magnetosphere Boundary", *Planet. and Space Sci.*, **15**, 787. [16]

Starrfield S.G., 1970, "The Rate of Mass Exchange in DQ Herculis", *Astrophys. J.*, **161**, 361. [11]

Starrfield S.G., Truran J.W., Sparks W.M. and Kutter G.S., 1972, "CNO Abundances in Hydrodynamic Models of the Nova Outburst", *Astrophys. J.*, **176**, 169. [11]

Strand K.Aa., 1948, "The Parallax of SS Cygni", *Astrophys. J.*, **107**, 106. [6]

Strittmatter P.A., Serkowski K., Carswell R., Stein W.A., Merril K.M. and Burbridge E.M., 1972, "Compact Extragalactic Nonthermal Sources", *Astrophys. J. Lett.*, **175**, L7. [17]

Struve O., 1950, *Stellar Evolution* (Princeton: Princeton University Press). [6, 8]

Sunyaev R.A., 1972, "Variability of X–Rays from Black Holes with Accretion Disks", *Astron. Zh.*, **49**, 1153 [*Sov. Astron.-AJ*, **16**, 941 (1973)]. [14, 19]

Sunyaev R.A. and Titarchuk L.G., 1980, "Comptonization of X–Rays in Plasma Cloud. Typical Radiation Spectra", *Astron. Astrophys.*, **86**, 121. [1]

Sunyaev R.A. and Truemper J., 1979, "Hard X-Ray Spectrum of Cyg X-1", *Nature*, **279**, 506. [1]

Swedlund J.B., Kemp J.C. and Wolstencroft R.D., 1975, "Periodic Circular Polarization Synchronous with the Rapid Light Variations", *Astrophys. J. Lett.*, **193**, L11. [21]

Syrovatskii S.I., 1967, "Direct Transformation of Magnetic-Field Energy into Energy of Fast Particle", in *Radio Astronomy and the Galactic System, I.A.U. Symposium No. 31*, ed. H. Van Woerden, p. 133 (Academic Press). [10]

Takahara F., Tsuruta S. and Ichimura S., 1981, "X-Rays from Active Galactic Nuclei", *Astrophys. J.*, **251**, 26. [1]

Tamazawa S., Toyama K., Kanako N. and Ono Y., 1975, "Optically Thick Accretion onto Black Holes", *Astroph. Space Sci.*, **32**, 403. [20]

Tananbaum H., Gursky H., Kellogg E.M., Levinson R., Schreier E. and Giacconi R., 1972, "Discovery of a Periodic Pulsating Binary X-Ray Source in Hercules from *UHURU*", *Astrophys. J. Lett.*, **174**, L143. [16]

Tarter C.B. and Salpeter E.E., 1969, "The Interaction of X-Ray Sources with Optically Thick Environments", *Astrophys. J.*, **156**, 953. [21]

Tarter C.B., Tucker W.H. and Salpeter E.E., 1969, "The Interaction of X-Ray Sources with Optically Thin Environments", *Astrophys. J.*, **156**, 943. [14, 21]

Tassoul J.L., 1978, *Theory of Rotating Stars* (Princeton: Princeton University Press). [24]

Taylor G.I., 1937, "Fluid Friction Between Rotating Cylinders. I. Torque Measurements", *Proc. Roy. Soc.*, **157**, 546. [14]

Thorne K.S., 1974, "Disk-Accretion onto a Black Hole. II. Evolution of the Hole", *Astrophys. J.*, **191**, 507. [27]

Thorne K.S. and Price R.H., 1975, "Cygnus X-1: An Interpretation of the Spectrum and Its Variability", *Astrophys. J. Lett.*, **195**, L101. [26]

Thorne K.S. and Zytkow A.N., 1977, "Stars with Degenerate Neutron Cores. I. Structure of Equilibrium Models", *Astrophys. J.*, **212**, 832. [28]

Thorne K.S., Flammang R.A. and Zytkow A.N., 1981, "Stationary Spherical Accretion into Black Holes. I. Equations of Structure", *Mon. Not. R. astr. Soc.*, **194**, 475. [1]

Tidman D.A., 1967, "Turbolent Shock Waves in Plasmas", *Phys. Fluids*, **10**, 547. [18]

Treves A., Belloni T., Chiappetti L., Maraschi L., Stella L., Tanzi E.G. and van der Klis M., 1988, "X–Ray Spectrum and Variability of the Black Hole Candidate LMCX–3", *Astrophys. J.*, **325**, 119. [1]

Trimble V.L. and Thorne K.S., 1969, "Spectroscopic Binaries and Collapsed Stars", *Astrophys. J.*, **156**, 1013. [12]

Trubnikov B.A., 1958, "Plasma Radiation in a Magnetic Field", *Soviet Phys.-Doklady*, **3**, 136. [20]

Trumper J., Pietsch W., Reppin C., Voges W., Staubert R. and Kendziorra E., 1978, "Evidence for Strong Cyclotron Line Emission in the Hard X–Ray Spectrum of Hercules X-1", *Astrophys. J. Lett.*, **219**, L105. [29]

Tucker W.H., 1967, "Physical Conditions in Sco X-1", *Astrophys. J. Lett.*, **149**, L105. [8]

Turolla R., Nobili L. and Calvani M., 1986, "On Hydrodynamics of Radiatively Driven Winds", *Astrophys. J.*, **303**, 573. [1]

van Citters G.W. and Morton D.C., 1970, "Model Atmospheres for B–Type Stars with Blanketing by Ultraviolet Lines", *Astrophys. J.*, **161**, 695. [16]

van de Hulst H.C., 1955, *Mém. Soc. R. Sci. Liège*, **15**, 393. [4]

van den Bergh, 1969, "Collapsed Objects in Cluster of Galaxies", *Nature*, **224**, 891. [12]

van den Heuvel E.P.J., 1977, "Evolutionary Processes in X–Ray Binaries and their Progenitor Systems", *Proc. 8th Texas Symposium on Relativistic Astrophysics* (*Ann. NY Acad. Sci.*, **302**, 14 ). [29]

Vedenov A.A., Velikhov E.P. and Sagdeev R.Z., 1961, "Stability of Plasma", *Usp. Fiz. Nauk*, **73**, 701 [*Sov. Phys.-Usp.*, **4**, 332 (1961)]. [9]

Waggett P.C., Warner P.J. and Baldwin J.E., 1977, "NGC 6251, a Very Large Radio Galaxy with an Exceptional Jet", *Mon. Not. R. astr. Soc.*, **181**, 465. [27]

Walker M.F., 1954, "Nova DQ Herculis (1934): an Eclipsing Binary with Very Short Period", *Publ. Astron. Soc. Pac.*, **66**, 230. [6]

Walker M.F., 1956, "A Photometric Investigation of the Short–Period Eclipsing Binary, Nova DQ Herculis (1934)", *Astrophys. J.*, **123**, 68. [6]

Walker M.F., 1957, in *I.A.U. Symposium, No. 3*, ed. G. H. Herbig (Cambridge: Cambridge University Press). [6]

Wallerstein G., 1959, "Three-Color Photometry of U Geminorum during an Outburst", *Publ. Astron. Soc. Pac.*, **71**, 316. [6]

Wallerstein G., 1961, "On the Colors of the U Geminorum Variables at Maximum Light", *Astrophys. J.*, **134**, 1020. [11]

Warner B., 1972, "Observations of Rapid Blue Variables. III. The Companion to Mira", *Mon. Not. R. astr. Soc.*, **159**, 95. [21]

Warner B., 1973, "On the Masses of Cataclysmic Variable Stars", *Mon. Not. R. astr. Soc.*, **162**, 189. [11]

Warner B., 1973, "More High-Speed Photometry of Cataclysmic Variables", *Sky Telesc.*, **46**, 298. [11]

Warner B. and Nather R.E., 1971, "Observations of Rapid Blue Variables. II. U Geminorum", *Mon. Not. R. astr. Soc.*, **152**, 219. [11]

Warner B. and Robinson E.L., 1972, "Non-Radial Pulsations in White Dwarf Stars", *Nature Phys. Sci.*, **239**, 2. [11]

Wasiutinski J., 1946, *Studies in Hydrodynamics and Structure of Stars and Planets*, Oslo. [14]

Webster B.L. and Murdin P., 1972, "Cygnus X-1. A Spectroscopic Binary with a Heavy Companion ?", *Nature*, **235**, 37. [13]

Weymann R., 1965, "Diffusion Approximation for a Photon Gas Interacting with a Plasma via the Compton Effect", *Phys. Fluids*, **8**, 2112. [9, 26]

Wheaton W.A., Ulmer M.P., Baity W.A., Datlowe D.W., Elcan M.J., Peterson L.E., Klebesadel R.W., Strong I.B., Cline T.L. and Desai U.D., 1973, "The Direction and Spectra Variability of a Cosmic Gamma-Ray Burst", *Astrophys. J. Lett.*, **185**, L57. [18]

White N.E., Mason K.O. and Sanford P.W., 1977, "Evidence for a 581-d Modulation in the Pulse Period of 3U0352+30", *Nature*, **267**, 229. [29]

White N.E., Stella L. and Parmar A.N., 1988, "The X-Ray Spectral Properties of Accretion Disks in X-Ray Binaries", *Astrophys. J.*, **324**, 363. [1]

White N.E., Mason K.O., Sanford P.W. and Murdin P., 1975, "The X-Ray Behaviour of 3U0352+30 (X Per)", *Mon. Not. R. astr. Soc.*, **176**, 201. [29]

Wickramasinghe D.T. and Whelan J.A.J., 1975, "The Periodic Transient X-Ray Sources", *Nature*, **258**, 502. [29]

Wilson D.B. and Rees M.J., 1978, "Induced Compton Scattering in Pulsar Winds", *Mon. Not. R. astr. Soc.*, **185**, 297. [27]

Wolfe A.M. and Burbridge G.R., 1970, "Black Holes in Elliptical Galaxies", *Astrophys. J.*, **161**, 419. [12]

Woltjer L., 1959, "Emission Nuclei in Galaxies", *Astrophys. J.*, **130**, 38. [10]

Wood P.R., 1977, "Mass Transfer Instabilities in Binary Systems", *Astrophys. J.*, **217**, 530. [22]

Wyckoff S., Wehinger P. and Gehren T., 1981, "The Resolution of Quasar Images", *Astrophys. J.*, **247**, 750. [27]

Wyller A.A., 1970, "Observational Aspects of Black Holes in Globular Clusters", *Astrophys. J.*, **160**, 443. [12]

Zdziarski A.A., 1985, "Power-Law X-Ray and Gamma-Ray Emission from Relativistic Thermal Plasmas", *Astrophys. J.*, **289**, 514. [1]

Zel'dovich Ya.B., 1963, *Voprosy Kosmog.*, **9**, 80. [5]

Zel'dovich Ya.B., 1964, "The Fate of a Star and the Evolution of Gravitational Energy upon Accretion", *Doklady Acad. Nauk SSSR*, **155**, 67 [*Soviet Phys.-Doklady*, **9**, 195 (1964)]. [1, 9, 12, 14, 18]

Zel'dovich Ya.B. and Gusejnov O.Kh., 1965, "Collapsed Stars in Binaries", *Astrophys. J.*, **144**, 840. [12]

Zel'dovich Ya.B. and Novikov I.D., 1964, "Mass of Quasi-Stellar Objects", *Doklady Acad. Nauk SSSR*, **158**, 811 [*Sov. Phys.-Doklady*, **9**, 834 (1965)]. [9]

Zel'dovich Ya.B. and Novikov I.D., 1965, "Relativistic Astrophysics. II.", *Usp. Fiz. Nauk*, **86**, 447 [*Sov. Phys-Usp.*, **8**, 522 (1966)]. [9]

Zel'dovich Ya.B. and Novikov I.D., 1971, *Relativistic Astrophysics*, **1** (Chicago: University Chicago Press). [1, 9, 12, 13, 16, 20]

Zel'dovich Ya.B. and Novikov I.D., 1971, *The Theory of the Gravitation and Stars Evolution* (Moskow: Nauka). [14]